P. A. Shaver L. DiLella A. Giménez (Eds.)

Astronomy, Cosmology and Fundamental Physics

Proceedings of the ESO/CERN/ESA
Symposium Held in Garching, Germany,
4-7 March 2002

W0193346

Springer

Volume Editors

Peter A. Shaver
European Southern Observatory
Karl-Schwarzschild-Str. 2
85748 Garching, Germany

Luigi DiLella
EP Division
CERN
1211 Geneva, Switzerland

Alvaro Giménez
Research and Scientific Support Department
ESTEC
Postbus 299
2200 AG Noordwijk,
The Netherlands

Series Editor

Bruno Leibundgut
European Southern Observatory
Karl-Schwarzschild-Strasse 2
85748 Garching, Germany

Library of Congress Cataloging-in-Publication Data applied for
Bibliographic information published by Die Deutsche Bibliothek

Die Deutsche Bibliothek lists this publication in the Deutsche Nationalbibliografie;
detailed bibliographic data is available in the Internet at http://dnb.ddb.de

ISBN 978-3-642-07281-9   e-ISBN 978-3-540-44851-8

Springer-Verlag Berlin Heidelberg New York
a member of BertelsmannSpringer Science+Business Media GmbH

http://www.springer.de

Cover design: Erich Kirchner, Heidelberg

Printed on acid-free paper   55/3141/du - 5 4 3 2 1 0

ESO ASTROPHYSICS SYMPOSIA
European Southern Observatory

Series Editor: Bruno Leibundgut

Springer
Berlin
Heidelberg
New York
Hong Kong
London
Milan
Paris
Tokyo

Physics and Astronomy

http://www.springer.de/phys/

Preface

Astronomy, cosmology and fundamental physics have thriving interfaces and modern research topics that bring together scientists from these different domains. Recent exciting developments in these fields include the structures in the cosmic background radiation, evidence for an accelerating Universe, searches for dark matter candidates, evidence for neutrino oscillations, space experiments on fundamental physics, and discoveries of extrasolar planets. ESO, CERN and ESA are thus involved in scientific endeavours and technologies which overlap considerably.

On 4–7 March 2002, a joint Symposium covering all of these fields was held in Garching, the first to be co-organized and co-sponsored by all three organizations. It was intended to give a broad overview of scientific areas of interest to the communities of the three organizations, and the future perspectives of the organizations themselves. The Symposium was attended by about 200 participants, and consisted largely of invited talks, with some contributed talks and a large number of posters on display. We are pleased that almost all of the talks and most of the posters are included in these Proceedings, and would like to thank all the participants for making the Symposium so stimulating and successful.

We are grateful to the Scientific Organizing Committee for their valuable contributions in putting the programme together. The members were: R. Battiston (Univ. of Perugia), R. Bender (Univ. of Munich), A. de Rujula (CERN), L. DiLella (CERN), C. Fabjan (CERN), A. Giménez (ESA), M. Jacob (CERN), F. Pacini (Arcetri), A. Renzini (ESO), P. Shaver (ESO; chair), M. Spiro (Saclay), B. Taylor (ESA), P. van der Kruit (Univ. of Groningen), S. Vitale (Univ. of Trento), S. Volonte (ESA), M. Ward (Univ. of Leicester).

We would like especially to thank Britt Sjöberg and Christina Stoffer for their essential roles in organizing the Symposium, a major effort that took months of preparation. Our thanks also go to Harald Kuntschner, Markus Kissler-Patig, and ESO's fellows and students for their assistance in making the Symposium a success. Finally, we would like to thank Pamela Bristow for her efficient work in preparing these proceedings for publication.

Garching, Geneva, Noordwijk
February 2003

Peter Shaver
Luigi DiLella
Alvaro Giménez

Contents

Poster Papers

List of Participants

Adams, Jenni
Univ. Canterbury, Christchurch
j.adams@phys.canterbury.ac.nz

Alkier, Bernd
Munich
bernd@alkier.de

Alves, Joao
ESO Garching
jalves@eso.org

Aufmuth, Peter
University of Hanover
pea@mpq.mpg.de

Banday, Anthony
MPI Astrophysik, Garching
banday@mpa-garching.mpg.de

Bardelli, Sandro
Osservatorio Astronomico di Bologna
bardelli@bo.astro.it

Barrientos, Felipe
P. Universidad Catolica de Chile
barrientos@astro.puc.cl

Battiston, Roberto
Università and INFN Perugia
battisto@krenet.it

Bauer, Florian
CEA Saclay
florian@hep.saclay.cea.fr

Bender, Ralf
University Observatory Munich
bender@usm.uni-muenchen.de

Benvenuti, Piero
ST-ECF Garching
pbenvenu@eso.org

Berat, Corinne
Institut des Sciences Nucléaires,
Grenoble
berat@isn.in2p3.fr

Bergeron, Jacqueline
Institut d'Astrophysique de Paris
bergeron@iap.fr

Berghöfer, Thomas
DESY, Hamburg
tberghoefer@epost.de

Binetruy, Pierre
APC Paris 7 / LPT Orsay
Pierre.Binetruy@th.u-psud.fr

Bluemer, Hans
Forschungszentrum Karlsruhe
bluemer@ik.fzk.de

Boehringer, Hans
MPI Extraterrestrische Physik,
Garching
hxb@mpe.mpg.de

Bork, Thomas
Munich
Dr.Thomas.M.Bork@t-online.de

Borthen, Peter
Technical University Munich
borthen@tep.ei.tum.de

Braine, Jonathan
Observatoire de Bordeaux
braine@observ.u-bordeaux.fr

Braxmaier, Claus
Astrium Friedrichshafen
claus.braxmaier
@astrium-space.com

Buschhorn, Gerd
MPI Physik, Munich
gwb@mppmu.mpg.de

Cabanero, Susana
ESO Garching
scabaner@eso.org

Cappi, Alberto
Osservatorio Astronomico di Bologna
cappi@bo.astro.it

Cardone, Vincenzo F.
University of Salerno
winny@na.infn.it

Caron, Christian
Springer-Verlag, Heidelberg
caron@Springer.de

Carretti, Ettore
IASF – CNR, Bologna
carretti@tesre.bo.cnr.it

Carrière, Jean-Claude
Paris

Cassé, Michel
CEA Saclay
casse@iap.fr

Catalano, Osvaldo
IFCAI / CNR, Palermo
catalano@ifcai.pa.cnr.it

Cavazzuti, Elisabetta
Italian Space Agency
elisabetta.cavazzuti@asi.it

Cesarsky, Catherine
ESO Garching
ccesarsk@eso.org

Chechelnitsky, Albert
Joint Institute for Nuclear Physics,
Dubna
ach@arcor.de

Chincarini, Guido
Osservatorio Astronomico di Brera
guido@merate.mi.astro.it

Chiosi, Cesare
University of Padova
chiosi@pd.astro.it

Coc, Alain
CSNSM, Orsay
coc@csnsm.in2p3.fr

Colacino, Carlo Nicola
MPI Gravitationsphysik, Hanover
cnc@mpq.mpg.de

Cristiani, Stefano
ST-ECF Garching
Stefano.Cristiani@eso.org

D'Odorico, Sandro
ESO Garching
sdodoric@eso.org

Daddi, Emanuele
ESO Garching
edaddi@eso.org

Danzmann, Karsten
MPI Gravitationsphysik, Hanover
kvd@mpq.mpg.de

de Bernardis, Paolo
Università di Roma "La Sapienza"
Paolo.deBernardis@roma1.infn.it

de Mello, Duilia
Onsala Space Observatory
duilia@oso.chalmers.se

de Padova, Thomas
Der Tagesspiegel
depadovat@aol.com

Demarco, Ricardo
ESO Garching
rdemarco@eso.org

Dietrich, Jörg
University of Bonn
jdietric@eso.org

DiLella, Luigi
CERN
luigi.di.lella@cern.ch

Dreibus, Gerlind
MPI Kosmochemie, Mainz
dreibus@mpch-mainz.mpg.de

Eckart, Andreas
University of Cologne, I. Phys. Inst.
eckart@ph1.uni-koeln.de

Ellis, John
CERN
John.Ellis@cern.ch

Enard, Daniel
EGO
enard@virgo.infn.it

Feigler, Bruce
MDD Research, Auckland
mdd@ihug.co.nz

Ferlet, Roger
Institut d'Astrophysique de Paris
ferlet@iap.fr

Fiorentini, Giovanni
University of Ferrara
fiorenti@fe.infn.it

Franck, Siegfried
Potsdam Institute
for Climate Impact Research
franck@pik-potsdam.de

Frutos, Francisco
University of Costa Rica
ffrutos@cariari.ucr.ac.cr

Fusi Pecci, Flavio
Osservatorio Astronomico di Bologna
flavio@bo.astro.it

Fuzfa, André
University of Namur
afu@math.fundp.ac.be

Galeotti, Piero
University of Torino
galeotti@to.infn.it

Gianotti, Fabiola
CERN
Fabiola.Gianotti@cern.ch

Giménez, Alvaro
ESA–ESTEC, Noordwijk
agimenez@rssd.esa.int

Giraud, Edmond
Université de Montpellier II
Edmond.Giraud@gamum2.in2p3.fr

Golden, Aaron
Computational Astrophysics Lab.,
Galway
agolden@physics.nuigalway.ie

Gong, Jin-Ook
Korea Advanced Institute
of Science & Technology, Daejon
devourer@muon.kaist.ac.kr

Görnitz, Thomas
J.W. Goethe-Universität, Frankfurt
goernitz@em.uni-frankfurt.de

Gorski, Krzysztof
ESO Garching
kgorski@eso.org

Grazian, Andrea
ESO Garching
agrazian@eso.org

Grimm, Oliver
ETH Zurich
grimm@particle.phys.ethz.ch

Haemmerle, Hannelore
University of Bonn
hanne@astro.uni-bonn.de

Halzen, Francis
University of Wisconsin-Madison
halzen@pheno.physics.wisc.edu

Hansen, Leif
Niels Bohr Institute, Copenhagen
leif@astro.ku.dk

Harris, Pauline
Univ. Canterbury, Christchurch
p.harris@phys.canterbury.ac.nz

Hartmann, Gernot
DLR, Bonn
gernot.hartmann@dlr.de

Hasinger, Günther
MPI Extraterrestrische Physik,
Garching
ghasinger@mpe.mpg.de

Heber, Ulrich
Dr. Remeis-Sternwarte, Bamberg
heber@sternwarte.uni-erlangen.de

Hernandez, Pilar
CERN
Pilar.Hernandez@cern.ch

Hillenbach, Mark
Springer-Verlag, Heidelberg
mark@th.physik.uni-bonn.de

Horvat, Raul
Rugjer Boskovic Institute, Zagreb
horvat@lei3.irb.hr

Ikebe, Yasushi
MPI Extraterrestrische Physik,
Garching
ikebe@mpe.mpg.de

Irgens-Jensen, Synnøve
Research Council of Norway, Oslo
sij@forskningsradet.no

Jacob, Maurice
CERN
maurice.jacob@cern.ch

Jagoutz, Emil
MPI Kosmochemie, Mainz
jagoutz@mpch-mainz.mpg.de

Jameux, David
ESA–ESTEC, Noordwijk
David.Jameux@esa.int

Janknecht, Eckart
Hamburger Sternwarte
ejanknecht@hs.uni-hamburg.de

Johann, Ulrich
Astrium Friedrichshafen
ulrich.johann@astrium-space.com

Jørgensen, Henning
Copenhagen Observatory
henning@astro.ku.dk

Käufl, Ulrich
ESO Garching
hukaufl@eso.org

Kellner, Gottfried
CERN
gottfried.kellner@cern.ch

Kellner, Marianne
CERN
marianne.kellner@cern.ch

Kilbinger, Martin
University of Bonn
kilbinge@astro.uni-bonn.de

Klose, Sylvio
Thüringer Landessternwarte
Tautenburg
klose@tls-tautenburg.de

Knöpfle, Karl-Tasso
MPI Kernphysik, Heidelberg
ktkno@mpi-hd.mpg.de

Koang, Dy-Holm
CNRS / UJF Grenoble
dy-holm.koang@isn.in2p3.fr

Komossa, Stefanie
MPI Extraterrestrische Physik,
Garching
skomossa@mpe.mpg.de

Königsmann, Kay
Universität Freiburg
k.konigsmann@cern.ch

Krauss, Lawrence
Case Western Reserve University
krauss@theory1.phys.cwru.edu

Krichbaum, Thomas
MPI Radioastronomie, Bonn
tkrichbaum@mpifr-bonn.mpg.de

Kryukov, Igor
Institute for Problems in Mechanics,
Moscow
kryukov@ipmnet.ru

Kuntschner, Harald
ESO Garching
hkuntsch@eso.org

Kurz, Richard
ESO Garching
rkurz@eso.org

Lavocat, Philippe
CEA Saclay
philippe.lavocat@cea.fr

Le Coultre, Pierre
CERN
Pierre.Le.Coultre@cern.ch

Leibundgut, Bruno
ESO Garching
bleibund@eso.org

Lorenzen, Dirk H.
Deutschlandfunk
DLorenzen@compuserve.com

Maccarone, Maria Concetta
IFCAI / CNR, Palermo
Cettina.Maccarone
@ifcai.pa.cnr.it

Madau, Piero
University of California at Santa Cruz
pmadau@ucolick.org

Mainieri, Vincenzo
ESO Garching
vmainier@eso.org

Marchetti, Enrico
ESO Garching
emarchet@eso.org

Mayor, Michel
Observatoire de Genève
Michel.Mayor@obs.unige.ch

Mellier, Yannick
Institut d'Astrophysique de Paris
mellier@iap.fr

Meylan, Georges
ESA / STScI
gmeylan@stsci.edu

Moorwood, Alan
ESO Garching
amoor@eso.org

Morselli, Aldo
INFN, Rome
Aldo.Morselli@roma2.infn.it

Müller, Thomas
Basler Zeitung
thomas.mueller@baz.ch

Nédélec, Patrick
LAPP, Annecy-le-Vieux
nedelec@lapp.in2p3.fr

Neuhäuser, Ralph
MPI Extraterrestrische Physik,
Garching
rne@mpe.mpg.de

Neumann, Hans-Werner
Sternwarte Neumarkt
hanswneumann@gmx.de

Noh, Hyerim
Korea Astronomical Observatory,
Taejon
hr@kao.re.kr

Pacini, Franco
Osservatorio Astrofisico di Arcetri
pacini@arcetri.astro.it

Palle, Davor
Rugjer Boskovic Institute, Zagreb
palle@mefisto.irb.hr

Paresce, Francesco
ESO Garching
fparesce@eso.org

Pasquali, Anna
ST-ECF Garching
apasqual@eso.org

Patat, Ferdinando
ESO Garching
fpatat@eso.org

Perryman, Michael
ESA–ESTEC, Noordwijk
mperryma@rssd.esa.int

Pich, Antonio
Univ. of Valencia, IFIC
antonio.pich@uv.es

Pirenne, Benoît
ESO Garching
bpirenne@eso.org

Pogorelov, Nikolai
Institute for Problems in Mechanics,
Moscow
pgrlv@ipmnet.ru

Preuss, Eugen
MPI Radioastronomie, Bonn
epreuss@mpifr-bonn.mpg.de

Primas, Francesca
ESO Garching
fprimas@eso.org

Puzia, Thomas H.
University Observatory Munich
puzia@usm.uni-muenchen.de

Raffelt, Georg
MPI Physik, Munich
raffelt@mppmu.mpg.de

Ranzan, Conrad
Cosmic Research Center,
Niagara Falls
CozmicResCenter@aol.com

Re, Virginia
University of Salerno
virginia.re@virgilio.it

Rees, Martin
IoA Cambridge
mjr@ast.cam.ac.uk

Reimelt, Nils
Burda Digital GmbH, Munich
reimelt@burdadigital.com

Reinhard, Ruedeger
ESA, Noordwijk
Ruedeger.Reinhard@esa.int

Renault, Cecile
Institut des Sciences Nucléaires,
Grenoble
rcecile@in2p3.fr

Renzini, Alvio
ESO Garching
arenzini@eso.org

Rich, James
CEA Saclay
rich@hep.saclay.cea.fr

Roeser, Siegfried
Astronomisches Rechen-Institut
roeser@ari.uni-heidelberg.de

Romaniello, Martino
ESO Garching
mromanie@eso.org

Roos, Matts
University of Helsinki
Matts.Roos@helsinki.fi

Roxburgh, Ian
Queen Mary College, Univ. London
I.W.Roxburgh@qmw.ac.uk

Rupprecht, Gero
ESO Garching
grupprec@eso.org

Rydbeck, Gustaf
Onsala Space Observatory
gustaf@oso.chalmers.se

Saavedra, Oscar
University of Torino
saavedra@to.infn.it

Sanchez-Conejo, Jorge
CERN
Jorge.Sanchez@cern.ch

Sbarra, Carla
ITESRE – CNR, Bologna
sbarra@tesre.bo.cnr.it

Schacher, Jürg
Lab. of High Energy Physics, Bern
juerg.schacher@lhep.unibe.ch

Schneider, Peter
University of Bonn
peter@astro.uni-bonn.de

Schulze, Michael P.
Astrophysikalisches Institut Potsdam
mps@physik.tu-berlin.de,
mpschulze@t-online.de

Schutz, Bernhard F.
MPI Gravitationsphysik, Golm
ute@aei-potsdam.mpg.de

Sereno, Mauro
Università di Napoli "Federico II"
sereno@na.infn.it

Shaver, Peter
ESO Garching
pshaver@eso.org

Siebenmorgen, Ralf
ESO Garching
rsiebenm@eso.org

Sikkema, Gert
ESO Garching
gsikkema@eso.org

Silva, David
ESO Garching
dsilva@eso.org

Southwood, David
European Space Agency
David.Southwood@esa.int

Sunyaev, Rashid
MPI Astrophysik, Garching
sunyaev@mpa-garching.mpg.de

Tao, Charling
CPPM, Marseille
tao@cppm.in2p3.fr

Tauscher, Ludwig
Institute for Physics, Basel
ludwig.tauscher@cern.ch

Taxil, Pierre
Centre de Physique Théorique,
Marseille
taxil@cpt.univ-mrs.fr

Teuscher, Richard
CERN / Enrico Fermi Institute
Richard.Teuscher@cern.ch

Thomsen, Sara
Universität Hamburg
spacesara@gmx.de

Turok, Neil
Cambridge University
N.G.Turok@damtp.cam.ac.uk

Ulbricht, Jürgen
Institut für Hochenergiephysik, Zürich
ulbricht@ihp.phys.ethz.ch

Vaas, Rüdiger
Bild der Wissenschaft /
Univ. Stuttgart
ruediger.vaas@dva.de

van de Bruck, Carsten
University of Cambridge
cv224@damtp.cam.ac.uk

van den Heuvel, Edward P.J.
University of Amsterdam
edvdh@astro.uva.nl

van der Kruit, Pieter
Kapteyn Astronomical Institute,
Groningen
vdkruit@astro.rug.nl

van Dishoeck, Ewine
Leiden Observatory
ewine@strw.leidenuniv.nl

van Eijndhoven, Nick
Utrecht University / Nikhef
nick@phys.uu.nl

Vangioni-Flam, Elisabeth
Institut d'Astrophysique de Paris
flam@iap.fr

Vanzella, Eros
ESO Garching
evanzell@eso.org

Vilenkin, Alexander
Tufts University
vilenkin@cosmos.phy.tufts.edu

Vitale, Stefano
Università di Trento
vitale@science.unitn.it

Völk, Heinrich J.
MPI Kernphysik, Heidelberg
Heinrich.Voelk@mpi-hd.mpg.de

Volonte, Sergio
European Space Agency
Sergio.Volonte@esa.int

von Rauchhaupt, Ulf
Frankfurter Allgemeine Zeitung
rauchhaupt@faz.de

Voûte, Lodie
University of Amsterdam
lodie@astro.uva.nl

Wallraff, Wolfgang
RWTH, Aachen
wallraff@physik.rwth-aachen.de

Walsh, Jeremy
ST-ECF Garching
jwalsh@eso.org

Warmels, Rein
ESO Garching
rwarmels@eso.org

Watson, Alan
University of Leeds
a.a.watson@leeds.ac.uk

Weghorn, Hans
University of Cooperative Education,
Stuttgart
weghorn@ba-stuttgart.de

Wiklind, Tommy
Onsala Space Observatory
tommy@oso.chalmers.se

Woltjer, Lodewijk
OHP / Osservatorio di Arcetri
woltjer@obs-hp.fr

Wrochna, Grzegorz
Soltan Institute for Nuclear Studies,
Warsaw
wrochna@fuw.edu.pl

Zaroubi, Saleem
MPI Astrophysik, Garching
saleem@mpa-garching.mpg.de

Zeilinger, Werner
Institute for Astronomy, Vienna
zeilinger@astro.univie.ac.at

Ziaeepour, Houri
Mullard Space Science Laboratory
hz@mssl.ucl.ac.uk

Zucca, Elena
Osservatorio Astronomico di Bologna
zucca@bo.astro.it

Cosmic Microwave Background Fluctuations

S. Masi[1], P.A.R. Ade[2], P. de Bernardis[1], J.J. Bock[3], J.R. Bond[4], J. Borrill[5],
A. Boscaleri[6], K. Coble[7], C.R. Contaldi[4], B.P. Crill[8], G. De Troia[1],
P. Ferreira[9], K. Ganga[10], M. Giacometti[1], E. Hivon[10], V.V. Hristov[8],
A. Iacoangeli[1], A.H. Jaffe[11], W.C. Jones[8], A.E. Lange[8], L. Martinis[12],
P. Mason[8], P.D. Mauskopf[2], A. Melchiorri[9], T. Montroy[7], F. Nati[1], P. Natoli[13],
C.B. Netterfield[14], E. Pascale[6], F. Piacentini[1], D. Pogosyan[4], G. Polenta[1],
F. Pongetti[15], S. Prunet[4], G. Romeo[15], J.E. Ruhl[7], F. Scaramuzzi[12]

[1] Dipartimento di Fisica, Universita' La Sapienza, Roma, Italy
[2] Dept. of Physics and Astronomy, Cardiff University, Cardiff CF24 3YB, Wales, UK
[3] Jet Propulsion Laboratory, Pasadena, CA, USA
[4] Canadian Institute for Theoretical Astrophysics, University of Toronto, Canada
[5] National Energy Research Scientific Computing Center, LBNL, Berkeley, CA, USA
[6] IROE-CNR, Firenze, Italy
[7] Dept. of Physics, Univ. of California, Santa Barbara, CA, USA
[8] California Institute of Technology, Pasadena, CA, USA
[9] Nuclear and Astrophysics Laboratory, University of Oxford, Keble Road, Oxford, OX 3RH, UK
[10] IPAC, California Institute of Technology, Pasadena, CA, USA
[11] Astrophysics Group, Blackett Laboratory, Imperial College, London
[12] ENEA, Frascati, Italy
[13] Dipartimento di Fisica, Seconda Universita' di Roma, Tor Vergata, Italy
[14] Depts. of Physics and Astronomy, University of Toronto, Canada
[15] Istituto Nazionale di Geofisica, Roma, Italy

Abstract. Several experiments have recently detected very low contrast, sub-horizon scale structures in the Cosmic Microwave Background (CMB). In the current cosmological model, these structures result from acoustic oscillations of the primeval plasma at recombination ($z \sim 1100$). In the framework of the Hot Big Bang theory with the inflation hypothesis, the statistical properties of the image of the CMB allow us to measure most of the cosmological parameters.

1 CMB Physics

In the framework of the Hot Big Bang model, when we look to the Cosmic Microwave Background we look back in time to the epoch when temperature decreased below 3000K for the first time: then hydrogen atoms formed from the primeval plasma and the universe became transparent (recombination). It is the end of the plasma era, at a redshift ~ 1000, when the universe was ~ 50000 times younger, ~ 1000 times hotter and $\sim 10^9$ times denser than today ([12], [58], [42]). Photons generated in the early Universe are last scattered at recombination. Thereafter they travel free to our telescopes. Density perturbations $\Delta\rho/\rho$ were oscillating in the plasma era (as a result of the opposite effects of gravity and photon pressure). After recombination, density perturbations can grow and create the hierarchy of structures we see in the nearby Universe. CMB temperature fluctuations are closely linked to density fluctuations at recombination through three effects: the density fluctuation of the plasma of photons and matter in thermal equilibrium, the gravitational redshift of photon escaping from a density perturbation, and the Doppler effect of photons scattered by electrons with infall velocity v. In formulas $\Delta T/T = \Delta\rho/\rho/3 + \Delta\phi/c^2/3 - (v/c)\cdot n$. Mapping the CMB temperature we map the density and velocity fields in the Universe at recombination, and sample the direct result of early physical processes. In the primeval plasma, photons/baryons density perturbations start to oscillate only when the sound horizon becomes larger than their linear size. Small wavelength perturbations oscillate faster than large ones. The result is that perturbations with different linear size arrive at recombination with different phase. The largest ones have about the same size as the horizon at recombination: these have just enough time to arrive at recombination maximally compressed (or rarified). They produce a network of cold and hot spots in the image, with the dimensions of the acoustic horizon (similar to the causal one, since the sound speed is $\sim c/\sqrt{3}$). The presence of a characteristic dimension of the spots results in a peak in the angular power spectrum of the CMB image. The projected angular size of the horizon can be computed by means of the Friedmann equation as a function of the cosmological parameters. We expect a strong dependence of this observable from the density parameter, which controls the curvature of the Universe. In a super-critical density, positive curvature Universe, light rays from opposite sides of the perturbation converge towards the observer along curved geodesics: the same density perturbations will appear larger than in an Euclidean Universe. The reverse happens in a low density Universe with negative curvature. The computation shows that the result is mainly a function of two cosmological parameters, Ω_Λ and Ω_{Mo}, but the typical size of the horizon is always of the order of 1°. If $\Omega_\Lambda + \Omega_{Mo} \simeq \Omega = 1$, then the dependence on the relative contributions of matter and vacuum is very weak, and the typical size of the horizon is $\theta_H \sim 0.85^o$. These spots produce a peak in the power spectrum at multipole $\ell_1 = \pi/\theta_H \sim 210$.

Smaller perturbations start to oscillate earlier and have more time to oscillate, so they can arrive at recombination with different phases. Also, they oscillate faster than large ones. Depending on the number of quarter periods they can undergo, they can arrive either with maximum compression (rarefaction) or with zero density contrast. We thus expect an approximately "harmonic" series of interleaved peaks and dips in the angular power spectrum of the CMB. In the adiabatic inflationary model, the relative amplitudes of the second to first peak is sensitive to the density of baryons Ω_{Bo} and to the spectral index n of the primordial density fluctuations spectrum, assumed of the form $P(k) \sim k^n$. Detection of the third peak allows to remove the degeneracy between n and Ω_{Bo} and measure both (see [18] for a review of the theory of cosmological acoustic oscillations and [9] for a discussion on the measurement of cosmological parameters from the angular power spectrum of CMB anisotropy).

The inflationary hypothesis (see e.g. [25], [19]) is very appealing, because it explains the large-scale smoothness of the CMB by solving the paradox of horizons; it stretches the space to flat, so that $\Omega = 1$ naturally; it inflates quantum fluctuations from microscopic scales to cosmological scales, creating scalar and tensor fluctuations responsible for CMB anisotropies and for galaxy formation. Scalar (density) fluctuations are expected to be gaussian, adiabatic, and to have a Harrison-Zeldovich power spectrum $P(k) \sim k^n$ with $n \lesssim 1$. In this framework, the image of the CMB is an image of quantum fluctuations at energies of 10^{19} GeV, seen through the powerful microscope of the inflationary Universe. Measurements of the CMB can test the inflationary hypothesis in three ways: by measuring Ω, by measuring n, and by measuring the peculiar CMB polarization pattern due to tensor fluctuations [20].

2 BOOMERanG

2.1 The Payload

The BOOMERanG experiment, a balloon-borne microwave telescope with cryogenic bolometric detectors, was flown on a long duration circum-Antarctic stratospheric balloon in 1998. Several aspects of the instrument, which was sensitive at 90, 150, 240, 410 GHz with beam resolution of 18′, 10′, 14′ and 13′ FWHM

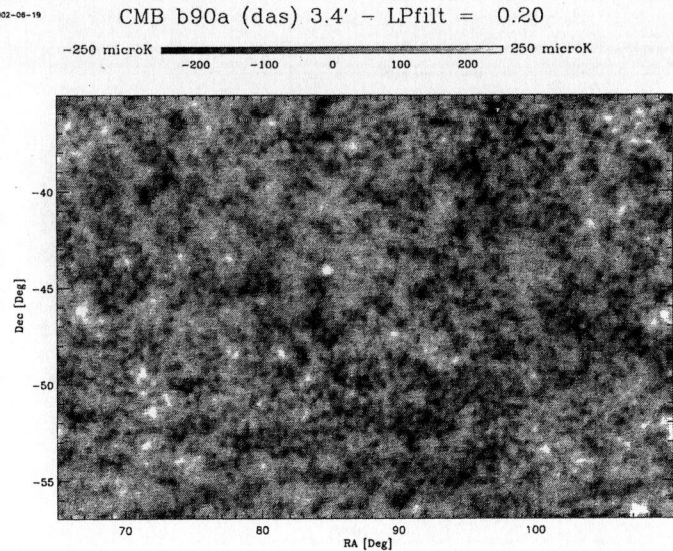

Fig. 1. Map of the sky brightness fluctuations measured by BOOMERanG at 90 GHz. The beam FWHM was 18′; the map has been smoothed with a 12′ FWHM gaussian

respectively, have been described in [33], [26], [27], [46], [41], [50], [6]. The payload was flown by NASA-NSBF on Dec.29, 1998, from McMurdo (Antarctica). It remained at float for 10.6 days, circumnavigating Antarctica at an average altitude of 37 km. About 57 million 16-bit samples of the signal were collected for each of the 16 detectors. The data were edited for known instrument glitches, temperature fluctuations, and cosmic rays events. Less than 5% of the bolometer data has been found to be contaminated. Constrained realizations of noise were substituted to the contaminated signals.

2.2   Sky Maps

The pointing has been reconstructed from the signals of the laser gyroscopes, of the differential GPS, and of the sun sensors. In the most recent pointing solution, repeated observations of compact sources show that the accuracy of the reconstruction is $\lesssim 2.5$ arcmin *rms*. Random errors in the pointing have the effect to smear-out the signals from small sources. This adds in quadrature to the intrinsic angular resolution of the telescope (9.5′ FWHM at 150 GHz). The finite size of the pixelization has a similar effect. In ℓ space these three effects are modelled as a low pass filter, with a shallow cutting-off at about $\ell \sim 600$. In order to remove the effect of 1/f noise and drifts we have high-pass filtered the data in the time domain. This process acts as a sharp high-pass at multipole ~ 30. The result is a window function $W(\ell)$, which has to be taken into account

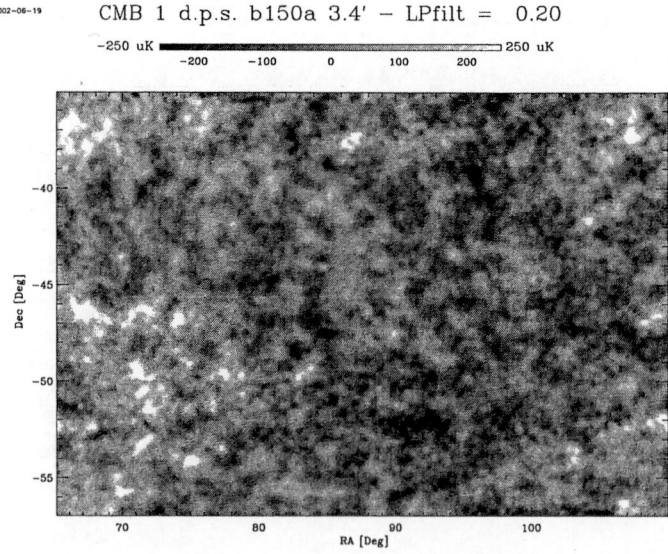

Fig. 2. Map of the sky brightness fluctuations measured by BOOMERanG at 150 GHz. The beam FWHM was 12′; the map has been smoothed with a 12′ FWHM gaussian

in the reconstruction of the angular power spectrum of the sky. These effects limit the sensitivity of our observations at low and high multipoles. The result is that BOOMERanG is sensitive to a range of multipoles from $\ell = 30$ to $\ell = 1000$.

Sky maps have been constructed from the time ordered data and pointing using four independent methods: naive maps (just coadding data on the same pixel); maximum likelihood maps obtained using the MADCAP package [3]; maximum likelihood maps obtained using the iterative method of [38]; suboptimal maps obtained using the fast map making method of [16]. Degree-scale structures with amplitude of the order of $\sim 100\ \mu K$ are evident in the map at 150 GHz. Consistent structure is also evident in the maps at 90 and 240 GHz (see Fig. 1, 2, 3). The similarity of the temperature maps obtained at different frequencies ([7]) is the best evidence for the CMB origin of the detected fluctuations. Foregrounds contamination can be constrained significantly in the center of the observed sky region [7], by comparison and correlation with dust templates [28]. Moreover, at variance with Galactic maps, the fluctuations detected at 150 GHz pass several tests for gaussianity [47].

2.3   Gaussianity

We have computed skewness, kurtosis and Minkowski functionals for the BOOMERanG maps at 150 GHz and 410 GHz, and for gaussian simulations of the CMB

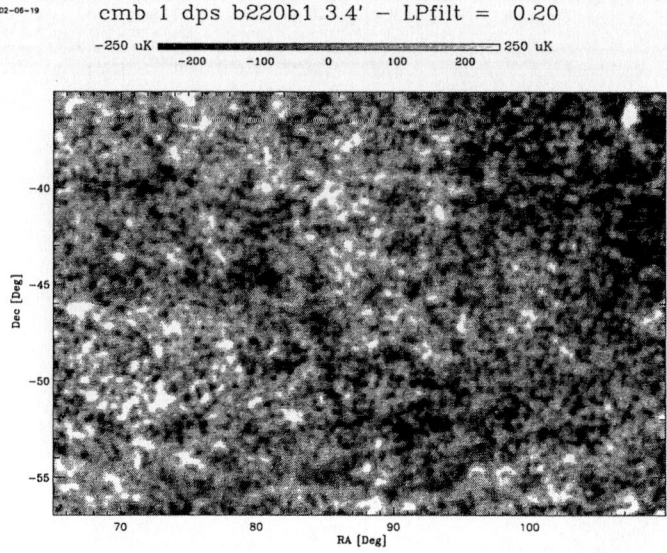

Fig. 3. Map of the sky brightness fluctuations measured by BOOMERanG at 240 GHz. The beam FWHM was 14′; the map has been smoothed with a 12′ FWHM gaussian

sky. For all these tests the measured data are fully consistent with the distributions expected from gaussian sky simulations in the high galactic latitude section of the 150 GHz map. We found deviations from gaussianity at lower latitudes, and at 410 GHz, and we ascribed them to Galactic dust contamination. For the 150 GHz map, where the bulk of the fluctuations is gaussian, it is important to estimate which is the maximum level of non gaussianity still consistent with the data. There are many types of non gaussianity. We started this analysis using two toy models.

In the first one a non gaussian sky (distributed as a χ^2 with DOF degrees of freedom) is obtained as the sum of DOF squared gaussian ΔT, with the constrain to produce the same power spectrum detected by BOOM-ERanG. BOOMERanG observations of that sky are simulated and a measured map, including noise, is obtained. Many realizations are simulated, and for each realization the three Minkowski functionals $v_M(\nu_i)$ are computed. Here ν_i are the selected temperature thresholds in units of the standard deviation of the sky temperature, and the index M ranges from 1 to 3 for Area, Length, and Genus functionals respectively. The simulations are compared to the functionals measured from the real sky $v_M^S(\nu_i)$, and the statistics $S_M = \sum_{ij}(v_M(\nu_i) - < v_M^S(\nu_i) >)M_{ij}^{-1}(v_M(\nu_i) - < v_M^S(\nu_i) >)$ is computed. Models with less than 80 DOF are excluded at 95% confidence (see Fig. 4; see [10] for the analogous analysis on the distribution of galaxies).

In the second toy model a slightly non gaussian sky was obtained as the sum of a gaussian ΔT plus a ΔT distributed as a χ_1^2. The same procedure as above

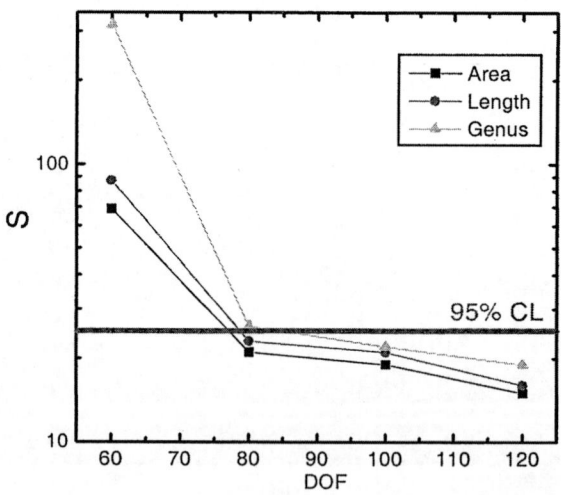

Fig. 4. The Statistics S, based on the Minkowski functionals, and defined in the text, is computed from the 150 GHz map of BOOMERanG and from non gaussian simulations distributed as a χ^2 with DOF degrees of freedom. The horizontal line is the 95% CL upper limit. Fluctuations with less than 80 DOF are excluded by the measured 150 GHz map

was followed. The rms value of the $\chi_1^2 \, \Delta T$ ranged from 1% to 15% of the rms of the gaussian ΔT. Models with a $\chi_1^2 \, \Delta T$ larger than 7% are excluded at 95% C.L. (see Fig. 5).

In a similar way we were able to exclude instrumental systematic effects and non gaussianity from foregrounds [47].

The gaussianity of the CMB temperature fluctuations is not trivial. As a matter of facts, if we do the gaussianity analysis on sky brightness data with the same angular resolution, but in different regions of the em spectrum, we do find important deviations from gaussianity. Our own 410 GHz channel is a good example of this behaviour [47]. The gaussian character of the fluctuations is telling us something important about the cosmological origin of the fluctuations, and is in good agreement with the predictions of the simplest inflationary models.

2.4   The Power Spectrum

Gaussianity also insures us that all the information encoded in the map is provided by the angular power spectrum of the map. We have computed the power spectrum of the sky by means of independent methods ([3], [16]), which rigorously take into account the effects of system noise, incomplete sky coverage, time-domain filtering of the data, beam shape etc. The current best spectrum of the sky is obtained from a combination of four channels at 150 GHz, and is shown in Fig. 6 (see [39] for all the details of the analysis). In this spectrum,

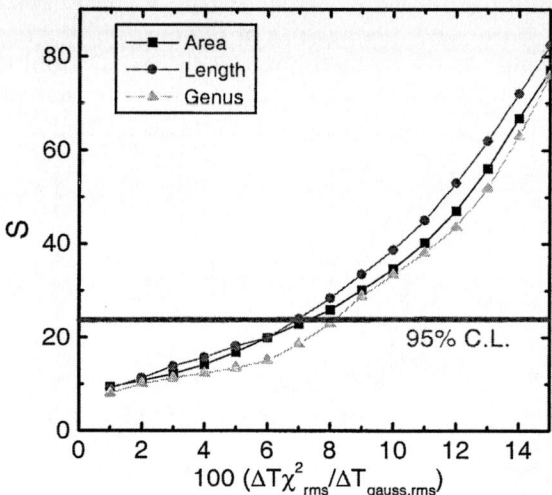

Fig. 5. The Statistics S, based on the Minkowski functionals, and defined in the text, is computed from the 150 GHz map of BOOMERanG and from a mix of non gaussian temperature fluctuations distributed as a χ^2 with one DOF and gaussian temperature fluctuations. The horizontal line is the 95% CL upper limit. The rms of the non gaussian component has to be less than 7% of the rms of the gaussian one to be consistent with the measured 150 GHz map

a peak at $\ell_{p1} \sim 213$ is evident at high statistical significance, and two further peaks at $\ell_{p2} \sim 541$, $\ell_{p3} \sim 845$ are also present at about 2σ significance [8]. They are interleaved with two dips at $\ell_{d1} \sim 416$ and $\ell_{d2} \sim 750$.

The power spectrum results from BOOMERanG have now been confirmed and extended by independent experiments. The measurements of MAXIMA, DASI, VSA, CBI ([23], [22], [51], [31]) are all consistent with the power spectrum detected by BOOMERanG. Moreover, all these independent experiments find consistent estimates for the cosmological parameters.

3   Impact of the Power Spectrum Results on Cosmology

The simplest interpretation is that peaks and dips in the power spectrum result from acoustic oscillations in the primeval plasma. The existence of these oscillations was predicted long time ago ([43],[56],[54]), and is expected in the standard inflationary scenario [2]. We derive the cosmological parameters within this theory framework, assuming adiabatic and gaussian initial density fluctuations. This a restrictive hypothesis (see e.g. [4]), but simplifies significantly the analysis; working with a more general set of initial conditions requires more data. With this assumption a full database of power spectra can be built by allowing each of the cosmological parameters to cover a wide range of values. The spectra are computed using the programs CMBFAST and CAMB ([52], [24]).

As explained in the introduction, the location of the first peak is directly related to the angular size of the acoustic horizon at recombination, and the comparison between linear and angular size of the horizon provides a clean way to measure Ω ([60] [35]). In [7] we preliminarily derived Ω from the location of

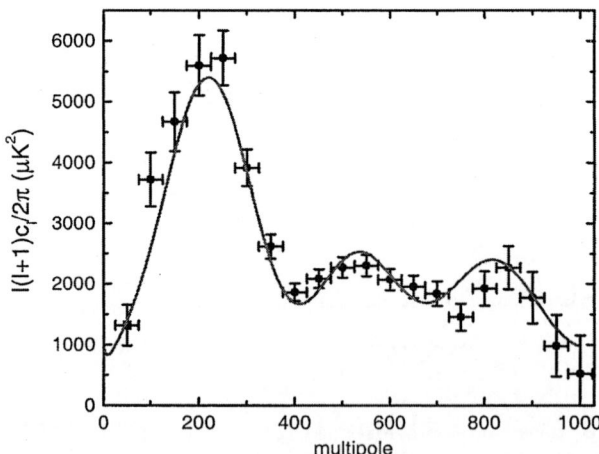

Fig. 6. Power spectrum of the CMB estimated from the combination of four independent 150 GHz maps of the sky measured by BOOMERanG. The best fit adiabatic inflationary model is shown as a continuous line

the first peak. We also derived a set of cosmological parameters by means of a bayesian analysis of the full power spectrum (for $\ell < 600$). While the results for Ω clearly indicated a flat geometry of the Universe, the results for other cosmological parameters (like $\Omega_b h^2$ and n_s) were still affected by calibration errors (effective beam and gain) and by the degeneracy of the spectra due to limited coverage of the multipoles space. In subsequent papers we derived the cosmological parameters from the full power spectrum of BOOMERanG (up to $\ell \sim 1000$). This was derived from the final BOOMERanG maps, featuring improved beam and pointing reconstruction ([39], [8]). The possibility to detect the third peak in the spectrum removed the degeneracy between $\Omega_b h^2$ and n_s. We find perfect consistency with a Euclidean Universe: $\Omega = 1.02^{+0.05}_{-0.06}$. This result is a significant improvement in precision, and is very important for Cosmology. If we trust General Relativity, the observation that $\Omega \sim 1$ today is very interesting. Our universe is following a very unstable solution of Einstein equations, and some mechanism (like inflation) has to be included in the standard Hot Big Bang scenario in order to explain this fact.

The second cosmological parameter well constrained by these data is the physical density of baryons. The density of baryons affects the symmetry of acoustic oscillations before recombination, thus controlling the ratio between amplitudes of even and odd order peaks in the power spectrum. We find $\Omega_b h^2 = 0.022^{+0.004}_{-0.003}$. The density of baryons also controls the nuclear reactions happening in the first minutes after the Big Bang and producing the light elements. Measurements of the primordial abundance of elements allow to estimate $\Omega_b h^2$ with high precision. Recent observations of primordial deuterium from

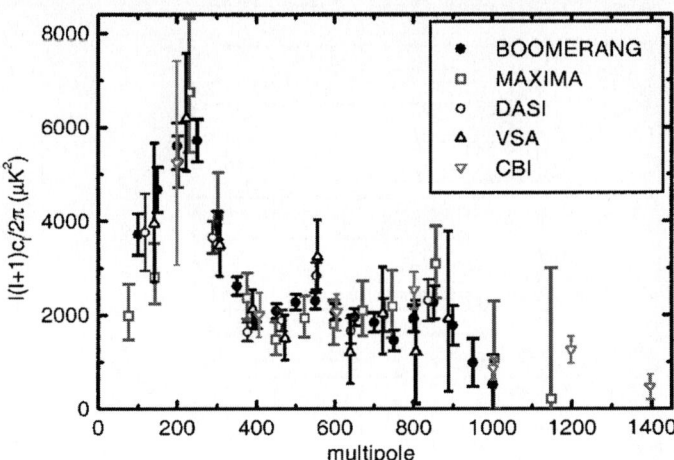

Fig. 7. Recent measurements of the angular power spectrum of the CMB. CBI data extend to $\ell \sim 3500$, so not all are shown. Due to the different binning and covariance properties of the data, this plot is not intended for a quantitative comparison. It displays, however, a remarkable consistency between independent and technically orthogonal experiments

quasar absorption line systems suggest a value $\Omega_b h^2 = 0.020 \pm 0.002$ at the 95% C.L. [5]. It is remarkable that such orthogonal techniques produce very consistent results: in our view it is a strong indication of the overall consistency of the hot big bang scenario.

The third parameter constrained is the spectral index of the power spectrum of primordial density perturbations. We find $n_s = 0.96 \pm 0.09$. In inflation models, the spectral index of the primordial fluctuations gives information about the shape of the primordial potential of the inflaton field which drove inflation. While there is no fundamental constraint on this parameter, the simplest models of inflation do give values that are just below unity.

With limited prior knowledge on the Hubble constant (i.e. $0.4 < h < 0.9$) and on the age of the Universe (i.e. $t > 10$ Gyrs)we are able to constrain significantly the density of dark matter and of dark energy. The result is a universe where ordinary matter accounts for only 4% of the total mass-energy density, while about 25% of the Universe is in the form of still undetected dark matter. A mysterious form of dark energy (i.e. something with negative pressure in the stress-energy tensor) accounts for about 70% of the mass-energy density. This apparently strange composition of the Universe is confirmed by independent cosmological observations (see e.g. [44], [45], [11], [49]) and remains one of the most important open questions in cosmology and fundamental physics.

Combining the results of the current CMB anisotropy experiments (BOOM-ERanG, CBI, MAXIMA, VSA, DASI and COBE) improves the estimates of the

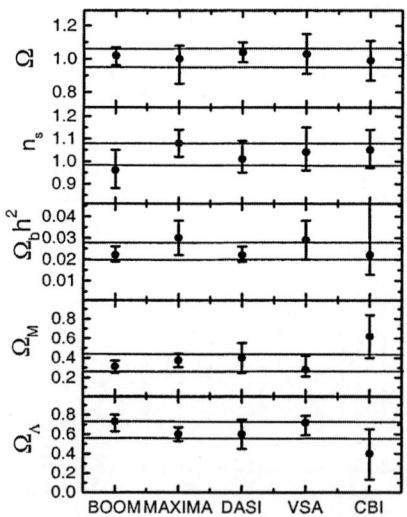

Fig. 8. Determination of the cosmological parameters from five independent experiments. All these experiment are analyzed with similar priors (see text), and in combination with the COBE-DMR data at large scales. The error bars correspond to 68% confidence intervals. The horizontal lines represent 95% confidence intervals for the parameters

cosmological parameters given above. (see Fig. 8). The best combined estimate of cosmological parameters from CMB experiments to date is in [53].

4   The Future

CMB experiments can improve our understanding of Nature in several ways. We will not even start to say how much the two CMB space mission (the ongoing MAP of NASA and the incoming Planck of ESA) will impact cosmology. They will provide full sky maps free of systematic effects, with high resolution, wide frequency coverage, and precise calibration, resulting in precision measurements of the CMB power spectrum and of the cosmological parameters. CMB Polarization measurements like the ones carried out by the forthcoming BOOMERanG payload (B2K, see e.g. [29], [30]) will confirm the origin of the anisotropy and provide evidence for the type of perturbations (adiabatic vs isocurvature, for example) responsible for the initial density fluctuations. A subsequent generation of polarization experiments will provide deep insight on the physics of inflation (see e.g. [20]). Meanwhile, interferometers like AMI, AmiBA, BIMA, CBI, SZA, and similar are/will extend our knowledge of the CMB power spectrum at very small angular scales, allowing the investigation of the early Universe at scales corresponding to clusters of galaxies and even smaller.

Acknowledgements

The BOOMERanG experiment is supported in Italy by Agenzia Spaziale Italiana, Programma Nazionale Ricerche in Antartide, Universita' di Roma La Sapienza; by PPARC in the U.K., by NASA, NSF OPP and NERSC in the U.S.A., and by CIAR and NSERC in Canada.

References

1. Benoit A., et al., Astroparticle Phys., in press, 2001, astro-ph/0106152
2. Bond, J.R. & Efstathiou, G. 1987, Mon. Not. R. Astron. Soc. 226, 655
3. Borrill, J., in 3K Cosmology astro-ph/9911389
4. Bucher M., Moodley K., Turok N., Phys.Rev.Lett. 87 (2001) 191301
5. Burles, S., Nollett, K.M. & Turner, M.S. 2000, astro-ph/0010171
6. Crill B., et al., astro-ph/0206254.
7. de Bernardis, P., et al. 2000, Nature, 404, 955-959
8. de Bernardis, P., et al. 2002, Ap.J. 564, 559-566, astro-ph/0105296
9. Efstathiou G., and Bond, J. R., 1999, MNRAS, 304, 75
10. Feldman H.A., et al., Phys. Rev. Lett. 86, 1434-1437, (2001), astro-ph/0010205
11. P.M. Garnavich et al. 1998, Ap.J. Letters 493, L53;
12. Gamow G., 1946, Phys.Rev., 70, 572
13. Gush H.P. et al., 1990, Phys. Rev. Lett., 65, 537
14. Hanany S., Jaffe A., Scannapieco E., 1998, MNRAS 299, 653-660, astro-ph/9801291
15. Hanany, S. et al., 2000, Ap.J., 545, L5-L9

16. Hivon E. et al., 2001, astro-ph/0105302
17. Hu W., Sugiyama N. & Silk J., 1997, Nature, 386, 37
18. Hu, W., & Dodelson, S., ARAA, 2002 (astro-ph/0110414).
19. Kolb E.W. and Turner M.S, 1990, The Early Universe, Addison-Wesley
20. Kamionkowski M., Kosowski A., Phys.Rev.D., 67, 685, (1998)
21. Lee A., et al., 1999, in "3K cosmology", AIP Conf. Proc. 476; astro-ph/9903249
22. Lee A., et al., 2002, submitted to Ap.J., astro-ph/0104459
23. Leitch E.M., et al., 2001, astro-ph/0104488
24. Lewis A., Challinor A., Lasenby A., astro-ph/9911177 (1999)
25. Linde A., 2002, Stephen Hawking's 60th birthday conference, Cambridge University, Jan. 2002 hep-th/0205259
26. Masi S., et al., Cryogenics, 38, 319-324 (1998)
27. Masi S., et al., Cryogenics, 39, 217-224 (1999)
28. Masi S., et al., Ap.J, 553, L93-L96, 2001, astro-ph/0101539
29. Masi S., et al., in Astrophysical Polarized Backgrounds, AIP Conf. Ser. 609, pg. 122-127; Cecchini, Cortiglioni, Sault and Sbarra eds. (2001)
30. Masi S., et al., in Experimental Cosmology at Millimetere Wavelengths, 2K1BC, AIP Conf. Ser. 616, 168-174, De Petris and Gervasi eds.(2001)
31. Mason B., et al., 2002, submitted to Ap.J., astro-ph/0205384; Pearson T.J., et al., 2002, submitted to Ap.J., astro-ph/0205388.
32. Mather J., et al., 1990, Ap.J., 354, L37
33. Mauskopf P., et al., Applied Optics 36, 765-771, (1997)
34. Mauskopf P., et al., 2000, Ap.J., 536, L59-L62
35. Melchiorri A. and Griffiths L.M., astro-ph/0011147, (2000)
36. Miller, A. et al. 1999, Ap.J., 524, L1
37. Miller, A. et al. 2001, astro-ph/0108030
38. Natoli P., et al. 2001, submitted to Ap.J., astro-ph/0101252
39. Netterfield B., et al., 2002, Ap.J., in press, astro-ph/0104460
40. Page L., 2000, Proc IAU Symposium 201 Eds A. Lasenby & A. Wilkinson, astro-ph/0012214
41. Pascale E., and Boscaleri A., in BC2K1, AIP Conf. Ser. 616, De Petris and Gervasi eds.
42. Peebles P.J.E, 1994, Principles of Physical Cosmology, Princeton Series in Physics
43. Peebles, P.J.E, and Yu, J.T. 1970, Ap.J. 162, 815
44. Perlmutter, S. et al. 1997, Ap. J. 483, 565;
45. S. Perlmutter et al. 1998, Nature 391, 51;
46. Piacentini F., et al., Ap.J.Suppl. 138, 315-336, (2002) astro-ph/0105148
47. Polenta G., et al., Ap.J.Letters in press, astro-ph/0201133
48. Pryke C., et al., astro-ph/0104490
49. A.G. Riess et al. 1998, Ap. J. 116, 1009
50. Romeo G., et al., in BC2K1, AIP Conf. Ser. 616, De Petris and Gervasi eds.
51. Scott P.F., et al., 2002, MNRAS, submitted, astro-ph/0205380
52. Seljak U., Zaldarriaga M., Astrophys.J., 469, 437, (1996)
53. Sievers J.L., et al., 2002, submitted to Ap.J., astro-ph/0205387
54. Silk J. & Wilson M. L. 1980, Physica Scripta, 21, 708
55. Smoot G., et al., 1992, Ap.J.,1992, 396, L1
56. Sunyaev, R.A. & Zeldovich, Ya.B., 1970, Astrophysics and Space Science 7, 3
57. Torbet E., et al., 1999, Ap.J., 521, L79-L82
58. Weinberg S., 1977, The first three minutes, Basic Books, ISBN-0465024378
59. White M., et al., astro-ph/9912422
60. Weinberg S., Phys.Rev. D62, 127302, (2000)

Evidence for an Accelerating Universe from Type Ia Supernovae*

Bruno Leibundgut

European Southern Observatory, Karl-Schwarzschild-Strasse 2,
D-85748 Garching, Germany

Abstract. The expansion history of the universe is revealed by measurements of cosmic distances. Since the various energy components in a Friedmann universe contribute differently to the evolution of the scale parameter it is possible to disentangle their importance as a function of time. Type Ia Supernovae are the only distance indicator available over a large enough range of the universal history. All other observations are based either on the assumption of a uniform evolution within a given cosmological model (e.g. Hubble constant, age of the universe, current matter density) or the comparison of exactly two epochs, the time of matter/radiation decoupling and now (CMB fluctuations).

Many questions concerning the systematics of type Ia supernovae remain unanswered, although their properties as distance indicators have been established securely. The small scatter around the expansion line in the Hubble diagram for supernovae in the nearby universe is evidence for the accurate distances. For objects in the distant universe this check does not work any longer and is replaced by the assumption of uniform peak luminosity. The most critical issue is evolution of the peak luminosity of the supernovae.

In the future, the equation of state parameter of the dominant form of energy in the universe can be established through observations of supernovae. The first experiments have started and will prepare the ground for possible future satellite missions.

1 Introduction

The application of particle physics within the largest available laboratory, the universe, has yielded exciting new insights. Cosmology allows us to combine the smallest and highest energies with measurements on the largest scales.

Many of the most pressing problems in physics today arise from astronomical observations. The dark matter problem, long referred to as the 'missing mass', has been in physics now for seventy years (Zwicky 1933). Recently, dark energy has been added leaving us to ponder what makes up 95% of the universe.

Tracing the expansion history of the universe requires acute knowledge of its constituents or a way to determine accurate distances. There are several ways how this can be achieved and the recent years have seen the exploration of angular size distances, luminosity distances and the summation of energy densities

---

* Partly based on the results obtained by the High-z Supernova Search Team.
  More information is available at
  http://cfa-www.harvard.edu/cfa/oir/Research/supernova/HighZ.html

in the various known forms. Several of the contributions to this conference deal with our current knowledge of the cosmological parameters (Krauss, de Bernardis, Mellier, this volume). Combinations of the various constraints have been presented in Garnavich et al. (1998), Turner and Tyson (1999) and Bahcall et al. (1999).

The measurement of cosmological distances plays a crucial role as it is the only direct way to investigate the dynamics of the expansion. By determining the distances at different lookback times it is possible to map out the change of the expansion rate over time, a significant difference between the knowledge gained from fluctuations in the Cosmic Microwave Background (CMB) and Type Ia Supernovae (SNe Ia). This quality turns the supernova result into the most direct evidence of the expansion history of the universe. SNe Ia have provided us with clear signatures of an accelerated expansion during the past $6 \cdot 10^9$ years. The source of this acceleration is strongly debated, but it is clear that the energy form dominating the expansion has to have a very strange equation of state and can not be attributed to any form of known energy (e.g. Garnavich et al. 1998, Perlmutter et al 1999b).

In the following, a summary of the current status of the SNe Ia result is given with an emphasis on possible loopholes in the argument. The astrophysical importance of SNe Ia goes beyond their cosmological applications (e.g. Leibundgut 2000), but the fact that we yet have to grasp the full understanding of the supernova physics means that systematic uncertainties could spoil the derived cosmology. It is these astrophysical limitations of our knowledge about SNe Ia, which currently limits the cosmological results most. Reviews of the supernova cosmology results are available from Riess (2000) and Leibundgut (2001).

The next section gives a brief summary of the cosmological model and the result derived from the supernovae (§2). In section 3 we discuss problems related to the nature of the SNe Ia and show how they could influence the inferred cosmology. The next steps in the cosmological exploration with SNe Ia, attempts to measure the equation of state parameter, are sketched in §4. The conclusions are presented in §5.

2   Cosmological Implications from Type Ia Supernovae

The expansion in a homogeneous and isotropic universe based on general relativity is governed by the energy densities of its constituents are encapsulated in the Friedmann equation (Friedmann 1922, Carroll et al. 1992). The luminosity distance depends on the total curvature (quadratic in redshift), the matter density (which scales with the volume) and a constant term (independent of redshift), the cosmological constant (Goobar and Perlmutter 1995).

2.1   Cosmology with Type Ia Supernovae

Several astrophysical issues have to be tackled by an observing cosmologist. First, a distance indicator has to be established in the nearby universe. Type Ia supernovae are often called standard candles, because they exhibit a small scatter around the linear expansion line of the local universe in a Hubble diagram (e.g. Tammann and Leibundgut 1990, Riess et al. 1996, Phillips et al. 1999, Parodi et al. 2000; Fig. 1). This can also be shown through painstakingly determining the distance to all objects, which are accessible to other distance indicators, most prominently Cepheids. These prove that almost all SNe Ia reach very similar luminosity at their maximum (Saha et al. 1999, Freedman et al. 2001, Ajhar et al. 1999).

SNe Ia are further refined by applying a normalisation based on their light curve shape (Riess et al. 1996, Phillips et al. 1999). This makes them exquisite distance indicators tested in the local universe. Most importantly, SNe Ia give extremely accurate relative distances even though the absolute calibration may still be contested. It is this quality, which makes them so important for cosmology.

Once the distance indicator can reliably be applied in the nearby universe attempts can be made to explore distant objects. As an experiment this poses the problem that the faint sources have to be securely identified through the correct classification procedure. With the decrease in flux the measurement errors increase. Since the distant objects also are at a much younger age of the universe it is very important to detect and control any evolutionary effects. These could influence the objects as a class or introduce selection biases, which have to be identified for a secure distance determination. Other problems are posed by any undetected interstellar or intergalactic dust components through which the light has travelled. A third problem is the fact that any gravitational light bending will magnify or de-magnify the light of the distant object (e.g. Mellier 2000). Gravitational lensing on average will de-magnify a set of objects distributed in randomly sampled sight lines (e.g. Holz and Wald 1998). The last problem is the redshifting of the light in the observer frame. This means that the observed light often corresponds to rest-frame wavelengths in the ultra-violet that are unobservable from the ground.

There are currently 52 objects published in the redshift range $0.3 < z < 1.0$ (Schmidt et al. 1998, Riess et al. 1998, Perlmutter et al. 1999a). Unfortunately, not all distant SNe have a spectroscopic classification and sometimes one has to resort to indirect classification criteria. The spectroscopy is also used to test for any evolutionary effects of the distant objects. In all cases, where this could be done in detail, there are no obvious discrepancies (Perlmutter et al. 1998, Riess et al. 1998, Coil et al. 2000).

Interstellar dust scatters and absorbs light. The scattering efficiency depends on the dust grain size and can be readily detected in colour changes for dust in our own Milky Way (Mathis 1990). Intergalactic dust with different properties has been proposed (Aguirre and Haimann 2000). So far, the supernovae have not shown any signatures of such dust (Riess et al. 2000), but the definitive experiment is under way only now.

Gravitational lensing does not introduce any changes in the observed supernova light, except magnification. Thus, it is not possible to assess the effect of

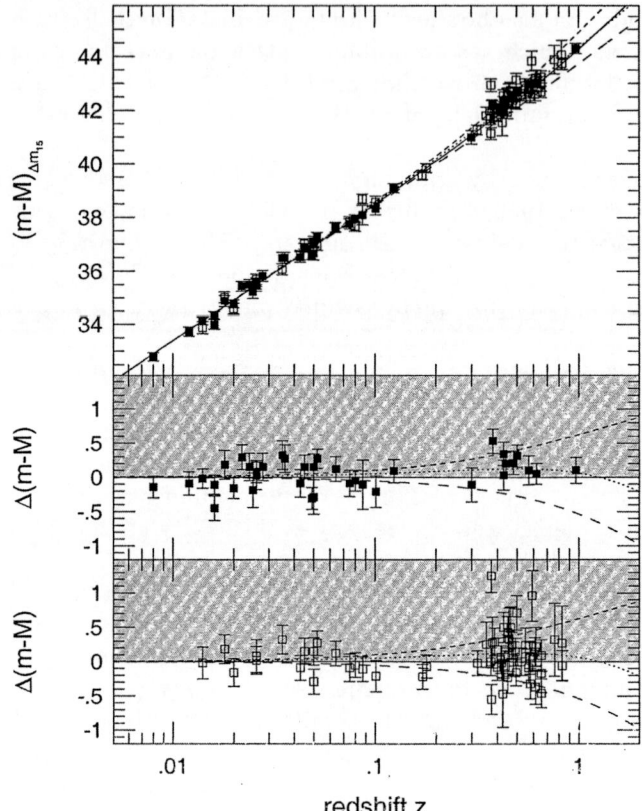

Fig. 1. Hubble diagram of SNe Ia. The lines are models of an empty universe (solid), an Einstein-de Sitter universe ($\Omega_M = 1$; long dashes), a Lambda-dominated universe ($\Omega_M = 0$, $\Omega_\Lambda = 1$; short dashes) and a 'concordance' flat universe ($\Omega_M = 0.3$, $\Omega_\Lambda = 0.7$; dotted)

gravitational lensing from the supernova observations alone and more detailed analyses of the mass distribution is required. For the redshifts concerned so far ($z \leq 1$), the systematic effect, however, is expected to be rather small and not important (Holz and Wald 1998). The correction from the observer frame to the rest frame is made fairly routinely now, but depends on the exact shape of the spectrum (Nugent et al. 2002, Tonry et al. 2003).

The result is summarised in Fig. 1. Note that only observed quantities are plotted. The redshift is defined as the ratio of the scale parameter between the time the light was emitted and now and is readily observed in the wavelength shift of spectral features. The normalised peak luminosity distance is plotted on the ordinate. The lower panels display the magnitude differences relative to an empty universe model. The middle panel shows the data of the High-z SN Search Team (Schmidt et al. 1998, Riess et al. 1998) and the bottom panel

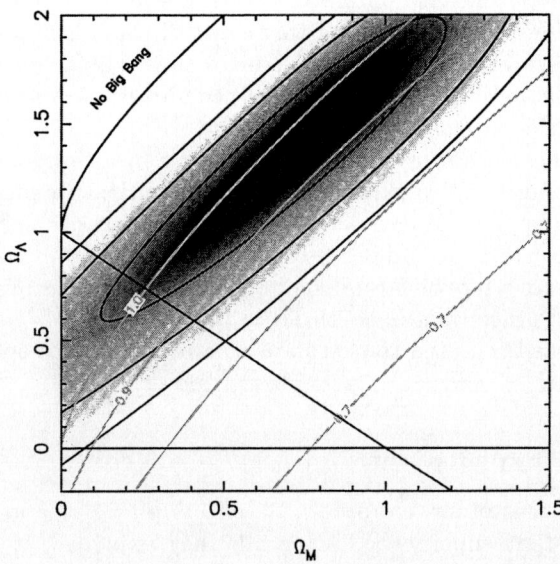

Fig. 2. Likelihood distribution for the cosmological parameters as derived from SNe Ia distances. The upper left corner is excluded for Big Bang models. The contours for 68.3%, 95.4% and 99.7% probability are shown. Isochrones for the dynamic age of the universe ($t_0 H_0$) are also indicated

the data from the Supernova Cosmology Project (Perlmutter et al. 1999a). The previous paradigm of a matter-dominated flat universe is plotted as the dashed line curving down in these diagrams. The shaded area is the parameter space where objects are fainter than they would appear in an empty universe. In the paradigm of the last decade this was an 'excluded' region. The distant supernovae fall into this 'forbidden' region. It is important to note that this is a result not only of the distant objects as the nearby supernovae 'anchor' the lines of the cosmological model. The scientific measurement is the magnitude difference between the nearby supernovae and the distant ones. The data sets of both groups clearly show the signature of an extra component driving the universe apart (Riess 2000, Leibundgut 2001).

The likelihood distribution for the cosmological parameters, Ω_Λ and Ω_M is shown in Fig. 2. This figure assumes that the observed signal is purely of cosmological nature. Other influences, which might alter some of this picture, will be discussed in the next section. The likelihood contours from luminosity distances are degenerate in the direction of $\Omega_M - \Omega_\Lambda$ (Goobar and Perlmutter 1995). From the contour lines it is clear that there is almost not overlap of the likelihood region with matter-dominated models ($\Omega_\Lambda = 0$). For reasonable values of the matter density ($0.3 < \Omega_M < 0.5$; Peacock et al. 2001, Turner 2002) and no other component the overlap with the likelihood from the supernova distances is vanishingly small. An additional attractive feature of the introduction of the Λ−component is that the dynamic age of the universe ($t_0 H_0$) approaches 1, which makes the universe a lot older than with the previously preferred parameters. In particular, the flat, matter-dominated (Einstein-de Sitter universe ($\Omega_M = 1, \Omega_A = 0$ giving $t_0 H_0 = \frac{2}{3}$) was falling way short of the age of the oldest stars independent of the specific choice of the value of the Hubble constant accepted by observations. The isochrones for the age of the universe are nearly parallel to the degeneracy of the supernova results.

Even more direct and stronger confirmation for a new component governing the dynamics of the universal expansion has come from the inference of an overall flat space-time geometry as derived from the measurements of the fluctuations in the cosmic microwave background (de Bernardis, this volume). For a flat universe ($\Omega_A + \Omega_M = 1$) the overlap with the likelihood from the supernova distances or the matter density is incompatible with $\Omega_A = 0$.

This evidence has been used to postulate a new force that is accelerating the universal expansion.

2.2   Alternative Explanations

It is quite possible that other, astrophysical, effects play a role in the apparent dimness of the distant supernovae. After all, they exploded at a time when the Solar System had not yet formed and, e.g., the metal enrichment of their progenitor stars could be significantly altered from the one of their counterparts we observe in the local universe. As outlined in the previous section an important part of the experiment is to ensure that the distant objects are not severely affected by these astrophysical influences. We will discuss the evidence in more detail in the following sections.

Dust Intervening dust could dim the distant object more strongly or in a different way than the nearer objects. The observations have attempted to closely monitor any possible effects dust could have. Most importantly, almost all distant objects in the High-z SN sample have been observed in two colours to detect any reddening, i.e. signature of dust. Dust with properties like the one we observe in the Milky Way can be excluded (Riess et al. 1998). In addition, one distant SN Ia has been observed covering a wavelength range nearly four times as large as all others (Riess et al. 2000). This gives a better leverage on any reddening, even from putative, gray dust, which has a reduced effect on the colour change. The best explanation is still that there is negligible intervening dust and it is not responsible for the faintness of the distant supernovae.

An even stronger argument is that the distant SNe Ia appear bluer in their rest frame colours than the nearby objects (Falco et al. 1999, Leibundgut 2001). This in itself argues that dust can not play a major role.

Evolution Claims for signatures of evolution in the distant SN Ia sample have been made (Drell et al. 2000). That study was based entirely on a parameter evaluation with evolution as a free parameter. More concrete indications of evolution of the distant supernovae are required. The most direct prediction for

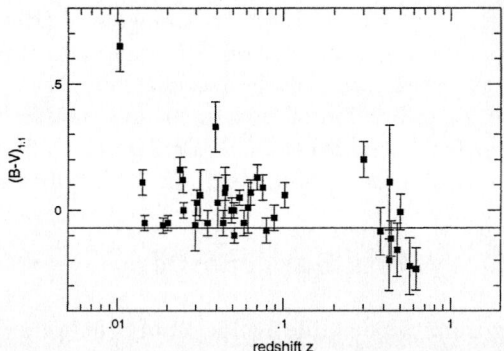

Fig. 3. Intrinsic colour of SNe Ia as a function of redshift. Data are from Phillips et al. (1999) for the nearby sample and from Riess et al. (1998) for the distant supernovae

evolution are subtle changes in the spectral appearance (Höflich et al. 1998). Even for drastic changes in the chemical composition of the progenitor star the spectrum changes only slightly in the blue region. High-quality spectroscopy of the distant supernovae is largely missing due to the difficulty in obtaining such data. Nevertheless, the available spectroscopy of the distant SNe Ia does not display obvious differences for objects out to $z = 1.2$ (Riess et al. 1998, Perlmutter et al. 1998, Coil et al. 2000, Leibundgut and Sollerman 2001).

The blue colour, however, does raise concerns (Leibundgut 2001). If the distant objects consistently are bluer than the nearby ones, then it is clear that some evolution must have occurred. This could be due to a selection bias so that we preferentially pick bluer objects, but it could also mean that most SNe Ia at large redshifts have intrinsically been bluer. In the first case, we should observe such blue objects also in the local universe, which is not the case (Fig. 3). Interestingly, the bluer colour and the apparent faintness do not follow the relations established in the local universe (Leibundgut 2001). Bluer colours normally go with more luminous objects (Phillips et al. 1999). If this relation were to hold for the distant objects as well, then we would have underestimated their distances and the acceleration would have been even larger. However, if the expansion did not accelerate then the distant SNe Ia behave differently from the nearby objects. It appears that we thus already have observed signatures of evolution in SNe Ia. Clearly, the import of the colour changes will need clarification and has been made the topic of a special investigation by the High-z SN team.

2.3   SN 1997ff

The most distant supernova ever discovered is SN 1997ff (Gilliland et al. 1999, Riess et al. 2001). At a redshift of $z = 1.7$ it probes the universe in its early phase, while it was still decelerating. Although no spectrum was obtained of this object and no good colour information is available, its light curve is fully consistent with a SN Ia at a redshift determined from multi-band photometry and a tentative detection of emission lines from the host galaxy (Riess et al. 2001). This supernova appears brighter at maximum light than the objects at smaller redshift, the signature expected for an early deceleration of the universe. Some discussion of a possible magnification of this supernova due to two foreground galaxies has taken place (Lewis and Ibata 2001, Benitez et al. 2002). It appears, however, that the amplification can not fully explain the brightness of the object.

3   Systematics of Type Ia Supernovae

The uncertainty arising from the possible evolution of SNe Ia limits their cosmological significance. This means we have to understand the underlying physics in these explosions better. So far, the explosion physics and the radiation transport have been formidable hurdles in our ability to link the observations with the theory. Most puzzling are the correlations between several distinct observational features. Of highest import for the cosmology is the relation of peak luminosity and the light curve shape originally discovered by Phillips (1993). This feature provides the normalisation of the peak luminosity and thus makes supernovae the superb distance indicators they are (Riess et al. 1998, Phillips et al. 1999, Perlmutter et al. 1999a). The correlations applied by different groups are, however, inconsistent (Leibundgut 2000). Some SNe Ia have recently been observed, which seem not to fit the one-parameter description at all (cf. SN 2000cx, Li et al. 2002; SN 2002cx, Li et al. 2003).

The main difficulty comes from the fact that we have no good idea about the progenitor stars and the way they explode. There still is a strong debate on which white dwarfs actually explode as SNe Ia (e.g. Livio 2000). Everybody accepts that they have to be in a binary star system, but the nature of their companion is not resolved. In some models, the companion is another white dwarf and the two stars will merge due to gravitational wave radiation (Iben and Tutukov 1985) in others the companion is either a main sequence or a giant star. Observations trying to detect them have so far been unsuccessful, although it appears that at least some of the super-soft X-ray sources have masses near the Chandrasekhar limit (Thoroughgood et al. 2001). The telltale signature of hydrogen around the explosion has, however, not been detected so far (Lundqvist et al. 2003).

Other problems arise from the lack of knowledge of the global parameters in the explosion. The total ejecta mass is not known nor is the exact composition of the white dwarf at the explosion. In the last few years we have been able to determine the energy source, the amount of radioactive nickel synthesised in the explosion. Through the construction of bolometric light curves the amount of nickel has been estimated using Arnett's rule (Arnett 1982, Pinto and Eastman 2000), which states that the energy released at maximum corresponds to the energy generated at that moment. This is a rather unique moment in the history of a SN Ia as during the rise phase more energy is generated at the centre than is released at the surface and after maximum not all γ−rays generated in the radioactive decays are converted into optical and infrared photons and some escape undetected. There is a large range of nickel masses produced in SNe Ia as is reflected in the peak luminosity (Contardo et al. 2000, Strolger et al. 2002). The reason for this is not entirely clear, but it could come from the initial conditions of the explosions (Reinecke et al. 2002).

The light curves themselves could provide additional information, but that still has to be explored. At late phases most of the energy generated in the γ−rays from the radioactive decay escapes the ejecta. The down-scattering probability depends on the column density, i.e. the amount of ejecta mass. Since this is determined by the total mass and the explosion energy, the rate of decline about 50 days after maximum light reflects the amount of energy converted into low-energy photons and hence the total mass (Leibundgut and Pinto 1992, Leibundgut and Suntzeff 2003). This works until about 120 days after maximum, when the positron annihilation channel in the decay starts to dominate (Milne et al. 2001).

A correlation of the late decline rates and the early decline rate has been observed (Vacca and Leibundgut 1997, Phillips et al. 1999). Interestingly, it means that slow declining objects remain so throughout the observable period. Since the more slowly declining objects are also the more luminous ones this implies that they emit also more energy at later phases. It turns out that these slow declining supernovae also display higher expansion velocities (Mazzali et al. 1998). Hence it appears that either they are more massive, they synthesise more nickel, or there are inhomogeneities in the explosions. It will be an important topic of supernova research to solve this problem in the future.

Theory has produced a gamut of interesting models. The problem is that supernova physics is a very complicated process (Hillebrandt and Niemeyer 2000). The flame dynamics in the explosions takes place in a region of parameter space that is extremely difficult to model. The radiation transport describing the escape of the photons from the explosion is just as complicated as it involves the treatment of non-thermal radiation in an expanding medium. The problem here is to find a theoretical description of the process where many excitation levels of many isotopes have to be treated statistically. Nevertheless, the progress in theoretical flame modelling and the radiation transport calculations have made the comparison with observations more favourable. It is encouraging to see that the theoretical models start to reproduce some of the observations.

4   The Equation of State Parameter

The universe is currently dominated by an unknown energy form, which has the strange property that it exerts a negative pressure. For a slowly changing dark energy and a Big Bang cosmology there must have been a transition from the matter-dominated epoch to the current epoch with the dark energy's domination. The time of this change can be determined with enough accurately measured distances mapping out the period in question. Interestingly, this transition was rather 'recently', i.e. between $0.5 < z < 1.0$ (Riess et al. 1998, 2001) or between 5 and 8 Gyr for a flat universe and a cosmological constant with $H_0 = 65$ km s^{-1} Mpc^{-1}. By accurately sampling this redshift range, it should be possible to evaluate the 'strength' of the dark energy. Since this is essentially measuring the turn-over of the model curve in Fig. 1, a pre-requisite for this measurement is that an independent estimate of the matter density becomes available. The measurement very sensitively depends on this parameter (Garnavich et al. 1998, Perlmutter et al. 1999b). Recent advances with the streaming motion of large samples of galaxies are already providing measurements with small errors (Peacock et al. 2001) and combining several independent methods it has been claimed that accuracies in the required range can be achieved (Turner 2002).

Essentially, the reason for the change comes from the different dependence of the dark energy component and the (pressure-less) matter component. The equation of state parameter is defined as $\omega \equiv \frac{p}{\rho c^2}$, with pressure p, density ρ and the speed of light c. The energy density scales with the universal scale parameter and the equation of state parameter $\rho \propto a^{-3(1+\omega)}$. For pressure-less matter $\omega = 0$ the density scales inversely with the volume. For a cosmological constant the density is scale independent and hence $\omega = -1$. From the acceleration measure we know that $\omega < -0.3$ for the dominant component governing the expansion of the universe. For many of the particle-physics motivated models the value of ω is different from -1 and this would be a clear way to distinguish between some of the models.

The High-z SN team has embarked on a large project to map the distances out to about half the age of the universe with many SNe Ia in the redshift range $0.3 < z < 0.8$. To achieve this we plan to obtain light curves and spectra of about 200 SNe Ia over a span of five years. Spectroscopy of all objects is absolutely essential to avoid contamination of the sample by other variable objects. The VLT (and Keck) will play a crucial role for the spectroscopy, while the light curves will be sampled with the CTIO 4m Blanco Telescope. The project starts in October 2002 and we plan to have a measurement of ω to within 10% by 2007.

It has to be noted that this will be an integral measurement of ω but is not capable of testing for time-dependence, which requires a larger SN sample and also a much wider redshift base. A proposal for a dedicated satellite is discussed in the US to achieve this measurement. SNAP, the SuperNova Acceleration Probe, is a satellite with a wide-field camera for optical and infrared observations and also spectroscopic capabilities. It should provide a huge sample of SNe Ia (about 2000 objects) out to $z = 1.7$ (e.g. Perlmutter et al. 2001).

5   Conclusions and Outlook

As distance indicators supernovae are currently the only way to map the history of the expansion of the universe over a significant redshift range. All other measures are either for a fixed lookback time, e.g. the CMB fluctuations, or combining different measurements, e.g. using the universal age and the value of the Hubble constant.

The next quest is to explore what the dark energy is and how it affects the expansion of the universe. After all, this is the major energy component in the universe today. It adds to the puzzle we already face with non-baryonic dark matter and makes the universe quite an enigmatic place. The next observational programs for Type Ia Supernovae should provide some indications what the evolution history of the expansion is, but these are major observational efforts and will take some time until the statistically significant samples will be collected.

These efforts are focused on determining the value of the equation of state parameter for the dominant energy form of the universe, the dark energy. They will profit from the combination with further, independent, determinations of the other cosmological parameters. The observations of the CMB and the mass estimations from lensing studies will contribute to this cosmic endeavour. Trying to find a time-dependent equation of state parameter requires to increase the redshift range out to beyond $z = 1.5$. This is out of reach of current ground-based observatories and will have to be achieved through satellite missions, like the proposed SNAP satellite.

These experiments are currently the only way to assess some aspects of the nature of the dark energy. The distinction between a decaying particle field (such as 'quintessence') and the cosmological constant is fundamental for progress in this field.

Nearer to home, we still have to work out the systematics of SNe Ia. Understanding the local SNe Ia is imperative for their cosmological applications. Evolution can not be tackled without a better physical understanding of these premium distance indicators. Without a physical underpinning of the explosions and their radiation we have to depend on empirical relations and the faith that evolution is not significant. Progress is being made and a European programme for a detailed study of nearby SNe Ia has started recently.

References

1. Aguirre, A.N., Haimann, Z., 2000, ApJ, 532, 28
2. Ajhar, E. A., Tonry, J. L., Blakeslee, J. P., Riess, A. G., Schmidt, B. P. 2001, ApJ, 559, 584
3. Arnett, W. D., 1982, ApJ, 253, 785
4. Bahcall, N.A., Ostriker, J.P., Perlmutter, S., Steinhardt, P.J., 1999, Science, 284, 1481
5. Benitez, N., Riess, A. G., Nugent, P. E., Dickinson, M., Chornock, R., Filippenko, A. V. 2002, ApJ, 577, L1
6. Carroll, S.M., Press, W.H., Turner, E.L 1992, ARA&A 30, 499
7. Coil, A.L., et al., 2000, ApJ, 544, L111
8. Contardo, G., Leibundgut, B., Vacca, W. D., 2000, A&A, 359, 876
9. Drell, P.S., Loredo, T.J., Wasserman, I., 2000, ApJ, 530, 593
10. Falco, E.E., et al., 1999, ApJ, 523, 617
11. Freedman, W.L., et al., 2001, ApJ, 553, 47
12. Friedmann, A. 1922, Z. Phys., 10, 377
13. Garnavich, P.M., et al., 1998, ApJ, 509, 74
14. Gilliland, R.L., Nugent, P.E., Phillips, M.M., 1999, ApJ, 521, 30
15. Goobar, A., Perlmutter, S., 1995, ApJ, 450, 14
16. Hillebrandt, W., Niemeyer, J. C.: 2000, ARA&A, 38, 191

17. Höflich, P., Wheeler, J.C., Thielemann, K.-F., 1998, ApJ 495, 617
18. Holz, D.E., Wald, R.M., 1998, Phys. Rev. D, 58, 063501
19. Iben, I., Tutukov, A.V, 1985, ApJS, 58, 661
20. Leibundgut, B., 2000, A&AR, 10, 179
21. Leibundgut, B., 2001, ARA&A, 39, 67
22. Leibundgut, B., Pinto, P. A., 1992, ApJ, 401, 49
23. Leibundgut, B., Sollerman, J., 2001, Europhysics News, 32, 4
24. Leibundgut, B., Suntzeff, N.B., 2003, in *Supernovae and Gamma-Ray Bursts*, ed. K. Weiler, Heidelberg:Springer, in press
25. Lewis, G. F., Ibata, R. A. 2001, MNRAS, 324, L25
26. Li, W., et al., 2001, PASP, 113, 1178
27. Li, W., et al., 2003, in preparation
28. Livio, M. 1999, in *Type Ia Supernovae: Theory and Cosmology*, eds. J. C. Niemeyer and J. W. Truran, Cambridge: Cambridge University Press, 33
29. Lundqvist, P., Sollerman, J., Baron, F. A , Fransson, C., Leibundgut, B., Nomoto, K., Spyromilio, J., 2003, in preparation
30. Mathis, J. S. 1990, ARA&A, 28. 37
31. Mazzali, P. A., Cappellaro, E., Danziger, I. J., Turatto, M., Benetti, S., 1998, ApJ, 499, L49
32. Mellier, Y., 1999, ARA&A, 37, 127
33. Milne, P. A., The, L.-S., Leising, M., 2001, ApJ, 559, 1019
34. Nugent, P., Kim, A., Perlmutter, S., 2002, PASP, 114, 803
35. Parodi, B.R., Saha, A., Sandage, A., Tammann, G.A., 2000, ApJ, 540, 634
36. Peacock, J.A., et al., 2001, Nature, 410, 169
37. Perlmutter, S., et al., 1998, Nature, 391, 51
38. Perlmutter, S., et al., 1999a, ApJ, 517, 565
39. Perlmutter, S., Turner, M.S., White, M., 1999b, Phys. Rev. Lett., 83, 670
40. Perlmutter, S., et al., 2001, BAAS, 33, 1480
41. Phillips, M. M., 1993, ApJ, 413, L105
42. Phillips, M.M., Lira, P., Suntzeff, N.B., Schommer, R.A., Hamuy, M., Maza, J., 1999, AJ, 118, 1766
43. Pinto, P. A., Eastman, R. G., 2000, ApJ, 530, 744
44. Reinecke, M., Hillebrandt, W., Niemeyer, J.C., 2002, A&A, 391, 1167
45. Riess, A.G., 2000, PASP, 112, 1284
46. Riess, A. G., Press, W. M., & Kirshner, R. P., 1996, ApJ, 473, 88
47. Riess, A. G., et al., 1998, AJ, 116, 1009
48. Riess, A.G., et al., 2000, ApJ, 536, 62
49. Riess, A.G., et al., 2001, ApJ, 560, 49
50. Saha, A, Sandage, A., Tammann, G.A., Labhardt, L., Macchetto, F.D., Panagia, N., 1999, ApJ, 522, 802
51. Schmidt, B.P., et al., 1998, ApJ, 507, 46
52. Strolger, L., et al., 2002, AJ, in press (astro-ph/0207409)
53. Tammann, G. A., Leibundgut, B., 1990, A&A, 236, 9
54. Thoroughgood, T.D., Dhillon, V.S., Littlefair, S.P., Marsh, T.R, Smith, D.A, 2001, MNRAS, 327, 1323
55. Tonry, J., et al., 2003, ApJ, submitted
56. Turner, M. S., 2002, ApJ, 576, L101
57. Turner, M. S., Tyson, J. A. 1999, Rev. Mod. Phys., 71, 145
58. Vacca, W. D., Leibundgut, B., 1997, in *Thermonuclear Supernovae*, eds. P. Ruiz-Lapuente, R. Canal, J. Isern, Kluwer, Dordrecht, 65
59. Zwicky, F., 1933, Helv. Phys. Acta, 6, 110
60. Zwicky, F., 1937, ApJ, 86, 217

Cosmological Weak Lensing

Yannick Mellier[1,2], Ludovic van Waerbeke[1,3], Francis Bernardeau[4], and
Ismael Tereno[1,5]

[1] IAP, 98bis boulevard Arago, 75014 Paris, France
[2] Observatoire de Paris, LERMA, 61 avenue de l'Observatoire, 75014 Paris, France
[3] CITA, 60 St George Str., M5S 3H8, Toronto, Canada,
[4] SPhT, CE Saclay, 91191 Gif-sur-Yvette Cedex, France
[5] Department of Physics, University of Lisbon, 1749-016 Lisboa, Portugal.

Abstract. We present the current status of cosmic shear studies and their implications on cosmological models. Theoretical expectations and observational results are discussed in the framework of standard cosmology and CDM scenarios. The potentials of the next generation cosmic shear surveys are discussed.

1 Introduction

The gravitational deflection of light beams by large scale structures of the universe (cosmological lensing) amplifies and modifies the shape of distant galaxies and quasars. Magnification produces correlation between the density of foreground lenses and the apparent luminosity of distant galaxies or quasars (magnification bias), whereas distortion induces a correlation of ellipticity distribution of lensed galaxies (cosmic shear). In both cases, the properties of cosmological lensing signals probe the matter content and the geometry of universe and how perturbations grew and clustered during the past Gigayears.

Albeit difficult to detect, the recent cosmic shear detections claimed by several groups demonstrate that it is no longer a technical challenge. It is therefore possible to study the universe through a new window which directly probes dark matter instead of light and allows cosmologists to measure cosmological parameters and dark matter power spectrum from weak gravitational distortion.

2 Weak Cosmological Lensing

Let us assume that the shape of galaxies can be simply characterize by their surface brightness second moments $I(\boldsymbol{\theta})$, (see [1], [2] and references therein):

$$M_{ij} = \frac{\int I(\boldsymbol{\theta})\,\theta_i\,\theta_j\,\mathrm{d}^2\theta}{\int I(\boldsymbol{\theta})\,\mathrm{d}^2\theta} \ . \tag{1}$$

Because of gravitational lensing, a galaxy with intrinsic ellipticity $\mathbf{e}$ is measured with an ellipticity $\mathbf{e} + \boldsymbol{\delta}$, where $\boldsymbol{\delta}$ is the gravitational distortion,

$$\boldsymbol{\delta} = 2\gamma\,\frac{(1-\kappa)}{(1-\kappa)^2 + |\gamma|^2} = \left(\delta_1 : \frac{M_{11} - M_{22}}{Tr(M)} \ ; \ \delta_2 : \frac{2M_{12}}{Tr(M)}\right) \ . \tag{2}$$

κ and γ are respectively the gravitational convergence and shear. Both depend on the second derivatives of the projected gravitational potential, φ:

$$\kappa(\boldsymbol{\theta}) = \frac{1}{2}(\varphi_{,11} + \varphi_{,22}); \quad \gamma_1(\boldsymbol{\theta}) = \frac{1}{2}(\varphi_{,11} - \varphi_{,22}); \quad \gamma_2(\boldsymbol{\theta}) = \varphi_{,12}. \quad (3)$$

In the case of weak lensing, $\kappa \ll 1$, $|\gamma| \ll 1$ and $\delta \approx 2\gamma$. Since large-scale structures have very low density contrast, this linear relation is in particular valid on cosmological scales.

Light propagation through an inhomogeneous universe accumulates weak lensing effects over Gigaparsec distances. Assuming structures formed from gravitational growth of Gaussian fluctuations, cosmological weak lensing can be predicted from Perturbation Theory at large scale. To first order, the convergence $\kappa(\boldsymbol{\theta})$ at angular position $\boldsymbol{\theta}$ is given by the line-of-sight integral

$$\kappa(\boldsymbol{\theta}) = \frac{3}{2}\Omega_0 \int_0^{z_s} n(z_s)\mathrm{d}z_s \int_0^{\chi(s)} \frac{D(z, z_s)\,D(z)}{D(z_s)}\delta(\chi, \boldsymbol{\theta})\,[1 + z(\chi)]\,\mathrm{d}\chi \quad (4)$$

where $\chi(z)$ is the radial distance out to redshift z, D the angular diameter distances, $n(z_s)$ is the redshift distribution of the sources. δ is the mass density contrast responsible for the deflection at redshift z. Its amplitude at a given redshift depends on the properties of the power spectrum and its evolution with look-back-time.

The cumulative weak lensing effects of structures induce a shear field which is primarily related to the power spectrum of the projected mass density, P_κ. Its statistical properties can be recovered by the shear top-hat variance [17,8,19],

$$\langle \gamma^2 \rangle = \frac{2}{\pi\theta_c^2} \int_0^\infty \frac{\mathrm{d}k}{k} P_\kappa(k)[J_1(k\theta_c)]^2, \quad (5)$$

the aperture mass variance [21,35]

$$\langle M_{\mathrm{ap}}^2 \rangle = \frac{288}{\pi\theta_c^4} \int_0^\infty \frac{\mathrm{d}k}{k^3} P_\kappa(k)[J_4(k\theta_c)]^2, \quad (6)$$

and the shear correlation function [17,8,19]:

$$\langle \gamma\gamma \rangle_\theta = \frac{1}{2\pi} \int_0^\infty \mathrm{d}k\, k P_\kappa(k) J_0(k\theta), \quad (7)$$

where J_n is the Bessel function of the first kind. Higher order statistics, like the skewness of the convergence, $s_3(\theta)$, can also be computed. They probe non Gaussian features in the projected mass density field, like massive clusters or compact groups of galaxies. (see [3]; [4] and references therein). The amplitude of cosmic shear signal and its sensitivity to cosmology can be illustrated in the fiducial case of a power law mass power spectrum with no cosmological constant and a background population at a single redshift z. In that case $< \kappa(\theta)^2 >$ and $s_3(\theta)$ write:

$$< \kappa(\theta)^2 >^{1/2} = < \gamma(\theta)^2 >^{1/2} \approx 1\%\; \sigma_8\; \Omega_m^{0.75}\; z_s^{0.8} \left(\frac{\theta}{1'}\right)^{-\left(\frac{n+2}{2}\right)}, \quad (8)$$

and

$$s_3(\theta) = \frac{\langle \kappa^3 \rangle}{\langle \kappa^2 \rangle^2} \approx 40 \; \Omega_m^{-0.8} \; z_s^{-1.35} \; , \qquad (9)$$

where n is the spectral index of the power spectrum of density fluctuations. Therefore, in principle the degeneracy between Ω_m and σ_8 can be broken when both the variance and the skewness of the convergence are measured.

3 Detection of Weak Distortion Signal

3.1 Observational Challenge

Eq. (8) shows that the amplitude of weak lensing signal is of the order of few percents, which is much smaller than the intrinsic dispersion of ellipticity distribution of galaxies. van Waerbeke et al ([5]) explored which strategy would be best suited to probe statistical properties of such a small signal. They have shown that the variance of κ can be measured with a survey covering about 1 deg^2, whereas for the skewness one needs at least 10 deg^2. Furthermore, more than 100 deg^2 must be observed in order to uncover information on Ω_Λ or the shape of the power spectrum over scales larger than 1 degree. For $\Omega_m = 0.3$ and $\sigma_8 = 1$, the limiting shear amplitude can be simply expressed as follows

$$< \gamma(\theta)^2 >^{1/2} = 0.3\% \left[\frac{A_T}{100 \; deg^2} \right]^{\frac{1}{4}} \times \left[\frac{\sigma_{\epsilon_{gal}}}{0.4} \right] \times \left[\frac{n}{20} \right]^{-\frac{1}{2}} \times \left[\frac{\theta}{10^i} \right]^{-\frac{1}{2}} , \qquad (10)$$

where A_T is the total sky coverage of the survey. The numbers given in the brackets correspond to a measurement at $3 - \sigma$ confidence level of the shear variance. Eq. (10) contains the specifications of a cosmic shear survey.

3.2 First Detection of Weak Distortion

Despite technical limitations discussed above, on scale significantly smaller than one degree, non-linear structures dominate and increase the amplitude of the lensing signal, making its measurement easier. Few teams started such surveys during the past years and succeeded to get a significant signal. Table 1 lists some published results. Since each group used different telescopes and adopted different observing strategy and data analysis techniques, one can figure out the reliability of the final results. Figure 1 shows that all these independent results are in very good agreement[1]. This is a convincing demonstration that the expected correlation of ellipticities is real.

---

[1] [26] data are missing because depth is different so the sources are at lower redshift and the amplitude of the shear is not directly comparable to other data plotted

Table 1. Present status of cosmic shear surveys with published results.

Telescope	Pointings	Total Area	Lim. Mag.	Ref.
CFHT	$5 \times 30' \times 30'$	1.7 deg^2	I=24.	[7][vWME+]
CTIO	$3 \times 40' \times 40'$	1.5 deg^2	R=26.	[9][WTK+]
WHT	$14 \times 8' \times 15'$	0.5 deg^2	R=24.	[10][BRE]
CFHT	$6 \times 30' \times 30'$	1.0 deg^2	I=24.	[12][KWL]
VLT/UT1	$50 \times 7' \times 7'$	0.6 deg^2	I=24.	[13][MvWM+]
HST/WFPC2	$1 \times 4' \times 42'$	0.05 deg^2	I=27.	[14]
CFHT	$4 \times 120' \times 120'$	6.5 deg^2	I=24.	[15][vWMR+]
HST/STIS	$121 \times 1' \times 1'$	0.05 deg^2	V≈ 26	[25]
CFHT	$5 \times 126' \times 140'$	24. deg^2	R=23.5	[26]
CFHT	$10 \times 126' \times 140'$	53. deg^2	R=23.5	[27]
CFHT	$4 \times 120' \times 120'$	8.5 deg^2	I=24.	[28]
HST/WFPC2	$271 \times 2.1 \times 2.1$	0.36 deg^2	I=23.5	[23]
Keck+WHT	$173 \times 2' \times 8'$	1.6 deg^2	R=25	[16]
	$+13 \times 16' \times 8'$			
	$7 \times 16' \times 16'$			

3.3 Nature of the Weak Distortion Signal

The detection of coherent signal is not a demonstration of its very nature. Even if a cosmological signal were expected, it could be contaminated by systematics, like optical and atmospheric distortions, which mix together with the gravitational shear. Contrary to lensing effects, these contributions are visible also on stars and can be corrected (using for example the KSB method, [20]). However, stars often show strong anisotropic shape with elongation much larger than the expected amplitude of the gravitational distortion. The reliability of artificial anisotropy corrections is therefore a critical step of the weak lensing analysis (see for example [6], [7] [10] and [11]). An elegant way to check whether corrections are correctly done and to confirm the gravitational nature of the signal is to decompose the signal into E- and B- modes. The E-mode contains signal produced by gravity-induced distortion whereas the B-mode is a pure curl-component, so it only contains intrinsic ellipticity correlation or systematics residuals. Both modes have been extracted using the aperture mass statistics by van Waerbeke et al ([15], [24]) and Pen et al ([28]) in the VIRMOS-DESCART survey as well as by Hoekstra et al ([27]) in the Red Cluster Sequence survey. In both samples, the E-mode dominates the signal, although a small residual is detected in the B-mode. This strongly supports the gravitational origin of the distortion.

An alternative to gravitational lensing effect could be an intrinsic correlations of ellipticities of galaxies produced by proximity effects. It could results from galaxy formation processes. Several recent numerical and theoretical studies (see for example [30]; [31]) show that intrinsic correlations are negligible on scales beyond one arc-minute, provided that the survey is deep enough. In that case, most lensed galaxies along a line of sight are spread over Gigaparsec scales and have no physical relation with its apparent neighbors. Hence, since most cosmic

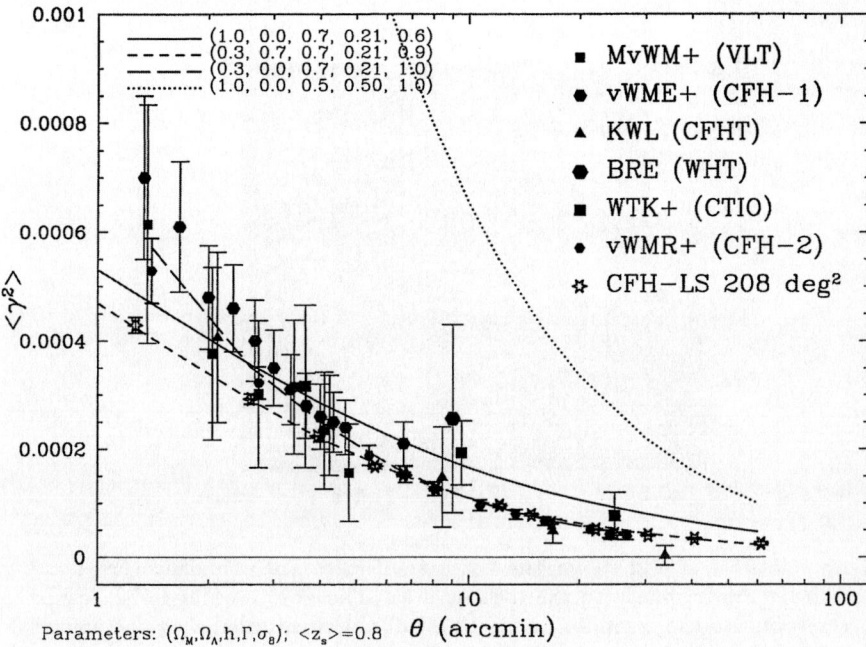

Fig. 1. Top hat variance of shear as function of angular scale from 6 cosmic shear surveys. The CFHT-LS (open black stars) illustrates the expected signal from a large survey covering 200 deg². For most points the errors are smaller than the stars

surveys are deep, they are almost free of intrinsic correlations. We therefore are confident that the signal measured by all teams is a genuine cosmic shear signal.

4   Cosmological Interpretations

4.1   2-Point Statistics and Variance

A comparison of the top-hat variance of shear with some realistic cosmological models is plotted in Fig. 1. The amplitude of the shear has been scaled using photo-z which gives $< z > \approx 1$. On this plot, we see that standard COBE-normalized CDM is ruled at a $10-\sigma$ confidence level. However, the degeneracy between Ω_m and σ_8 discussed in the previous section still hampers a strong discrimination among most popular cosmological models. The present-day constraints resulting from independent analyses by Maoli et al ([13]), Rhodes et al ([14]), van Waerbeke et al ([15,24]), Hoekstra et al ([26,27]) and Réfrégier et al ([23]) can be summarized by the following conservative boundaries (90% confidence level):

$$0.05 \leq \Omega_m \leq 0.8 \quad \text{and} \quad 0.5 \leq \sigma_8 \leq 1.2 , \tag{11}$$

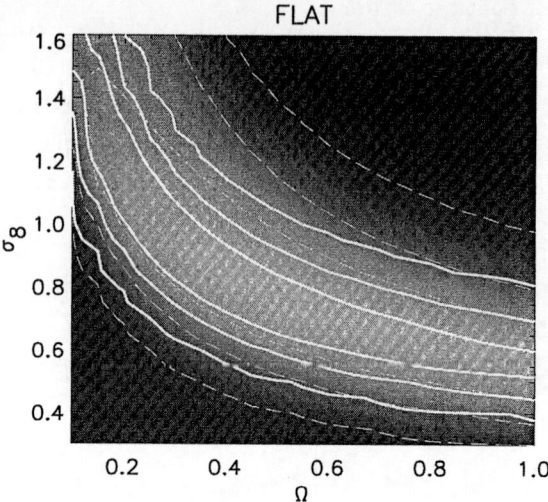

Fig. 2. Constraints on Ω and σ_8 for the flat cosmologies. The confidence levels are [68, 95, 99.9] from the brightest to the darkest area. The gray area and the dashed contours correspond to the computations with a full marginalisation over the default prior $\Gamma \in [0.05, 0.7]$ and $z_s \in [0.24, 0.64]$. The thick solid line contours are obtained from the prior $\Gamma \in [0.1, 0.4]$ and $z_s \in [0.39, 0.54]$ (which is a mean redshift $\bar{z}_s \in [0.8, 1.1]$). From van Waerbeke et al. (2002)

and, in the case of a flat-universe with $\Omega_m = 0.3$, they lead to $\sigma_8 \approx 0.9$.

4.2   The 3-Point Shear Correlations Function: Breaking the $\Omega_m - \sigma_8$ Degeneracy

The measurement of non-Gaussian features needs informations on higher order statistics than variance. Although the afore mentioned skewness of κ looks a promising quantity for this purpose, its measurements suffers from a number a practical difficulties which are not yet fixed. Recently, Bernardeau, van Waerbeke & Mellier ([37]) have proposed an alternative method using some specific patterns in the shear three-point function. Despite the complicated shape of the three-point correlation pattern, they uncovered it can be used for the measurement of non-Gaussian features. Their detection strategy based on their method has been tested on ray tracing simulations and turns out to be robust, usable in patchy catalogs, and quite insensitive to the topology of the survey.

Bernardeau, Mellier & van Waerbeke ([38]) used the analysis of the 3-point correlations function on the VIRMOS-DESCART data. Their results on Fig. 3 show a 2.4σ signal over four independent angular bins, or equivalently, a 4.9-σ confidence level detection with respect to measurements errors on scale of about 2 to 4 arc-minutes. The amplitude and the shape of the signal are consistent with theoretical expectations obtained from ray-tracing simulations. This result supports the idea that the measure corresponds to a cosmological signal due to

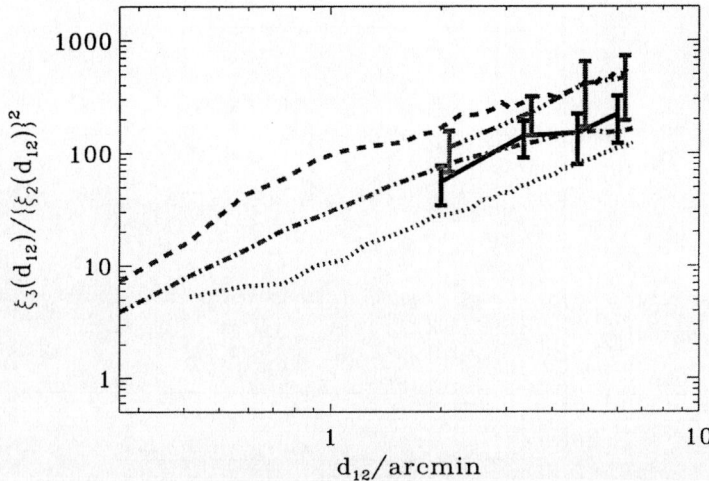

Fig. 3. Results for the VIRMOS-DESCART survey of the reduced three point correlation function ([38]). The solid line with error bars shows the raw results, when both the E and B contributions to the two-point correlation functions are included. The dot-dashed line with error bars corresponds to measurements where the contribution of the B mode has been subtracted out from the two-point correlation function. These measurements are compared to results obtained in τCDM, OCDM and ΛCDM simulations (dashed, dotted and dot-dashed lines respectively)

the gravitational instability dynamics. Moreover, its properties could be used to put constraints on the cosmological parameters, in particular on the density parameter of the Universe. Although the errors are still large to permit secure conclusions, one clearly see that the amplitude and the shape of the 3-point correlations function match the most likely cosmological models. Remarkably, the ΛCDM scenario perfectly fit the data points.

The Bernardeau et al. ([38]) result is the first detection of non-Gaussian features in a cosmic shear survey and it opens the route to break the $\Omega_m - \sigma_8$ degeneracy. Furthermore, this method is weakly dependent on other parameters, like the cosmological constant or the properties of the power spectrum. However, there are still some caveats which may be considered seriously. One difficulty is the source clustering which could significantly perturb high-order statistics (Hamana et al 2000, [32]). If so, multi-lens plane cosmic shear analysis will be necessary which implies a good knowledge of the redshift distribution. For very deep cosmic shear surveys, this could be could be a challenging issue.

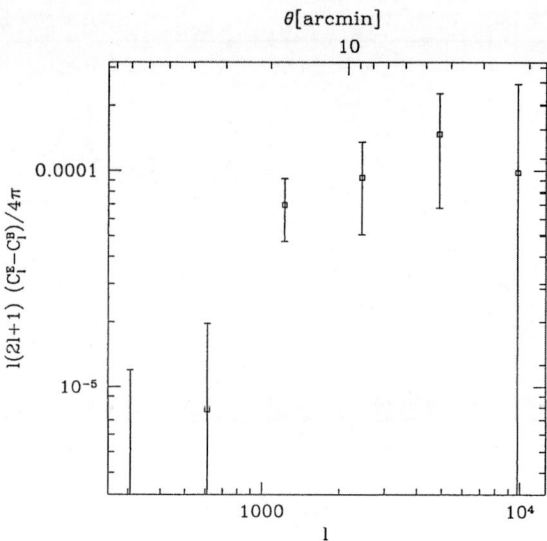

Fig. 4. Cosmological results from cosmic shear surveys: The angular power spectrum of the convergence field from the VIRMOS-DESCART survey is plotted (From Pen et al 2001). These are the first $C(l)$ of dark matter ever measured in cosmology

5   What Next?

Because on going surveys increase both in solid angle and in number of galaxies, they will quickly improve the accuracy of cosmic shear measurements, at a level where Ω_m and σ_8 will be known with a 10% accuracy. Since it is based on gravitational deflection by intervening matter spread over cosmological scales, the shape of the distortion field also probes directly the shape of the projected power spectrum of the (dark) matter. Pen et al ([28]) already explored its properties measuring for the first time the $C(l)$ of the dark matter (see Fig. 4). We therefore know this is feasible with present data. However, we expect much more within the next decade. Surveys covering hundreds of degrees, with multi-bands data in order to get redshift of sources and possibly detailed information of their clustering properties, are scheduled. The CFHT Legacy Survey[2] will cover 200 deg^2 and is one of those next-generation cosmic shear survey. Figures 1 and 5 shows it potential for cosmology. On Fig. 1 we simulated the expected signal to noise of the shear variance as function of angular scale for a ΛCDM cosmology. The error bars are considerably reduced as compared to present-day survey. On Fig. 5, we compare the expected signal to noise ratio of the CFHT Legacy Survey with the expected amplitude of the angular power spectrum for several theoretical quintessence fields models. It shows that even with 200 deg^2

---

[2] http://www.cfht.hawaii.edu/Science/CFHLS/

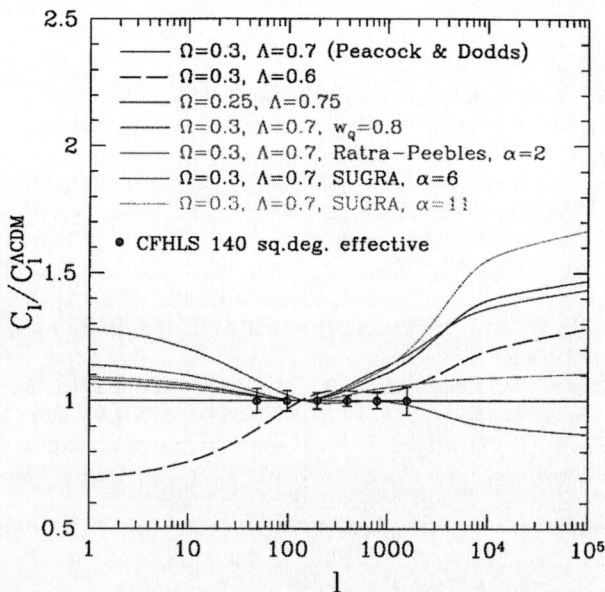

Fig. 5. The future of from cosmic shear surveys: Theoretical expectations of the CFHT Legacy Survey. The dots with error bars are the expected measurements of cosmic shear data from the 208 deg² of the CFHT Legacy Survey. The lines shows various models discussed by Benabed & Bernardeau ([39])

which include multi-color informations in order to get redshift of sources, one can already obtain interesting constraints on cosmology beyond standard models.

The use of cosmic shear data can be much more efficient if they are used together with other surveys, like CMB (Boomerang, MAP, Planck), SNIa surveys, or even galaxy surveys (2dF, SDSS). For example, SDSS will soon provide the 100, 000 quasars with redshifts. Ménard & Bartelmann ([40]) have recently explored the interest of this survey in order to cross-correlate the foreground galaxy distribution with the quasar population. The expected magnification bias generated by dark matter associated with foreground structures as mapped by galaxies depends on Ω_m and the biasing σ_8. In principle magnification bias in the SDSS quasar sample can provide similar constrains as cosmic shear.

Acknowledgements

We thank M. Bartelmann, K. Benabed, D. Bond, T. Hamana, H. Hoekstra, B. Ménard, S. Prunet and P. Schneider for useful discussions. This work was supported by the TMR Network "Gravitational Lensing: New Constraints on Cosmology and the Distribution of Dark Matter" of the EC under contract No. ERBFMRX-CT97-0172. YM thanks the organizers of the meeting for financial support.

References

1. Mellier, Y.; 1999 ARAA 37, 127.
2. Bartelmann, M.; Schneider, P.; 2001 Phys. Rep. 340, 292.
3. Bernardeau, F.; van Waerbeke, L.; Mellier, Y.; 1997 A&A 322, 1.
4. Jain, B.; Seljak, U.; 1997 ApJ 484, 560.
5. van Waerbeke, L.; Bernardeau, F.; Mellier, Y.; 1999 A&A 342, 15.
6. Erben, T.; van Waerbeke, L.; Bertin, E.; Mellier, Y.; Schneider, P.; 2001 A&A 366, 717.
7. van Waerbeke, L.; Mellier, Y.; Erben, T.; et al.; 2000 A&A 358, 30 [vWME+].
8. R. Blandford, A. Saust, T. Brainerd, J. Villumsen; 1991 MNRAS 251, 600
9. Wittman, D.; Tyson, J.A.; Kirkman, D.; Dell'Antonio, I.; Bernstein, G. 2000a Nature 405, 143 [WTK+].
10. Bacon, D.; Réfrégier; A., Ellis, R.S.; 2000 MNRAS 318, 625 [BRE].
11. Bacon, D.; Réfrégier; A., Clowe, D., Ellis, R.S.; 2000 MNRAS 325, 1065.
12. Kaiser, N.;, Wilson, G.;, Luppino, G. 2000 preprint, astro-ph/0003338 [KWL].
13. Maoli, R.; van Waerbeke, L.; Mellier, Y.; et al.; 2001 A&A 368, 766 [MvWM+].
14. Rhodes, J.; Réfrégier, A., Groth, E.J.; 2001 ApJ 536, 79.
15. van Waerbeke, L.; Mellier, Y.; Radovich, M.; et al.; 2001 A&A 374, 757 [vWMR+].
16. Bacon, D., Massey, R., Réfrégier, A., Ellis, R. astro-ph/0203134
17. J. Miralda-Escudé; 1991 ApJ380,1
18. Pen, U-L., Van Waerbeke, L., Mellier, Y.; 2002 ApJ 567, 31
19. Kaiser, N. 1992 ApJ 388, 272
20. kaiser, N., Squires, G., Broadhurst, T. 1995, ApJ 449, 460.
21. Kaiser, N. et al., 1994, in Durret et al., *Clusters of Galaxies*, Eds Frontières
22. P. Schneider, L. Van Waerbeke, B. Jain, G. Kruse; 1998 ApJ333, 767.
23. A. Réfrégier, J. Rhodes, E. Groth, ApJL, in press, astro-ph/0203131
24. L. Van Waerbeke, Y. Mellier, R. Pello et al., A& A, in press, astro-ph/0202503
25. Hämmerle, H.; Miralles, J.-M.; Schneider, P.; Erben, T.; Fosbury, R.A.E.; Freudling, W.; Pirzkal; N., Jain, B.; White, S.D.M.; 2002 A&A385 , 743
26. Hoekstra, H.; Yee, H.;, Gladders, M.D. 2001 ApJ 558, L11
27. H. Hoekstra, H. Yee, M. Gladders, ApJ, in press, astro-ph/0204295
28. Pen, U.; van Waerbeke, L.; Mellier, Y.; 2001 ApJ in press astro-ph/0109182.
29. Pierpaoli, E., Scott, D., White, M. 2001 MNRAS 325, 77.
30. Crittenden, R.G.; Natarajan, P.; Pen, U.; Theuns, T. 2001 ApJ 559 , 552.
31. Mackey, J.; White, M.; Kamionkowski, M.; 2001 preprint, astro-ph/0106364.
32. Hamana, T. et al. 2000. preprint astro-ph/0012200
33. Davis, M.; Newman, J.; Faber, S.; Phillips, A.; 2000 Proc. ESO/ECF/ESA on Deep Fields   Springer.
34. Le Fèvre, O.; Saisse, M.; Mancini, M.; 2000 SPIE 4008, 546.
35. Schneider, P. 1998 ApJ 498, 43.
36. van Waerbeke, L. 1998 A&A 334, 1.
37. Bernardeau, F., van Waerbeke, L., Mellier, Y. 2002 preprint astro-ph/0201029.
38. Bernardeau, F., Mellier, Y., van Waerbeke, L. 2002 A&A in press. preprint astro-ph/0201032
39. Benabed, K.; Bernardeau, F.; 2001 preprint, astro-ph/0104371.
40. Ménard, B., Bartelmann, M. 2002 preprint astro-ph/0203163

Constraining Cosmological Parameters with Observations of Galaxy Clusters

Hans Böhringer[1], Peter Schuecker[1], Luigi Guzzo[2], and Chris A. Collins[3]

[1] Max-Planck-Institut für Extraterrestrische Physik, D-85748 Garching, Germany
[2] Osservatorio Astronomico di Brera, via Bianchi, 22055 Merate (LC), Italy
[3] Astrophysics Research Institute, Liverpool John Moores University, Birkenhead CH41 1LD, U.K.

Abstract. Galaxy clusters are ideal probes of the large-scale structure of the Universe and for the tests of cosmological models. Based on the REFLEX redshift survey of X-ray selected clusters of galaxies we determine statistical properties of the galaxy cluster population, their spatial correlation, and the density fluctuation power spectrum of the cosmic matter distribution on large scales up to about 1 Gpc. Comparing these results with predictions of cosmological models we obtain tight constrains for the matter density parameter of the Universe, consistent with the combined results from observations of the microwave background anisotropies and distant type Ia supernovae.

1 Introduction

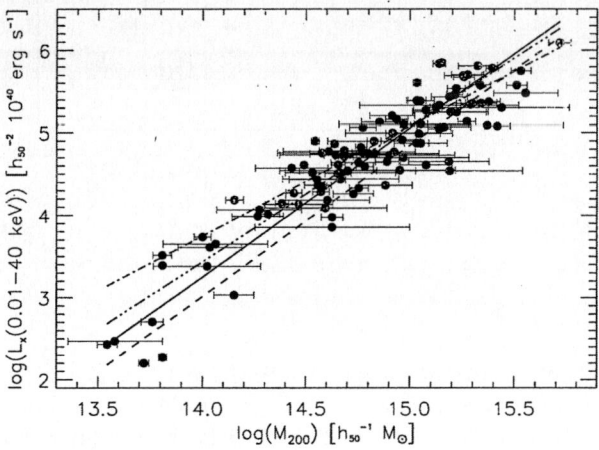

Fig. 1. Mass-X-ray luminosity relation determined from the 106 brightest galaxy clusters found in the ROSAT All-Sky Survey by Reiprich & Böhringer (2001)

Galaxy clusters with masses from about 10^{14} to over 10^{15} $M_\odot$ are the largest clearly defined building blocks of our Universe. Their formation and evolution is tightly connected to the evolution of the cosmic large-scale structure. Clusters are therefore ideal probes for the study of the large-scale matter distribution. The hot, intracluster plasma with temperatures of several 10 Million degrees makes galaxy clusters the most luminous X-ray emitters next only to quasars. The hot gas and its X-ray emission allows us to obtain mass estimates of the clusters and to detect clusters as gravitationally bound entities out to very large distances (Sarazin 1986). Systematic studies show that clusters have within a first order description, a quite standardized, self-similar appearance. This is reflected in the tight correlation of X-ray luminosity and cluster mass enabling us to construct interesting cosmological samples of clusters above a certain mass limit based on the detection of their X-ray luminosities. Figure 1 shows the X-ray luminosity-mass relation constructed from the observation of the X-ray brightest clusters found in the ROSAT All-Sky Survey (RASS).

2   Measurement of the Large-Scale Structure

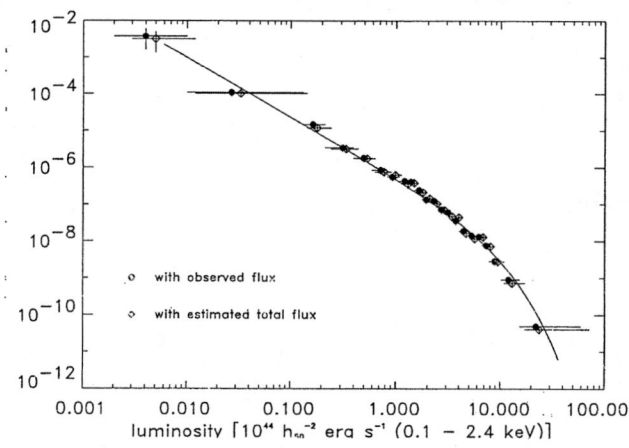

Fig. 2. X-ray luminosity function of the REFLEX cluster Survey (Böhringer et al. 2002). The filled and open data point refer to observed and corrected total luminosities, respectively

We are exploiting this relation and the data base of the RASS (Trümper 1993) with our X-ray selected cluster redshift surveys: the northern NORAS Survey (Böhringer et al. 2000) and the southern REFLEX (ROSAT-ESO Flux-Limited X-ray) Survey (Böhringer et al. 2001a). The latter project is currently more complete comprising a sample of 452 clusters. Therefore we describe the results of this project in the following. Most of the redshifts for this survey have been obtained within the frame of an ESO key program (Böhringer et al. 1998, Guzzo et al. 1999). The basic census of the REFLEX cluster survey is the X-ray luminosity function shown in Fig. 2 (Böhringer et al. 2002).

The essential goal of this survey is the assessment of the statistics of the large-scale structure. The most fundamental statistical description of the spatial structure is based on the second moments on the distribution, characterized either by the two-point-correlation function or its Fourier transform, the density fluctuation power spectrum. The two-point correlation function of REFLEX shows a power law shaped correlation function with a slope of 1.83, a correlation length of $18.8h_{100}^{-1}$ Mpc and a possible zero crossing at $\sim 45h_{100}^{-1}$ Mpc (Collins et al. 2000). The density fluctuation power spectrum (Fig. 3; Schuecker et al. 2001a is characterized by a power law at large values of the wave vector, k, with a slope of $\propto k^{-2}$ for $k \leq 0.1h$ Mpc^{-1} and a maximum around $k \sim 0.03h$ Mpc^{-1} (corresponding to a wavelength of about $200h^{-1}$ Mpc). This maximum reflects the size of the horizon when the Universe featured equal energy density in radiation and matter and is a sensitive measure of the mean density of the Universe, Ω_0, providing approximately the following constraints $h\Omega_0 = 0.12$ to 0.26 (Schuecker et al. 2001).

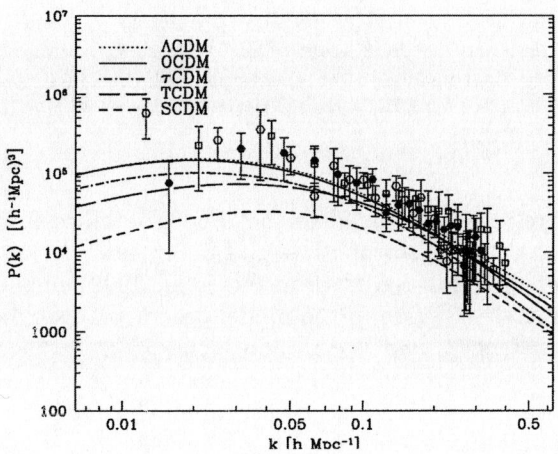

Fig. 3. Power spectra of the density fluctuations in the REFLEX cluster sample together with predictions from various popular cosmological models taken from the literature. For details see Schuecker et al. (2001a)

3   Cosmological Tests with Galaxy Clusters

For a more stringent cosmological test we are using a statistical analysis based on an eigenmode decomposition of the spatial structure of the REFLEX cluster distribution with the Karhunen-Loéve method in comparison with semi-analytic cosmological models of structure formation (Schuecker 2002a,b). With this method the density fluctuations of the distribution as well as the redshift dependent cluster number density has been taken into account simultaneously for the first time. The analysis of the REFLEX data provides the very important constraints on the cosmic matter density shown in Fig. 4.

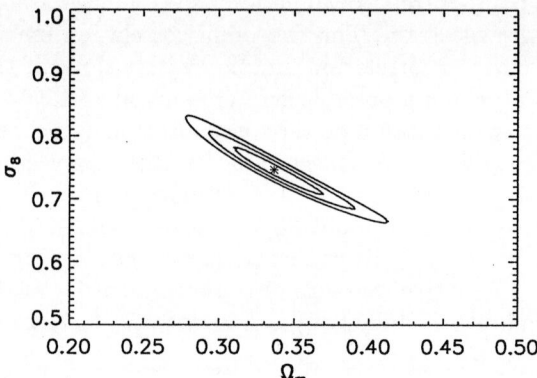

Fig. 4. Constraints on the cosmological density parameter, Ω_m, and the amplitude of the matter density fluctuations on a scale of $8h^{-1}$ Mpc, σ_8 obtained from the comparison of the density fluctuation power spectrum and the clusters abundance as a function of redshift in a Karhunen-Loéve statistical analysis of the REFLEX Survey data (Schuecker et al. 2002b)

These results are to be compared to the results obtained from observations of the cosmic microwave anisotropies (e.g. De Bernardis et al. 2002) and of the study of distant SN Ia (e.g. Perlmutter et al. 1999) providing combined constraints that encircle a region in the model parameter space spanned by the cosmological parameters Ω_0 and Ω_Λ around values of $\Omega_0 = 0.3$ and $\Omega_\Lambda = 0.7$. The galaxy cluster results provide a different cut through this parameter space crossing the other two results at their intersection. Thus, the evidence for a low density universe is solidifying.

References

1. Böhringer, H., Guzzo, L., Collins, C.A., et al. 1998, The Messenger, No. **94**, 21
2. Böhringer, H., Voges, W., Huchra, J.P., et al., 2000, ApJS, **129**, 435
3. Böhringer, H., Schuecker, P., Guzzo, L., et al., 2001, A&A, **369**, 826
4. Böhringer, H., Collins, C.A., Guzzo, L., 2002, ApJ, **566**, 1
5. Collins, C.A., Guzzo, L., Böhringer, H., 2000, MNRAS, **319**, 939
6. De Bernardis et al., 2002, ApJ, **564**, 556
7. Guzzo, L., Böhringer, H., Schuecker, P., et al., 1999, The Messenger, No. **95**, 27
8. Perlmutter, N., 1999, ApJ, **517**, 565
9. Reiprich, T.H. & Böhringer, H., 2001, ApJ, **567**, 716
10. Sarazin, C.L., 1986, Rev. Mod. Phys., **58**, 1
11. Schuecker, P., Böhringer, H., Guzzo, et al., 2001, A&A, **368**, 86
12. Schuecker, P., Guzzo, L., Collins, C.A., Böhringer, H., 2002a, MNRAS (submitted), astro-ph/0205342
13. Schuecker, P. et al. 2002b (in preparation)
14. Trümper, J., 1993, Science, **260**, 1769

First Light and the Reionization of the Universe*

Piero Madau

University of California Santa Cruz, CA 95064, USA

Abstract. The development of primordial inhomogeneities into the non-linear regime and the formation of the first bound objects mark the transition from a simple cooling universe – described by just a few parameters – to a very messy hot one – the realm of radiative, hydrodynamic and star formation processes. In popular cold dark matter cosmological scenarios, stars and quasars may have first appeared in significant numbers around a redshift of 10 or so, as the gas within subgalactic halos cooled rapidly due to atomic processes and fragmented. This early generation of sources generated the ultraviolet radiation and kinetic energy that ended the cosmic "dark ages" and reheated and reionized most of the hydrogen in the cosmos by a redshift of 6. The detailed thermal history of the intergalactic medium – the main repository of baryons at high redshift – during and soon after these crucial formative stages depends on the power-spectrum of density fluctuations on small scales and on a complex network of poorly understood 'feedback' mechanisms, and is one of the missing links in galaxy formation and evolution studies.

1 Introduction

The past few years have witnessed great progress in our understanding of the high-redshift universe. The pace of observational cosmology and extragalactic astronomy has never been faster. *Hubble Space Telescope* deep imaging surveys and spectroscopic follow-up from 8–10 m telescopes are together elucidating the history of star formation and galaxy clustering back to redshift $z = 4$. The measurement of the far-IR/sub-mm background by DIRBE and FIRAS on board the *COBE* satellite and the detection of distant ultraluminous sub-mm sources by the SCUBA camera have shed new light on the 'optically hidden' side of galaxy formation, and shown that a significant fraction of the energy released by stellar nucleosynthesis is re-emitted as thermal radiation by dust. Ongoing X-ray studies with the *Chandra* and *XMM Newton* satellites may be discovering the population of highly absorbed quasistellar objects predicted to be responsible for the hard X-ray background. Big surveys currently in progress such as the Sloan Digital Sky Survey, together with the use of novel instruments have led to the discovery of galaxies and quasars at redshifts $z \gtrsim 6$ [9] (the current record holder appears to be a $z = 6.56$ galaxy gravitationally lensed by a foreground cluster [20]), when the universe was about 6% of its current age. These observations may be at the

---

* Invited talk at the ESO-CERN-ESA Symposium on Astronomy, Cosmology, and Fundamental Physics, March 4-7 2002, Garching, Germany.

threshold of probing the epoch of first light and cosmological reionization. *Keck* and *VLT* observations of redshifted H I Lyα ('forest') absorption in the spectra of distant quasars are becoming an increasingly sensitive probe of the distribution of gaseous matter in the universe, and are revealing the topology, thermal and ionization state, and chemical composition of the intergalactic medium (IGM) – the main repository of baryons at high redshift – in unprecedented detail. From all these data, an empirical picture is beginning to emerge of the processes and timescales whereby ordinary matter in the universe has been transformed as it is subjected to radiative and chemical interactions and processed through stars.

On the theoretical side progress has been equally significant. The key idea, that primordial density fluctuations grow by gravitational instability driven by cold dark matter (CDM), has been elaborated upon and explored in detail through large-scale numerical simulations on supercomputers, leading to a hierarchical ('bottom-up') scenario of structure formation. In this model, the first objects to form are on subgalactic scales, and merge to make progressively bigger structures ('hierarchical clustering'). Ordinary matter in the universe follows the dynamics dictated by the dark matter until radiative, hydrodynamic, and star formation processes take over. Perhaps the most remarkable success of this theory has been the prediction of anisotropies in the temperature of the cosmic microwave background (CMB) radiation at about the level subsequently measured by the *COBE* satellite and most recently by the *BOOMERANG, MAXIMA, DASI,* and *CBI* experiment.

In spite of the remarkable progress achieved in recent years, many fundamental questions (beside, of course, the nature of the dark matter and dark energy components) remain only partially answered. How and when was the universe reheated? We know that at least some galaxies and quasars were already shining when the universe was less than 10^9 yr old. But when exactly did the first luminous structures form and how bright were they? We believe there is a strong coupling between the thermodynamic state of the IGM and the process of galaxy formation, and we suspect complex feedback mechanisms are at work in this interaction. The detailed astrophysics of these processes is, however, unclear. Finally, the precise location and degree of metal enrichment of most the baryons in the universe remains an open question.

2   The Dark Ages

At epochs corresponding to $z \sim 1200$, about half a million years after the big bang, the universe recombined, became optically thin to Thomson scattering, and entered, in the words of Sir Martin Rees [30], a "dark age". The primordial radiation then cooled below 3000 K, shifting first into the infrared and then into the radio. We understand the microphysics of the post-recombination universe well. The fractional ionization freezed out to the value $10^{-5}\Omega_M/(h\Omega_b)$:[1]

---

[1] Throughout this talk I will denote with $\Omega_M, \Omega_\Lambda,$ and Ω_b the matter, vacuum, and baryon density parameters today, and with h the dimensionless Hubble constant, $h = H_0/100 \text{ km s}^{-1} \text{ Mpc}^{-1}$.

these residual electrons were enough to keep the matter in thermal equilibrium with the radiation via Compton scattering until a thermalization redshift $z_t \simeq 800(\Omega_b h^2)^{2/5} \simeq 160$, i.e. well after the universe became transparent [28]. Thereafter, the matter temperature dropped as $(1+z)^2$ due to adiabatic expansion (Fig. 1) until primordial inhomogeneities in the density field evolved into the non-linear regime. The minimum mass scale for the gravitational aggregation of cold dark matter particles is negligibly small. One of the most popular CDM candidates is the neutralino: in neutralino CDM, collisional damping and free streaming smear out all power of primordial density inhomogeneities only below $\sim 10^{-7}$ M$_\odot$ [19]. Baryons, however, respond to pressure gradients and do not fall into dark matter clumps below the cosmological Jeans mass (in linear theory this is the minimum mass-scale of a perturbation where gravity overcomes pressure)

$$M_J = \frac{\pi \bar{\rho}}{6} \left(\frac{5\pi kT}{3G\bar{\rho}m_p \mu} \right)^{3/2} \approx 3 \times 10^4 \, h^{-1} \, \text{M}_\odot (aT/\mu)^{3/2} \Omega_M^{-1/2}. \tag{1}$$

Here $a = (1+z)^{-1}$ is the scale factor, $\bar{\rho}$ the total mass density, μ the mean molecular weight, and T the gas temperature. The evolution of M_J is shown in Fig. 1. In the post-recombination universe, the baryon-electron gas is thermally coupled to the CMB, $T \propto a^{-1}$, and the Jeans mass is independent of redshift and comparable to the mass of globular clusters, $M_J \approx 10^5 \, h^{-1}$ M$_\odot$. For $z < z_t$, the temperature of the baryons drops as $T \propto a^{-2}$, and the Jeans mass decreases with time, $M_J \propto a^{-3/2}$. This trend is reversed by the reheating of the IGM. The energy released by the first collapsed objects drives the Jeans mass up to galaxy

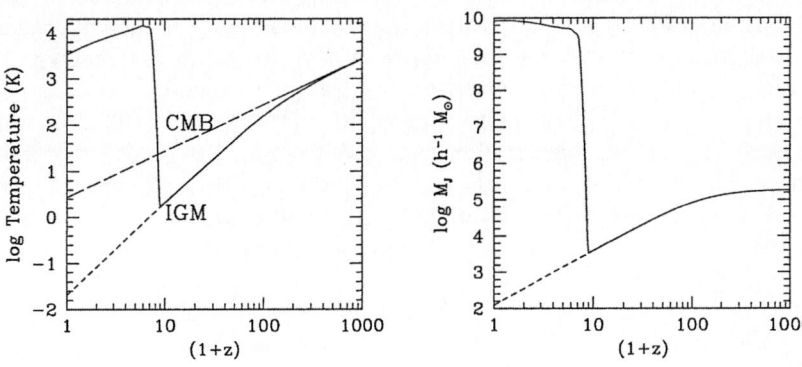

Fig. 1. *Left:* Evolution of the radiation (*long-dashed line*, labeled CMB) and gas (*solid line*, labeled IGM) temperatures after recombination. The universe is assumed to be reionized by ultraviolet radiation at $1+z = 10$. The *short-dashed line* is the extrapolated gas temperature in the absence of any reheating mechanism. *Right:* Cosmological (gas + dark matter) Jeans mass

scales (Fig. 1): previously growing density perturbations decay as their mass drops below the new Jeans mass. Photoionization by the ultraviolet radiation from the first stars and quasars will heat the IGM to temperatures of $\approx 10^4$ K (corresponding to a Jeans mass $M_J \approx 4 \times 10^9 \, h^{-1}$ M$_\odot$ at $1+z = 10$), suppressing gas infall into low mass halos and preventing new (dwarf) galaxies from forming.[2]

3   The Epoch of Reionization

Hierarchical clustering theories provide a well-defined cosmological framework in which the history of baryonic material can in principle be tracked through the development of cosmic structure. Probing the reionization epoch may then provide a means for constraining competing models for the formation of cosmic structures. For example, popular modifications of the CDM paradigm that attempt to improve over CDM by suppressing the primordial power-spectrum on small scales, like warm dark matter (WDM), are known to reduce the number of collapsed halos at high redshifts and make it more difficult to reionize the universe [2].

In practice, we are unable to predict when reionization actually occurred. While N-body+hydrodynamical simulations have convincingly shown that intergalactic gas is expected to fragment into structures at early times in CDM cosmogonies, the same simulations are much less able to predict the efficiency with which the first gravitationally collapsed objects lit up the universe at the end of the dark age. The crucial processes of star formation, feedback (e.g. the effect of the heat input from the first generation of sources on later ones), and assembly of massive black holes in the nuclei of galaxies are poorly understood. Consider the following illustrative example:

Hydrogen photoionization requires more than one photon above 13.6 eV per hydrogen atom: of order $t/\bar{t}_{rec} \sim 10$ (where $\bar{t}_{rec}$ is the volume-averaged hydrogen recombination timescale) extra photons appear to be needed to keep the gas in overdense regions and filaments ionized against radiative recombinations [14]. A 'typical' stellar population produces during its lifetime about 4000 Lyman continuum (ionizing) photons per stellar proton. A fraction $f \sim 0.25\%$ of cosmic baryons must then condense into stars to supply the requisite ultraviolet flux. This estimate assumes a standard (Salpeter) initial mass function (IMF), which determines the relative abundances of hot, high mass stars versus cold, low mass ones.

The very first generation of stars ('Population III') must have formed, however, out of unmagnetized metal-free gas: numerical simulations of the fragmentation of pure H and He molecular clouds [4] [1] have shown that these

---

[2] It has been pointed out by [16] that, when the Jeans mass itself varies with time, linear gas fluctuations tend to be smoothed on a (filtering) scale that depends on the full thermal history of the gas instead of the instantaneous value of the sound speed: after reheating, this filtering scale is actually smaller than the Jeans scale. Numerical simulations of cosmological reionization show that the characteristic suppression mass is typically lower than the linear-theory Jeans mass [15].

characteristics likely led to a 'top-heavy' IMF biased towards very massive stars (VMSs, i.e. stars a few hundred times more massive than the Sun), quite different from the present-day Galactic case. Metal-free VMSs emit about 10^5 Lyman continuum photons per stellar baryon [5], approximately 25 times more than a standard stellar population. A corresponding smaller fraction of cosmic baryons would have to collapse then into VMSs to reionize the universe, $f \sim 10^{-4}$. There are of course further complications. Since, at zero metallicity, mass loss through radiatively-driven stellar winds is expected to be negligible [22], Population III stars may actually die losing only a small fraction of their mass. If they retain their large mass until death, VMSs with masses $100 \lesssim m \lesssim 250$ M$_\odot$ will encounter the electron-positron pair instability and disappear in a giant nuclear-powered explosion [12], leaving no compact remnants and polluting the universe with the first heavy elements [32]. In still heavier stars, however, oxygen and silicon burning is unable to drive an explosion, and complete collapse to a black hole will occur instead [3]. Thin disk accretion onto a Schwarzschild black hole releases about 50 MeV per baryon. The conversion of a trace amount of the total baryonic mass into early black holes, $f \sim 3 \times 10^{-6}$, would suffice to reionize the universe.

Quite apart from the uncertainties in the IMF, it is also unclear what fraction of the cold gas will be retained in protogalaxies after the formation of the first stars (this will affect the global efficiency of star formation at these epochs) and whether an early input of mechanical energy will play a role in determining the thermal and ionization state of the IGM on large scales. The same massive stars that emit ultraviolet light also explode as supernovae (SNe), returning most of the metals to the interstellar medium of pregalactic systems and injecting about 10^{51} ergs per event in kinetic energy. For a standard IMF, one has about one SN every 150 M$_\odot$ of baryons that forms stars. A complex network of feedback mechanisms is likely at work in these systems, as the gas in shallow potential is more easily blown away [7], thereby quenching star formation. Furthermore, as the blastwaves produced by supernova explosions sweep the surrounding intergalactic gas and enrich it with newly formed heavy elements [23], they may inhibit the formation of surrounding low-mass galaxies due to 'baryonic stripping' [31], and drive vast portions of the IGM to a significantly higher adiabat than expected from photoionization.

4   Detecting Protogalaxies Before Reionization Breakthrough

By a redshift of 7 no large galaxies should have assembled yet, but hierarchical clustering theories predict the existence of a large population of subgalactic fragments with masses comparable to present-day dwarf ellipticals. What are the prospects for detecting these early star-forming systems before reionization breakthrough? Prior to complete reionization at redshift z_r, sources of ultraviolet radiation will be seen behind a large column of intervening gas that is still neutral. Photons with energies between 10.2 and 13.6 eV will propagate

freely into the IGM until, redshifted by the Hubble expansion, they will ultimately reach the Lyα transition energy of 10.2 eV, resonantly scatter off neutral
hydrogen, and be removed from the line-of-sight. The spectrum of a source at
$z > z_r$ should then show a Gunn-Peterson [17] absorption trough at wavelengths
shorter than $\lambda_\alpha (1 + z)$, where $\lambda_\alpha = 1216$ Å. In fact, the Gunn-Peterson optical
depth would be so large that the damping wing of this absorption trough would
spill over into the red of Lyα [27]. This characteristic feature extends for more
than 1500 km s^{-1} to the red of the resonance, and may significantly suppress
the Lyα emission line in the spectra of the first generation of cosmic sources. A
disappearing Lyα could then be used in principle as a flag of the observation of
the universe before reionization.

In practice, to generate a Lyα emission line by recombination these objects
have to produce ionizing photons. If the escape fraction for Lyman continuum
photons into the intergalactic space is significant, a photoionized region of IGM
will surround each individual source [25], increasing the transmission of photons
redward of the Lyα resonance. If the source lifetime is shorter than the cosmic
expansion and gas recombination timescales, the volume of ionized gas will be
proportional to the total number of photons produced above 13.6 eV that leak
into intergalactic space; the effect of this local photoionization is to greatly reduce
the scattering opacity between the redshift of the source and the boundary of
its H II region. In Fig. 2 I have plotted the expected ultraviolet spectrum and
Lyα line profile for a luminous star-forming galaxy observed at $z = 7$ before

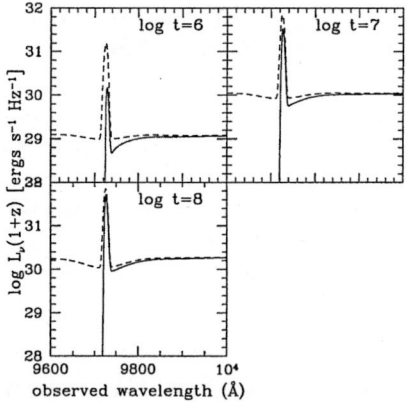

Fig. 2. *Dashed lines*: Synthetic ultraviolet intrinsic spectra of a luminous galaxy at
$z = 7$ as a function of age t (in yr). The stellar population is formed at a constant rate
of $\dot{M} = 10$ M$_\odot$ yr^{-1}, with a Salpeter IMF and metallicity $Z = 0.05$ Z$_\odot$. The Lyα line is
computed in Case B recombination assuming an escape fraction of Lyman continuum
photons of 50% and a Doppler width of 150 km s^{-1}. *Solid lines*: Same spectra transmitted through a uniform IGM, assuming the source is being observed prior to the
reionization epoch

reionization breakthrough (assumed to occur at $z_r = 6$), as a function of stellar age. The suppression effect of the Lyα emission line by the red wing of the Gunn-Peterson trough, while clearly visible at early times (i.e. in the absence of a local H II region, top-left panel), weakens significantly with time as the size of the photoionized zone increases. In this figure the escape fraction of ionizing photons into the IGM was assumed to be 50%. In general (depending on the intrinsic line width), the emission line appears to remain observable at late times if the escape fraction of Lyman continuum photons exceeds 5–10% (see also [18]).

This should be taken just as an illustrative example, as in some numerical simulations [14] reionization was already well in progress prior to redshift 6, and the form of the damping profile would be different in the case of patchy ionization along the line of sight. Yet, the figure shows that the detection of Lyα emission in the spectra of luminous galaxies cannot then be used, by itself, as a constraint on the reionization epoch. Even before reionization, the line is unlikely to be obscured from view; early star-forming protogalaxies may still be best detected in the near-IR via their Lyα emission.

5   First Light at 21 cm

Recent progress with cosmological hydro-simulations based on hierarchical structure formation models has led to important insight into the topology of intergalactic baryons. According to these calculations, a truly inter- and proto-galactic medium collapses under the influence of dark matter gravity into flattened and filamentary structures, which are seen in absorption against background quasi stellar sources. Balloon experiments have produced convincing detections of fluctuations in the temperature of the microwave background radiation on angular scales of a few arcminutes, providing a direct link between the state of the universe at recombination and the high contrast structures visible at later times.

Prior to the epoch of full reionization, the intergalactic medium and gravitationally collapsed systems may be detectable in emission or absorption against the CMB at the frequency corresponding to the redshifted 21 cm line (associated with the spin-flip transition from the triplet to the singlet state of neutral hydrogen.) In emission, the contribution of a patch of neutral IGM with overdensity δ to the background spectrum at $21(1+z)$ cm amounts to a brightness temperature at Earth of

$$\delta T_b \simeq 0.02\,\mathrm{K}\ h^{-1}(1+\delta)\left(\frac{\Omega_b h^2}{0.02}\right)\left[\left(\frac{1+z}{10}\right)\left(\frac{0.3}{\Omega_M}\right)\right]^{1/2}, \qquad (2)$$

very small compared to the CMB. Nonetheless, 21 cm spectral features are expected to display structure in redshift space as well as angular structure due to inhomogeneities in the gas density field, hydrogen ionized fraction, and spin temperature. Several different signatures have been investigated in the recent literature: (a) the fluctuations in the redshifted 21 cm emission induced by the gas density inhomogeneities that develop at early times in CDM-dominated cosmologies [24] [34] and by virialized "minihalos" with virial temperatures below the threshold for atomic cooling [21]; (b) the global feature ('reionization step') in the continuum spectrum of the radio sky that may mark the abrupt overlapping phase of individual intergalactic H II regions [33]; (c) and the 21 cm narrow lines generated in absorption against very high-redshift radio sources by the neutral IGM [6] and by intervening minihalos and protogalactic disks [13].

A quick summary of the physics of 21 cm radiation will illustrate the basic ideas and unresolved issues behind these studies. The emission or absorption of 21 cm photons from a neutral IGM is governed by the hydrogen spin temperature T_S defined by $n_1/n_0 = 3 \exp(-T_*/T_S)$, where n_0 and n_1 are the number densities of atoms in the singlet and triplet $n = 1$ hyperfine levels and $T_* = 0.068$ K is the excitation temperature of the 21 cm transition. In the presence of only the CMB radiation with $T_{CMB} = 2.73\,(1 + z)$ K, the spin states will reach thermal equilibrium with the CMB on a timescale of $T_*/(T_{CMB}A_{10}) \approx 3 \times 10^5\,(1+z)^{-1}$ yr $(A_{10} = 2.9 \times 10^{-15}\,s^{-1}$ is the spontaneous decay rate of the hyperfine transition of atomic hydrogen), and intergalactic H I will produce neither an absorption nor an emission signature. A mechanism is required that decouples T_S and T_{CMB}, e.g. by coupling the spin temperature instead to the kinetic temperature T_K of the gas itself. Two mechanisms are available, collisions between hydrogen atoms [29] and scattering by Lyα photons [10]. The collision-induced coupling between the spin and kinetic temperatures is dominated by the spin-exchange process between the colliding hydrogen atoms. The rate, however, is too small

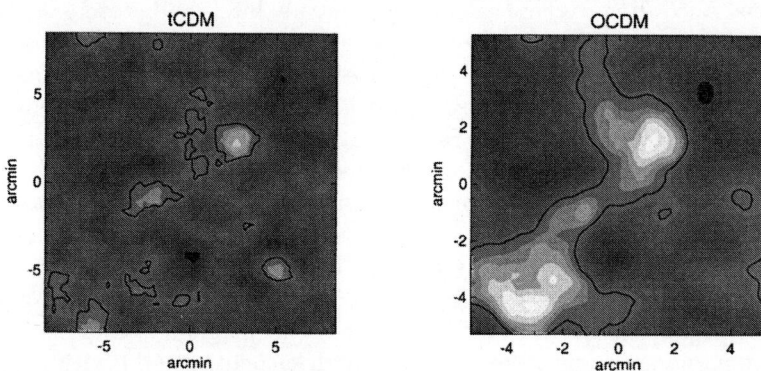

Fig. 3. *Left:* Radio map of redshifted 21 cm emission against the CMB in a 'tilted' CDM (tCDM) cosmology at $z = 8.5$. Here a collisionless N-body simulation with 64^3 particles has been performed with box size $20h^{-1}$ comoving Mpc, corresponding to 17 (11) arcmin in tCDM (OCDM). The baryons are assumed to trace the dark matter distribution without any biasing, the spin temperature to be much greater than the temperature of the CMB everywhere, and the gas to be fully neutral. The frequency window is 1 MHz around a central frequency of 150 MHz. The contour levels outline regions with signal greater than $4\,\mu$Jy per beam. *Right:* Same for a open cosmology (OCDM). Since the growth of density fluctuations ceases early on in an open universe (and the power spectrum is normalized to the abundance of clusters today), the signal at a given angular size is much larger in OCDM than in tCDM at these early epochs [34]

for realistic IGM densities at the redshifts of interest, although collisions may be important in dense regions with $\delta\rho/\rho \gtrsim 30[(1+z)/10]^{-2}$, like virialized minihalos.

In the low density IGM, the dominant mechanism is the scattering of continuum ultraviolet photons redshifted by the Hubble expansion into local Lyα photons. The many scatterings mix the hyperfine levels of neutral hydrogen in its ground state via intermediate transitions to the $2p$ state, the Wouthuysen-Field process. As the neutral IGM is highly opaque to resonant scattering, the shape of the continuum radiation spectrum near Lyα will follow a Boltzmann distribution with a temperature given by the kinetic temperature of the IGM [11]. In this case the spin temperature of neutral hydrogen is a weighted mean between the matter and CMB temperatures. There exists then a critical value of the background flux of Lyα photons which, if greatly exceeded, would drive the spin temperature away from T_{CMB}. In [34] we used N-body cosmological simulations and, assuming a fully neutral medium with $T_S \gg T_{CMB}$, showed that prior to reionization the same network of sheets and filaments (the 'cosmic web') that gives rise to the Lyα forest at $z \sim 3$ should lead to fluctuations in the 21 cm brightness temperature at higher redshifts (Fig. 3). At 150 MHz ($z = 8.5$), for observations with a bandwidth of 1 MHz, the root mean square fluctuations should be ~ 10 mK at $1'$, decreasing with scale. Because of the smoothness of the CMB sky, fluctuations in the 21 cm radiation will dominate the CMB fluctuations by about 2 orders of magnitude on arcmin scales.

While the microphysics is well understood, our understanding of the astrophysics of 21 cm tomography is still poor. As mentioned above, it is the presence of a sufficient flux of Lyα photons which renders the neutral IGM 'visible'. Without heating sources, the adiabatic expansion of the universe will lower the kinetic temperature of the gas well below that of the CMB, and the IGM will be detectable through its absorption. If there are sources of radiation that preheat the IGM, it may be possible to detect it in emission instead. The energetic demand for heating the IGM above the CMB temperature is meager, only ~ 0.004 eV per particle at $z \sim 10$. Consequently, even relatively inefficient heating mechanisms may be important warming sources well before the universe was actually reionized. Perhaps more importantly, prior to full reionization the IGM will be a mixture of neutral, partially ionized, and fully ionized structures: low-density regions will be fully ionized first, followed by regions with higher and higher densities. Radio maps at $21(1+z)$ cm will show a patchwork of emission/absorption signals from H I zones modulated by H II regions where no signal is detectable against the CMB.

6   Summary

Despite much recent progress in our understanding of the formation of early cosmic structure and the high-redshift universe, the end of the cosmic "dark ages" remains one of missing link in galaxy formation and evolution studies. We are left very uncertain about the whole era from 10^7 to 10^9 yr – the epoch of the first galaxies, stars, supernovae, and massive black holes. Some of the issues discussed above are likely to remain a topic of lively controversy until the launch of the *Next Generation Space Telescope*, ideally suited to image the earliest generation of stars in the universe. If the first massive black holes form in pregalactic systems at very high redshifts, they will be incorporated through a series of mergers into larger and larger halos, sink to the center owing to dynamical friction, accrete a fraction of the gas in the merger remnant to become supermassive, and form a binary system [35]. Their coalescence would be signalled by the emission of low-frequency gravitational waves detectable by the planned *Laser Interferometer Space Antenna*. The search at 21 cm for the epoch of first light has become one of the main science drivers of the *LOw Frequency ARray*. While remaining an extremely challenging project due to foreground contamination from extragalactic radio sources [8], the detection and imaging of these small-scale structures with *LOFAR* is a tantalizing possibility within range of the thermal noise of the array.

I would like to thank all my collaborators for discussions on the topics described here, and Martin Rees for many insightful suggestions. Support for this work was provided by NASA through grant NAG5-11513 and by NSF grant AST-0205738.

References

1. Abel, T., Bryan, G., & Norman, M. 2000, ApJ, 540, 39
2. Barkana, R., Haiman, Z., & Ostriker, J. P. 2001, ApJ, 558, 482
3. Bond, J. R., Arnett, W. D., & Carr, B. J. 1984, ApJ, 280, 825
4. Bromm, V., Coppi, P. S., & Larson, R. B. 2002, ApJ, 564, 23
5. Bromm, V., Kudritzki, R. P., & Loeb, A. 2001, ApJ, 552, 464
6. Carilli, C., Gnedin, N. Y., & Owen, F. 2002, ApJ, 577, 22
7. Dekel, A., & Silk, J. 1986, ApJ, 303, 39
8. Di Matteo, T., Perna, R., Abel, T., & Rees, M. J. 2002, ApJ, 564, 576
9. Fan, X., et al. 2001, AJ, 122, 2833
10. Field, G. B. 1958, Proc. IRE, 46, 240
11. Field, G. B. 1959, ApJ, 129, 551
12. Fryer, C. L., Woosley, S. E., & Heger, A. 2001, ApJ, 550, 372
13. Furlanetto, S., & Loeb, A. 2002, ApJ, submitted (astro-ph/0201313)
14. Gnedin, N. Y. 2000, ApJ, 535, 530
15. Gnedin, N. Y. 2000, ApJ, 542, 535
16. Gnedin, N. Y., & Hui, L. 1998, MNRAS, 296, 44
17. Gunn, J. E., & Peterson, B. A. 1965, ApJ, 142, 1633
18. Haiman, Z. 2002, ApJ, submitted (astro-ph/0205410)
19. Hofmann, S., Schwarz, D. J., & Stocker, H. 2001, PhRvD 64, 083507
20. Hu, E. M., et al. 2002, ApJ, 568, L75
21. Iliev, I. T., Shapiro, P. R., Ferrara, A., & Martel, H. 2002, ApJ, 572, L123
22. Kudritzki, R. P. 2000, in The First Stars, ed. A. Weiss, T. Abel, & V. Hill (Heidelberg: Springer), 127
23. Madau, P., Ferrara, A., & Rees, M. J. 2001, 555, 92
24. Madau, P., Meiksin, A., & Rees, M. J. 1997, ApJ, 475, 492
25. Madau, P., & Rees, M. J. 2000, ApJ, 542, L69

26. Madau, P., & Rees, M. J. 2001, ApJ, 551, L27
27. Miralda-Escudé, J., & Rees, M. J. 1998, ApJ, 497, 21
28. Peebles, P. J. E. 1993, Principles of Physical Cosmology (Princeton: Princeton University Press)
29. Purcell, E. M., & Field, G. B. 1956, ApJ, 124, 542
30. Rees, M. J. 2000, RSPTA, 358, 1989
31. Scannapieco, E., Ferrara, A., & Madau, P. 2002, ApJ, 574, 590
32. Schneider, R., Ferrara, A., Natarajan, P., & Omukai, K. 2002, ApJ, 571, 30
33. Shaver, P., Windhorst, R., Madau, P., & de Bruyn, G. 1999, A&A, 345, 380
34. Tozzi, P., Madau, P., Meiksin, A., & Rees, M. J. 2000, ApJ, 528, 597
35. Volonteri, M., Haardt, F., & Madau, P. 2002, ApJ, in press

The State of the Universe: Cosmological Parameters 2002

Lawrence M. Krauss

Departments of Physics and Astronomy, Case Western Reserve University, 10900 Euclid Ave., Cleveland, OH 44106-7079, USA

Abstract. In the past decade, observational cosmology has had one of the most exciting periods in the past century. The precision with which we have been able to measure cosmological parameters has increased tremendously, while at the same time, we have been surprised beyond our wildest dreams by the results. I review here recent measurements of the expansion rate, geometry, age, matter content, and equation of state of the universe, and discuss the implications for our understanding of cosmology.

1 Introduction

As early as a decade ago, the uncertainties in the measurement of cosmological parameters was such that few definitive statements could be made regarding cosmological models. That situation has changed completely. Instead all cosmological observables have now converged on a single cosmological model. Unfortunately, or perhaps fortunately for theorists, the "standard model" of cosmology from the 1980's is now dead. Instead, the model that has survived the test of observation is completely inexplicable at the present time, producing many more questions than answers. At the very least, our vision of the future of the Universe has completely changed, and the long-tauted connection between geometry and destiny is now dead.

I have been asked here to review the current status of our knowledge of cosmological observables. Following previous reviews I have prepared, it seems reasonable to divide this into three subsections, Space, Time, and Matter. Specifically, I shall concentrate on the following observables:

Space:

- Expansion Rate
- Geometry

Time:

- Age of the Universe

Matter:

- Baryon Density
- Large Scale Structure
- Matter Density
- Equation of State

2   Space: The Final Frontier

2.1   The Hubble Constant

Arguably the most important single parameter describing the physical universe today is the Hubble Constant. Since the discovery in 1929 that the Universe is expanding, the determination of the rate of expansion dominated observational cosmology for much of the rest of the 20th century. The expansion rate, given by the Hubble Constant, sets the overall scale for most other observables in cosmology.

The big news, if any, is that by the end of the 20th century, almost all measurements have converged on a single range for this all important quantity. (I say *almost* all, because to my knowledge Alan Sandage still believes the claimed limits are incorrect [4].)

Recently, the Hubble Space Telescope Key Project has announced its final results. This is the largest scale endeavor carried out over the past decade with a goal of achieving a 10 % absolute uncertainty in the Hubble constant. The goal of the project has been to use Cepheid luminosity distances to 25 different galaxies located within 25 Megaparsecs in order to calibrate a variety of secondary distance indicators, which in turn can be used to determine the distance to far further objects of known redshift. This in principle allows a measurement of the distance-redshift relation and thus the Hubble constant on scales where local peculiar velocities are insignificant. The five distance indicators so constrained are: (1) the Tully Fisher relation, appropriate for spirals, (2) the Fundamental plane, appropriate for ellipticals, (3) surface brightness fluctuations, and (4) Supernova Type 1a distance measures, and (5) Supernovae Type II distance measures.

The Cepheid distances obtained from the HST project include a larger LMC sample to calibrate the period-luminosity relation, a new photometric calibration, and corrections for metallicity. As a result they determined a new LMC distance modulus, of $\mu_o = 18.50 \pm 0.10$ mag. The number of Cepheid calibrators used for the secondary measures include 21 for the Tully-Fisher relation, and 6 for each of the Type Ia and surface fluctuation measures.

The HST-Key project reported measurements for each of these methods is present below [1]. (While I shall adopt these as quoted, it is worth pointing out that some critics have stressed that this involves utilizing data obtained by other groups, who themselves sometimes report different values of H_0). The first quoted uncertainty is statistical, the second is systematic (coming from such things as LMC zero point measurements, photometry, metallicity uncertainties, and remnant bulk flows).

$$H_O^{TF} = 71 \pm 3 \pm 7$$

$$H_O^{FP} = 82 \pm 6 \pm 9$$

$$H_O^{SBF} = 70 \pm 5 \pm 6$$

$$H_O^{SN1a} = 71 \pm 2 \pm 6$$

$$H_O^{SNII} = 72 \pm 9 \pm 7$$

On the basis of these results, the Key Project reports a weighted average value:

$$H_O^{WA} = 72 \pm 3 \pm 7 \ kms^{-1}Mpc^{-1}(1\sigma)$$

and a final combined average of

$$H_O^{WA} = 72 \pm 8 \ kms^{-1}Mpc^{-1}(1\sigma)$$

The Hubble Diagram obtained from the HST project [1] is reproduced here.

In the weighted average quoted above, the dominant contribution to the 11% one sigma error comes from an overall uncertainty in the distance to the Large Magellanic Cloud. If the Cepheid Metallicity were shifted within its allowed 4% uncertainty range, the best fit mean value for the Hubble Constant from the HST-Key project would shift downward to 68 ± 6.

S-Z Effect:

The Sunyaev-Zeldovich effect results from a shift in the spectrum of the Cosmic Microwave Background radiation due to scattering of the radiation by electrons as the radiation passes through intervening galaxy clusters on the way to our receivers on Earth. Because the electron temperature in Clusters exceeds that in the CMB, the radiation is systematically shifted to higher frequencies, producing a deficit in the intensity below some characteristic frequency, and

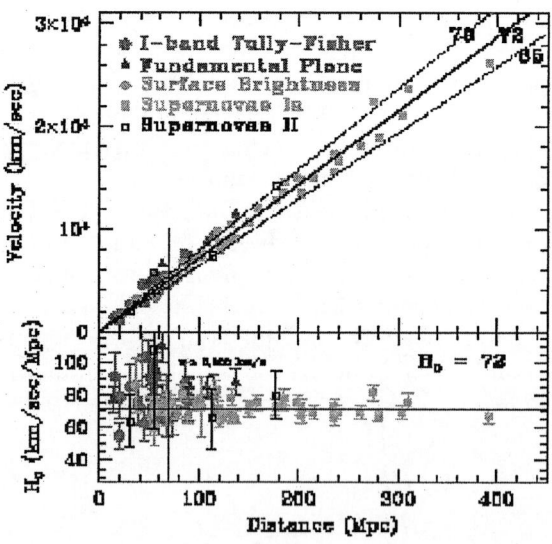

Fig. 1. HST Key Project Hubble Diagram

an excess above it. The amplitude of the effect depends upon the Thompson scattering cross section, and the electron density, integrated over the photon's path:

$$SZ \approx \int \sigma_T n_e dl$$

At the same time the electrons in the hot gas that dominates the baryonic matter in galaxy clusters also emits X-Rays, and the overall X-Ray intensity is proportional to the *square* of the electron density integrated along the line of sight through the cluster:

$$X - Ray \approx \int n_e^2 dl$$

Using models of the cluster density profile one can then use the the differing dependence on n_e in the two integrals above to extract the physical path-length through the cluster. Assuming the radial extension of the cluster is approximately equal to the extension across the line of sight one can compare the physical size of the cluster to the angular size to determine its distance. Clearly, since this assumption is only good in a statistical sense, the use of S-Z and X-Ray observations to determine the Hubble constant cannot be done reliably on the basis of a single cluster observation, but rather on an ensemble.

A recent preliminary analysis of several clusters [2] yields:

$$H_0^{SZ} = 60 \pm 10 \ ks^{-1} Mpc^{-1}$$

Type 1a SN (non-Key Project):

One of the HST Key Project distance estimators involves the use of Type 1a SN as standard candles. As previously emphasized, the Key Project does not perform direct measurements of Type 1a supernovae but rather uses data obtained by other groups. When these groups perform an independent analysis to derive a value for the Hubble constant they arrive at a smaller value than that quoted by the Key Project. Their most recent quoted value is [3]:

$$H_0^{1a} = 64_{-6}^{+8} \ ks^{-1} Mpc^{-1}$$

At the same time, Sandage and collaborators have performed an independent analysis of SNe Ia distances and obtain [4]:

$$H_0^{1a} = 58 \pm 6 \ ks^{-1} Mpc^{-1}$$

Surface Brightness Fluctuations and The Galaxy Density Field:

Another recently used distance estimator involves the measurement of fluctuations in the galaxy surface brightness, which correspond to density fluctuations allowing an estimate of the physical size of a galaxy. This measure yields a slightly higher value for the Hubble constant [5]:

$$H_0^{SBF} = 74 \pm 4 \ ks^{-1} Mpc^{-1}$$

Time Delays in Gravitational Lensing:

One of the most remarkable observations associated with observations of multiple images of distant quasars due to gravitational lensing intervening galaxies has been the measurement of the time delay in the two images of quasar $Q0957 + 561$. This time delay, measured quite accurately to be 417 ± 3 days is due to two factors: The path-length difference between the quasar and the earth for the light from the two different images, and the Shapiro gravitational time delay for the light rays traveling in slightly different gravitational potential wells. If it were not for this second factor, a measurement of the time delay could be directly used to determine the distance of the intervening galaxy. This latter factor however, implies that a model of both the galaxy, and the cluster in which it is embedded must be used to estimate the Shapiro time delay. This introduces an additional model-dependent uncertainty into the analysis. Two different analyses yield values [6]:

$$H_0^{TD1} = 69^{+18}_{-12}(1 - \kappa) \ ks^{-1} Mpc^{-1}$$

$$H_0^{TD2} = 74^{+18}_{-10}(1 - \kappa) \ ks^{-1} Mpc^{-1}$$

where κ is a parameter which accounts for a possible deviation in cluster parameters governing the overall induced gravitational time delay of the two signals from that assumed in the best fit. It is assumed in the analysis that κ is small.

Summary:

It is difficult to know how to best incorporate all of the quoted estimates into a single estimate, given their separate systematic and statistical uncertainties. Assuming large number statistics, where large here includes the quoted values presented here, I perform a simple weighted average of the individual estimates, and find an approximate average value:

$$H_0^{Av} \approx 70 \pm 5 \ ks^{-1} Mpc^{-1} \tag{1}$$

2.2 Geometry

Again, for much of the 20th century the effort to determine the geometry of the Universe involved a very indirect route. Einstein's Equations yield a relationship between the Hubble constant, the energy density, and the curvature of the Universe. By attempting to determine the first two quantities, one hoped to constrain the third. The problem is that until the past decade the uncertainty in the Hubble constant was at least 20-30% and the uncertainty in the average energy density of the universe was even greater. As a result, almost any value for the net curvature of the universe remained viable.

It has remained a dream of observational cosmologists to be able to directly measure the geometry of space-time rather than infer the curvature of the universe by comparing the expansion rate to the mean mass density. While several such tests, based on measuring galaxy counts as a function of redshift, or the variation of angular diameter distance with redshift, have been attempted in the past, these have all been stymied by the achilles heel of many observational measurements in cosmology, evolutionary effects.

Recently, however, measurements of the cosmic microwave background have finally brought us to the threshold of a direct measurement of geometry, independent of traditional astrophysical uncertainties. The idea behind this measurement is, in principle, quite simple. The CMB originates from a spherical shell located at the surface of last scattering (SLS), at a redshift of roughly $z \approx 1000$:

COSMIC MICROWAVE BACKGROUND

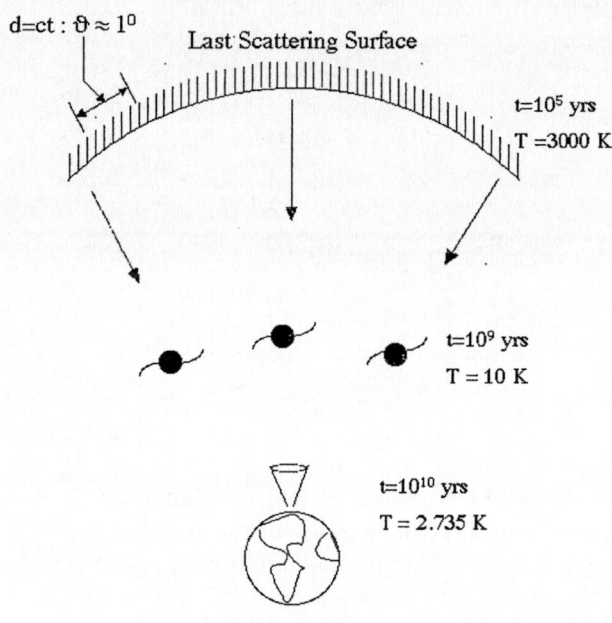

Fig. 2. A schematic diagram of the surface of last scattering, showing the distance traversed by CMB radiation

If a fiducial length could unambiguously be distinguished on this surface, then a determination of the angular size associated with this length would allow a determination of the intervening geometry:

Angular Size of a Fixed Scale in
Open, Closed, and Flat Universes:

First Scale to Collapse after
Recombination (≈distance spanned
by light ray =horizon size)

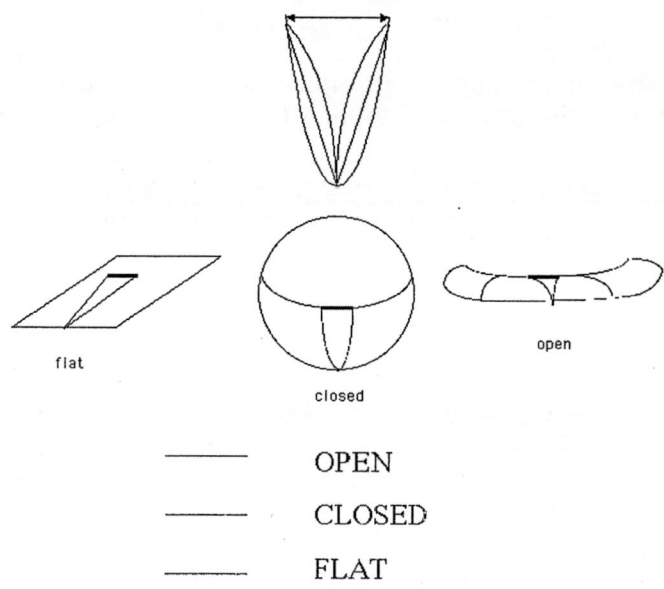

flat

closed

open

—————— OPEN

— — — — CLOSED

- - - - - FLAT

Fig. 3. The geometry of the Universe and ray trajectories for CMB radiation

Fortunately, nature has provided such a fiducial length, which corresponds roughly to the horizon size at the time the surface of last scattering existed (In this case the length is the "sound horizon", but since the medium in question is relativistic, the speed of sound is close to the speed of light.) The reason for this is also straightforward. This is the largest scale over which causal effects at the time of the creation of the surface of last scattering could have left an imprint. Density fluctuations on such scales would result in acoustic oscillations of the matter-radiation fluid, and the doppler motion of electrons moving along with this fluid which scatter on photons emerging from the SLS produces a characteristic peak in the power spectrum of fluctuations of the CMBR at a wavenumber corresponding to the angular scale spanned by this physical scale. These fluctuations should also be visually distinguishable in an image map of the CMB, provided a resolution on degree scales is possible.

Recently, a number of different ground-based balloon experiments, launched in places such Texas and Antarctica have resulted in maps with the required resolution [7–10]. Shown below is a comparison of the actual Boomerang map with several simulations based on a gaussian random spectrum of density fluctuations in a cold-dark matter universe, for open, closed, and flat cosmologies. Even at this qualitative level, it is clear that a flat universe provides better agreement to between the simulations and the data than either an open or closed universe.

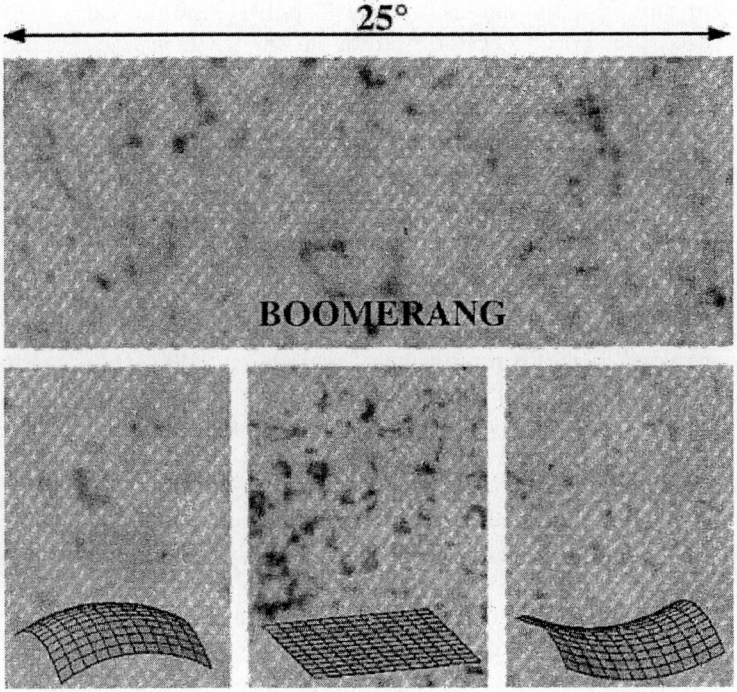

Fig. 4. Boomerang data visually compared to expectations for an open, closed, and flat CDM Universe

On a more quantitative level, one can compare the inferred power spectra with predicted spectra [11]. Such comparisons for the most recent data [12] yields a constraint on the density parameter:

$$\Omega = 1.03^{+.05}_{-.06}(68\%CL) \tag{2}$$

For the first time, it appears that the longstanding prejudice of theorists, namely that we live in a flat universe, may have been vindicated by observation! However, theorists can not be too self-satisfied by this result, because the source of this energy density appears to be completely unexpected, and largely inexplicable at the present time, as we will shortly see.

3 Time

3.1 Stellar Ages

Ever since Kelvin and Helmholtz first estimated the age of the Sun to be less than 100 million years, assuming that gravitational contraction was its prime energy source, there has been a tension between stellar age estimates and estimates of the age of the universe. In the case of the Kelvin-Helmholtz case, the age of the sun appeared too short to accommodate an Earth which was several billion years old. Over much of the latter half of the 20th century, the opposite problem dominated the cosmological landscape. Stellar ages, based on nuclear reactions as measured in the laboratory, appeared to be too old to accommodate even an open universe, based on estimates of the Hubble parameter. Again, as I shall outline in the next section, the observed expansion rate gives an upper limit on the age of the Universe which depends, to some degree, upon the equation of state, and the overall energy density of the dominant matter in the Universe.

There are several methods to attempt to determine stellar ages, but I will concentrate here on main sequence fitting techniques, because those are the ones I have been involved in. For a more general review, see [13].

The basic idea behind main sequence fitting is simple. A stellar model is constructed by solving the basic equations of stellar structure, including conservation of mass and energy and the assumption of hydrostatic equilibrium, and the equations of energy transport. Boundary conditions at the center of the star and at the surface are then used, and combined with assumed equation of state equations, opacities, and nuclear reaction rates in order to evolve a star of given mass, and elemental composition.

Globular clusters are compact stellar systems containing up to 10^5 stars, with low heavy element abundance. Many are located in a spherical halo around the galactic center, suggesting they formed early in the history of our galaxy. By making a cut on those clusters with large halo velocities, and lowest metallicities (less than 1/100th the solar value), one attempts to observationally distinguish the oldest such systems. Because these systems are compact, one can safely assume that all the stars within them formed at approximately the same time.

Observers measure the color and luminosity of stars in such clusters, producing color-magnitude diagrams of the type shown in Fig. 5 (based on data from [15].

Next, using stellar models, one can attempt to evolve stars of differing mass for the metallicities appropriate to a given cluster, in order to fit observations. A point which is often conveniently chosen is the so-called main sequence-turnoff (MSTO) point, the point in which hydrogen burning (main sequence) stars have exhausted their supply of hydrogen in the core. After the MSTO, the stars quickly expand, become brighter, and are referred to as Red Giant Branch (RGB) stars. Higher mass stars develop a helium core that is so hot and dense that helium fusion begins. These form along the horizontal branch. Some stars along this branch are unstable to radial pulsations, the so-called RR Lyrae stars mentioned earlier, which are important distance indicators. While one in principle

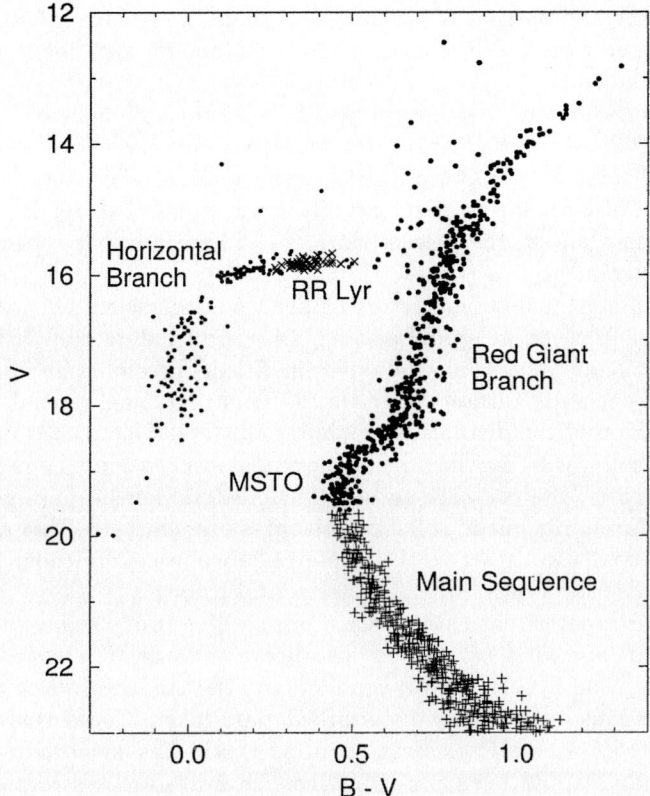

Fig. 5. Color-magnitude diagram for a typical globular cluster, M15. Vertical axis plots the magnitude (luminosity) of the stars in the V wavelength region and the horizontal axis plots the color (surface temperature) of the stars

could attempt to fit theoretical isochrones (the locus of points on the predicted CM curve corresponding to different mass stars which have evolved to a specified age), to observations at any point, the main sequence turnoff is both sensitive to age, and involves minimal (though just how minimal remains to be seen) theoretical uncertainties.

Dimensional analysis tells us that the main sequence turnoff should be a sensitive function of age. The luminosity of upper main sequence stars is very roughly proportional to the third power of solar mass. Hence the time it takes to burn the hydrogen fuel is proportional to the total amount of fuel (proportional to the mass M), divided by the Luminosity – proportional to M^3. Hence the lifetime of stars on the main sequence is roughly proportional to the inverse square of the stellar mass.

Of course the ability to go beyond this rough approximation depends completely on the on the confidence one has in one's stellar models. What is most important for the comparison of cosmological predictions with inferred age estimates is the uncertainties in stellar model parameters, and not merely their best fit values.

Over the course of the past several years, I and my collaborators have tried to incorporate stellar model uncertainties, along with observational uncertainties into a self consistent Monte Carlo analysis which might allow one to estimate a reliable range of globular cluster ages. Others have carried out independent, but similar studies, and at the present time, rough agreement has been obtained between the different groups (i.e. see [18]).

I will not belabor the detailed history of all such efforts here. The most crucial insight has been that stellar model uncertainties are small in comparison to an overall observational uncertainty inherent in fitting predicted main sequence luminosities to observed turnoff magnitudes. This matching depends crucially on a determination of the distance to globular clusters. The uncertainty in this distance scale produces by far the largest uncertainty in the quoted age estimates.

In many studies, the distance to globular clusters can be parametrized in terms of the inferred magnitude of the horizontal branch stars. This magnitude can, in turn, be presented in terms of the inferred absolute magnitude, M_v(RR) of RR Lyrae variable stars located on the horizontal branch.

In 1997, the Hipparcos satellite produced its catalogue of parallaxes of nearby stars, causing an apparent revision in distance estimates. The Hipparcos parallaxes seemed to be systematically smaller, for the smallest measured parallaxes, than previous terrestrially determined parallaxes. Could this represent the unanticipated systematic uncertainty that David has suspected? Since all the detailed analyses had been pre-Hipparcos, several groups scrambled to incorporate the Hipparcos catalogue into their analyses. The immediate result was a generally lower mean age estimate, reducing the mean value to 11.5-12 Gyr, and allowing ages of the oldest globular clusters as low as 9.5 Gyr. However, what is also clear is that there is now an explicit systematic uncertainty in the RR Lyrae distance modulus which dominates the results. Different measurements are no longer consistent. Depending upon which distance estimator is correct, and there is now better evidence that the distance estimators which disagree with Hipparcos-based main sequence fitting should not be dismissed out of hand, the best-fit globular cluster estimate could shift up perhaps 1σ, or about 1.5 Gyr, to about 13 Gyr.

Within the past two years, Brian Chaboyer and I have reanalyzed globular cluster ages, incorporating new nuclear reaction rates, cosmological estimates of the ^{4}He abundance, and most importantly, several new estimates of M_v(RR), shown below.

The result is that while systematic uncertainties clearly still dominate, we argue that the best fit age of globular clusters is now $12.6^{+3.4}_{-2.4}$ (95%) Gyr, with a 95 % confidence range of about 11-16 Gyr [13].

If we are to turn this result into a lower limit on the age of the Universe we must add to this estimate the time after the Big Bang that it took for the first globular clusters in our galaxy to form. Here there is great uncertainty.

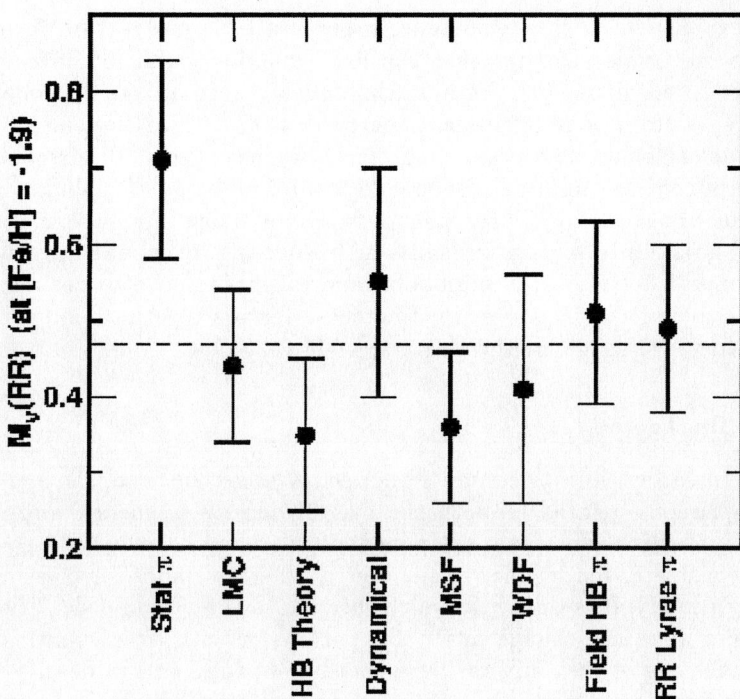

Fig. 6. Different estimates of the inferred magnitude of horizontal branch RR Lyrae stars, with uncertainties

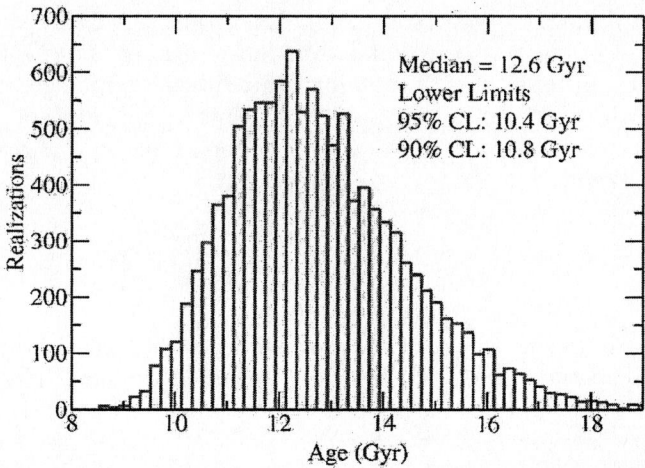

Fig. 7. Histogram showing range of age fits to old globular clusters using Monte Carlo analysis

However a robust lower limit comes from observations of structure formation in the Universe, which suggest that the first galaxies could not have formed much before a redshift of 6-7. Turning this redshift into an age depends upon the equation of state of the dominant energy density at that time (see below). However, one can show that at such high redshifts, the effects of a possible dark energy component are minimal, leading to a minimum age of globular cluster formation of about .8 Gyr. The maximum age is much less certain, as it is possible for galaxies to form at redshifts as low as 1-2. Thus, one must add an age of perhaps 3.5-4 Gyr to the globular age estimate above to get an upper limit on the age of the Universe. Putting these factors together, one derives a 95% confidence age range for the Universe of 11.2-20 Gyr.

3.2   Hubble Age

As alluded to earlier, in a Friedman-Robertson-Walker Universe, the age of the Universe is directly related to both the overall density of energy, and to the equation of state of the dominant component of this energy density. The equation of state is parameterized by the ratio $\omega = p/\rho$, where p stands for pressure and ρ for energy density. It is this ratio which enters into the second order Friedman equation describing the change in Hubble parameter with time, which in turn determines the age of the Universe for a specific net total energy density.

The fact that this depends on two independent parameters has meant that one could reconcile possible conflicts with globular cluster age estimates by altering either the energy density, or the equation of state. An open universe, for example, is older for a given Hubble Constant, than is a flat universe, while a flat universe dominated by a cosmological constant can be older than an open matter dominated universe.

If, however, we incorporate the recent geometric determination which suggests we live in a flat Universe into our analysis, then our constraints on the possible equation of state on the dominant energy density of the universe become more severe. If, for existence, we allow for a diffuse component to the total energy density with the equation of state of a cosmological constant ($\omega = -1$), then the age of the Universe for various combinations of matter and cosmological constant is given by:

$$H_0 t_0 = \int_0^\infty \frac{dz}{(1+z)[(\Omega_m)(1+z)^3 + (\Omega_X)(1+z)^{3(1+w)}]^{1/2}} \qquad (3)$$

This leads to ages as shown in the table below.

The existing limits on the age of the universe from globular clusters are thus already are *incompatible* with a flat matter dominated universe. This is a very important result, as it implies that now all three classic tests of cosmology, including geometry, large scale structure, and age of the Universe now support the same cosmological model, which involves a universe dominated by dark energy. We can provide limits on the equation of state for dark energy as well. Shown in Fig. 8, is the constraint on w, assuming a Hubble constant of 72 [13].

Table 1. Hubble Ages for a Flat Universe, $H_0 = 70 \pm 8$,

Ω_M	Ω_x	t_0
1	0	9.7 ± 1
0.2	0.8	15.3 ± 1.5
0.3	0.7	13.7 ± 1.4
0.35	0.65	12.9 ± 1.3

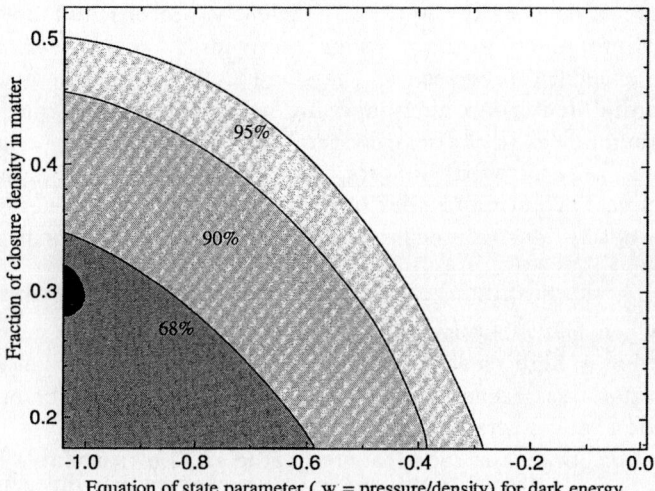

Fig. 8. Constraint on the equation of state parameter for dark energy as a function of the fraction of closure density in matter resulting from age constraint described here

At the same time, it is worth noting that unfortunately the upper limit on the age of the universe coming from globular cluster ages cannot provide a useful limit on the equation of state parameter w, because there is an upper limit on the Hubble Age, independent of w, if the contribution of matter to the total density is greater than 20% [14].

4   Matter

Having indirectly probed the nature of matter in the Universe using the previous estimates, it is now time to turn to direct constraints that have been derived in the past decade.

4.1   The Baryon Density: A Re-Occuring Crisis?

The success of Big Bang Nucleosynthesis in predicting in the cosmic abundances of the light elements has been much heralded. Nevertheless, the finer the ability to empirically infer the primordial abundances on the basis of observations, the greater the ability to uncover some small deviation from the predictions. Over the past five years, two different sets of observations have threatened, at least in some people's minds, to overturn the simplest BBN model predictions. I believe it is fair to say that most people have accepted that the first threat was overblown. The concerns about the second have only recently subsided.

i. Primordial Deuterium: The production of primordial deuterium during BBN is a monotonically decreasing function of the baryon density simply because the greater this density the more efficiently protons and neutrons get processed to helium, and deuterium, as an intermediary in this reactions set, is thus also more efficiently processed at the same time. The problem with inferring the primordial deuterium abundance by using present day measurements of deuterium abundances in the solar system, for example, is that deuterium is highly processed (i.e. destroyed) in stars, and no one has a good enough model for galactic chemical evolution to work backwards from the observed abundances in order to adequately constrain deuterium at a level where this constraint could significantly test BBN estimates.

Five years ago, the situation regarding deuterium as a probe of BBN changed dramatically, when David Tytler and Scott Burles convincingly measured the deuterium fraction in high redshift hydrogen clouds that absorb light from even higher redshift quasars. Because these clouds are at high redshift, before significant star formation has occurred, little post BBN deuterium processing is thought to have taken place, and thus the measured value gives a reasonable handle on the primordial BBN abundance. The best measured system [20] yields a deuterium to hydrogen fraction of

$$(D/H) = (3.3. \pm 0.5) \times 10^{-5} \quad (2\sigma) \tag{4}$$

This, in turn, leads to a constraint on the baryon fraction of the Universe, via standard BBN,

$$\Omega_B h^2 = .0190 \pm .0018 \quad (2\sigma) \tag{5}$$

where the quoted uncertainty is dominated by the observational uncertainty in the D/H ratio, and where $H_0 = 100h$. Thus, taken at face value, we now know the baryon density in the universe today to an accuracy of about 10%!

When first quoted, this result sent shock waves through some of the BBN community, because this value of Ω_B is only consistent if the primordial helium fraction (by mass) is greater than about 24.5%. However, a number of previous studies had claimed an upper limit well below this value. However, recent studies, for example, place an upper limit on the primordial helium fraction closer to 25%.

In any case, even if somehow the deuterium estimate is wrong, one can combine all the other light element constraints to produce a range for $\Omega_b h^2$ consistent with observation:

$$\Omega_B h^2 = .016 - 0.025 \tag{6}$$

ii. CMB constraints: Beyond the great excitement over the observation of a peak in the CMB power spectrum at an angular scale corresponding to that expected for a flat universe lay some excitement/concern over the small apparent size of the next peak in the spectrum, at higher multipole moment (smaller angular size). The height of the first peak in the CMB spectrum is related to a number of cosmological parameters and thus cannot alone be used to constrain any one of them. However, the relative height of the first and second peaks is strongly dependent on the baryon fraction of the universe, since the peaks themselves arise from compton scattering of photons off of electrons in the process of becoming bound to baryons. Analyses of the two first small-scale CMB results originally produced a constraint which was in disagreement with the BBN estimate. However, more recent data indicates $\Omega_B h^2 = 0.021$, precisely where one would expect it to be based on BBN predictions.

Most recently reported measurements of ^3He in the Milky Way Galaxy give the constraint, $^3He/H = (1.1.\pm0.2) \times 10^{-5}$, which in turn implies $\Omega_B h^2 = 0.02$. Thus, all data is now consistent with the assumption that the Burles and Tytler limit on $\Omega_B h^2$ is correct, adding further confidence in the predictions of BBN. Taking the range for H_0 given earlier, one derives the constraint on Ω_B of

$$\Omega_B = .045 \pm 0.15 \tag{7}$$

4.2 Ω_{matter}

Perhaps the second greatest change in cosmological prejudice in the past decade relates to the inferred total abundance of matter in the Universe. Because of the great intellectual attraction Inflation as a mechanism to solve the so-called Horizon and Flatness problems in the Universe, it is fair to say that most cosmologists, and essentially all particle theorists had implicitly assumed that the Universe is flat, and thus that the density of dark matter around galaxies and clusters of galaxies was sufficient to yield $\Omega = 1$. Over the past decade it became more and more difficult to defend this viewpoint against an increasing number of observations that suggested this was not, in fact, the case in the Universe in which we live.

The earliest holes in this picture arose from measurements of galaxy clustering on large scales. The transition from a radiation to matter dominated universe at early times is dependent, of course, on the total abundance of matter. This transition produces a characteristic signature in the spectrum of remnant density fluctuations observed on large scales. Making the assumption that dark matter dominates on large scales, and moreover that the dark matter is cold (i.e. became non-relativistic when the temperature of the Universe was less than about a keV), fits to the two point correlation function of galaxies on large scales yielded [21,22]:

$$\Omega_M h = .2 - .3 \tag{8}$$

Unless h was absurdly small, this would imply that Ω_M is substantially less than 1.

New data from the Sloan and 2DF surveys refine this limit further, with reported values of [23,24]

$$\Omega_M = 0.23 \pm 0.09 (2DF) \tag{9}$$

$$\Omega_M h \approx 0.14^{+.11}_{-.06} \ (2\sigma) \ (Sloan) \tag{10}$$

The second nail in the coffin arose when observations of the evolution of large scale structure as a function of redshift began to be made. Bahcall and collaborators [25] argued strongly that evidence for any large clusters at high redshift would argue strongly against a flat cold dark matter dominated universe, because in such a universe structure continues to evolve with redshift up to the present time on large scales, so that in order to be consistent with the observed structures at low redshift, far less structure should be observed at high redshift. Claims were made that an upper limit $\Omega_B \leq 0.5$ could be obtained by such analyses.

A number of authors have questioned the systematics inherent in the early claims, but it is certainly clear that there appears to be more structure at high redshift than one would naively expect in a flat matter dominated universe. Future studies of X-ray clusters, and use of the Sunyaev-Zeldovich effect to measure cluster properties should be able to yield measurements which will allow a fine-scale distinction not just between models with different overall dark matter densities, but also models with the same overall value of Ω and different equations of state for the dominant energy [26].

One of the best overall constraint on the total density of clustered matter in the universe comes from the combination of X-Ray measurements of clusters with large hydrodynamic simulations. The idea is straightforward. A measurement of both the temperature and luminosity of the X-Rays coming from hot gas which dominates the total baryon fraction in clusters can be inverted, under the assumption of hydrostatic equilibrium of the gas in clusters, to obtain the underlying gravitational potential of these systems. In particular the ratio of baryon to total mass of these systems can be derived. Employing the constraint on the total baryon density of the Universe coming from BBN, and assuming that galaxy clusters provide a good mean estimate of the total clustered mass in the Universe, one can then arrive at an allowed range for the total mass density in the Universe [27–29]. Many of the initial systematic uncertainties in this analysis having to do with cluster modelling have now been dealt with by better observations, and better simulations (i.e. see [30]), so that now a combination of BBN and cluster measurements yields:

$$\Omega_M = 0.35 \pm 0.1 \ (2\sigma) \tag{11}$$

Combining these results, one derives the constraint:

$$\Omega_M \approx 0.3 \pm 0.05 \ (2\sigma) \tag{12}$$

4.3 Equation of State of Dominant Energy

The above estimate for Ω_M brings the discussion of cosmological parameters full circle, with consistency obtained for a flat 13 billion year old universe, but not one dominated by matter. As noted previously, a cosmological constant dominated universe with $\Omega_M = 0.3$ has an age which nicely fits in the best-fit range. However, based on the data discussed thus far, there was no direct evidence that the dark energy necessary to result in a flat universe actually has the equation of state appropriate for a vacuum energy. Direct motivation for the possibility that the dominant energy driving the expansion of the Universe violates the Strong Energy Condition actually came somewhat earlier, in 1998, from two different sets of observations of distant Type 1a Supernovae. In measuring the distance-redshift relation [31,32] these groups both came to the same, surprising conclusion: the expansion of the Universe seems to be accelerating! This is only possible if the dominant energy is "cosmological-constant-like", namely if $\omega < -0.5$ (recall that $\omega = -1$ for a cosmological constant).

In order to try and determine if the dominant dark energy does in fact differ significantly from a static vacuum energy – as for example may occur if some background field that is dynamically evolving is dominating the expansion energy at the moment – one can hope to search for deviations from the distance-redshift relation for a cosmological constant-dominated universe. To date, none have been observed. In fact, existing measurements already put a (model dependent) limit of approximately $-1.7 \leq \omega \leq -0.7$ [33]. Recent work [34] suggests that the best one might be able to do from the ground using SN measurements would be to improve this limit to $\omega \leq -0.7$. Either other measurements, such as galaxy cluster evolution observations, or space-based SN observations would be required to further tighten the constraint.

5 Conclusions: A Cosmic Uncertainty Principle

I list the overall constraints on cosmological parameters discussed in this review in the table below. It is worth stressing how completely remarkable the present situation is. After 20 years, we now have the first direct evidence that the Universe might be flat, but we also have definitive evidence that there is not enough matter, including dark matter, to make it so. We seem to be forced to accept the possibility that some weird form of dark energy is the dominant stuff in the Universe. It is fair to say that this situation is more mysterious, and thus more exciting, than anyone had a right to expect it to be.

The new situation changes everything about the way we think about cosmology. In the first place, it demonstrates that Geometry and Destiny are no longer linked. Previously, the holy grail of cosmology involved determining the density parameter Ω, because this was tantamount to determining the ultimate future of our universe. Now, once we accept the possibility of a non-zero cosmological constant, we must also accept the fact that any universe, open, closed, or flat, can either expand forever, or reverse the present expansion and end in a big crunch [35]. But wait, it gets worse, as my colleague Michael Turner and I have

Table 2. Cosmological Parameters 2001

Parameter	Allowed range	Formal Conf. Level (where approp.)
H_0	70 ± 5	2σ
t_0	$13^{+7}_{-1.8}$	2σ
$\Omega_B h^2$	$.02 \pm .004$	2σ
Ω_B	0.045 ± 0.015	2σ
Ω_M	0.3 ± 0.1	2σ
Ω_{TOT}	1.03 ± 0.1	2σ
Ω_X	0.7 ± 0.1	2σ
ω	≤ -0.7	2σ

also demonstrated, there is no set of cosmological measurements, no matter how precise, that will allow us to determine the ultimate future of the Universe. In order to do so, we would require a theory of everything.

On the other hand, if our universe is in fact dominated by a cosmological constant, the future for life is rather bleak [36]. Distant galaxies will soon blink out of sight, and the Universe will become cold and dark, and uninhabitable...

This bleak picture may seem depressing, but the flip side of all the above is that we live in exciting times now, when mysteries abound. We should enjoy our brief moment in the Sun.

References

1. W.L. Freedman *et al*, Ap. J. **553**, 47 (2001)
2. M. Birkinshaw, Phys. Rep., **310**, 97 (1999)
3. J. Saurabh *et al*, Ap. J. Suppl. **125**, 73 (1999)
4. B. R Parodi *et al*, Ap. J. **540**, 634 (2000)
5. J. P. Blakeslee *et al*, Ap. J. Lett. **527**, 73 (1999)
6. K-H. Chae, Ap. J. **524**, 582 (1999)
7. P. de Bernardis *et al*, Nature **404**, 995 (2000)
8. S. Hanany *et al*, Ap. J. Lett. **545**, 5 (2000)
9. P.F. Scott *et al* astro-ph/0205380 (2002)
10. N.W. Halverson *et al*, Ap. J. Lett. bf 568, 38 (2001)
11. A. H. Jaffe *et al*, astro-ph/0007333 (2000)
12. J. E. Ruhl *et al*, astro-ph/0212229 (2002)
13. L.M. Krauss and B. Chaboyer, Science, Jan 3 2003 issue
14. L.M. Krauss, astro-ph/0212369, Ap. J., submitted.
15. P.R. Durrell and W. E. Harris, AJ, **105**, 1420 (1993)
16. B. Chaboyer, P. Demarque, and A. Sarajedini, Ap. J. **459**, 558 (1996)
17. B. Chaboyer, and Y.-C. Kim, Ap.J. **454**, 76 (1995)
18. L. M. Krauss, Phys Rep.**333-334**, 33 (2000)
19. B. Chaboyer and L.M. Krauss, to appear.
20. S. Burles and D. Tytler, Ap. J. **499**, 699 (1998)
21. J. A. Peacock and S. J. Dodds, MNRAS **280**, 19 (1996)
22. A. Liddle *et al*, MNRAS **278**, 644 (1996); **282**, 281 (1996)
23. E. Hawkins *et al*, astro-ph/0212375.

24. S. Dodelson *et al* astro-ph/0107421
25. N. A. Bahcall *et al* Ap. J. Lett. **485**, 53 (1997)
26. Z. Haiman *et al*, astro-ph/0002336 (2000)
27. S. D. M. White *et al*, MNRAS **262**, 1023 (1993)
28. L. M. Krauss, Ap. J. **501**, 461 (1998)
29. A. E. Evrard, MNRAS **292**, 289 (1997)
30. J. Mohr *et al* astro-ph/0004244 (2000)
31. S. Perlmutter *et al*, Ap. J. **517**, 565 (1999)
32. B. Schmidt *et al*, Ap. J. **507**, 46 (1998)
33. A. Melchiorri *et al*, astro-ph/0211522
34. L.M. Krauss, E. Linton, D. Davis, M. Grugel, to appear.
35. L. M. Krauss and M. S. Turner, J. Gen. Rel. Grav. **31**, 1453 (1999)
36. L. M. Krauss and G. Starkman, Ap. J. **531**, 22 (2000)

Cosmological Constant Problems and Their Solutions

Alexander Vilenkin

Institute of Cosmology, Department of Physics and Astronomy,
Tufts University, Medford, MA 02155, USA

Abstract. There are now two cosmological constant problems: (i) why the vacuum energy is so small and (ii) why it comes to dominate at about the epoch of galaxy formation. Anthropic selection appears to be the only approach that can naturally resolve both problems. The challenge presented by this approach is that it requires scalar fields with extremely flat potentials or four-form fields coupled to branes with an extremely small charge. Some recent suggestions are reviewed on how such features can arise in particle physics models.

1   The Problems

The cosmological constant is (up to a factor) the vacuum energy density, ρ_V. Particle physics models suggest that the natural value for this constant is set by the Planck scale M_P,

$$\rho_V \sim M_P^4 \sim (10^{18} \ GeV)^4, \tag{1}$$

which is some 120 orders of magnitude greater than the observational bound,

$$\rho_V \lesssim (10^{-3} \ \text{eV})^4. \tag{2}$$

In supersymmetric theories, one can expect a lower value,

$$\rho_V \sim \eta_{SUSY}^4, \tag{3}$$

where η_{SUSY} is the supersymmetry breaking scale. However, with $\eta_{SUSY} \gtrsim 1\,\text{TeV}$, this is still 60 orders of magnitude too high. This discrepancy between the expected and observed values is the first cosmological constant problem. I will refer to it as the old CCP.

Until recently, it was almost universally believed that something so small could only be zero, due either to some symmetry or to a dynamical adjustment mechanism. (For a review of the early work on CCP, see [1].) It therefore came as a surprise when recent observations [2] provided evidence that the universe is accelerating, rather than decelerating, suggesting a non-zero cosmological constant. The observationally suggested value is

$$\rho_V \sim \rho_{M0} \sim (10^{-3} \ \text{eV})^4, \tag{4}$$

where ρ_{M0} is the present density of matter. This brings yet another puzzle. The matter density ρ_M and the vacuum energy density ρ_V scale very differently with the expansion of the universe, and there is only one epoch in the history of the universe when $\rho_M \sim \rho_V$. It is difficult to understand why we happen to live in this special epoch. Another, perhaps less anthropocentric statement of the problem is why the epoch when the vacuum energy starts dominating the universe ($z_V \sim 1$) nearly coincides with the epoch of galaxy formation ($z_G \sim 1-3$), when the giant galaxies were assembled and the bulk of star formation has occurred:

$$t_V \sim t_G. \tag{5}$$

This is the so-called cosmic coincidence problem, or the second CCP.

2 Proposed Solutions

2.1 Quintessence

Much of the recent work on CCP involves the idea of quintessence [3]). Quintessence models require a scalar field ϕ with a potential $V(\phi)$ approaching zero at large values of ϕ. A popular example is an inverse power law potential,

$$V(\phi) = M^{4+\beta}\phi^{-\beta}, \tag{6}$$

with a constant $M \ll M_P$. It is assumed that initially $\phi \ll M_P$. Then it can be shown that the quintessence field ϕ approaches an attractor "tracking" solution, $\phi(t) \propto t^{2/(2+\beta)}$, in which its energy density grows relative to that of matter, $\rho_\phi/\rho_M \sim \phi^2/M_P^2$. When ϕ becomes comparable to M_P, its energy dominates the universe. At this point the nature of the solution changes: the evolution of ϕ slows down and the universe enters an epoch of accelerated expansion. The mass parameter M can be adjusted so that this happens at the present epoch.

A nice feature of the quintessence models is that their evolution is not sensitive to the choice of the initial conditions. However, I do not think that these models solve either of the two CCPs. The potential $V(\phi)$ is assumed to vanish in the asymptotic range $\phi \to \infty$. This assumes that the old CCP has been solved by some unspecified mechanism. The coincidence problem also remains unresolved, because the time of quintessence domination depends on the choice of the parameter M, and there seems to be no reason why this time should coincide with the epoch of galaxy formation.

2.2 A Small Cosmological Constant from Fundamental Physics

One possibility here is that some symmetry of the fundamental physics requires that the cosmological constant should be zero. A small value of ρ_V could then arise due to a small violation of this symmetry. One could hope that ρ_V would be given by an expression like

$$\rho_V \sim M_W^8/M_P^4 \sim (10^{-3}\ \text{eV})^4, \tag{7}$$

where $M_W \sim 10^3$ GeV is the electroweak scale. There have been attempts in this direction [4], but no satisfactory implementation of this program has yet been developed. And even if we had one, the time coincidence $t_V \sim t_G$ would still remain a mystery.

Essentially the same remarks apply to the braneworld [5] and holographic [6] approaches to CCPs.

2.3   Anthropic Approach

According to this approach, what we perceive as the cosmological constant is in fact a stochastic variable which varies on a very large scale, greater than the present horizon, and takes different values in different parts of the universe. We shall see that situation of this sort can naturally arise in the context of the inflationary scenario.

The key observation here is that the gravitational clustering that leads to galaxy formation effectively stops at $z \sim z_V$. An anthropic bound on ρ_V can be obtained by requiring that it does not dominate before the redshift z_{max} when the earliest galaxies are formed. With $z_{max} \sim 5$ one obtains [7]

$$\rho_V \lesssim 200 \rho_{M0}. \tag{8}$$

For negative values of ρ_V, a lower bound can be obtained by requiring that the universe does not recollapse before life had a chance to develop [8],

$$\rho_V \gtrsim - \rho_{M0}. \tag{9}$$

The bound (8) is a dramatic improvement over (1) or (3), but it still falls short of the observational bound by a factor of about 50. If all values in the anthropic range (8) were equally probable, then $\rho_V \sim \rho_{M0}$ would still be ruled out at a 95% confidence level. However, the values in this range are *not* equally probable. The anthropic bound (8) specifies the value of ρ_V which makes galaxy formation barely possible. Most of the galaxies will be not in regions characterized by these marginal values, but rather in regions where ρ_V dominates after the bulk of galaxy formation has occured, that is $z_V \lesssim 1$ [9,10].

This can be made quantitative by introducing the probability distribution as [9]

$$dP(\rho_V) = \mathcal{P}_*(\rho_V)\nu(\rho_V)d\rho_V. \tag{10}$$

Here, $\mathcal{P}_*(\rho_V)d\rho_V$ is the prior distribution, which is proportional to the volume of those parts of the universe where ρ_V takes values in the interval $d\rho_V$, and $\nu(\rho_V)$ is the average number of galaxies that form per unit volume with a given value of ρ_V. The calculation of $\nu(\rho_V)$ is a standard astrophysical problem; it can be done, for example, using the Press-Schechter formalism [11].

The distribution (10) gives the probability that a randomly selected galaxy is located in a region where the effective cosmological constant is in the interval $d\rho_V$. If we are typical observers in a typical galaxy, then we should expect to observe a value of ρ_V somewhere near the peak of this distribution.

The prior distribution $\mathcal{P}_*(\rho_V)$ should be determined from the inflationary model of the early universe. Weinberg [12,13] has argued that a flat distribution,

$$\mathcal{P}_*(\rho_V) = const, \tag{11}$$

should generally be a good approximation. The reason is that the function $\mathcal{P}_*(\rho_V)$ is expected to vary on some large particle physics scale, while we are only interested in its values in the tiny anthropically allowed range (8). Analysis shows that this Weinberg's conjecture is indeed true in a wide class of models, but one finds that it is not as automatic as one might expect [14,15].

Martel, Shapiro and Weinberg [16] (see also [10,12]) presented a detailed calculation of $d\mathcal{P}(\rho_V)$ assuming a flat prior distribution (11). They found that the peak of the resulting probability distribution is close to the observationally suggested values of ρ_V.

The cosmic time coincidence is easy to understand in this approach [17,18]. Regions of the universe where $t_V \ll t_G$ do not form any galaxies at all, whereas regions where $t_V \gg t_G$ are suppressed by "phase space", since they correspond to a very tiny range of ρ_V. It was shown in Ref. [17] that the probability distribution for t_G/t_V is peaked at $t_G/t_V \approx 1.5$, and thus most observers will find themselves in galaxies formed at $t_G \sim t_V$.

We thus see that the anthropic approach naturally resolves both CCPs. All one needs is a particle physics model that would allow ρ_V to take different values and an inflationary cosmological model that would give a more or less flat prior distribution $\mathcal{P}_*(\rho_V)$ in the anthropic range (8).

3   Models with a Variable ρ_V

3.1   Scalar Field with a Very Flat Potential

One possibility is that what we perceive as a cosmological constant is in fact a potential $V(\phi)$ of some field $\phi(x)$ [14]. The slope of the potential is assumed to be so small that the evolution of ϕ is slow on the cosmological time scale. This is achieved if the slow roll condition

$$M_P V' \ll V \lesssim \rho_{M0}, \qquad (12)$$

is satisfied up to the present time. This condition ensures that the field is over-damped by the Hubble expansion, and that the kinetic energy is negligible compared with the potential energy (so that the equation of state is basically that of a cosmological constant term.) The field ϕ is also assumed to have negligible couplings to all fields other than gravity.

Let us now suppose that there was a period of inflation in the early universe, driven by the potential of some other field. The dynamics of the field ϕ during inflation are strongly influenced by quantum fluctuations, causing different regions of the universe to thermalize with different values of ϕ. Spatial variation of ϕ is thus a natural outcome of inflation.

The probability distribution $\mathcal{P}_*(\phi)$ is determined mainly by the interplay of two effects. The first is the "diffusion" in the field space caused by quantum fluctuations. The dispersion of ϕ over a time interval Δt is $\Delta\phi \sim H(H\Delta t)^{1/2}$, where H is the inflationary expansion rate. The effect of diffusion is to make all values of ϕ equally probable over the interval $\Delta\phi$. The second effect is the differential expansion. Although $V(\phi)$ represents only a tiny addition to the inflaton potential, regions with larger values of $V(\phi)$ expand slightly faster, and thus the probability for higher values of $V(\phi)$ is enhanced. The effect of differential expansion is negligible if $\Delta t_{anth} \ll \Delta t_{de}$. The corresponding condition on $V(\phi)$ is [15]

$$V'^2 \gg \rho_{M0}^3/H^3 M_P^2. \tag{13}$$

In this case, the probability distribution for ϕ is flat in the anthropic range, $\mathcal{P}_*(\phi) = const$. The probability distribution for the effective cosmological constant $\rho_V = V(\phi)$ is given by

$$\mathcal{P}_*(\rho_V) = \frac{1}{V'}\mathcal{P}_*(\phi),$$

and it will also be very flat, since V' is typically almost constant in the anthropic range.

As we discussed in Section II, a flat prior distribution for the effective cosmological constant in the anthropic range entails an automatic explanation for the two cosmological constant puzzles. On the other hand, if the condition (13) is not satisfied, then the prior probability for the field values with a higher $V(\phi)$ would be exponentially enhanced with respect to the field values at the lower anthropic end. This would result in a prediction for the effective cosmological constant which would be too high compared with observations.

A simple example is given by a potential of the form

$$V(\phi) = \rho_\Lambda + \frac{1}{2}\mu^2\phi^2, \tag{14}$$

where ρ_Λ represents the "true" cosmological constant. We shall assume that $\rho_\Lambda < 0$, so that the two terms in (14) partially cancel one another in some parts of the universe. With $|\rho_\Lambda| \sim (1 \text{ TeV})^4$, the slow roll condition (12) gives

$$\mu \lesssim 10^{-90} M_P. \tag{15}$$

Thus, an exceedingly small mass scale must be introduced. The condition (13) yields a lower bound on μ,

$$\mu \gtrsim 10^{-137} M_P. \tag{16}$$

(Here, I have used the upper bound on the expansion rate at late stages of inflation, $H \lesssim 10^{-5} M_P$, which follows from the CMB observations.)

We thus see that models with a variable ρ_V can be easily constructed in the framework of inflationary cosmology. The challenge here is to explain the very small mass scale (15) in a natural way.

3.2   Four-Form Models

Another class of models, first discussed by Brown and Teitelboim [19], assumes that the cosmological constant is due to a four-form field, $F^{\alpha\beta\gamma\delta} = F\epsilon^{\alpha\beta\gamma\delta}$. The field equation for F is $\partial_\mu F = 0$, so F is a constant, but it can change its value through nucleation of bubbles bounded by domain walls, or branes. The total vacuum energy density is given by

$$\rho_V = \rho_\Lambda + F^2/2 \tag{17}$$

and once again it is assumed that $\rho_\Lambda < 0$. The change of the field across the brane is $\Delta F = q$, where the "charge" q is a constant fixed by the model. Thus, F takes a discrete set of values, and the resulting spectrum of ρ_V is also discrete. The four-form model has recently attracted much attention [20–23,15] because four-form fields coupled to branes naturally arise in the context of string theory.

In the range where the bare cosmological constant is almost neutralized, $|F| \approx |2\rho_\Lambda|^{1/2}$, the spectrum of ρ_V is nearly equidistant, with a separation

$$\Delta\rho_V \approx |2\rho_\Lambda|^{1/2}q. \tag{18}$$

In order for the anthropic explanation to work, $\Delta\rho_V$ should not exceed the present matter density,

$$\Delta\rho_V \lesssim \rho_{m0} \sim (10^{-3}\ eV)^4. \tag{19}$$

With $\rho_\Lambda \gtrsim (1\ TeV)^4$, it follows that

$$q \lesssim 10^{-90}M_P^2. \tag{20}$$

Once again, the challenge is to find a natural explanation for such very small values of q.

4   Explaining the Small Parameters

Both scalar field and four-form models discussed above have some seemingly unnatural features. The scalar field models require extremely flat potentials and the four-form models require branes with an exceedingly small charge. The models cannot be regarded satisfactory until the smallness of these parameters is explained in a natural way. Here I shall briefly review some possibilities that have been suggested in the literature.

4.1   Scalar Field Renormalization

Let us start with the scalar field model. Weinberg [13] suggested that the flatness of the potential could be due to a large field renormalization. Consider the Lagrangian of the form

$$L = \frac{Z}{2}(\nabla\phi)^2 - V(\phi). \tag{21}$$

The potential for the canonically normalized field $\phi' = \sqrt{Z}\phi$ will be very flat if the field renormalization constant is very large, $Z \gg 1$.

4.2   A Discrete Symmetry

Another approach attributes the flatness of the potential to a spontaneously broken discrete symmetry [24]. The main ingredients of the model are: (1) a four-form field $F_{\mu\nu\sigma\tau}$ which can be obtained from a three-form potential, $F_{\mu\nu\sigma\tau} = \partial_{[\mu}A_{\nu\sigma\tau]}$, (2) a complex field X which develops a vacuum expectation value, $\langle X \rangle = \eta e^{ia}$, and whose phase a becomes a Goldstone boson, and (3) a scalar field Φ which is used to break a discrete Z_{2N} symmetry,

$$Z_{2N}: \quad \Phi \to \Phi e^{i\pi/N}, \qquad a \to -a \quad (\text{or } X \to X^\dagger), \tag{22}$$

Below the symmetry breaking scales of X and Φ, the effective Lagrangian for the model includes a mixing term of the Goldstone a with the three-form potential,

$$g\eta^2 \frac{\langle \Phi \rangle^N}{M_P^N} \epsilon^{\mu\nu\sigma\tau} A_{\nu\sigma\tau} \partial_\mu a. \tag{23}$$

Here, $g \lesssim 1$ is a dimensionless coupling and it is assumed that the Planck scale M_P plays the role of the ultraviolet cutoff of the theory.

The effect of the mixing term (23) is to give a mass

$$\mu = g\eta \frac{\langle \Phi \rangle^N}{M_P^N} \tag{24}$$

to the field a. This mass can be made very small if $\langle \Phi \rangle \ll M_P$ and N is sufficiently large. For example, with $\langle \Phi \rangle \sim 1$ TeV, $\eta \ll M_P$ and $N \geq 6$, we have $\mu \ll 10^{-90} M_P$, as required.

Models of this type can also be used to generate branes with a very small charge. In this case a is assumed to be a pseudo-Goldstone boson, like the axion, and the theory has domain wall solutions with a changing by 2π across the wall. The mixing of a and A couples these walls to the four-form field, and it can be shown that the corresponding charge is

$$q = 2\pi g\eta^2 \frac{\langle \Phi \rangle^N}{M_P^N}. \tag{25}$$

Once again, the anthropic constraint on q is satisfied for $\langle \Phi \rangle \sim 1$ TeV, $\eta \ll M_P$ and $N \geq 6$.

The central feature of this approach is the Z_{2N} symmetry (22). What makes this symmetry unusual is that the phase transformation of Φ is accompanied by a charge conjugation of X. It can be shown, however, that such a symmetry can be naturally embedded into a left-right symmetric extension of the standard model [24].

4.3   String Theory Inspired Ideas

Feng *et al.* [22] have argued that branes with extremely small charge and tension can naturally arise due to non-perturbative effects in string theory. A potential problem with this approach is that the small brane tension and charge appear to be unprotected against quantum corrections below the supersymmetry breaking scale [24]. The cosmology of this model is also problematic, since it is hard to stabilize the present vacuum against copious brane nucleation [15].

A completely different approach was taken by Bousso and Polchinski [20]. They assume that several four-form fields F_i are present so that the vacuum energy is given by

$$\rho_V = \rho_\Lambda + \frac{1}{2} \sum_i F_i^2. \tag{26}$$

The corresponding charges q_i are not assumed to be very small, but Bousso and Polchinski have shown that with multiple four-forms the spectrum of the allowed values of ρ_V can be sufficiently dense to satisfy the anthropic condition (19) in the range of interest. However, the situation here is quite different from that in the single-field models. The vacua with nearby values of ρ_V have very different values of F_i, and there is no reason to expect the prior probabilities for these vacua to be similar. Moreover, the low energy physics in different vacua is likely to be different, so the process of galaxy formation and the types of life that can evolve will also differ. It appears therefore that the anthropic approach to solving the cosmological constant problems cannot be applied to this case [23].

5   Conclusions

In conclusion, it appears that the only approach that can solve both cosmological constant problems is the one that attributes them to anthropic selection effects. In this approach what we perceive as the cosmological constant is in fact a stochastic variable which varies from one part of the universe to another. A typical observer then finds himself in a region with a small cosmological constant which comes to dominate at about the epoch of galaxy formation. Cosmological models of this sort can easily be constructed in the framework of inflation. What one needs is either a scalar field with a very flat potential, or a four-form field coupled to branes with a very small charge. Some interesting suggestions have been made on how such features can arise; the challenge here is to implement these suggestions in well motivated particle physics models. (One attempt in this direction has been made in [24].)

Acknowledgements

This work was supported in part by the National Science Foundation and by the Templeton Foundation.

References

1. S. Weinberg, Rev. Mod. Phys. **61**, 1 (1989).
2. S. Perlmutter et al., Ap.J. **483**, 565 (1997); S. Perlmutter et al., astro-ph/9812473 (1998); B. Schmidt et al., Ap.J. **507**, 46 (1998); A. J. Riess et al., A.J. **116**, 1009 (1998); astro-ph/0104455.
3. P.J.E. Peebles and B. Ratra, Ap. J. Lett. **325**, L17 (1988); C. Wetterich, Nucl. Phys. **B302**, 668 (1988); R.R. Caldwell, R. Dave and P.J. Steinhardt, Phys. Rev. Lett. **80**, 1582 (1998); I. Zlatev, L. Wang and P.J. Steinhardt, Phys. Rev. Lett. **82**, 896 (1999).
4. S. Kachru, J. Kumar and E. Silverstein, Phys. Rev. **D59**, 106004 (1999); E. Guendelman and A. Kaganovich, Phys. Rev. **D60**, 065004 (2000); P.H. Frampton, hep-th/0002053; N. Arkani-Hamed, L.J. Hall, C. Colda and H. Murayama, Phys. Rev. Lett. **85**, 4434 (2000).
5. N. Arkani-Hamed, S. Dimopoulos, N. Kaloper and R. Sundrum, Phys. Lett. **B480**, 193 (2000); S. Kachru, M. Schulz and E. Silverstein, Phys. Rev. **D62**, 045021 (2000); G. Dvali and G. Gabadadze, Phys. Rev. **D63**, 065007 (2001); S.-H. Tye and I. Wasserman, Phys. Rev. Lett. **86**, 1682 (2001).
6. T. Banks, hep-th/9601151, 0007146; A.G. Cohen, D.B. Caplan and A.E. Nelson, Phys. Rev. Lett. **82**, 4971 (1999); P. Horava and D. Minic, Phys. Rev. Lett. **85**, 1610 (2000); S. Thomas, hep-th/0010145.
7. S. Weinberg, Phys. Rev. Lett. **59**, 2607 (1987).
8. J.D. Barrow and F.J. Tipler, *The Anthropic Cosmological Principle*, Clarendon, Oxford (1986).
9. A. Vilenkin, Phys. Rev. Lett. **74**, 846 (1995).
10. G. Efstathiou, M.N.R.A.S. **274**, L73 (1995).
11. W.H. Press and P. Schechter, Astrophys. J. **187**,425 (1974).
12. S. Weinberg, in *Critical Dialogues in Cosmology*, ed. by N. G. Turok (World Scientific, Singapore, 1997).
13. S. Weinberg, Phys. Rev. **D61** 103505 (2000); "The cosmological constant problems," Talk given at 4th International Symposium on Sources and Detection of Dark Matter in the Universe (DM 2000), Marina del Rey, California astro-ph/0005265.
14. J. Garriga and A. Vilenkin, Phys. Rev. **D61**, 083502 (2000).
15. J. Garriga and A. Vilenkin, "Solutions to the cosmological constant problems", hep-th/0011262.
16. H. Martel, P. R. Shapiro and S. Weinberg, Ap.J. **492**, 29 (1998).
17. J. Garriga, M. Livio and A. Vilenkin, Phys. Rev. **D61**, 023503 (2000).
18. S. Bludman, Nucl. Phys. **A663-664**,865 (2000).
19. J.D. Brown and C. Teitelboim, Nucl. Phys. **279**, 787 (1988).
20. R. Bousso and J. Polchinski, JHEP 0006:006 (2000).
21. J. Donoghue, JHEP 0008:022 (2000).
22. J. L. Feng, J. March-Russell, S. Sethi and F. Wilczek, hep-th/0005276.
23. T. Banks, M. Dine and L. Motl, hep-th/0007206.
24. G. Dvali and A. Vilenkin, "Field theory models for variable cosmological constant", hep-th/0102142.

Searches for Dark Matter

Charling Tao

CPPM, 163 av. Luminy
13288 Marseille, Cedex, France

Abstract. Only 4% of the energy density in the Universe consists of ordinary matter. 2/3 could be in the form of unknown Dark Energy, and 1/3 is Dark Matter (DM). Is this DM due to yet unidentified baryons (in molecular hydrogen, cosmic dust, brown dwarves, black holes, ...) or to non-baryonic elementary particles, such as neutrinos, or other unknown particles like the axion or the supersymmetric LSP version of the Weakly Interacting Massive Particles (WIMPs)? If WIMPs are the Dark Matter in the Galaxy, how can they be detected? This paper briefly reviews the current status of some searches for this dark component of the Universe.

1 Introduction

One of the great fascinations of the field of Dark Matter, is that it presents so many possible exploration paths. There is enough for everyone's taste and creativity and physicists have let their imagination flourish! Dark Matter candidates of every possible nature have been proposed and searched for with every conceivable detector. Such a short summary cannot do justice to the whole field, and the reader should refer to more complete lectures on the different topics addressed here.

2 Evidence for Dark Matter

The existence of a large amount of Dark Matter in the Universe is inferred by a wide variety of observations[1]. Two of the most convincing evidences are the rotation curves of spiral galaxies, and the observation of arclets due to gravitational lensing. The average density of the Universe is usually expressed by the parameter $\Omega = \frac{\rho}{\rho_c}$ in reference to the critical density, ρ_c, needed to halt the expansion of the Universe ($\rho_c = 3H^2/8\pi G_N = 1.0610^{-29} g/cm^3$, where H is the Hubble velocity of the expansion of the Universe and G_N is the Newtonian gravitational constant). Cosmic Microwave Background (CMB) measurements (cf De Bernardis[2]) give a value of Ω=1 ± 0.1, of which 2/3 is in the form of unknown Dark Energy, and 1/3 consists of matter (cf Krauss[3]).

Big Bang Nucleosynthesis (or BBN) models[10] calculate the abundances of light nuclei (H^1, H^2, He^3, He^4 and Li^7) as a function of the baryon to photon ratio η in the Universe with simple assumptions based on the Big Bang cosmology. The comparison with measured abundances leads to serious constraints on the density of baryons: Ω_b is 0.04 ± 0.004. So most of the unknown 1/3 matter in the Universe is in a yet unknown form, that of Dark Matter.

3   The Nature of Dark Matter

There is room in galaxies for baryonic DM, since $\Omega_b= 0.04 \pm 0.004$ is much larger than $\Omega_v isible$ (0.002 to 0.006). We also know of the existence of some non-baryonic matter, like electrons and neutrinos. But there are many other possibilities.

3.1   Baryonic Dark Matter

The pioneers in the field of DM had in mind several forms of dark stars, planets, asteroids and comets [12]. Zwicky[13] thought of intergalactic gas left from an inefficient galaxy formation process. Neutral hydrogen can be observed through the characteristic 21 cm line; star remnants should have many heavy elements which are not observed and comets are not frequent enough to be good candidates. True stars, even the faintest, would irradiate more infrared than is observed from the haloes of nearby galaxies. Neutron stars and stellar mass black holes would produce too many X-rays because they accrete gas from their surroundings. Several candidates have been searched for in recent years. Details can be found in [11].

- **Massive black holes** might have been formed in the early Universe. In addition to having contributed perhaps to the formation of helium and photons in the past, they would now stir up the disk stars and may contribute to the known increase of stellar velocity dispersion with age that is generally blamed on giant molecular clouds or spiral arms. Massive black holes are excluded since they would be too efficient at this, but a 10^{12} solar mass halo consisting of 10^6 solar mass objects is just right. If similar black holes make up the dark matter in dwarf ellipticals, then they must be clustered towards the center, predicting an outward decrease of stellar velocity dispersion as a test of the model. In addition, 10^6 solar mass black holes in halo should reveal themselves as gravitational lenses. Resolution of radio components may be possible with VLBI, and optical observations could be obtained with the space optical interferometer.

- **Molecular hydrogen** could provide much more dark matter density than previously estimated, if they clump in hierarchical fractal types of structures. F.Combes and D.Pfenniger[14] propose a spheroidal distribution of cold fractal molecular clouds, with a central bar in the Galaxy. Salati et al[15] have shown that it is difficult for these molecular clouds to exist in large quantities (more than 4%) inside the region where cosmic rays diffuse (inner part of the halo). Cosmic rays diffuse in the magnetic fields of the galaxy. A reliable calculation of their flux is now possible, in good agreement with measurements. The spectrum of γ-ray diffuse spectrum was measured with high precision by the EGRET instrument on board the Compton Gamma Ray Observatory (CGRO) in the range 100 MeV to 10 GeV. The analysis performed by Salati et al. did not evidence the existence of a large additional component of the γ ray diffuse emission. Molecular hydrogen could still contribute in the outer parts of the halo.

- **Brown dwarves** are substellar objects whose only energy source is contraction. They could be present in the halo of our Galaxy, and would be evidenced by microlensing experiments[16]. Although some microlensing events have indeed been observed by several experiments[18,17,19,20], they are not necessarily MACHOs (MAcroscopic Compact Halo Objects) from the Halo of our Galaxy. Observations seem to exclude that the observed candidates could be brown dwarfs, since the masses that could explain the microlensing candidates are quite large (around 0.2 solar masses). In any case, microlensing experiments do not exclude the need for other types of DM.

3.2 Why WIMPS?

Cosmology relates the present density of Dark Matter particles to their annihilation cross-sections. If Dark Matter was once in thermal equilibrium with quarks and leptons, the present density in the Universe is a function of the annihilation rate at the time the Dark Matter went out of equilibrium. If the annihilation rate is much faster than the rate of expansion of the Universe, the particles are suppressed by a huge Boltzmann factor and would not constitute the present Dark Matter. If the annihilation rate is too small, the expansion of the Universe dilutes these particles, but their abundance remains too large. In order to explain Ω around 1, the particles should have cross sections of the order of those of the neutrinos.

$$\sigma \approx \frac{10^{-38} cm^2}{\Omega \, h^2}$$

Is this "weak" scale a coincidence or an evidence that the Dark Matter is made of neutrino-like particles? In any case, the name has remained: WIMPs for *Weakly Interacting Massive Particles*!

3.3 Hot and Cold Dark Matter

The observations of complex structures (bubbles, voids, filaments, etc...) are not easy to reproduce from homogeneous initial conditions with baryons alone, and initial particles have been included in the scenarii of structure formations. Cosmologists distinguish between Hot and Cold Dark Matter.

Hot Dark Matter consists of particles that are still relativistic at the time of decoupling (masses below 100 eV), eg neutrinos (discussed by P. Hernandez[4]). While this kind of Dark Matter may be particularly useful in creating large scale structures (superclusters and beyond) first, the formation of the (smaller) galaxies is not well understood in this scenario.

Cold Dark Matter (CDM) are WIMPs that become non-relativistic well above $10^4 K$ (with large rest masses up to 10^{16} to 10^{19} GeV for some candidates). They favour the formation of small scale structures. Larger objects then arise from gravitational interactions and from biasing, an effect allowing protogalaxies to form only when very large fluctuations of masses are reached. When electrons and protons recombine into hydrogen, the Universe becomes transparent to photons which escape and baryons are trapped in the potential wells created by the WIMPs. WIMPs are capable of enhancing the fluctuations at the earlier stage, when baryons and photons are locked together and have been the favorite candidates of galaxy formationists.

4   WIMP Candidates

The number of candidates for WIMPs is large and has lead to a large number of experimental searches. The leading candidate for Dark Matter has long been the neutrino, since it is known to exist. It is not preferred in present scenarii of galaxy formations, but might still contribute a substantial fraction of the DM.

The axion[23]is a light pseudoscalar boson postulated to solve the strong CP problem. It couples to pairs of photons. A discussion on axion searches is given by G. Raffelt[5].

The rest of this review will concentrate on the search for Supersymmetric WIMPs.

4.1   Supersymmetric Particles

There seems to be no escape from Supersymmetry in particle physics, yet no decisive experimental evidence exists to date. Supersymmetry is the latest Graal in the field. In simple supersymmetric models (SUSY) [24], a quantum number called R parity, which takes a value R=+1 for particles and -1 for their SUSY partners, is conserved. R parity conservation is linked to baryon number B and lepton number L conservation: $R = (-1)^{3B+L+2S}$, where S is the spin. R parity can be violated by explicit couplings which violate either B or L. The conservation of R parity has three important implications: Sparticles are produced in pairs, heavy sparticles decay into lighter sparticles, and the lightest supersymmetric particle (**LSP**) is stable because it has no available decay mode.

In the Minimal Supersymmetric extension of the Standard Model (MSSM) the **neutralino** χ is the lowest mass particle (LSP). It is assumed to be stable, and is a linear combination of the two gauginos (photino $\tilde{\gamma}$ and zino $\tilde{Z}$) and the two higgsinos ($\tilde{H}_1^0$ and $\tilde{H}_2^0$) : $\chi = a_1\tilde{\gamma} + a_2\tilde{Z} + a_3\tilde{H}_1^0 + a_4\tilde{H}_2^0$ The two gauginos are linear combinations of the U(1) and SU(2) neutral gauginos $\tilde{B}$ and $\tilde{W}_3$. The neutralino mass m_χ and the coefficients a_i depend on the $\tilde{W}_3$ mass, M_2, on the Higgs mixing parameter μ and on $\tan\beta = v_u/v_d$, the ratio of the vacuum expectations values of the Higgs giving masses to the u and d quarks. Since even constrained MSSM models have more than 7 parameters, SUSY has very little predictive power. Most experimental constraints come from accelerator experiments (LEP, $p\bar{p}$, b $\rightarrow$ s γ, ...)

4.2   Dark Matter Distribution in the Galaxy

It is commonly assumed that our Galaxy is a standard spiral galaxy, with a visible flat disk, a central bulge and a spherical halo of dark matter. The hypothesis that the particles are trapped in the Galaxy constrains the particle to virial velocities of order $10^{-3}c$. The dark matter is in thermal equilibrium, with a Maxwellian velocity distribution of rms 270 km/s, with an escape velocity of 600 to 800 km/s, and a local density ρ_{loc}, around 0.3 GeV/cm^3. The solar system is at around 8 kpc from the Galactic Center and moves through this distribution on a circular galactic orbit with v_c about 220 km/sec. Because of the Earth movement about the Sun, the Earth velocity with respect to the dark matter halo has an annual modulation of ±30 km/sec. The distribution of matter closer to the Galactic center is much less known and gives rise to controversies that have implications for Indirect Detection predictions from the Galaxy center.

This "Simplified Model" of Matter in Galaxy will be called SMMG in this review. It is commonly used by most experimentalists to estimate event numbers or to compare different experimental results. This simplified model was an acceptable hypothesis, ten years ago, due to the lack of data and information on the Dark Matter in the Galaxy. Today, there are some reasons to doubt its exact validity. First, microlensing experiments (OGLE and MACHO[17,18]) seem to have measured 3 times the number of microlensing events in the bulge than would be expected with a simple description of the matter density in our Galaxy. (It is questioned by the EROS experiment[21]).

In addition, refined N-body CDM simulations for the Low Surface Brightness galaxies[25], confirm previous studies in dwarf galaxies, that the CDM hypothesis does not fit well the measured rotation curves. Some other components or hypothesis might be needed, which affects the amount of local neutralino contribution. Furthermore, these hierarchical simulations seem to imply the existence of possible large density enhancements (or clumps) in galaxies[26].

Those large uncertainties add to the difficulty of making precise predictions for experiments. Although results show that the simple description of the Dark Matter distribution is more complex than was previously assumed, precise quantitative assumptions are still lacking.

5   Direct Detection of Dark Matter

Goodman and Witten[27] were the first to point out that the Dark Matter in our Galaxy would interact with laboratory detectors, with rates that are small, but possibly within the reach of experiments. The challenges facing these experiments did not frighten the experimental community: How can one detect low energy nuclear recoils of a few keV, with reasonable resolutions, when predicted rates are very low (a few events/day) in the presence of radioactive backgrounds?

5.1   Kinematics

In a collision of a WIMP of mass m and velocity $v(=\beta c)$ with a nucleus of mass M, the nucleus receives a kinetic energy T given by

$$T = Mv^2(1 - \cos\theta)\left[\frac{m}{(m+M)}\right]^2$$

where θ is the center of mass scattering angle. For a given M, the observed distribution of recoil energy is then determined by m, the distribution of v and the differential cross-section $d\sigma/d\cos\theta$.

Since the velocity distribution is roughly Maxwellian (exp^{-Cv^2}), and T is proportional to v^2, the observed distribution of T will be roughly an exponential.

As for the angular distribution, most dark matter candidates scatter via the exchange of heavy particles, so the low energy behaviour is independent of v and isotropic in the center of mass. For large nuclei, the assumption of isotropic scattering has to be corrected for the nuclear elastic form factor.

5.2   DM Direct Detection Detectors

Nuclear recoils in the keV range can be observed in ionization detectors such as semiconductor diodes made of Germanium or Silicon or Time projection Chambers, scintillators of all sorts, solid or liquid, cryogenic bolometers, superconducting granules or moderately heated liquids.

One experiment, the "DAMA" collaboration[28], has reported the observation of an annual modulation of 100 kg NaI rate, which they claim could come from a 60 GeV neutralino with a cross-section on nucleon of about 10^5 pb. Many criticisms have been formulated against this observation (cf [29]). In addition, the claimed region is being excluded now by the EDELWEISS[30] bolometer experiment, (CDMS[31] excludes it partially).

Cryogenic bolometer detectors are measuring phonons, the thermal component of the ionizing event. So they are fully sensitive to the low energy nuclear recoils, while scintillation and ionization detectors are only sensitive to a fraction of the deposited energy. Hybrid detectors, capable of recording both heat and ionization (or heat and scintillation) use this difference of response to discriminate between a nuclear recoil and an electromagnetic interaction in the detector. The hybrid bolometers developed by CDMS or EDELWEISS aim at rejecting more than 99% of the electromagnetic background. This means a 100 times potential improvement on the sensitivity of the existing Germanium experiments.

EDELWEISS is installed in the deep underground Frejus laboratory (LSM) and is currently running with several 300 g Ge. A 10 kg project is planned. The CDMS group has measured several Ge and Si crystals in a shallow site (10 m underground) in Stanford and is moving this year to the deeper SOUDAN mine (690 m deep).

Another promising development is the use of LXe, which interests many groups in the world. The installation of 1 ton detector is foreseen by the UKDM group[32] in the Boulby mine.

5.3   DM Direct Detection Signatures

Experiments are able to set limits with relatively small masses (eg, 300 g as in the case of bolometers). Should a signal be observed, evidence that it is not due to spurious detector effects should be given, and the Galactical origin must be proven. This will require much larger detectors for statistics and the study of systematic effects. 100 kg of NaI(Tl) have been installed and measured in LNGS. Much larger detectors (eg, 100 tons of NaI(Tl)) have already been proposed.

1- The most unambiguous signature would be a change in the event rate and the spectrum of energy deposition with the time of the year. The Sun goes around the Galaxy and therefore through the halo at 220 km/sec and the earth is either adding (or subtracting) half of its velocity to the Sun velocity in the summer (or winter). Both the mean energy deposition and the rate should vary by about 7%. In order to observe such an effect, some 5000 events are needed and therefore large mass detectors are required.

2- The direction of the scattered nucleus is another interesting signature. Because of the rotation of the Sun inside the halo, dark matter particles come preferentially from one direction. This is very difficult since few detectors are sensitive to the directionality of a very small recoil energy. Low pressure TPC with hydrogenous gas have been studied and will probably be revived if an indication of Dark matter is shown.

3- The shape of the energy spectrum will be measured by high resolution detectors, such as Germanium and hybrid bolometers. This will test the velocity distribution of the Dark Matter in the Galaxy.

4- A comparison of the rates and spectra on targets of different materials allows a determination of the masses of the WIMP. The use of different nuclei is important to prove the galactic nature of the signal.

6 Indirect Detection of Dark Matter

Astroparticle detectors as different as antimatter detectors[34,6], gamma-ray telescopes[7] or neutrino telescopes[8], share a common possibility to detect indirectly the Dark Matter of the Universe, if WIMPS exist and are present in our Galaxy, forming its invisible halo. These massive particles accumulate and annihilate almost at rest in the centre of celestial bodies, such as the Galactic Center(GC)[68], the Sun or the Earth[35,36]. The capture rate is proportional to the elastic scattering cross sections on the nuclei of the celestial body. As the number of WIMPs in the body gradually increases, a steady state will eventually be reached where the capture rate is balanced by the annihilation rate. The WIMP annihilations gives secondary leptons and hadrons that can decay and the products can be detected. The most up to date derivation of such effects and the estimates of the fluxes and spectra due to SUSY neutralino annihilations can be found using the DarkSUSY[33] package.

6.1 Positron Measurements

The balloon experiments CAPRICE and HEAT[37] have observed a structure in the positron spectrum around 7 GeV. This structure first needs confirmation from other experiments (PAMELA[38] soon, or AMS[34] in 2005). In addition, SUSY calculations, would require clumps to explain this structure[39].

6.2   Antiproton Measurements

A few years ago, the BESS balloon experiment detected antiprotons at low energies below 1 GeV [40]. The size of the signal was slightly above the expected calculations available at the time, of the interaction of cosmic rays with interstellar gas and the Earth atmosphere. Several authors[41,42] have reevaluated the effect of helium interactions, as well as collective nuclear effects and proton and antiproton secondary interactions. The consequences are an increase of the expectations of antiprotons in the energy range below 1 GeV, with the main uncertainty coming from the parametrization of the primary proton spectrum. In parallel, the BESS experiment also improved[43] its measurements, and the measured antiproton yields are now smaller. With those developments, there is today no indication for new physics in the antiproton signal, and more severe constraints on the SUSY parameters can be set.

Studies on antideuterons[61] seem to provide a clear signal for SUSY.

6.3   Gamma Rays

Gamma rays produced in neutralino annihilations give a continuous spectrum, but also a distinct signal of monoenergetic photons. Existing limits on Dark Matter detection come from analysis of data from the EGRET[44] satellite below 10 GeV, and the Whipple[45] ground based telescope above 100 GeV. Ongoing STACEE[46] and CELESTE[47] should cover the intermediate interesting region from 10 to 100 GeV.

The next generation of gamma ray telescopes with the ground based experiments such as MAGIC[48] or the even bigger VERITAS[49], should give 1 to 2 orders of magnitude improvement in sensitivity. The space particle detector GLAST[50], with 1 m^2 track detector should be able to improve over EGRET by more than 2 orders of magnitude, and extend the CELESTE and STACEE ground based telescopes energy region.

Compared to the Dark Matter Direct Searches, the main interest of the gamma ray detection is the complementarity of the parameter space. The largest cross-sections for gamma production come from higgsino-like neutralinos, while more favourable to Direct Detection are the gaugino-like neutralinos[51]. In addition, [26] concluded that gamma ray detectors would be the best instruments to test for clumpiness scenarii.

6.4   Neutrinos

Neutrinos are also produced in neutralino annihilations either directly but most frequently by secondary decay of heavy quarks or gauge bosons. The neutrinos produced in the center of the Earth, the Sun or the Galaxy, interact in and around neutrino telescopes and produce muons that can be detected, by looking in the searched direction.

The difficulty of identifying these neutrinos is the presence of the atmospheric neutrinos, produced by cosmic ray interactions in the Earth atmosphere. The first limits were published by Kamioka[52] in 1992 with 770 m^{2}yr, and the best limits to date are given by MACRO[55] and SuperK[53] for roughly 10^3 to 10^4 m^{2}yr, with a detection sensitivity for muon fluxes of about 10^{-14}/cm^2/s or roughly 3000/km^2/yr. The AMANDA neutrino telescope in the South pole ice is now reaching this sensitivity[56]. Several neutrino telescopes are being developed in water, one in the lake BAIKAL[58], and the others in the Mediterranean Sea (ANTARES[57], NEMO[59], NESTOR[60]).

Very large telescopes of order 1 km^2 are necessary to improve substantially the present sensitivity and probe more SUSY parameter space. The planned sea neutrino telescope projects, will be able to reach 0.1km^{2}yr in the near future, and in the future aim at 10 km^{2}yr, as is the case of ICECUBE[8,56] in the South Pole.

6.5   Comparisons with Direct Detection

The expected flux of muons from this process depends on the annihilation rate of WIMPs in the Sun (or GC or Earth), and on the spectrum of neutrinos produced in the annihilations. For most interesting (ie detectable) dark matter candidates, the capture cross sections are (low but) high enough for the annihilation and capture rates to have reached equilibrium. In this case, and assuming the SMMG is valid,the flux of muons is proportional to the elastic scattering cross sections on the nuclei in the Sun (or GC or Earth). The Sun is mainly hydrogen, and helium, but coherent couplings induce a non negligible contribution for heavy nuclei. For coherent couplings in the Earth, Fe is the main contributor. In addition, in the case of the Earth, the capture rates are much lower so that a steady state may or may not have been reached, depending on the mass of the WIMP and the annihilation cross section.

For a given spectrum of annihilation neutrinos, the flux limits on upward going muons can be interpreted in terms of a limit on the effective elastic scattering cross section. These can be then compared with the limits from direct searches if one makes a hypothesis on how the cross section depends on the interaction with the nucleus[54].

Damour and Krauss[63] have suggested that planet perturbations would induce the existence of new orbits for WIMPs scattered from the surface layers of the Sun. This new population could intersect the Earth and increase the predicted neutrino fluxes up by a large amount (up to 2 orders of magnitude[65,66]. This population would affect much less the Direct Detection experiments, since this population has low velocity, thus giving low energy recoils, below most detectors thresholds. However, it has been pointed out by Gould and Khairul Alam[64] that this population could be very similar to asteroids. It is known that Earth crossing asteroids are systematically driven towards the Sun in timescales of a few million years, much less than the WIMP diffusion times. The consequences of such asteroid-like behavior would be a reduction of the contribution for the center of the Earth[65,66]. If these effects are to be taken seriously, it is difficult to extract exact exclusion plots for the center of the Earth. Only discovery of a signal would be meaningful...

The planet perturbations hypothesis and/or the asteroid-like subsequent behaviour do not concern the neutralinos rates in the Sun or the Galactic Center.

In the Galactic Center(GC), much is unknown still, but it does seem that there is a massive black hole Sgr A* with a mass about 2.6 +/- 0.2 million times that of the Sun([9,67], as in most galaxies. Gondolo and Silk[68] first estimated that the presence of CDM in the center of the Galaxy, would produce an extra large neutrino flux, if the black hole produced a central spike. In that case, the MACRO experiment could set very stringent limits on SUSY parameters. Later on, Gondolo[69]realized that synchrotron radiation would also be produced in large quantity, and he uses the measured values at 408 MHz to conclude that either no neutralinos are in the halo, or the matter density in the center of the Galaxy is not cuspy. These conclusions have been revisited by [70], who would still allow massive neutralinos of 1 TeV to contribute. Others[71] have questioned the existence of the spike in the GC, thus rejecting the dramatic conclusions of Gondolo.

7   Conclusions

The field is in a very interesting period. Many ideas are discussed and enormous signals can appear and disappear depending on the predictions. For experimentalists, who need a longer time scale to build an experiment, what is important is to know that surprises can happen, and to design experiments with an open mind. Historically, Kamioka was first built to measure the proton lifetime predicted by SU(5). It did not see the proton decay but opened a new area for neutrino physics. It provided a strong confirmation of the long standing puzzle of solar neutrino deficit and has discovered the atmospheric neutrino anomaly. Evidence for neutrino oscillations are probably the experimental results of a proton decay experiment. What is Nature preparing for next generation telescopes?

References

1. Trimble V., Ann. Rev. A. A. **25** (1987) 425.
   Trimble V., Contemp. Phys.**19** (1988) 373.
   Primack J. R., Sadoulet B. and Seckel D., Ann. Rev. Nucl. Sci. **38** (1988) 751
   Smith P.F. and Lewin J. D., Phys. Rep. **187** (1990) 203
   Bernabei R., Nuovo Cimento **18** (1995) N. 5
   Jungman G., Kamiokowski M., Griest K., Phys. Rep. **267** (1996) 195-373.
2. de Bernardis P., "Cosmic Microwave Background Fluctuations", these proceedings.
3. Krauss, L., "The Cosmological Parameters", these proceedings.
4. Hernandez P., "Neutrino Masses and Oscillations", these proceedings.
5. Raffelt G., "Stars and Fundamental Physics", these proceedings.
6. Battiston R., "Astroparticle Physics from Space", these proceedings.
7. Voelk H., "Astrophysics with High Energy Gamma Rays", these proceedings.
8. Halzen F., "High Energy Neutrinos from Astrophysical Sources", these proceedings.
9. Eckart A., "The Center of the Milky Way", these proceedings.

10. Yang J. et al., Ap. J. **281** (1984) 493.
    Olive K.A., Schramm D.N., Steigman G. and Walker T., Phys. Lett. **B426** (1990)
    Malaney R.A. and Mathews G.J., Phys. Rep. **229** (1993) 145.
11. Carr, B., Ann.Rev.Astron.Astrophys. 32 (1994) 531-490.
12. Oort J.H., Bull. astr. Inst. Neth.6 (1932) 249.
13. Zwicky F., Helv. Phys. Acta **6** (1933) 110.
14. Combes F., Pfenniger D., Martinet L., A&A (1994) **285** 79, 83.
15. Salati P. et al, A&A (1995), ENSLAPP-A-528/95.
16. Paczynski B, Ap.J. **304** (1986) 1.
17. Udalski C.A. et al., Act. astr.43 (1993) 283, and A&A, 343(1999) 10.
18. Alcock C. et al., Nature **365** (1993) 621, ApJS, 124 (1999) 171. Alcock C. et al.,
    Phys. Rev. Lett. **74** (1995) 2867.
19. www.lal.in2p3.fr/recherche/eros/publif.html Lasserre Th. et al., Astron. As-
    trophys. 355, P L39-L42 (2000), A&A, 348 (1999) 175.
20. www.vuw.ac.nz/scps/moa/moapapers.html
21. Afonso C., PhD. Thesis 2001, U. Paris XI.
22. Alard C. et al., Messenger **80** (1995) 31.
23. Sikivie P., Phys. Rev. Lett.51 (1983) 1415.
24. Fayet P., in Unifications of the Fundamental Particle Interactions, eds. Ferrara S.,
    Ellis J., and Van Nieuwenhuizen, P. (Plenum Press, New York) (1980) 587.
25. Moore B. et al., ApJ. 499 (1998) L5, and astro-ph/9903164.
26. Bergstrom L. et al., astro-ph/9806072, Phys. Rev. D59 (1999) 043506, and astro-
    ph/0012346.
    Stiff et al., astro-ph/0106048.
27. Goodman M. W. and Witten E., Phys. Rev. D **31**, (1985) 3059
28. INFN/AE-00/01, also Phys.Rev D61 (2000) 023512.
29. Gerbier et al., hep-ph/9710181, astro-ph/9902194
    Also gaitskell.brown.edu/physics/dm/DAMA summary.html
30. edelweiss.in2p3.fr and Mosca, L., Status Report Conseil Scientifique de Modane,
    May 2002, to be published soon.
31. cdms.berkeley.edu/ and Perera T. et al., AIP Conf.Proc., July 22-27, 2002.
32. hepwww.rl.ac.uk/ukdmc/pub/publications.html
33. Darksusy, Gondolo P., Edsjo J., Bergstrom L., Ullio P., Baltz E.A., astro-
    ph/0012234, Proc. 3rd Int. Workshop on the Identification of Dark Matter
    (IDM2000).
34. pierre.mit.edu/~eluc/AMS/
35. Press W.H. and Spergel D.N., Ap. J. **296** (1985) 679.
36. Gould A., Ap. J. **321** (1987) 571.
37. CAPRICE2, ida1.physik.uni-siegen.de/caprice2.html
    HEAT, pooh.physics.lsa.umich.edu/heat/
38. Bonvicini V. et al., NIM A461 (2001) 262.
39. Ullio P., Phys.Rev. D59 (1999) Baltz et al., Phys Rev D65 (2002) 063511.
40. BESS95, astro-ph/9809326, Phys.Rev.Lett. 81 (1998) 4052-4055.
41. Bergstrom,L., Edsjo,J., and Ullio, P., ApJ 526 (1999) 215
42. Bieber et al., Phys.Rev..Lett. 83 (1999) 674
43. BESS Collaboration, cf www.phys.kobe-u.ac.jp/~nozaki/BESS/
    And astro-ph/9906426, Phys.Rev.Lett. 84(2000) 1078-1081.
44. cossc.gsfc.nasa.gov/cossc/egret/
45. egret.sao.arizona.edu/
46. hep.uchicago.edu/~stacee/ and astro-ph/0010341, Proc. Heidelberg Symp. on
    High-Energy Gamma-Ray Astronomy.

47. wwwcenbg.in2p3.fr/Astroparticule/celeste/e-index.html
    and astro-ph/0010265, Proc. Heidelberg Sympos. on High-Energy Gamma-Ray Astronomy.
48. hegra1.mppmu.mpg.de/MAGICWeb/
49. veritas.sao.arizona.edu/veritas/index.shtml and astro-ph/9908135, proc. 26th ICRC (Salt Lake City, 1999).
50. www-glast.stanford.edu/ and astro-ph/9912139
51. Bergstrom L., hep-ph/0002126, Rept.Prog.Phys. 63 (2000) 793
52. Kamiokande collaboration: Mori M. et al., Phys. Lett. B **289** (1992) 463.
    Mori M. et al., Phys. Rev. D **48** (1993) 5505.
53. www-sk.icrr.u-tokyo.ac.jp/index.html Habig A., 27th ICRC Hamburg 2001, hep-ex/0106024
54. Rich J. and Tao C., DAPNIA/SPP 95-01, Saclay.
    Kamionkowski M. et al., Phys. Rev. Lett. **74** (1995) 5174.
55. www.lngs.infn.it/lngs/htexts/macro.htmlx
    For WIMPs, Montaruli,T., hep-ex/9905021, or hep-ex/9812020, Phys. Rev. D60 (1999) 082002.
56. amanda.berkeley.edu/
57. antares.in2p3.fr/
58. www.ifh.de/baikal/
59. nemoweb.lns.infn.it/
60. www.nestor.org.gr/
61. Donato F., Fornengo N., Salati P., hep-ph/9904481, Phys.Rev.D 62(2000) 043003.
62. Bergstrom L., Edsjo J., Gondolo P., hep-ph/9806293, Phys.Rev. D58 (1998) 103519
63. Damour Th. and Krauss L., astro-ph/9807099, Phys.Rev. D59 (1999) 063509
64. Gould A. and Khairul Alam S.M., astro-ph/9911288, Astrophys.J. 549 (2001) 72.
65. Bergstrom L., et al, hep-ph/9905446, JHEP 9908 (1999) 010,
66. Bottino A. et al., hep-ph/0001309, Phys. Rev.D62 (2000) 056006.
67. Eckart A. and Genzel R., Nature 383 (1996) 415.
    Klein B.M. et al. Astrophys.J. 509(1998) 678.
    Genzel R. et al., astroph/0001428 in MNRAS.
68. Gondolo P. and Silk J., astro-ph/9906391, Phys.Rev.Lett. 83 (1999) 1719-1722.
69. Gondolo P., hep-ph/0002226, Phys.Lett. B494 (2000) 181-186
70. Bertone G., Sigl G., Silk J., astro-ph/0011553, Proc. 3rd Int. Workshop on the Identification of Dark Matter (IDM2000).
71. Ullio P., Zhao H. and Kamionkowski M., astro-ph 0101481, Phys. Rev. D64 (2001) 043504.

Neutrino Masses and Oscillations

Pilar Hernández

Theory Division, CERN, 1211 Genève 23, Switzerland

1   Introduction

LEP era has established the validity of the Standard Model (SM) at better than the 1% level [1]. This is a gauge theory based on the group $SU(3) \times SU(2) \times U(1)$, which is spontaneously broken to the subgroup describing the strong and electromagnetic interactions $SU(3) \times U(1)_{\text{em}}$. In this process the boson mediating the weak interactions adquire a mass. The recent discovery of the last missing particles ν_τ and t, which were predicted a long time ago, has confirmed the family structure shown in Table 1.

All the fermions in the model can be classified according to irreducible representations of the gauge group. There are two features that are particularly striking in this table. One is the family structure: there is a three-fold degeneracy of multiplets with the same quantum numbers, they simply differ in their mass. The second is that the $SU(2)$ is left-handed: it only couples to left-handed particles. A relativistic spin 1/2 particle in the limit $v \to c$ has a well defined helicity: either the spin points in the same direction as the momentum (right-handed) or in the opposite one (left-handed). In the Standard Model only the second type of spin 1/2 particles are charged under the $SU(2)$ group. In particular, the neutrinos ν_i being neutral under any the strong and electromagnetic interactions, do not have right-handed counterparts.

The origin of these features is one of the fundamental open questions in particle physics.

$(1,2)_{-\frac{1}{2}}$	$(3,2)_{\frac{1}{6}}$	$(1,1)_{-1}$	$(3,1)_{\frac{2}{3}}$	$(3,1)_{-\frac{1}{3}}$
$\begin{pmatrix} \nu_e \\ e \end{pmatrix}_L$	$\begin{pmatrix} u^i \\ d^i \end{pmatrix}_L$	e_R	u_R^i	d_R^i
$\begin{pmatrix} \nu_\mu \\ \mu \end{pmatrix}_L$	$\begin{pmatrix} c^i \\ s^i \end{pmatrix}_L$	μ_R	c_R^i	s_R^i
$\begin{pmatrix} \nu_\tau \\ \tau \end{pmatrix}_L$	$\begin{pmatrix} t^i \\ b^i \end{pmatrix}_L$	τ_R	t_R^i	b_R^i

Table 1. Fermion content of the Standard Model, classified according to the dimension of the SU(3) and SU(2) representations as well as the U(1) charge: $(d_{SU(3)}, d_{SU(2)})_{U(1)}$

2 Fermion Masses

A mass can be thought as a left $\rightarrow$ right transition:

$$m\,\bar{\psi}_R\psi_L + h.c. \qquad (1)$$

In the SM fermion masses originate in the interaction with the Higgs field by a coupling of the type:

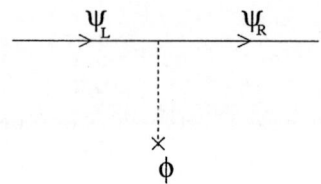

Upon spontaneous symmetry breaking the Higgs field acquires a vev, $\langle\phi\rangle = v$, generating a mass for the fermions of the form:

$$m_f = \lambda_f\,v. \qquad (2)$$

The different values of the fermion masses are obtained by adjusting the Yukawa couplings, λ_f, which are fundamental parameters in the SM.

When the SM was invented there were only upper bounds on the ν masses from weak-decay kinematics and they were conjectured to be zero. It is however easy to include ν masses. One possibility is to add a right-handed singlet ν for each family so that Yukawa couplings to the Higgs induce neutrino masses, in analogy with the remaining leptons. However the problem of doing this is that we are faced with a new "hierarchy" problem: why are neutrinos so much lighter than the remaining fermions?

It was realized a long time ago [2] that a different alternative is possible. Since neutrinos are neutral under the unbroken group, the right-handed neutrino can be identified with the antiparticle of the left-handed one. In this case, it is said that the neutrino is a Majorana particle and a new type of coupling to the Higgs field exists, which by simple dimensional analysis involves a new scale M:

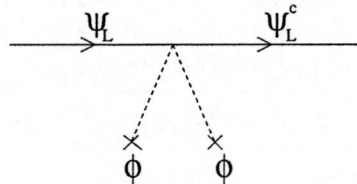

Upon spontaneous symmetry breaking the neutrino mass is of the form:

$$m_\nu = \lambda_\nu \frac{v^2}{M}. \tag{3}$$

If $v \ll M$, neutrino masses are naturally small.

The addition of this interaction to the SM has far reaching consequences, since there is a new physics scale M and the SM is just valid at $E \ll M$. Furthermore this interaction breaks total lepton number (i.e. the number of leptons - antileptons), which is satisfied in the SM with only Yukawa interactions. This is in fact extremely interesting, because a mechanism to explain the baryon asymmetry in the Universe, called leptogenesis, emerges naturally [3]. This shows that measuring neutrino masses could give us a clue of what lies beyond the Standard Model. Indeed if we consider values of the neutrino masses in the range implied by recent data and values of their Yukawa couplings close to those of the charged leptons, we find a value for the scale $M \sim O(10^{15})$ GeV, which is intriguingly close to the scale at which the gauge couplings seem to unify [4].

3   Neutrino Masses and Mixing

If the fermions have a mass, the charged current interactions are not flavour diagonal in the mass eigenbasis: there is generically mixing among the families. The weak and mass bases are related by a unitary matrix, called the mixing matrix. There are two such matrices in the SM:

$$\begin{pmatrix} d \\ s \\ b \end{pmatrix} = V_{CKM} \begin{pmatrix} d_m \\ s_m \\ b_m \end{pmatrix}, \quad \begin{pmatrix} \nu_e \\ \nu_\mu \\ \nu_\tau \end{pmatrix} = U_{MNS} \begin{pmatrix} \nu_1 \\ \nu_2 \\ \nu_3 \end{pmatrix}. \tag{4}$$

The CKM matrix [5] in the quark sector and the MNS matrix [6] in the lepton sector. These matrices are in general complex and in this case, the discrete symmetry CP (a combination of parity and charge conjugation) is broken. The violation of CP that has been observed to date in the SM can be explained by the complex entries of V_{CKM} matrix [7]. The MNS matrix has four physical parameters, three angles and one CP-violating phase, if neutrinos are Dirac particles, and two additional phases if they are Majorana.

4   Neutrino Oscillations

Neutrino mixing is responsible for the beautiful mechanism of neutrino oscillations [8]. Neutrinos are necessarily produced in weak interactions, i.e. in flavour eigenstates. Due to neutrino mixing, this state is a combination of the mass eigenstates, which evolve differently in their propagation. The state at a distance L is no longer the original flavour eigenstate

$$x = 0 \ |\nu_\alpha\rangle = \sum_j U_{\alpha j}|\nu_j\rangle \Rightarrow \quad |\nu_\alpha(L)\rangle = \sum_j U_{\alpha j} \ e^{iEt} \ e^{iLp_j}|\nu_j\rangle \qquad (5)$$

with $p_j \equiv \sqrt{E^2 - m_j^2}$. There is a non-zero probability that what its measured at L (again via a weak interaction) is a different flavour. In the case of two families, the mixing matrix depends only on one angle. The probability for the transition between the two flavours is given by the well-known formula [8]:

$$P_{\alpha\beta} = \sin^2 2\theta \ \sin^2\left(\frac{\Delta m^2 L}{4E_\nu}\right), \qquad (6)$$

where θ is the mixing angle and $\Delta m^2 = m_2^2 - m_1^2$ is the splitting between the mass eigenstates.

Neutrino oscillation experiments can be classified in two types. Those in which the *appearance* of a different flavour to the one originally produced is measured, or the case in which the *disappearance* probability of a certain flavour is measured, $P_{\alpha\alpha} = 1 - P_{\alpha\beta}$ in the case of two families.

The most characteristic features of neutrino oscillations are:

- Flavour transitions
- The peculiar L/E_ν dependence of the oscillation pattern.

The optimal experiment to observe these two features, is an appearance experiment with

$$\langle E_\nu\rangle/L \sim |\Delta m^2|. \qquad (7)$$

At much smaller values of $\langle E_\nu\rangle/L$, the oscillation is very fast (see Fig. 1)and given the finite resolution of the detector, the transition probability is then averaged

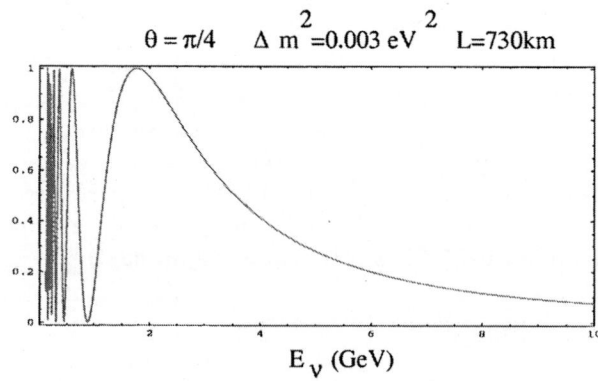

$$\theta = \pi/4 \quad \Delta m^2 = 0.003 \ eV^2 \quad L = 730km$$

$$E_\nu \ (GeV)$$

Fig. 1. Oscillation probability as a function of the ν energy

to:

$$P_{\alpha\beta} \simeq \frac{1}{2}\sin^2 2\theta. \tag{8}$$

No energy dependence and no sensitivity to the mass splitting. At much larger values of this ratio, neutrinos have not yet oscillated.

5 Evidence for Neutrino Oscillations

5.1 Solar Neutrinos

The first evidence for neutrino oscillations was found in the pioneer experiment of R. Davis and collaborators who measured for the first time the solar neutrino flux and found a deficit with respect to the expectations. Several experiments with different experimental techniques and different thresholds for the neutrino energy have followed [9,10], as shown in Table 2, and consistently measured a deficit of $40 - 70\%$ with respect to the standard solar model, which predicts the neutrino spectra shown in Fig. 2.

Recently impressive progress has been achieved in this field. There is now evidence that neutrinos coming from the sun do oscillate into a different flavour, independent of the input of the standard solar model. Since the conference, the SNO experiment [10] has established, with a $\sim 5\sigma$ confidence level, that

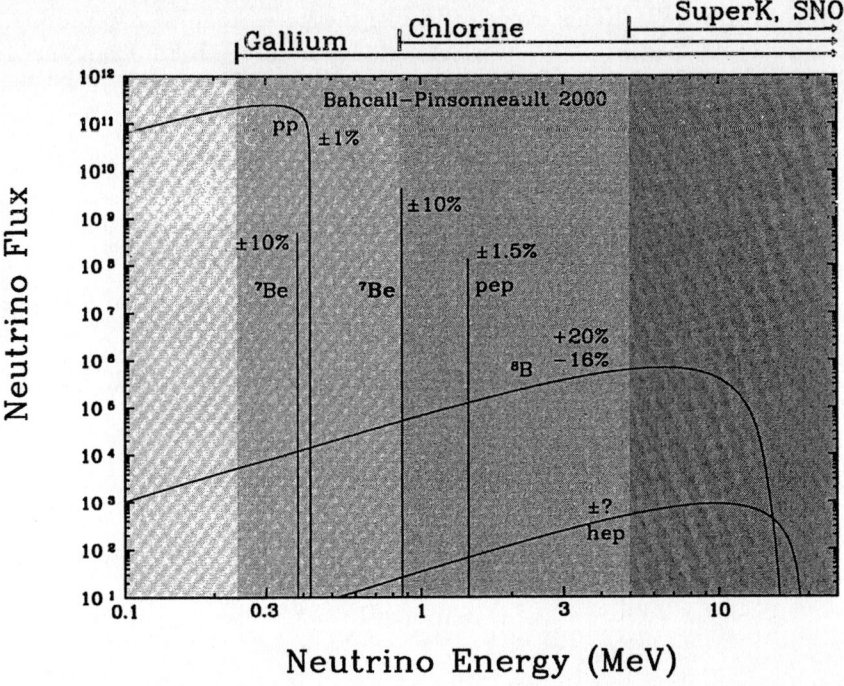

Fig. 2. Solar neutrino fluxes from the standard solar model of [11]

Exp.	Tech.	E_ν^{th}	data/SSM
Homestake ('68)	$^{37}\text{Cl}(\nu, e^-)\ ^{37}\text{Ar}$	> 0.81 MeV	0.34(3)
Gallex⊕Sage ('90)	$^{71}\text{Ga}(\nu, e^-)\ ^{71}\text{Ge}$	> 0.23 MeV	0.56(4)
K→SK ('95)	Cerenkov	> 5 MeV	0.46(2)
SNO-CC ('99)	Cerenkov	> 5 MeV	0.35(2)

Table 2. Rates of solar nu_e measured in different experiments.

the neutrinos that arrive from the sun are not only ν_e but also ν_μ/ν_τ. This experiment is able to measure solar neutrinos with two reactions:

$$\text{CC} : \nu_e + d \to p + p + e^- \qquad \text{NC} : \nu_x + d \to p + n + \nu_x, x = e, \mu, \tau. \qquad (9)$$

In the first one only ν_e can contribute, while in the second one all neutrino species that couple to the Z^0 can intervene. The different flux measured with the two reactions is an unambiguous evidence that the sun also shines ν_μ and ν_τ.

After this experiment, the global fits of all solar neutrino data to the hypothesis of neutrino oscillations, give a much more consistent picture. See Fig. 3. At 90% confidence level there is only one solution allowed: the so-called large mixing angle (LMA) one with a $\Delta m^2 \geq 10^{-5}\text{eV}^2$ and close-to-maximal mixing. The other solutions shown in this figure have just been excluded by the impressive results of KamLAND. This experiment can measure the flux of $\bar{\nu}_e$ produced in several reactors located at an average distance of $L = 175$km from the detector, in such a way that they are sensitive to the range of the LMA solution in Fig. 3. The first results of this experiment [13] indicate that there is indeed

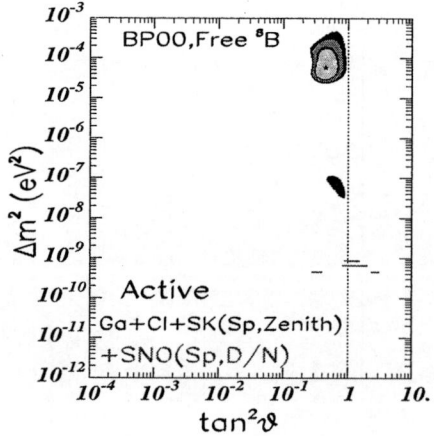

Fig. 3. Allowed neutrino oscillation solutions to the solar ν deficit [12]

an oscillation signal, which excludes with a very high confidence level the other islands in Fig. 3, selecting the LMA solution as the right one.

5.2   Atmospheric Neutrinos

Atmospheric neutrinos are produced in the atmosphere when primary cosmic rays impinge on it producing kaons and pions, which subsequently decay. The flux of atmospheric neutrinos is mainly composed of $\nu_\mu, \bar{\nu}_\mu, \nu_e, \bar{\nu}_e$. Although there is a large uncertainty in the absolute normalization of this flux, the uncertainty in the ratio of the $\nu_\mu(\bar{\nu}_\mu)$ over the $\nu_e(\bar{\nu}_e)$ is smaller than $\sim 5\%$. This ratio has been measured by several experiments to be about half of what is expected [14].

That this effect is the result of neutrino oscillations was demonstrated by the impressive data obtained by the SuperKamiokande detector [15]. This detector can distinguish electrons and muons, and thus the charged current reactions produced by $\nu_\mu(\bar{\nu}_\mu)$ or $\nu_e(\bar{\nu}_e)$. They can also measure the energy and zenith angle of these leptons, which are strongly correlated (especially for the largest energies) with the energy and distance travelled by the parent neutrino. The left plot of Fig. 4 shows that the distance travelled by the neutrino, since it is produced at the higher levels of the atmosphere, can vary from $L \sim 13,000$ km for the neutrinos moving upwards ($\cos\theta = -1$) to $L \sim 15$ km, for those moving downwards ($\cos\theta = 1$). Based on the final lepton energy the data are separated in several samples (sub-gev and multi-gev for electron and muon events, and for the latter there are also stopping and through-going events), which correspond to the different distributions of the neutrino parent energies shown in the right figure of Fig. 4.

The measured distributions in zenith angle within these samples are shown in Fig. 5. The solid curves are the expected events in the absence of oscillations. The electron events agree with this expectation, while the muon events show a very strong suppression for $\cos\theta < 0$. The suppression disappears as we approach $\cos\theta = 1$. This is precisely the expected L dependence in the case that

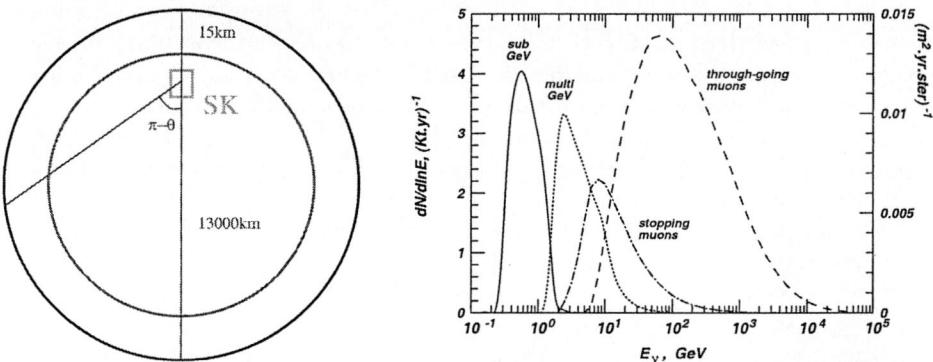

Fig. 4. Left: correlation between the zenith angle and the distance. Right: parent ν energy of the different data samples

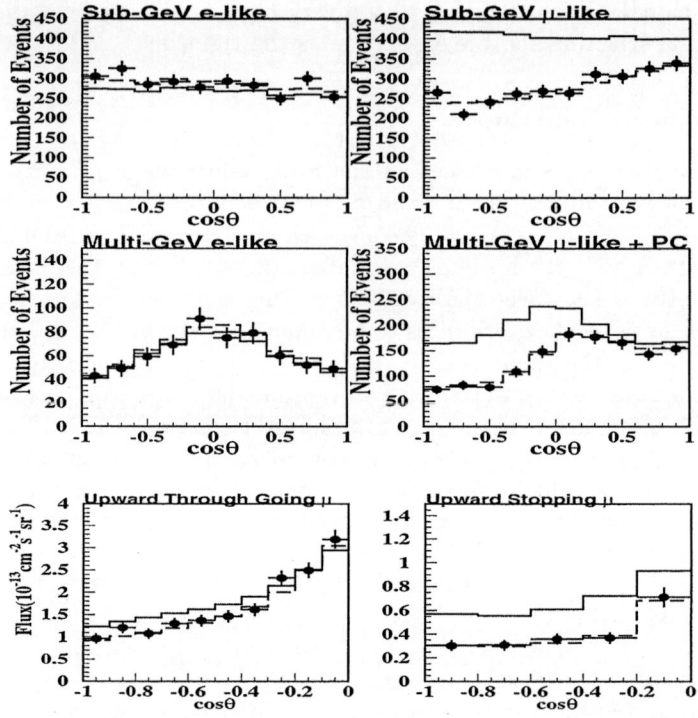

Fig. 5. Zenith angle distributions of data in the different samples [15]

these neutrinos have oscillated into a different type other than electron or muon. The most favoured hypothesis of $\nu_\mu \to \nu_\tau$ gives the dashed curves, which agree extremely well with the data.

Figure 6 shows the values of Δm^2 ($\geq 10^3$ eV^2) and $\sin^2 2\theta$ (close to 1) allowed by the atmospheric data if interpreted in terms of $\nu_\mu \to \nu_\tau$ oscillations. The atmospheric mass splitting is a factor of 10 or more larger than the one observed in solar data. The fact that the ν_e do not seem to oscillate in the atmospheric range is mostly the result of a reactor experiment, Chooz, which has measured the disappearance probability of reactor neutrinos in this range, finding a probability of one within errors [16]. This implies that the electron component of the two mass eigenstates responsible for the atmospheric splitting is very small.

The atmospheric oscillation will be further confirmed in man-made sources. It turns out that

$$|\Delta m^2_{\text{atmos}}| \sim \frac{E_\nu(1 - 10GeV)}{L(10^2 - 10^3 km)} \tag{10}$$

and ν_μ beams in this energy range can be easily produced in accelerators:

$$p \;\; \to \;\; \text{Target} \to \pi^+, K^+ \to \nu_\mu \;(\% \; \nu_e, \bar{\nu}_\mu, \bar{\nu}_e)$$
$$\nu_\mu \to \nu_x$$

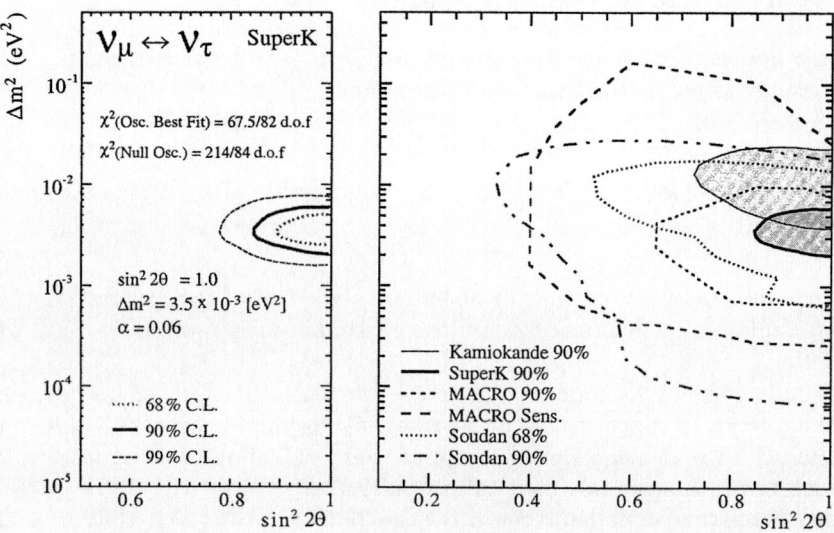

Fig. 6. Allowed neutrino oscillation solutions to explain the atmospheric ν flux [15]

Three such "conventional" beams have been designed: from KEK to Kamioka (L=235 km), Fermilab to Soudan (L = 730 km) and CERN to Gran Sasso (L=730 km) and will search for the disappearance of ν_μ or the appearance of ν_τ. The first of these experiments has already observed the disappearance of ν_μ in agreement with the atmospheric oscillation, with a confidence level above $2 - 3\sigma$ [17].

5.3   LSND

The third evidence for neutrino oscillations was obtained in the laboratory from a ν_μ beam produced in the decay-in-flight of a pion beam and the decay-at-rest of the secondary muons [18]. The signal is the appearance of more electron neutrinos in the beam than those expected from the background. If interpreted in terms of neutrino oscillations, the best fit values for the oscillation parameters are $\Delta m^2 \sim 1.2$ eV^2, which is very much smaller than the splitting in solar and atmospheric data. The mixing angle required is much smaller.

This result has not yet been confirmed by an independent experiment. In fact the experiment KARMEN [19], which has a sensitivity range which covers a large fraction of the allowed oscillation solutions of LSND, has observed no effect. This unsatisfactory situation will soon be resolved by a new dedicated experiment, MiniBooNE.

6   Neutrino Masses in the Standard Model

The three bodies of evidence for neutrino oscillations discussed in the previous section require three distinct mass splittings since:

$$|\Delta m^2_{solar}| \quad \ll \quad |\Delta m^2_{atmos}| \quad \ll \quad |\Delta m^2_{LSND}|. \tag{11}$$

On the other hand we know from LEP that the number of light standard neutrinos is

$$N_\nu = 3.00 \pm 0.06. \tag{12}$$

With three-neutrino mixing we can at best explain two of the anomalies. If we insist on explaining the three, an sterile neutrino species is needed. But, who ordered that?

Fortunately the hypothesis of an active $\to$ sterile oscillation has become very improbable recently, since it cannot fit solar, atmospheric nor LSND data. Although solutions with some sterile component are still allowed at a conservative confidence level, there is not compelling theoretical reason to believe in these solutions. If on the other hand we drop the LSND data (given that it is the only result that has not been confirmed independently), the remaining results fit wonderfully in the mixing of the three standard neutrinos: ν_e, ν_μ, ν_τ.

Let us analyze what is known in this scenario. Although we do not know the absolute neutrino mass scale, we know that there is a hierarchy between the solar and atmospheric mass splittings. The neutrino spectrum can be of the following two types, depending on the sign of the atmospheric mass splitting:

The mixing matrix is roughly the following:

$$U_{MNS} \sim \begin{pmatrix} \frac{1}{\sqrt{2}} & \frac{1}{\sqrt{2}} & s_{13} \\ -\frac{1}{2}(e^{i\delta} + s_{13}) & \frac{1}{2}(e^{i\delta} - s_{13}) & \frac{1}{\sqrt{2}} \\ \frac{1}{2}(e^{i\delta} - s_{13}) & -\frac{1}{2}(e^{i\delta} + s_{13}) & \frac{1}{\sqrt{2}} \end{pmatrix} \begin{pmatrix} 1 & 0 & 0 \\ 0 & e^{i\alpha_1} & 0 \\ 0 & 0 & e^{i\alpha_2} \end{pmatrix}, \tag{13}$$

for the choice $\Delta m^2_{12} = \Delta m^2_{solar}$, $\Delta m^2_{13} = \Delta m^2_{atmos}$. $\theta_{13} < 10°$, while all the CP-violating phases: $\delta, \alpha_1, \alpha_2$ are unconstrained. At present we do not know if there is CP violation in the lepton sector of the SM. Another important feature to notice is the large difference between this matrix and the one in the quark sector, which is close to the identity matrix:

$$V_{CKM} \simeq \begin{pmatrix} 1 & O(\lambda) & O(\lambda^3) \\ O(\lambda) & 1 & O(\lambda^2) \\ O(\lambda^3) & O(\lambda^2) & 1 \end{pmatrix}, \tag{14}$$

where $\lambda \sim 0.2$. This is another indication that the dynamics of the mass generation in the lepton sector might well differ from that in the quark sector.

7 Completing the Picture

After the next generation of neutrino experiments we will probably remain far from having the complete picture. Eventually, we would like to answer the following questions:

- Fill in the unknowns of the mixing matrix: θ_{13} and the phases
- Stablish if there is leptonic CP-violation: e.g. $\delta \neq 0, \pi$
- Find the correct spectrum: $\Delta m^2_{atmos} > $ or < 0
- Establish the Majorana nature of neutrinos
- Determine the absolute neutrino mass scale

In order to address the first three questions, we need more precise neutrino oscillation experiments in the atmospheric range:

$$\langle E_\nu \rangle / L \sim |\Delta m^2_{atmos}| \tag{15}$$

which can measure the subleading transition: $\nu_e \leftrightarrow \nu_{\mu(\tau)}$ and its CP-conjugated $\bar{\nu}_e \leftrightarrow \bar{\nu}_{\mu(\tau)}$.

The last two problems on the other hand require other techniques, since neutrino oscillations are not sensitive, neither to the absolute mass scale nor to the "majorananess" of the neutrinos. One possibility is the direct measurement of the neutrino mass by studying the end-point spectrum of Tritium β-decay [20]. These experiments have reached a sensitivity to $m_{ee} \equiv \sum_i m_i |U^2_{ei}|$ of 2.2 eV at 95% CL. There is a proposal to reduce this to ~ 0.3 eV. Another possibility is the search for rare processes, such as neutrinoless double β decay. This rare process can only occur if neutrinos are Majorana and the amplitude is proportional to the combination:

$$\langle m \rangle = \left| \sum_\alpha U^2_{e\alpha} m_\alpha \right|. \tag{16}$$

The present limit on this quantity is $0.34 eV$ [21], but there are proposals to arrive at a limit of $\sim 0.01 eV$. This would be already in an interesting range, even if the spectrum is hierarchical, since the heaviest neutrino mass eigenstate should be smaller than $\sqrt{\Delta m^2_{atmos}} \sim 0.05$ eV.

Coming back to the possibility of answering the first three questions, in the last few years there have been several interesting proposals for more precise neutrino oscillation experiments. The challenge is to measure small oscillation probabilities for the first time, such as those mediated by the small angle θ_{13}. Systematic errors in conventional accelerator neutrino beams are too large to achieve this goal satisfactorily. Not only more intense but also purer sources are required. There are two ideas in the market:

- Neutrino factory: ν-beams from accelerated and stored muon decays

$$\mu^- \to e^- \; \nu_\mu \; \bar{\nu}_e \, ;$$
$$\bar{\nu}_e \to \bar{\nu}_\mu \to \mu^+$$
$$\nu_\mu \to \nu_\mu \to \mu^-$$

Muons decay in the same number of ν_μ as $\bar{\nu}_e$. Provided the neutrino detector is capable of measuring the muon charge, the signal of the transition $\bar{\nu}_e \to \bar{\nu}_\mu$ is the appearance of a wrong-sign muon [22]. There is no intrinsic background. Furthermore the total flux and spectrum of this beam at some far away location can be known to be at the per mil level. Intensities up to 10^{21} muon decay per year are foreseeable and the machine can be run in both polarities, so that the CP-conjugated transitions can be measured sequentially. Figure 7 shows the possible layout of such a facility in the CERN site. 2GeV protons impinge on a target producing kaons and pions, which decay into muons. The latter are cooled, accelerated to some reference energy and stored in a ring with a triangular or bow tie shape, such that muons which decay most of the time in the long straight sections, produce neutrinos pointing into two convenient far away locations.

Studies have shown that the optimal location to study CP violation is in the range of a few thousand kilometers [23]. Indeed there are some interesting underground locations in Europe at the right distance from CERN, such as the Canary Islands!

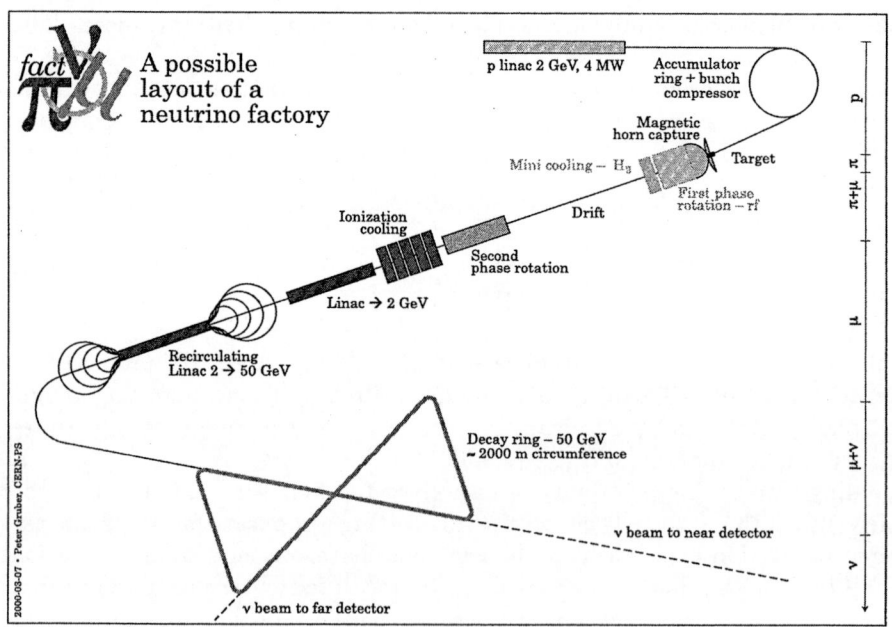

Fig. 7. Possible layout of a CERN-based Neutrino Factory

- β-beams: ν-beams from boosted heavy ion decays [24]

$$^6He^{++} \to\, ^6_3 Li^{+++}\ e^- \qquad\qquad \bar{\nu}_e$$

$$\bar{\nu}_e \to \bar{\nu}_\mu \to \mu^+ \tag{17}$$

In this case, the interesting channel is also the appearance of muons, but the charge does not need to be measured, because there is absolutely no intrinsic background, not even with the opposite charge. Although the foreseeable intensities of this machine are not yet comparable to those of the neutrino factory, sensitivity to CP-violation is achievable if the angle θ_{13} is above a few degrees.

8   Conclusions

An ingenious experimental effort is being devoted to prove neutrino oscillations beyond possible doubt. Recently two smoking guns for neutrino oscillations have indeed been demonstrated: the up-down asymmetry in the atmospheric neutrino flux and the presence of ν_μ and/or ν_τ in the solar flux. These signals are starting to be confirmed by "man-made" neutrino sources.

ν masses almost certainly imply a new physics scale beyond the Standard Model of particle physics, which might actually be related to a GUT scale. The determination of all the parameters in the neutrino mass matrices will be very valuable information to understand the dynamics of this new physics scale. This challenging task requires however more precise neutrino experiments, for which new ideas are not lacking.

Even if neutrinos are no longer good candidates for dark matter, they are for visible matter, since they can explain the observed baryon asymmetry in the Universe!

References

1. F. Gianotti, these proceedings.
2. M. Gell-Mann, P. Ramond and R. Slansky, in *Supergravity*, edited by P. Nieuwenhuizen and D. Freedman, North-Holland 1979, pag. 315. T. Yanagida, 1979, in *Proceedings of Workshop on Unified Theory and Baryon Number in the Universe*, edited by O. Sawada and A. Sugamoto (KEK); R.N. Mohapatra, and G. Senjanovic, 1980, Phys. Rev. Lett. **44**, 912.
3. M. Fukugita and T. Yanagida, Phys. Lett. **B174** (1986) 45. For an introductory review on the subject see W. Buchmuller, hep-ph/0204288.
4. J. Ellis, these proceedings.
5. Kobayashi, M., and T. Maskawa,, Prog. Theor. Phys. **49**, 652 (1973).
6. Z. Maki, M. Nakagawa, and S. Sakata, Prog. Theor. Phys. **28**, 870 (1962).
7. A. Pich, these proceedings.
8. B. Pontecorvo, J. Exptl. Theoret. Phys. **33**, 549 (1957).

9. B.T. Cleveland *et al.*, Astrophys. J. **496**, 505 (1998). GALLEX Coll., W. Hampel *et al.*, Phys. Lett. **B447**, 127 (1999). GNO Coll., M. Altmann *et al.*, Phys. Lett.**B490**, 16 (2000). SAGE Coll., J.N. Abdurashitov *et al.*, J. Exp. Theor. Phys. **95**, 181 (2002). Kamiokande Coll., Y. Fukuda *et al.*, Phys. Rev. Lett. **77**, 1683 (1996). Super-Kamiokande Coll., S. Fukuda *et al.*, Phys. Lett. B **539**, 179 (2002).
10. SNO Coll., Q.R. Ahmad *et al.* Phys. Rev. Lett. **87** (2001) 071301 and **89** (2002) 011301; ibid 011302.
11. J.N. Bahcall, H.M. Pinsonneault, and S. Basu, Astrophys. J. **555**, 990 (2001).
12. J.N. Bahcall, M.C. Gonzalez-Garcia, C. Peña-Garay, JHEP **0207**, 054 (2002).
13. KamLAND Coll., K. Eguchi *et al*, hep-ex/0212021.
14. IMB Coll., R. Becker-Szendy, *et al.*, Phys. Rev. D **46**, 3720 (1992). Kamiokande Coll., Y. Fukuda, *et al.*, Phys. Lett. B **335**, 237 (1994); Soudan Coll., W.W.M. Allison, *et al.*, Phys. Lett. B **449**, 137 (1999); MACRO Coll., M. Ambrosio, *et al.*, Phys. Lett. B **517**, 59 (2001).
15. SuperKamiokande Coll., S. Fukuda *et al*, Phys. Rev. Lett. **81** (1998) 1562.
16. Chooz Coll., M. Apollonio *et al.*, Phys. Lett. **B420**, 397 (1998).
17. K2K Coll., S.H. Ahn, *et al.*, Phys. Lett. B **511**, 178 (2001).
18. LSND Coll., C. Athanassopoulos, *et al.*, Phys. Rev. Lett. **81**, 1774 (1998) 1774.
19. KARMEN Coll., M. Steidl, Nucl. Phys. Proc. Suppl. **110** (2002) 417.
20. J. Bonn, *et al.*, Nucl. Phys. Proc. Suppl. **91**, 273 (2001); V.M. Lobashev, *et al.*, Nucl. Phys. Proc. Suppl. **91**, 280 (2001).
21. L. Baudis *et al*, Phys. Rev. Lett. **83** (1999) 41.
22. S. Geer, Phys. Rev. **D57** 6989 (1998). A. De Rújula, M. B. Gavela and P. Hernández, Nucl. Phys. B **547**, 21 (1999).
23. A. Cervera *et al*, Nucl. Phys. **B579** (2000) 17; Erratum,ibid. **B593** (2001) 731.
24. P. Zucchelli, hep-ex/0107006.

The Standard Model of Particle Physics: Status and Low-Energy Tests

Antonio Pich

IFIC, Universitat de València – CSIC, Apt. 22085, E-46071 València, Spain

Abstract. Precision measurements of low-energy observables provide stringent tests of the Standard Model structure and accurate determinations of its parameters. An overview of the present experimental status is presented. The main topics discussed are the muon anomalous magnetic moment, the asymptotic freedom of strong interactions, the lepton universality of gauge couplings, the quark flavour structure and CP violation.

1 Standard Model Structure

The Standard Model (SM) [1,2] is a gauge theory, based on the group $SU(3)_C \otimes SU(2)_L \otimes U(1)_Y$, which describes strong, weak and electromagnetic interactions, via the exchange of the corresponding spin-1 gauge fields: 8 massless gluons and 1 massless photon for the strong and electromagnetic forces, respectively, and 3 massive bosons, $W^\pm$ and Z, for the weak interaction. The gauge symmetry determines the dynamics in terms of the three couplings g_s, g and g', associated with the $SU(3)_C$, $SU(2)_L$ and $U(1)_Y$ subgroups. Strong interactions are governed by the first group factor, while the other two provide a unified description of the electroweak forces, their gauge parameters being related through $g \sin\theta_W = g' \cos\theta_W = e$.

The fermionic matter content is given by the known leptons and quarks, which are organized in a 3-fold family structure:

$$\begin{bmatrix} \nu_e & u \\ e^- & d' \end{bmatrix} , \qquad \begin{bmatrix} \nu_\mu & c \\ \mu^- & s' \end{bmatrix} , \qquad \begin{bmatrix} \nu_\tau & t \\ \tau^- & b' \end{bmatrix} , \tag{1}$$

where (each quark appears in 3 different *colours*)

$$\begin{bmatrix} \nu_l & q_u \\ l^- & q_d \end{bmatrix} \equiv \begin{pmatrix} \nu_l \\ l^- \end{pmatrix}_L , \quad \begin{pmatrix} q_u \\ q_d \end{pmatrix}_L , \quad l_R^- , \quad (q_u)_R , \quad (q_d)_R , \tag{2}$$

plus the corresponding antiparticles. Thus, the left-handed fields are $SU(2)_L$ doublets, while their right-handed partners transform as $SU(2)_L$ singlets. The three fermionic families in (1) appear to have identical properties (gauge interactions); they only differ by their mass and their flavour quantum number.

The gauge symmetry is broken by the vacuum, which triggers the Spontaneous Symmetry Breaking (SSB) of the electroweak group to the electromagnetic subgroup:

$$SU(3)_C \otimes SU(2)_L \otimes U(1)_Y \xrightarrow{\text{SSB}} SU(3)_C \otimes U(1)_{QED} . \tag{3}$$

The SSB mechanism generates the masses of the weak gauge bosons, and gives rise to the appearance of a physical scalar particle, the so-called *Higgs*. The fermion masses and mixings are also generated through the SSB mechanism.

The SM constitutes one of the most successful achievements in modern physics. It provides a very elegant theoretical framework, which is able to describe all known experimental facts in particle physics. A detailed description of the SM and its impressive phenomenological success can be found in [3,4].

2   Quantum Corrections

The high accuracy achieved by the most recent experiments allows to make stringent tests of the SM structure at the level of quantum corrections. The following discussion concentrates on Quantum Electrodynamics (QED) and Quantum Chromodynamics (QCD). Electroweak effects are covered in [5].

2.1   Running Couplings

Let us consider the electromagnetic interaction between two electrons. At lowest order, the scattering amplitude $T(q^2) \sim \alpha/q^2$ with $\alpha = e^2/(4\pi)$. The leading quantum corrections are generated by the photon self-energy contribution:

$$T(Q^2) \sim \frac{\alpha}{Q^2} \left\{ 1 - \Pi(Q^2) + \Pi(Q^2)^2 + \cdots \right\} = \frac{\alpha}{Q^2} \frac{1}{1 + \Pi(Q^2)} \sim \frac{\alpha(Q^2)}{Q^2} .$$

This defines an effective *running* coupling,

$$\alpha(Q^2) = \frac{\alpha(Q_0^2)}{1 - \frac{\beta_1}{2\pi} \alpha(Q_0^2) \ln(Q^2/Q_0^2)} , \tag{4}$$

where $Q^2 \equiv -q^2$ and $\alpha(m_e^2) = \alpha$. The e^+e^- loop induces a logarithmic correction with $\beta_1 = 2/3 > 0$. Therefore, the effective QED running coupling increases with the energy scale: $\alpha(Q^2) > \alpha(Q_0^2)$ if $Q^2 > Q_0^2$, i.e., the electromagnetic charge decreases at large distances. This can be intuitively understood as the screening due to virtual e^+e^- pairs generated, through quantum effects, around the electron charge. The physical QED vacuum behaves as a polarized dielectric

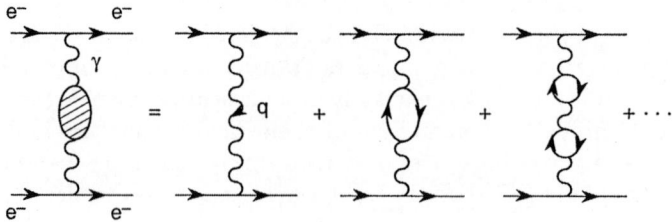

Fig. 1. Photon self-energy contribution to e^-e^- scattering

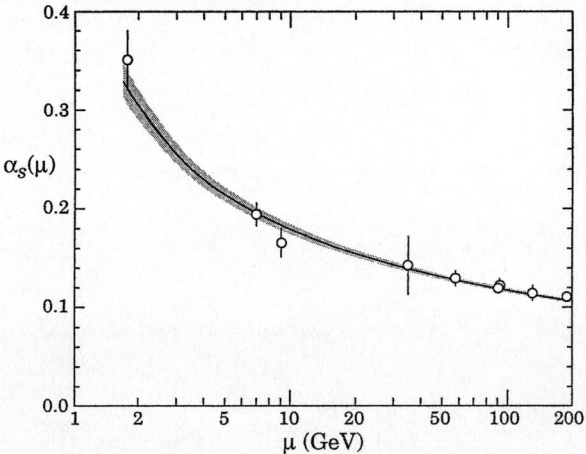

Fig. 2. Energy dependence [10] of the strong coupling α_s

medium. The huge difference between the electron and Z mass scales makes this quantum correction relevant at LEP energies [6,7]:

$$\alpha(m_e^2)^{-1} \; = \; 137.03599976\,(50) \; > \; \alpha(M_Z^2)^{-1} \; = \; 128.95 \pm 0.05 \,. \qquad (5)$$

The strong interaction between two quarks can be analyzed in a similar way. Owing to the non-abelian character of the $SU(3)_C$ group, QCD leads to cubic and quartic self-interactions among gluons. This results in a strong running coupling $\alpha_s(Q^2)$ with the same Q^2 dependence (4), but with a negative β_1 [8]:

$$\beta_1 \; = \; \frac{2\,N_f - 11\,N_C}{6} \; < \; 0 \,. \qquad (6)$$

The contribution proportional to the number of quark flavours N_f is generated by the q-$\bar{q}$ loop corrections to the gluon self-energy. The gluonic self-interactions introduce the additional negative term proportional to the number of quark colours N_C. Since $\beta_1 < 0$, $\alpha_s(Q^2)$ decreases at short distances. Thus, QCD has the required property of *asymptotic freedom*: quarks behave as free particles when $Q^2 \to \infty$. The predicted running of α_s, known to four loops [9], agrees very well with the experimental determinations at different energies. Normalizing all measurements at the Z mass scale, the present world average is [10]:

$$\alpha_s(M_Z^2) \; = \; 0.118 \pm 0.002 \,. \qquad (7)$$

2.2   Lepton Anomalous Magnetic Moments

The most stringent QED test [11,12] comes from the high-precision measurements of the e [10] and μ [13] anomalous magnetic moments $a_l^\gamma \equiv (g_l - 2)/2$:

$$a_e^\gamma = \begin{cases} (115\,965\,215.35 \pm 2.40) \times 10^{-11} & \text{(Theory)} \\ (115\,965\,218.69 \pm 0.41) \times 10^{-11} & \text{(Experiment)} \end{cases}, \qquad (8a)$$

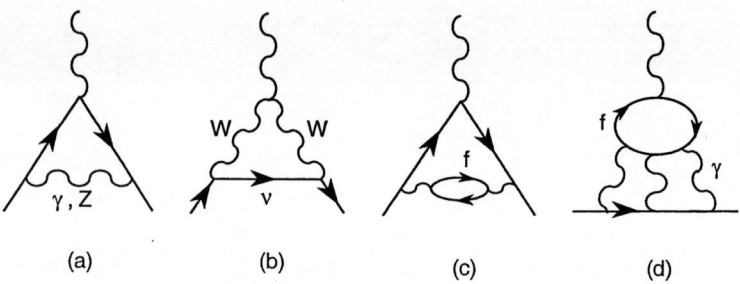

Fig. 3. Some Feynman diagrams contributing to a_l^γ

$$a_\mu^\gamma = \begin{cases} (1\,165\,917.9 \pm 1.0) \times 10^{-9} & \text{(Theory)} \\ (1\,165\,920.3 \pm 1.5) \times 10^{-9} & \text{(Experiment)} \end{cases} . \tag{8b}$$

The impressive agreement between theory and experiment promotes QED to the level of the best theory ever build by the human mind to describe nature.

To a measurable level, a_e^γ arises entirely from virtual electrons and photons; these contributions are known [11] to $O(\alpha^4)$. The theoretical error is dominated by the uncertainty in the input value of the QED coupling α. Turning things around, a_e^γ provides the most precise determination of the fine structure constant.

The anomalous magnetic moment of the muon is sensitive to virtual contributions from heavier states; compared to a_e^γ, they scale as m_μ^2/m_e^2. The main theoretical uncertainty on a_μ^γ has a QCD origin. Since quarks have electric charge, virtual quark-antiquark pairs induce *hadronic vacuum polarization* corrections to the photon propagator (Fig. 3.c). Owing to the non-perturbative character of QCD at low energies, the light-quark contribution cannot be reliably calculated at present; fortunately, this effect can be extracted from the measurement of the cross-section $\sigma(e^+e^- \to \text{hadrons})$ and from the invariant-mass distribution of the final hadrons in τ decays [6]. Additional QCD uncertainties stem from the smaller *light-by-light scattering* contributions (Fig. 3.d); a recent reevaluation of these corrections [14] has detected a sign mistake in previous calculations [15], improving the agreement with the experimental measurement [13].

The Brookhaven E821 experiment [13] is expected to push its sensitivity to at least 4×10^{-10}, and thereby observe the contributions from virtual $W^\pm$ and Z bosons [12,16]. This would require a better control of the QCD corrections.

3   Lepton Universality

In the SM all lepton doublets have identical couplings to the W boson:

$$\mathcal{L} = \frac{g}{2\sqrt{2}} W_\mu^\dagger \sum_l \bar{\nu}_l \gamma^\mu (1 - \gamma_5) l + \text{h.c.} \qquad (l = e, \mu, \tau) . \tag{9}$$

Comparing the measured decay widths of leptonic or semileptonic decays which

Table 1. Experimental determinations of the ratios $g_l/g_{l'}$

	$\Gamma_{\tau \to \nu_\tau \mu \bar\nu_\mu}/_{\nu_\tau e \bar\nu_e}$	$\Gamma_{\pi \to \mu \bar\nu_\mu}/_{e \bar\nu_e}$	$\Gamma_{W \to \mu \bar\nu_\mu}/_{e \bar\nu_e}$
$|g_\mu/g_e| =$	1.0006 ± 0.0021	1.0017 ± 0.0015	1.000 ± 0.011

	$\Gamma_{\tau \to \nu_\tau e \bar\nu_e}/\Gamma_{\mu \to \nu_\mu e \bar\nu_e}$	$\Gamma_{\tau \to \nu_\tau \pi}/\Gamma_{\pi \to \mu \bar\nu_\mu}$	$\Gamma_{\tau \to \nu_\tau K}/\Gamma_{K \to \mu \bar\nu_\mu}$	$\Gamma_{W \to \tau \bar\nu_\tau}/\mu \bar\nu_\mu$
$|g_\tau/g_\mu| =$	0.9995 ± 0.0023	1.005 ± 0.007	0.977 ± 0.016	1.026 ± 0.014

	$\Gamma_{\tau \to \nu_\tau \mu \bar\nu_\mu}/\Gamma_{\mu \to \nu_\mu e \bar\nu_e}$	$\Gamma_{W \to \tau \bar\nu_\tau}/_{e \bar\nu_e}$
$|g_\tau/g_e| =$	1.0001 ± 0.0023	1.026 ± 0.014

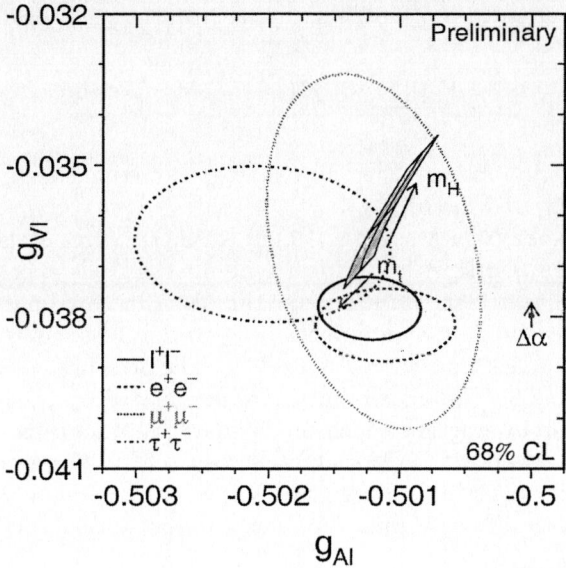

Fig. 4. Contours of 68% probability in the a_l-v_l plane from LEP and SLD measurements [17]. The solid contour assumes lepton universality. The shaded region corresponds to the SM prediction for $m_t = 174.3 \pm 5.1$ GeV and $m_H = 300 \,^{+700}_{-186}$ GeV

only differ by the lepton flavour, one can test experimentally that the W interaction is indeed the same, i.e. that $g_e = g_\mu = g_\tau \equiv g$. As shown in Table 1, the present data [4,10,17] verify the universality of the leptonic charged-current couplings to the 0.2% level.

The interactions of the neutral Z boson are diagonal in flavour. Moreover, all fermions with equal electric charge have identical axial-vector, $a_f = T_3^f = \pm 1/2$, and vector, $v_f = T_3^f \left(1 - 4 |Q_f| \sin^2 \theta_W\right)$, couplings to the Z. This has been accurately tested at LEP and SLD through a precise analysis of $e^+e^- \to \gamma, Z \to f\bar{f}$ data. Figure 4 shows the 68% probability contours in the a_l-v_l plane, obtained from leptonic observables [17]. The universality of the leptonic Z couplings is now verified to the 0.15% level for a_l, while only a few per cent precision has been achieved for v_l due to the smallness of the leptonic vector coupling. The measured leptonic asymmetries provide an accurate determination of the electroweak mixing angle [17]:

$$\sin^2 \theta_W = 0.23113 \pm 0.00021 \,. \tag{10}$$

4   Flavour Mixing

In the SM, all mass scales are generated through the Higgs mechanism. After the SSB, the Yukawa couplings to the Higgs scalar doublet give rise to non-diagonal fermionic mass terms. The mass eigenstates are then different from the weak eigenstates, which leads to flavour mixing in the charged-current interaction:

$$\mathcal{L} = \frac{g}{2\sqrt{2}} W_\mu^\dagger \sum_{ij} \bar{u}_i \gamma^\mu (1 - \gamma_5) \mathbf{V}_{ij} d_j + \text{h.c.} \,. \tag{11}$$

With non-zero neutrino masses, there are analogous mixing effects in the lepton sector, which are covered in [18].

The Cabibbo-Kobayashi-Maskawa [19,20] (CKM) matrix $\mathbf{V}$ is unitary and couples any up-type quark with all down-type quarks. It is a priori unknown, because the gauge symmetry does not fix the Yukawa couplings. The matrix element $\mathbf{V}_{ij}$ can be obtained experimentally from semileptonic weak processes associated with the quark transition $d_j \to u_i l^- \bar{\nu}_l$. The present determinations are summarized in Table 2. The uncertainties are dominated by theoretical errors, related to the strong interaction which binds quarks into hadrons.

The most precisely known CKM matrix element is $\mathbf{V}_{ud}$. The weighted average of the two determinations in Table 2 gives $\mathbf{V}_{ud} = 0.9738 \pm 0.0008$. Taking for $\mathbf{V}_{us}$ the more reliable K_{e3} determination, one obtains

$$|\mathbf{V}_{ud}|^2 + |\mathbf{V}_{us}|^2 + |\mathbf{V}_{ub}|^2 = 0.9965 \pm 0.0019 \,. \tag{12}$$

The unitarity of $\mathbf{V}_{ij}$ appears to be slightly violated by 1.8σ. At this level of precision, a small underestimate of some uncertainties seems plausible. A less accurate unitarity test is provided by the hadronic width of the W boson [17]:

$$\sum_{j=d,s,b} \left(|\mathbf{V}_{uj}|^2 + |\mathbf{V}_{cj}|^2 \right) = 2.039 \pm 0.025 \,. \tag{13}$$

The CKM matrix shows a hierarchical pattern, with its diagonal elements being very close to one, the ones connecting the two first generations having a size $\lambda \equiv |\mathbf{V}_{us}|$, the mixing between the second and third families being of order

Table 2. Direct V_{ij} determinations.

CKM entry	Value	Source
$|V_{ud}|$	0.9740 ± 0.0010	Nuclear β decay [10]
	0.9733 ± 0.0015	$n \to p\,e^-\bar{\nu}_e$ [10]
$|V_{us}|$	0.2196 ± 0.0023	K_{e3} [10]
	0.2176 ± 0.0026	Hyperon decays [10]
$|V_{cd}|$	0.224 ± 0.016	$\nu d \to c\,X$ [10]
$|V_{cs}|$	1.04 ± 0.16	$D \to \bar{K}\,e^+\nu_e$ [10]
$|V_{cb}|$	0.0421 ± 0.0022	$B \to D^*l\bar{\nu}_l$ [21]
	0.0404 ± 0.0011	$b \to c\,l\,\bar{\nu}_l$ [21]
$|V_{ub}|$	0.0033 ± 0.0006	$B \to \rho\,l\,\bar{\nu}_l$ [22]
	0.0041 ± 0.0006	$b \to u\,l\,\bar{\nu}_l$ [23,24]
$|V_{tb}|/\sqrt{\sum_q |V_{tq}|^2}$	$0.97^{+0.16}_{-0.12}$	$t \to b\,W/q\,W$ [25]

λ^2, and the mixing between the first and third quark flavours having a much smaller size of about λ^3. It is convenient to use the parameterization [26]:

$$\mathbf{V} = \begin{bmatrix} 1 - \lambda^2/2 & \lambda & A\lambda^3(\varrho - i\eta) \\ -\lambda & 1 - \lambda^2/2 & A\lambda^2 \\ A\lambda^3(1 - \varrho - i\eta) & -A\lambda^2 & 1 \end{bmatrix} + O(\lambda^4) . \qquad (14)$$

Imposing the unitarity constraint, the CKM determinations in Table 2 imply

$$\lambda = 0.223 \pm 0.003 , \qquad A = 0.83 \pm 0.04 , \qquad \sqrt{\varrho^2 + \eta^2} = 0.40 \pm 0.07 . \quad (15)$$

4.1 B^0-$\bar{B}^0$ Mixing

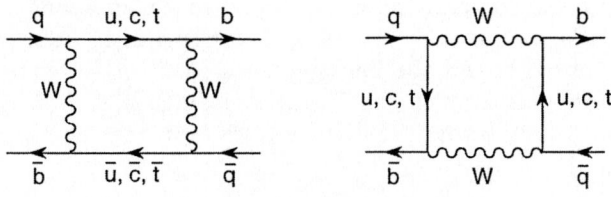

Fig. 5. B^0-$\bar{B}^0$ mixing diagrams

Additional information on the CKM parameters is obtained from flavour-changing neutral-current transitions, occurring at the 1-loop level. An important example is provided by the mixing between the B^0 meson and its antiparticle:

$$\langle \bar{B}_d^0 | \mathcal{H}_{\Delta B=2} | B_0 \rangle \sim \left\{ \sum_{ij=u,c,t} V_{id} V_{ib}^* V_{jd}^* V_{jb} \, S(r_i, r_j) \right\} \left(\frac{2}{3} M_B^2 \xi_B^2 \right) , \qquad (16)$$

with $S(r_i, r_j)$ a loop function of $r_i \equiv m_i^2 / M_W^2$. Owing to the unitarity of the CKM matrix, the mixing amplitude vanishes for equal (up-type) quark masses. Thus the effect is proportional to the mass splittings between the u, c and t quarks. Since all CKM factors have a similar size, $V_{ud} V_{ub}^* \sim V_{cd} V_{cb}^* \sim V_{td} V_{tb}^* \sim A\lambda^3$, the top contribution dominates completely. This transition can then be used to perform an indirect determination of $|V_{td}|$. The main uncertainty stems from the hadronic matrix element of the four-quark operator generated by the box diagrams in Fig. 5, which is characterized through the non-perturbative parameter $\xi_B \equiv \sqrt{2\hat{B}_B} f_B = 230 \pm 45$ MeV [27,28]. The measured mixing between the B_d^0 and $\bar{B}_d^0$ mesons, $\Delta M_{B_d^0} = 0.496 \pm 0.007$ ps^{-1} [23], implies:

$$|V_{td}| = 0.0077 \pm 0.0011 , \qquad \sqrt{(1-\varrho)^2 + \eta^2} \approx \left| \frac{V_{td}}{\lambda V_{cb}} \right| = 0.84 \pm 0.12 . \quad (17)$$

A similar analysis can be applied to the B_s^0-$\bar{B}_s^0$ mixing. The non-perturbative uncertainties are reduced to the level of $SU(3)$ breaking through the ratio

$$\frac{\Delta M_{B_s^0}}{\Delta M_{B_d^0}} \approx \frac{M_{B_s^0} \, \xi_{B_s^0}^2}{M_{B_d^0} \, \xi_{B_d^0}^2} \left| \frac{V_{ts}}{V_{td}} \right|^2 \equiv \Omega^2 \left| \frac{V_{ts}}{V_{td}} \right|^2 . \qquad (18)$$

Taking $\Omega \approx 1.15 \pm 0.08$ [27,28], the experimental bound $\Delta M_{B_s^0} > 14.9$ ps^{-1} (95% CL) [23] implies

$$\left| \frac{V_{ts}}{V_{td}} \right| \approx \frac{1}{\lambda \sqrt{(1-\varrho)^2 + \eta^2}} > 4.2 . \qquad (19)$$

5   CP Violation

With N_G fermion generations, the matrix $\mathbf{V}$ is characterized by $N_G(N_G - 1)/2$ moduli and $(N_G-1)(N_G-2)/2$ phases. In the simpler case of two fermion families $\mathbf{V}$ is determined by a single parameter, the so-called Cabibbo angle [19], while for $N_G = 3$ the CKM matrix is described by 3 angles and 1 phase [20]. This is the only complex phase in the SM Lagrangian; thus, it is a unique source for violations of the CP symmetry. It was for this reason that the third generation was assumed to exist [20], before the discovery of the τ and the b.

5.1   Kaon Physics

For many years, the only experimental evidence of CP-violation phenomena came from the kaon system. The ratios,

$$\frac{A(K_L \to \pi^+ \pi^-)}{A(K_S \to \pi^+ \pi^-)} \approx \varepsilon_K + \varepsilon_K' , \qquad \frac{A(K_L \to \pi^0 \pi^0)}{A(K_S \to \pi^0 \pi^0)} \approx \varepsilon_K - 2\varepsilon_K' , \qquad (20)$$

involve final 2π states which are even under CP. Therefore, they measure a CP-violating amplitude which can originate either from a small CP-even admixture in the initial K_L state (indirect CP violation), parameterized by ε_K, or from direct CP violation in the decay amplitude. This latter effect, parameterized by ε'_K, requires the interference between the two $K \to 2\pi$ isospin $(I = 0, 2)$ amplitudes, with different weak and strong phases.

The parameter ε_K is well determined [10]:

$$\varepsilon_K = (2.271 \pm 0.017) \times 10^{-3} \ e^{i\phi(\varepsilon_K)} \,, \qquad \phi(\varepsilon_K) = 43.5° \pm 0.5° \,. \tag{21}$$

ε_K has been also measured [10,29] through the CP asymmetry between the two $K_L \to \pi_{\text{pl}}^m l^{\pm} \overset{(-)}{\nu_l}$ decay widths, which implies $\mathrm{Re}\,(\varepsilon_K) = (1.654 \pm 0.032) \times 10^{-3}$, in good agreement with (21).

The value of ε'_K has been established very recently. The present experimental world average [30–33],

$$\mathrm{Re}\,(\varepsilon'_K / \varepsilon_K) = (1.72 \pm 0.18) \times 10^{-3} \,, \tag{22}$$

provides clear evidence for the existence of direct CP violation.

The CKM mechanism generates CP-violation effects both in the $\Delta S = 2$ K^0-$\bar{K}^0$ transition (box diagrams) and in the $\Delta S = 1$ decay amplitudes (penguin diagrams). The theoretical analysis of K^0-$\bar{K}^0$ mixing is quite similar to the one applied to the B system. This time, however, the charm loop contributions are non-negligible. The main uncertainty stems from the calculation of the hadronic matrix element of the four-quark $\Delta S = 2$ operator, which is usually parameterized through the non-perturbative parameter $\hat{B}_K$.

The experimental value of ε_K specifies a hyperbola in the (ϱ, η) plane. This is shown in Fig. 6 [34], together with the constraints obtained from $|V_{ub}/V_{cb}|$, B_d^0-$\bar{B}_d^0$ mixing and the experimental bound on $\Delta M_{B_s^0}$. This figure assumes [34] $\hat{B}_K = 0.87 \pm 0.06 \pm 0.13$, $\xi_B = 230 \pm 25 \pm 20$ MeV and $\Omega = 1.15 \pm 0.04 \pm 0.05$.

The theoretical estimate of $\varepsilon'_K / \varepsilon_K$ is more involved [35], because several four-quark operators need to be considered in the analysis. Moreover, the strong rescattering of the final pions generates an important enhancement through infrared logarithms [36]. Taking into account all large logarithmic corrections at short and long distances, the SM prediction for ε'/ε is found to be [36]:

$$\mathrm{Re}\,(\varepsilon'/\varepsilon) = \left(1.7 \pm 0.2 {}^{+0.8}_{-0.5} \pm 0.5\right) \times 10^{-3} \,, \tag{23}$$

in excellent agreement with the measured experimental value (22).

5.2 B Physics

The unitarity tests in (12) and (13) involve only the moduli of the CKM parameters, while CP violation has to do with their phases. The most interesting off-diagonal unitarity condition is

$$\mathbf{V}_{ub}^* \mathbf{V}_{ud} + \mathbf{V}_{cb}^* \mathbf{V}_{cd} + \mathbf{V}_{tb}^* \mathbf{V}_{td} = 0 \,, \tag{24}$$

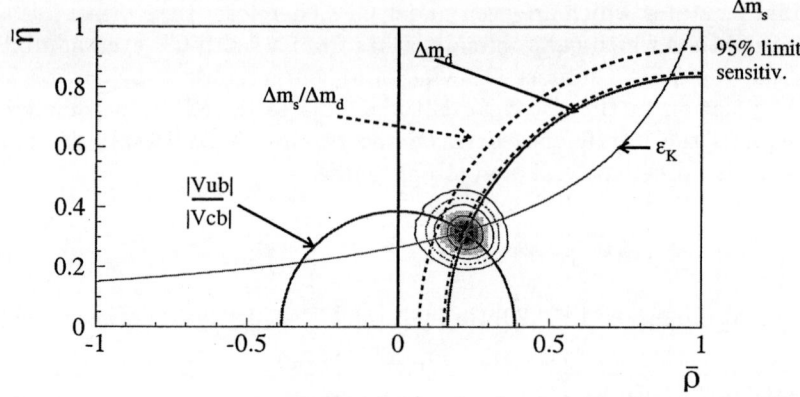

Fig. 6. Constraints on the (ϱ, η) vertex of the unitarity triangle [34]

which has three terms of a similar size. This relation can be visualized by a triangle in the complex plane, which is usually scaled by dividing its sides by $\mathbf{V}_{cb}^* \mathbf{V}_{cd}$. This aligns one side of the triangle along the real axis and makes its length equal to 1; the coordinates of the 3 vertices are then $(0,0)$, $(1,0)$ and (ϱ, η). In the absence of CP violation ($\eta = 0$), this unitarity triangle would degenerate into a segment along the real axis.

The length of the sides and the angles (α, β, γ) of the unitarity triangle can be directly measured. In fact, we have already determined its sides from $\Gamma(b \to u)/\Gamma(b \to c)$ and B_d^0-$\bar{B}_d^0$ mixing, and the position of the (ϱ, η) vertex has been further pinned down in Fig. 6 with ε_K. This gives [34]:

$$\varrho = 0.224 \pm 0.038 , \qquad \eta = 0.317 \pm 0.040 , \qquad \sin 2\beta = 0.698 \pm 0.066 , \quad (25)$$

where $\beta \equiv -\arg(\mathbf{V}_{cd}\mathbf{V}_{cb}^*/\mathbf{V}_{td}\mathbf{V}_{tb}^*)$.

B^0 decays into CP self-conjugate final states provide independent ways to determine the angles [37]. The B^0 (or $\bar{B}^0$) can decay directly to the given final state f, or do it after the meson has been changed to its antiparticle via the mixing process. CP-violating effects can then result from the interference of these two contributions. The time-dependent CP-violating rate asymmetries contain direct information on the CKM parameters. The gold-plated decay mode is $B_d^0 \to J/\psi K_S$, which gives a clean measurement of β [38] without strong-interaction uncertainties, in good agreement with (25):

$$\sin 2\beta = 0.80 \pm 0.10 . \tag{26}$$

Additional tests of the CKM matrix are underway. The B factories should accomplish an approximate determination of $\alpha \equiv -\arg(\mathbf{V}_{td}\mathbf{V}_{tb}^*/\mathbf{V}_{ud}\mathbf{V}_{ub}^*)$, from $B_d^0 \to \pi^+\pi^-$, and many other interesting studies with B decays. Complementary and very valuable information could be also obtained from the kaon decay modes $K^{\pm} \to \pi^{\pm}\nu\bar{\nu}$, $K_L \to \pi^0\nu\bar{\nu}$ and $K_L \to \pi^0 e^+ e^-$ [3,35].

6   Summary

The SM provides a beautiful theoretical framework which is able to accommodate all our present knowledge on electroweak and strong interactions. It is able to explain any single experimental fact and, in some cases, it has successfully passed very precise tests at the 0.1% to 1% level [5]. In spite of this impressive phenomenological success, the SM leaves too many unanswered questions to be considered as a complete description of the fundamental forces. We do not understand yet why fermions are replicated in three (and only three) nearly identical copies? Why the pattern of masses and mixings is what it is? Are the masses the only difference among the three families? What is the origin of the SM flavour structure? Which dynamics is responsible for the observed CP violation?

The fermionic flavour is the main source of arbitrary free parameters in the SM. The problem of fermion-mass generation is deeply related with the mechanism responsible for the electroweak SSB. Thus, the origin of these parameters lies in the most obscure part of the SM Lagrangian: the scalar sector. Clearly, the dynamics of flavour appears to be "terra incognita" which deserves a careful investigation.

The SM incorporates a mechanism to generate CP violation, through the single phase naturally occurring in the CKM matrix. Although the present laboratory experiments are well described, this mechanism is unable to explain the matter-antimatter asymmetry of our universe. A fundamental explanation of the origin of CP-violating phenomena is lacking.

Many interesting experimental signals are expected to be seen in the near future. Large surprises may well be discovered, probably giving the first hints of new physics and offering clues to the problems of mass generation, fermion mixing and family replication.

Acknowledgements

This work has been supported by MCYT, Spain (Grant FPA-2001-3031) and by the EU TMR network EURODAPHNE (Contract ERBFMX-CT98-0169).

References

1. S.L. Glashow: Nucl. Phys. **22**, 579 (1961). S. Weinberg: Phys. Rev. Lett. **19**, 1264 (1967). A. Salam: in *Elementary Particle Theory*, ed. by N. Svartholm (Almquist and Wiksells, Stockholm 1969) p. 367. S.L. Glashow, J. Iliopoulos, L. Maiani: Phys. Rev. D **2**, 1285 (1970)
2. H. Fritzsch, M. Gell-Mann, H. Leutwyler: Phys. Lett. B **47**, 365 (1973)
3. A. Pich: 'The Standard Model of Electroweak Interactions', hep-ph/9412274; 'Flavourdynamics', hep-ph/9601202; 'Rare Kaon Decays', hep-ph/9610243; 'Aspects of Quantum Chromodynamics', hep-ph/0001118
4. A. Pich: Nucl. Phys. B (Proc. Suppl.) **66**, 456 (1998); **81**, 183 (2000); **98**, 385 (2001)
5. F. Gianotti: these proceedings

6. F. Jegerlehner: hep-ph/0104304. J.F. de Trocóniz, F.J. Yndúrain: Phys. Rev. D **65**, 093001, 093002 (2002). S. Narison: Phys. Lett. B **513**, 53 (2001); **526**, 414 (2002). A. Pich, J. Portolés: Phys. Rev. D **63**, 093005 (2001). M. Davier, A. Höcker: Phys. Lett. B **435**, 427 (1998). R. Alemany et al.: Eur. Phys. J. C **2**, 123 (1998)

7. H. Burkhardt, B. Pietrzyk: Phys. Lett. B **513**, 46 (2001); Eur. Phys. J. C **19**, 681 (2001). A.D. Martin et al.: Phys. Lett. B **492**, 69 (2000)

8. D.J. Gross, F. Wilczek: Phys. Rev. Lett. **30**, 1343 (1973). H.D. Politzer: Phys. Rev. Lett. **30**, 1346 (1973)

9. T. van Ritbergen et al.: Phys. Lett. B **400**, 379 (1997)

10. Particle Data Group: Eur. Phys. J. C **15**, 1 (2000); http://pdg.lbl.gov/

11. T. Kinoshita: Rep. Prog. Phys. **59**, 1459 (1996)

12. A. Czarnecki, W.J. Marciano: Phys. Rev. D **64**, 013014 (2001). J. Prades: hep-ph/0108192

13. H.N. Brown et al.: Phys. Rev. Lett. **86**, 2227 (2001)

14. M. Knecht, A. Nyffeler: Phys. Rev. D **65**, 073034 (2002). M. Knecht et al.: Phys. Rev. Lett. **88**, 071802 (2002). I. Blokland et al.: Phys. Rev. Lett. **82**, 071803 (2002). J. Bijnens et al.: Nucl. Phys. B **626**, 410 (2002). M. Hayakawa, T. Kinoshita: hep-ph/0112102. M.J. Ramsey-Musolf, M. Wise: hep-ph/0201297.

15. M. Hayakawa, T. Kinoshita: Phys. Rev. D **57**, 465 (1998). J. Bijnens et al.: Nucl. Phys. B **474**, 379 (1996); Phys. Rev. Lett. **75**, 1447, 3781 (1995)

16. M. Knecht et al.: hep-ph/0205102. S. Peris et al: Phys. Lett. B **355**, 523 (1995). A. Czarnecki et al.: Phys. Rev. Lett. **76**, 3267 (1996); Phys. Rev. D **52**, 2619 (1995). T.V. Kukhto et al.: Nucl. Phys. B **371**, 567 (1992)

17. The LEP Collaborations ALEPH, DELPHI, L3, OPAL, the LEP Electroweak Working Group and the SLD Heavy Flavour Group: LEPEWWG/2002-01

18. P. Hernández: these proceedings

19. N. Cabibbo: Phys. Rev. Lett. **10**, 531 (1963)

20. M. Kobayashi and T. Maskawa: Progr. Theor. Phys. **42**, 652 (1973)

21. M. Artuso, E. Barberio: hep-ph/0205163

22. CLEO: Phys. Rev. D **61**, 052001 (2000)

23. LEP Heavy Flavour Steering Group: http://lephfs.web.cern.ch/LEPHFS/

24. CLEO: hep-ex/0202019

25. CDF: Phys. Rev. Lett. **86**, 3233 (2001)

26. L. Wolfenstein: Phys. Rev. Lett. **51**, 1945 (1983)

27. M. Ciuchini et al.: JHEP **0107**, 013 (2001)

28. K. Hagiwara, S. Narison, D. Nomura: hep-ph/0205092. A. Pich, J. Prades: Phys. Lett. B **346**, 342 (1995). A. Pich: Phys. Lett. B **206**, 322 (1988)

29. KTeV: Phys. Rev. Lett. **88**, 181601 (2002)

30. NA48: Eur. Phys. J. C **22**, 231 (2001); Phys. Lett. B **465**, 1999 (335)

31. NA31: Phys. Lett. B **317**, 1993 (233); **206**, 169 (1988)

32. R. Kessler: hep-ex/0110020; KTeV: Phys. Rev. Lett. **83**, 1999 (22)

33. E731: Phys. Rev. Lett. **70**, 1993 (1203)

34. ALEPH, CDF, DELPHI, L3, OPAL, SLD, hep-ex/0112028. F. Parodi, A. Stocchi: CKM Workshop (CERN, February 2002)

35. A.J. Buras: hep-ph/0101336

36. E. Pallante, A. Pich, I. Scimemi: Nucl. Phys. B **617**, 441 (2001). E. Pallante, A. Pich: Nucl. Phys. B **592**, 2000 (294); Phys. Rev. Lett. **84**, 2000 (2568)

37. A.B. Carter, A.I. Sanda: Phys. Rev. Lett. **45**, 952 (1980); Phys. Rev. D **23**, 1567 (1981). I.I. Bigi, A.I. Sanda: Nucl. Phys. B **193**, 85 (1981)

38. BABAR: hep-ex/0201020. BELLE: hep-ex/0202027. CDF: Phys. Rev. D **61**, 072005 (2000). ALEPH: Phys. Lett. B **492**, 259 (2000). OPAL: Eur. Phys. J. C **5**, 379 (1998)

High-Energy Tests of the Standard Model and Beyond

Fabiola Gianotti

CERN, EP Division, 1211 Genève 23, Switzerland

Abstract. Over the past decade, particle physics experiments carried out at high-energy machines, like the CERN LEP collider, have performed very precise tests of the predictions of the Standard Model, the theory which describes the interactions between elementary particles. The goals and strategy of these precise measurements and the major achievements are discussed. Then, the present status of the searches for the Higgs boson and for new particles predicted by theories beyond the Standard Model is discussed briefly. Prospects at future machines are also mentioned.

1 Introduction

The theory which describes the electroweak and strong interactions between elementary particles, the so-called Standard Model (SM), is discussed elsewhere in these Proceedings [1]. This paper summarizes the precise experimental tests of the predictions of this theory performed over the last ten years by high-energy physics experiments. A few examples are presented (Sect. 2) out of an enormous amount of results produced by several machines and experiments.

Experimental searches for the Higgs boson, the only fundamental particle predicted by the SM which has not been observed yet, are also discussed; the present status and future prospects are described in Sect. 3.

Although the above-mentioned precise measurements represent a great success for the SM and have given the theory a wide and undisputable experimental support, there are numerous indications today that the SM is not the ultimate theory of the elementary particles and their interactions. The main motivations for looking for physics beyond the SM are summarized in Sect. 4, and the experimental investigations of a specific scenario, Supersymmetry, are discussed briefly.

2 Precision Measurements of the Standard Model

The discoveries of the $W^{\pm}$ and Z bosons at the CERN $p\bar{p}$ collider in 1982-1983 [2] mark a step of fundamental importance for the Standard Model. These particles, which mediate the weak interactions, had been predicted by the theory [3] about 15 years before their direct observation.

A few years later, in 1989, the LEP $e^{+}e^{-}$ collider at CERN started to operate at a centre-of-mass energy $\sqrt{s} \sim 90$ GeV, which corresponds to the mass of the Z boson, with the aim of performing detailed studies of the SM interactions. At the same time, the Tevatron collider at Fermilab (Chicago) was delivering first $p\bar{p}$ collisions at $\sqrt{s} \sim 2$ TeV. One of the Tevatron main goals was the search for the top quark, the only fermion predicted by the SM which had not been observed yet. Some of the beautiful tests of the SM performed at these machines are discussed below[1].

2.1   Measurements at the Z Peak

The CERN LEP collider is a 27 km ring equipped with four experiments: ALEPH, DELPHI, L3, OPAL. In the period 1989-1995, the so-called LEP1 phase, it has operated at a centre-of-mass energy around 90 GeV. As shown

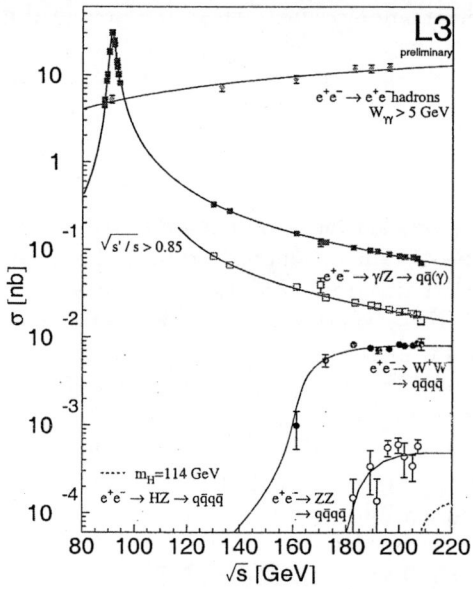

Fig. 1. Production cross-sections of several Standard Model processes as a function of the centre-of-mass energy, as measured by the L3 experiment at LEP (dots and squares). The full lines show the theoretical predictions. The dotted line shows the expected cross-section for a Higgs boson of mass 114 GeV

in Fig. 1, the main process occurring in e^+e^- collisions at these energies is the production of a Z particle, through the diagrams depicted in Fig. 2. The region below the Z resonance has been addressed by machines which operated before LEP. The region above the Z resonance has been studied by LEP in 1996-2000,

---

[1] The Stanford e^+e^- SLC collider has also run at $\sqrt{s} \sim m_Z$ in the period 1989-1998. With the Mark II and SLD experiments it has contributed in an important way to the SM precise tests. It is not specifically discussed in this paper, but general information can be found in Ref. [4].

the so-called LEP2 phase, when the centre-of-mass energy was increased up to $\sqrt{s} \simeq 209$ GeV. This has allowed the production of W pairs, and therefore the measurement of the W mass (Sect. 2.3), as well as powerful searches for the Higgs boson (Sect. 3.1) and other new heavy particles predicted by theories beyond the SM (Sect. 4.1).

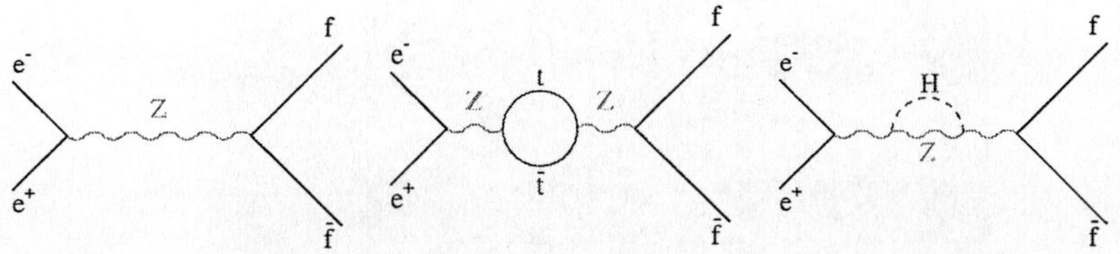

Fig. 2. Z-boson production processes in e^+e^- collisions: dominant diagram (left) and diagrams with radiative corrections involving a top loop (middle) and a Higgs loop (right)

In the period 1989-1995, the four LEP experiments have collected a total of about 20 million events at the Z peak, and have therefore made a large number of precise measurements of the Z particle properties. These measurements include the Z mass and width, the Z production cross-section, all the features of the Z couplings to quarks and leptons, etc. The precision reached is better than 0.1% in most cases. A complete summary of all the results can be found in Ref. [5].

Two main goals have been achieved:

- Detailed and extensive tests of the SM predictions have been performed, in particular of the weak sector of the theory (the Z boson has only weak interactions) which was much less known than the electromagnetic sector.
- Weak radiative quantum corrections [1], which provide indirect sensitivity to heavy physics, have been probed. The Standard Model says that (tiny) contributions to the Z production come also from processes, like the middle and right diagrams in Fig. 2, involving loops of heavy particles such as the top quark and the Higgs boson. As a consequences, all the observables at the Z peak are predicted to depend also on the top mass (quadratic dependence) and on the Higgs mass (logarithmic dependence). Therefore, by measuring as many observables as possible and by comparing them to the theoretical predictions it should be possible to deduce indirectly the masses of the top quark and Higgs boson, which are too heavy to be directly produced at LEP. However, since the radiative corrections are small and modify the electroweak observables by only a few percent, experimental precisions at the level of a few permil are needed to be sensitive to these effects. This study requires in parallel theoretical predictions accurate enough to match the experimental precision. A remarkable demonstration that the weak radiative corrections exist, and that the needed experimental and theoretical sensitivities had been achieved, was obtained in 1994, when the LEP experiments were able

to predict the mass of the top quark a few months before this quark was discovered at the Tevatron.

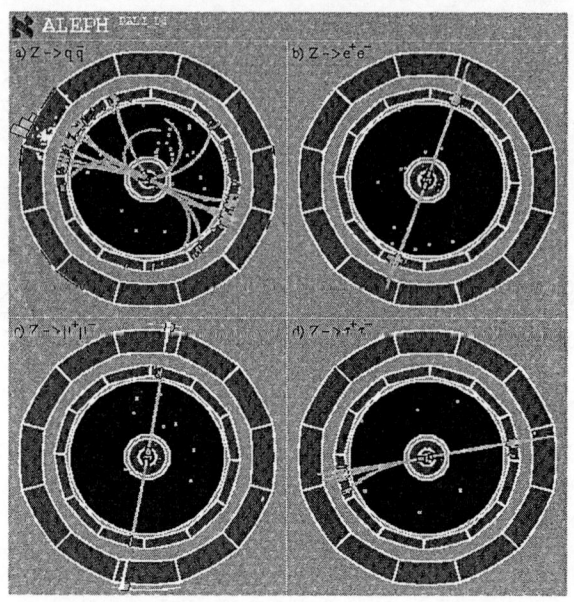

Fig. 3. Graphic displays of four Z events collected by the ALEPH experiment, where the Z has decayed into **(a)** a $q\bar{q}$ pair, **(b)** an e^+e^- pair, **(c)** a $\mu^+\mu^-$ pair, and **(d)** a $\tau^+\tau^-$ pair. The ALEPH detector, shown here in the view transverse to the beam, consists (from the collision centre outwards) of inner tracking devices (up to the end of the dark region) followed by electromagnetic and hadron calorimeters and by muon chambers (the latter are shown only in panel **c**)

Figure 3 shows four Z events recorded by the ALEPH experiment, illustrating all visible final states resulting from the production and decay of a Z boson. A $Z \rightarrow e^+e^-$ event is characterized by an electron and a positron emitted in opposite directions ("back-to-back") and carrying each an energy of about 45 GeV (i.e. half of the Z mass), for reasons of energy and momentum conservation. The electron and the positron are recognized mainly by the ionization tracks in the inner tracking devices pointing to electromagnetic showers in the electromagnetic calorimeter. A $Z \rightarrow \mu^+\mu^-$ decay gives rise to two back-to-back muons leaving ionization tracks in the inner tracking devices and hits in the external muon chambers. A $Z \rightarrow q\bar{q}$ event produces two jets of hadrons coming from the fragmentation and hadronization of the original quarks [1], and therefore two clusters of tracks in the inner tracking devices (due e.g. to $\pi^{\pm}$) and showers in the calorimeters (due to e.g. $\pi^0 \rightarrow \gamma\gamma$). Finally an hadronic $Z \rightarrow \tau^+\tau^-$ decay appears as two low-multiplicity jets of hadrons (coming from the τ decays).

The branching ratios (i.e. the fractions of times the various decay modes occur) of the Z for the above final states are about 3% for each charged lepton species and about 70% for $q\bar{q}$. They have been measured with experimental precisions of about 0.1% and found to be in agreement with the theoretical prediction. The Z boson can also decay into $\nu\bar{\nu}$ pairs (with branching ratio ~20%), a final state which is not observable because neutrinos have only weak interactions and therefore do not leave signals in the detector.

To determine the shape and features of the Z resonance, the LEP centre-of-mass energy was varied between 88 GeV and 94 GeV, and the experiments have measured the Z production cross-section as a function of energy (Fig. 1). These data have allowed the measurements of e.g. the Z mass and width.

The measured value of the Z mass is $m_Z = 91187.5 \pm 2.1$ MeV. Thanks to the spectacular precision of this result (0.002%), m_Z is today one of the best known parameters of the SM (together with e.g. the Fermi constant G_F [1]). The dominant uncertainty in the above measurement (~ 1.7 MeV) comes from the knowledge of the LEP beam energy. The latter can be determined with an intrinsic precision of 100 keV (over 45 GeV !) by using a method called resonant depolarization [6]. However, some subtle effects increased the uncertainty from 100 keV to about 1.7 MeV. These effects include the moon tidal forces, which modify the size of the accelerator and therefore induce a day-night variation of the beam energy, and the impact on the magnetic field of the machine of vagabond currents produced by the French TGV trains transiting in the Geneva area.

The Z width has been determined by the LEP experiments with an overall precision of better than 0.1%: $\Gamma_Z = 2495.2 \pm 2.3$ MeV. From this measurement and from the measurements of the Z branching ratios for the decays into visible final states it has been possible to deduce the branching ratio into invisible final states and therefore the number of light neutrinos (i.e. neutrinos with masses smaller than about half the mass of the Z). The result is $N_\nu = 2.984 \pm 0.008$, i.e. compatible with three families of light neutrinos, as also indicated by nucleosynthesis arguments. Figure 4 shows how the Z lineshape is expected to change when the number of light neutrinos is varied from two to four.

The measured value of Γ_Z can be compared to the prediction of the SM as a function of the top mass, as shown in Fig. 5. The SM prediction depends on the top mass because of the already-mentioned radiative corrections (see middle diagram in Fig. 2). From Fig. 5 it is possible to derive an indirect measurement of the top mass. A higher precision can be obtained by using all electroweak observables and not just Γ_Z. The result is $m_{top} = 180.4 \pm 9.7$ GeV, in remarkable agreement with the direct measurement performed by the Tevatron experiments ($m_{top} = 174.3 \pm 5.1$ GeV, see Sect. 2.2). This is a spectacular demonstration that the radiative corrections exist and have the size predicted by the Standard Model.

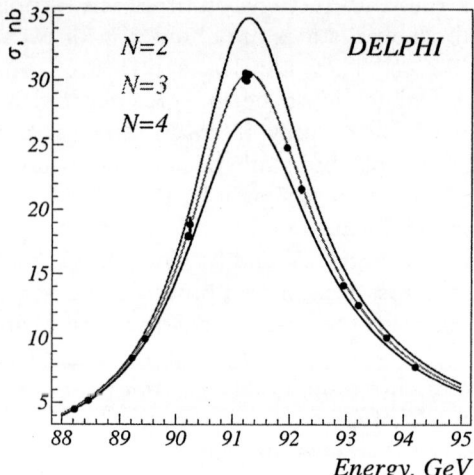

Fig. 4. Z-boson production cross-section as a function of the centre-of-mass energy (Z "lineshape") as measured by the DELPHI experiment at LEP (dots) and as predicted by the Standard Model for two, three and four families of light neutrinos (full lines)

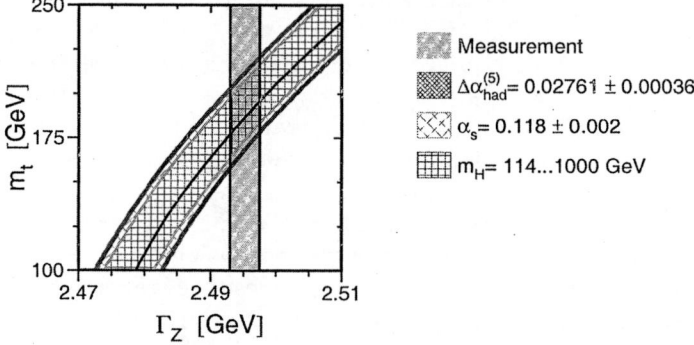

Fig. 5. Z-boson width as measured by the LEP experiments (vertical band) and as predicted by the Standard Model as a function of the top mass (curved band). The width of the SM band is due to the uncertainties on the values of the electromagnetic coupling constant, of the strong coupling constant and of the Higgs mass. From Ref. [5]

2.2   Measurement of the Top Mass

As of today the top quark has only been observed at the Fermilab Tevatron collider, a 6.5 km ring in the Chicago area providing $p\bar{p}$ collisions at $\sqrt{s} \sim 2$ TeV to two experiments, CDF and D0. This machine has also operated in two phases: Run 1 (1989-1996) has been characterized by the very important discovery of the top quark [7]; Run 2 started in Spring 2001 with upgraded machine and detectors and should accumulate up to 100 times more statistics than Run 1, therefore allowing improved measurements of the top and W masses and more effective searches for the Higgs bosons and other new particles to be performed.

At the Tevatron, pairs of top quarks are produced in the strong interaction between two colliding quarks through the exchange of a gluon: $q\bar{q} \to g \to t\bar{t}$. Both top quarks then decay fast (i.e. as free quarks, before hadronising) into a W and a b quark: $t \to Wb$. This gives rise for instance to the topology shown in Fig. 6. In this event one of the W's has decayed into two jets and the other one into an electron-neutrino pair. The jets produced by the two b quarks do not originate from the primary vertex of the interaction. This is because the b-hadrons coming from the fragmentation and hadronization of b quarks have a lifetime of $\sim$1.5 ps, and therefore decay at a few mm from the production point. This gives rise to secondary displaced vertices (with respect to the primary interaction vertex), which can be reconstructed by using high-granularity detectors like Silicon strips located close to the beam pipe.

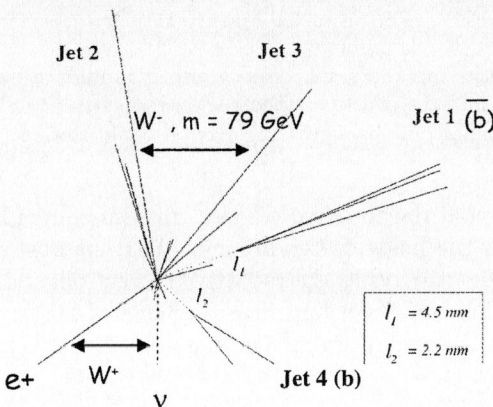

Fig. 6. Schematic view of a $t\bar{t}$ event collected by the CDF experiment. One top has decayed as $\bar{t} \to W^-b \to jj\,b$ (where j indicates a jet) and the other one as $t \to W^+b \to e^+\nu\,b$. The reconstructed W mass for the $W^- \to jj$ decay is indicated, as well as the two displaced b-jet vertices (l_1, l_2)

Figure 7 shows the reconstructed top mass spectrum obtained by the CDF experiment for a sample of $t\bar{t} \to bjj\,b\ell\nu$ candidate events. The distribution shows a peak at $m_{top} \sim$ 175 GeV; it is in agreement with the expectation for a top signal plus background from other processes and inconsistent with the expectation for background only. The top mass can be obtained from such distributions, and the combined measurement from the CDF and D0 experiments gives $m_{top} = 174.3 \pm 5.1$ GeV. The uncertainty of about 3% is mainly due to the limited size of the $t\bar{t}$ data sample, which today consists of only $\sim$200 events. Nevertheless the top mass is today the best known of all quark masses.

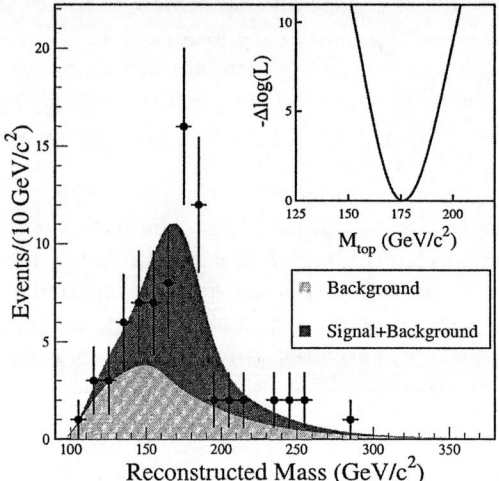

Fig. 7. The reconstructed top mass spectrum from $t\bar{t}$ candidate events collected by CDF in Run 1 (dots). The light (dark) shaded region indicates the SM expectation for background only (top signal plus background) events. From Ref. [8]

It should be noted that the top quark is the most intriguing fermion, because of its large mass (it is the heaviest fundamental particle observed so far), and therefore the special rôle it plays in the radiative corrections, and its large width (which causes its fast decay). Precision measurements of the top properties may therefore give clues to the origin of particle masses.

2.3   Measurement of the W Mass

The W mass is another fundamental parameter of the Standard Model. It is related to the electromagnetic coupling constant α_{EM}, the Fermi constant G_F and the Weinberg angle $\sin\theta_W$ [1] through the following relation

$$m_W = \left(\frac{\pi\alpha_{EM}}{\sqrt{2}G_F}\right)^{1/2} \frac{1}{\sin\theta_W \sqrt{1-\Delta r}} \qquad (1)$$

As any other observable, m_W receives radiative corrections (Δr), which depend as usual on the top mass and Higgs mass and which amount to about 3%. Since α_{EM}, G_F and $\sin\theta_W$ are all known with high accuracy, precise measurements of both the top mass and the W mass allow the only unknown parameter in the above relation, the Higgs mass, to be constrained. This constraint is unfortunately relatively loose because of the logarithmic dependence of Δr on m_H.

The W mass has been measured at LEP and at the Tevatron. In its phase two LEP has run at a centre-of-mass energy above 160 GeV, i.e. above $2 \times m_W$, and therefore was able to produce W pairs through the processes shown in Fig. 8. Two of the diagrams in Fig. 8 involve the interaction of three bosons

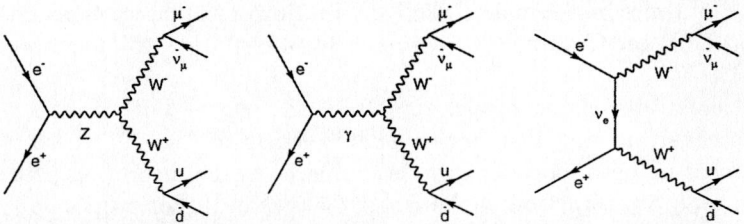

Fig. 8. Processes contributing to WW production at LEP

($WW\gamma$ and WWZ). The existence of these triple-boson vertices in the SM is a consequence of the intrinsic structure of the theory [1]. Figure 9 shows the WW production cross-section as a function of the LEP energy, as measured by the four experiments together. The data points are in agreement with the SM prediction based on all the processes shown in Fig. 8, and clearly inconsistent with the contribution of the ν-exchange diagram only, thereby confirming the existence of triple-boson vertices.

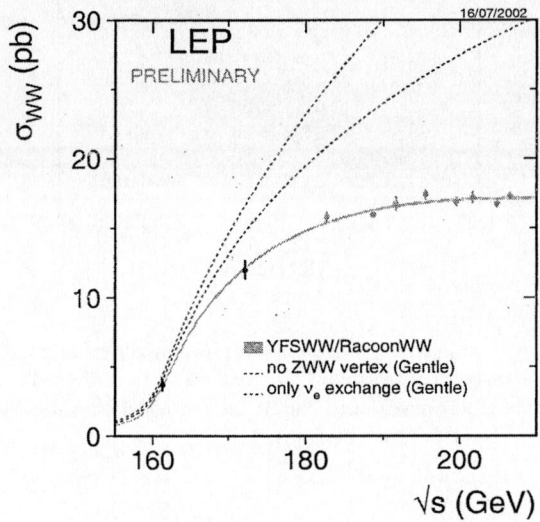

Fig. 9. WW production cross-section measured at LEP [5] as a function of the centre-of-mass energy (dots). The curves show the cross-section predicted by the theory if all production processes are taken into account (lower curve), if only the ν-exchange process is taken into account (upper curve), and if only the ν-exchange and γ-exchange processes are taken into account (middle curve)

The W mass is reconstructed from the energy and momenta of the decay products measured in the final state (the W boson decays 70% of times in a $q\bar{q}$

pair and 30% of times in a $\ell\nu$ pair). The combination of the direct measurements performed at LEP and Tevatron gives $m_W = 80.449 \pm 0.034\,\text{GeV}$, i.e. a precision of about 0.4%.

As mentioned above, the measurements of the top and W masses can be combined to constrain the Higgs mass. This is shown in Fig. 10, where the experimental measurement (dotted curve) is compared to the theoretical prediction for m_W and m_{top} as a function of the Higgs mass (band). For a fixed top mass, the predicted value for the W mass decreases when the Higgs mass increases. It can be seen that the data favour a light Higgs boson, with mass of order 100 GeV. This is discussed in more detail in the next Section.

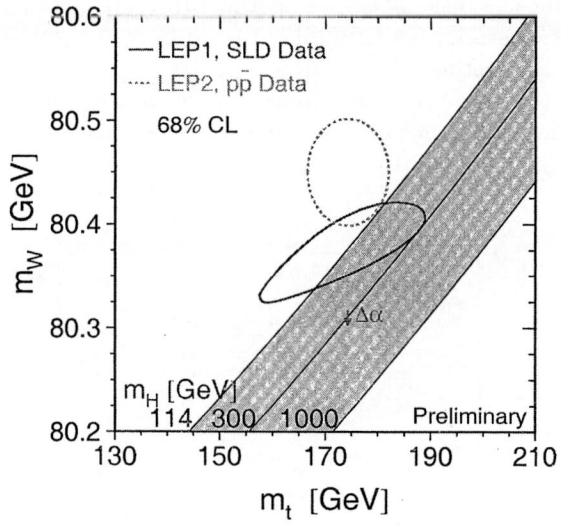

Fig. 10. In the plane (m_W, m_{top}), the 68% C.L. region obtained from the direct measurements at LEP and Tevatron (dotted curve), the 68% C.L. region obtained indirectly from the electroweak data (full curve), and the Standard Model prediction as a function of the Higgs boson mass (band). From Ref. [5]

2.4   Global Fit of the Standard Model to the Data

The various electroweak measurements, i.e. the Z peak observables, the W and top masses, can be combined to test the consistency of the SM and to constrain the unknown parameters of the theory.

Figure 11 shows the list of electroweak observables, some of which are discussed in the previous sections, the measured values, and the difference between the measured values and the values obtained from a fit of the Standard Model to the ensemble of electroweak data. In most cases the difference between the

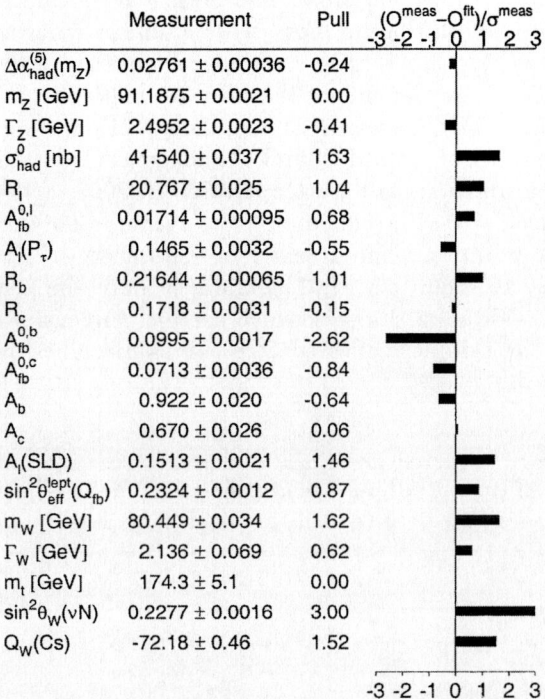

Summary 2002

	Measurement	Pull	$(O^{meas} - O^{fit})/\sigma^{meas}$
$\Delta\alpha^{(5)}_{had}(m_Z)$	0.02761 ± 0.00036	-0.24	
m_Z [GeV]	91.1875 ± 0.0021	0.00	
Γ_Z [GeV]	2.4952 ± 0.0023	-0.41	
σ^0_{had} [nb]	41.540 ± 0.037	1.63	
R_l	20.767 ± 0.025	1.04	
$A^{0,l}_{fb}$	0.01714 ± 0.00095	0.68	
$A_l(P_\tau)$	0.1465 ± 0.0032	-0.55	
R_b	0.21644 ± 0.00065	1.01	
R_c	0.1718 ± 0.0031	-0.15	
$A^{0,b}_{fb}$	0.0995 ± 0.0017	-2.62	
$A^{0,c}_{fb}$	0.0713 ± 0.0036	-0.84	
A_b	0.922 ± 0.020	-0.64	
A_c	0.670 ± 0.026	0.06	
$A_l(SLD)$	0.1513 ± 0.0021	1.46	
$\sin^2\theta^{lept}_{eff}(Q_{fb})$	0.2324 ± 0.0012	0.87	
m_W [GeV]	80.449 ± 0.034	1.62	
Γ_W [GeV]	2.136 ± 0.069	0.62	
m_t [GeV]	174.3 ± 5.1	0.00	
$\sin^2\theta_W(\nu N)$	0.2277 ± 0.0016	3.00	
$Q_W(Cs)$	-72.18 ± 0.46	1.52	

Fig. 11. The list of electroweak observables (first column), their measured values (second column), and the difference between the measured values and the values deduced from the fit of the Standard Model to the ensemble of electroweak data divided by the experimental uncertainty (third and fourth columns). From Ref. [5]

measured and the fitted values is within $\sim 1\sigma$ of the experimental error, thereby indicating a very good internal consistency of the theory. Out of 20 measurements only two, one from LEP and one from NuTeV (a fixed-target neutrino scattering experiment at Fermilab [9]), show larger deviations (2.5-3σ).

In this global fit the Higgs mass is a free parameter, since its value is not known. A prediction for the Higgs mass can therefore be derived from the fit as that mass which gives the best χ^2 of the fit. The result is $m_H = 81^{+52}_{-33}$ GeV. The big error is due to the already-mentioned logarithmic dependence of the radiative corrections on the Higgs mass. The 95% C. L. upper limit obtained from this fit is $m_H < 193$ GeV. Thus, as anticipated, the electroweak data favour a light Higgs boson.

Finally, an illustration of the improved experimental knowledge of the electroweak observables over the last ten years is given in Table 1, which shows the achieved precision in the measurements of the top and W masses as a function of time, as well as future prospects. Ten years ago, in 1991, the top quark had not been discovered yet and only a mass lower limit of about 90 GeV existed. The W mass was known to about 300 MeV. The present uncertainties, 34 MeV on the W mass and 5.1 GeV on the top mass, will be further reduced in the future. First by the Tevatron Run 2, which in particular should halve the present error on the top mass. Then, by the end of this decade, experiments at the CERN Large Hadron Collider (LHC), presently under construction, should measure the W mass with a precision of ~15 MeV and the top mass to ~1 GeV. The LHC [10] is a pp machine which should start operation in 2007, and should be able to collect 10^7 Z events, 10^8 W events and 10^7 $t\bar{t}$ events in only one year of data taking, thanks to the unprecedented centre-of-mass energy ($\sqrt{s}$=14 TeV) and luminosity (up to ~ 10^{34} cm^{-2} s^{-1}). Therefore, in only one year of operation the LHC should provide data samples which are in many cases orders of magnitude larger than the samples collected at previous machines over their whole life.

Table 1. Achieved and expected precisions in the measurements of the top and W masses as a function of time. The approximate uncertainty on the indirect determination of the Higgs mass is also given (assuming a light Higgs boson)

Year	Δm_{top} (GeV)	Δm_W (MeV)	$\Delta m_H/m_H$
1991	$m_{top} > 91$ GeV	270	–
1996	6.0	125	~80%
2002	5.1	34	~50%
~2006 (Tevatron Run 2)	~2	~25	~35%
~2008 (LHC)	~1	~15	~25%

Table 1 shows also the evolution with time of the error on the Higgs mass obtained indirectly from the electroweak measurements. The present uncertainty of about 50% should be reduced to 25% at the LHC. It should be noticed that if the Higgs boson is discovered, a comparison between the measured value of its mass and the indirect value derived from the electroweak measurements will provide a necessary consistency test of the theory, in particular of the sector responsible for the mass generation.

3   Where Is the Higgs Boson?

The results discussed in the previous sections testify to the success of the Standard Model in describing the interactions of elementary particles at the energies tested so far.

One crucial part of the Standard Model, however, remains largely untested as of today. This is the Higgs sector, which is needed to explain the origin of the particle masses. As a consequence of the so-called "electroweak symmetry breaking" [1], which is the mechanism by which fermions and bosons acquire mass in the SM, the theory predicts the existence of an elementary scalar particle, the Higgs boson, for which there is no direct experimental evidence so far.

The value of the Higgs mass is not predicted by the theory, which gives only an upper bound of ~1 TeV. Direct searches performed so far have set a mass lower limit of about 114 GeV (see Sect. 3.1). In addition, and as already mentioned, out of the allowed mass range ~115-1000 GeV the electroweak data prefer the region below 200 GeV.

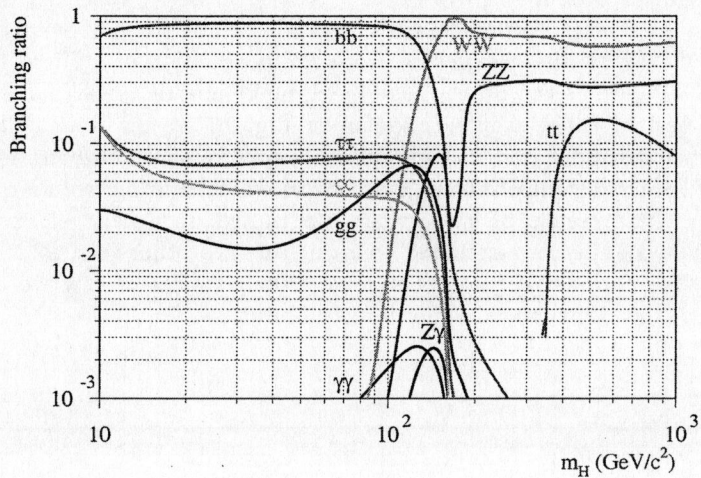

Fig. 12. Decay branching ratios of the SM Higgs boson as a function of mass

In contrast, for a given mass the Higgs boson decay modes are known (Fig. 12), because the SM predicts that the Higgs particle couples to fermions and bosons with strength proportional to their masses. Therefore, if $m_H < 120$ GeV the Higgs boson is expected to decay mainly into $b\bar{b}$ pairs, since b quarks are the most massive particles which are accessible in this m_H range. For larger Higgs masses, decays into heavier particles, like W pairs and Z pairs, open and dominate.

The present experimental status of Higgs searches and the future prospects are reviewed in the next two sections.

3.1 Higgs Searches at LEP

So far LEP has been the most powerful machine for Higgs searches. At LEP2 the Higgs particle would be produced mainly in association with a Z boson through the process $e^+e^- \to Z^* \to Z^{(*)}H$. In the mass range accessible to LEP (up to $m_H \sim 116$ GeV), the Higgs boson is expected to decay mainly into $b\bar{b}$ and, to a lesser extent, into $\tau\tau$ pairs (see Fig. 12). Therefore Higgs production at LEP2 should give rise to four main topologies: four jets final states, if $H \to b\bar{b}$ and

$Z \to q\bar{q}$ or $Z \to b\bar{b}$, which is the channel with the largest branching ratio; two jets and missing energy final states, if $H \to b\bar{b}$ and $Z \to \nu\bar{\nu}$ (the missing energy is due to the two neutrinos escaping detection); two jets and two leptons final states, if $H \to b\bar{b}$ and $Z \to ee, \mu\mu$; and two jets and two taus final states, if either $H \to b\bar{b}$ and $Z \to \tau\tau$ or $H \to \tau\tau$ and $Z \to q\bar{q}$.

Backgrounds to these topologies arise for instance from ZZ production and WW production, which have larger cross-sections than the Higgs signal (Fig. 1) and can also give rise to events containing jets, jets and leptons, jets and missing energy. However, these backgrounds can be effectively rejected by exploiting for instance the fact that signal events are expected to contain two b-jets (from the Higgs decay) and a Z boson.

In 2000, the last year of operation of the LEP collider, a few events with features compatible with those expected from the production of a Higgs boson of mass ~115 GeV (i.e. at the limit of the machine kinematic reach) have been observed. The best of these events is shown in Fig. 13. It has been collected by the ALEPH experiment and is a candidate event for HZ production with $H \to b\bar{b}$ and $Z \to q\bar{q}$. It contains four jets, two of which coming from b quarks as suggested by the presence of two well-reconstructed displaced vertices (see bottom right panel). The reconstructed mass of the two other jets, 92.1 GeV, is very close to the nominal Z mass. The reconstructed mass of the two b-jets, which come from the Higgs boson candidate, is about 114 GeV.

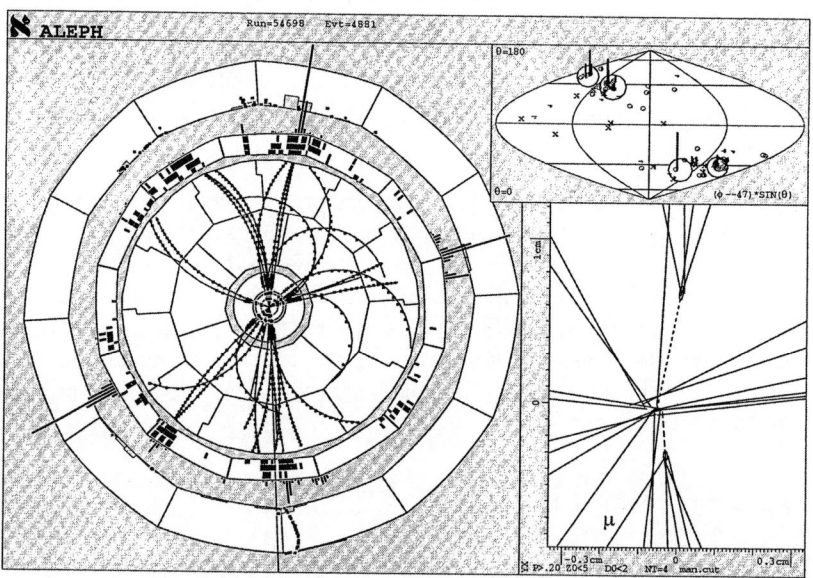

Fig. 13. Display of the best LEP Higgs boson candidate in the ALEPH detector [11]. The event is shown in the view transverse to the beam direction (left), in the $\theta - \phi \sin\theta$ view (top right), and in a zoom of the vertex region (bottom right)

In spite of a few nice events, the probability that the LEP observation is due to a fluctuation of the background is ∼10%, i.e. the excess of Higgs-like events is only 1.7σ larger than the expected background (where σ is the background uncertainty) [12]. This is not enough to claim a discovery, for which an excess of at least 5σ, corresponding to a background fluctuation probability of only 10^{-7}, is required. As a consequence, the LEP experiments could only set a lower limit on the Higgs mass, which is $m_H > 114.4$ GeV at the 95% C.L. [12], and it is up to present and future machines to clarify whether or not the LEP observation is the first hint of a Higgs signal.

3.2   Future Prospects for Higgs Searches

The Higgs question should be definitely clarified at the LHC and maybe already at the Tevatron Run 2.

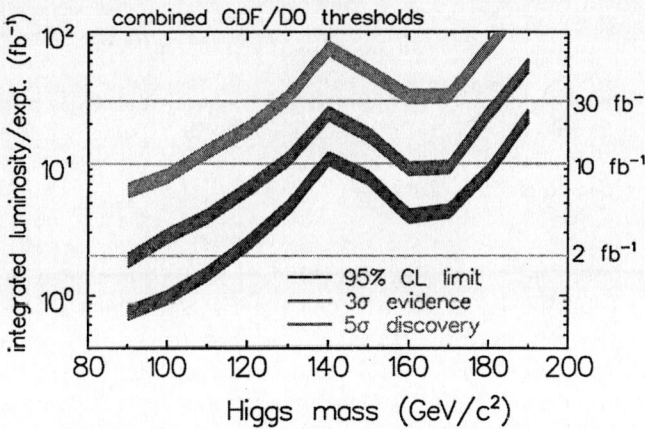

Fig. 14. The integrated luminosity per experiment needed at the Tevatron Run 2 to exclude a SM Higgs boson at the 95% C.L. (lower band), to observe it at the 3σ level (middle band) and to discover it at the 5σ level (upper band), as a function of mass [13]. The width of each band indicates the systematic uncertainty

Figure 14 shows the integrated luminosity (a parameter proportional to the amount of collected data) needed at the Tevatron for a 95% C.L. exclusion, a 3σ observation and a 5σ discovery of a Higgs signal, as a function of m_H. It can be seen that if the Higgs boson has a mass around 115 GeV, i.e. close to the present lower limit and to the LEP hint, CDF and D0 should be able to discover it (at 5σ) with 15 fb^{-1} of data per experiment, which may be collected by 2008. On the other hand with only 2 fb^{-1} of data, i.e. by ∼2005, the Tevatron experiments should be able to exclude a signal at the 95% C.L. if there is no Higgs of such a mass. If the Higgs boson is heavier than ∼115 GeV, discovery will be very difficult at the Tevatron.

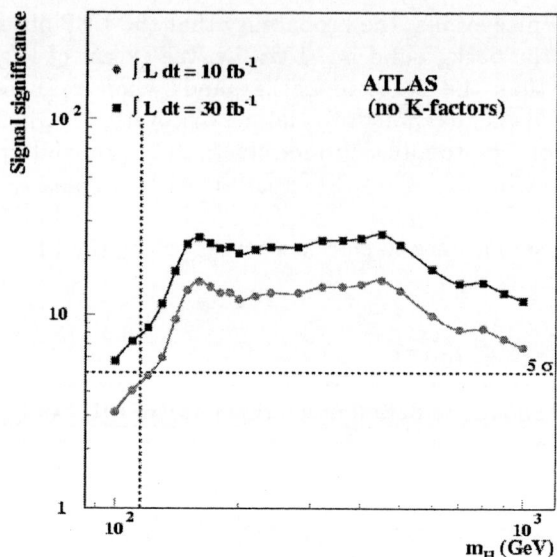

Fig. 15. Expected signal significance for a SM Higgs boson in ATLAS as a function of mass, for integrated luminosities of 10 fb^{-1} (dots) and 30 fb^{-1} (squares). The vertical line shows the mass lower limit from LEP. The horizontal line indicates the minimum significance (5σ) needed to claim a discovery

The discovery potential of the next machine, the LHC, is shown in Fig. 15. With an integrated luminosity of 10 fb^{-1} per experiment, which should be collected in only one year of operation, the two general-purpose LHC experiments ATLAS [14] and CMS [15] should be able to discover a Higgs signal with a significance of 5σ or larger over the full allowed mass range up to 1 TeV. Therefore the LHC should be able to say the final word about the SM Higgs mechanism, i.e. if nothing is found at the LHC this mechanism is wrong.

The years 2007-2008 may be very exciting for the Higgs hunt, because if the Higgs mass is indeed close to 115 GeV both machines, the Tevatron and the LHC, may have collected enough luminosity for a 5σ discovery.

4   Beyond the Standard Model

Despite most of the SM predictions have been verified experimentally with a spectacular accuracy, particle physicists have good reasons today to believe that new (and presently unknown) physics exists beyond the Standard Model [16].

Figure 16 shows the energy scale Λ up to which the Standard Model is valid, as a function of the Higgs mass. It can be seen that if the Higgs mass is close to the LEP limit, i.e. ~115 GeV, the SM works only up to an energy scale of ~ 10^6 GeV, which means that some new physics should appear at this scale or below. If, on the other hand, the Higgs mass is in the range 130-180 GeV, then

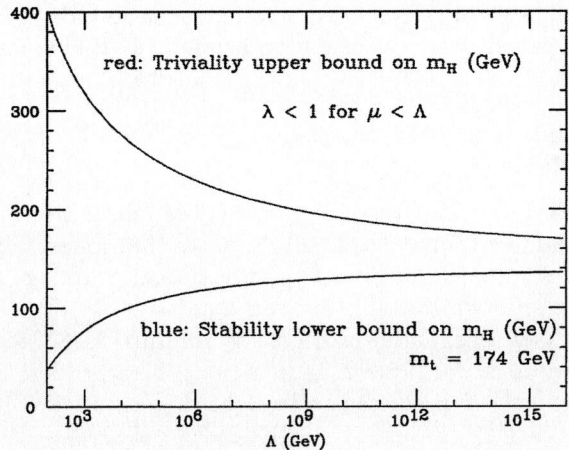

Fig. 16. Theoretical bounds on the Higgs boson mass (in GeV) as a function of the energy scale Λ up to which the SM works. The region above the top curve and the region below the bottom curve are theoretically forbidden (the top region because the Higgs self-couplings diverge, the bottom region because the electroweak vacuum is unstable)

the Standard Model technically works up to the Planck scale. This scenario, not particularly attractive from the experimental point of view, introduces two problems: the so-called "hierarchy problem", i.e. the huge difference between the electroweak scale ($\sim$100 GeV) and the Planck scale ($\sim 10^{19}$ GeV); and the so-called "naturalness problem", i.e. the fact that in the Standard Model the radiative corrections to the Higgs mass (and therefore the Higgs mass itself) increase as Λ^2, and some tricks are needed to keep it down to the electroweak scale.

In addition, many other questions of fundamental importance, such as the existence of three fermion families with apparently random mass patterns, the magnitude of the cosmological constant, etc., remain without fully satisfactory answers in the SM.

The above arguments, discussed in more details in [16], indicate that the Standard Model is most likely not the ultimate theory of particle interactions, but rather an approximation, valid at the energies tested so far, of a more fundamental theory. There are today several candidate scenarios beyond the Standard Model [16], among which Supersymmetry, Technicolour and theories with Extra-dimensions. They have a difficult task, since they are asked to solve the SM problems without contradicting the electroweak data, which are very precise and therefore very constraining. It should be noted that all these theories predict that something new, new particles or new phenomena, should manifest at the TeV scale, and this strongly motivates the LHC which has a direct discovery potential for particle masses of up to $\sim$ 5 TeV.

Supersymmetry, probably the most motivated scenario today for physics beyond the SM, is discussed briefly in the next Section, with emphasis on the experimental status.

4.1 Supersymmetry

Supersymmetry (SUSY) is an appealing symmetry relating fermions and bosons, i.e. matter fields and force fields [16]. It states that for each SM particle p there exists a supersymmetric partner $\tilde{p}$, with identical couplings and quantum numbers except the spin which differs by half a unit.

The SUSY particles (sparticles) predicted by minimal SUSY models, such as the MSSM [16], are listed in Table 2. In addition, there are five Higgs bosons, h, H, Λ, $H^{\pm}$, of which h is predicted to be light ($m_h < 135$ GeV).

Table 2. Standard Model particles and their supersymmetric partners

SM particles	SUSY partners	Spin of SUSY partners
quarks	squarks $\tilde{q}$	0
leptons	sleptons $\tilde{\ell}$	0
gluon	gluino $\tilde{g}$	1/2
$W^{\pm}, H^{\pm}$-field	charginos $\chi_{1,2}^{\pm}$	1/2
Z, γ, H-field	neutralinos $\chi_{1,2,3,4}^{0}$	1/2

No experimental evidence for Supersymmetry has been found as of today. Therefore, either Supersymmetry does not exist, or the SUSY particles are too heavy (well above 100 GeV) to be produced at present machines. However, the above-mentioned problem of the divergence of the Higgs mass can be solved in Supersymmetry only if the sparticle masses are ~ 1 TeV or lighter, so sparticles should not escape discovery at the LHC.

Important phenomenological consequences come from the fact that the theory contains a multiplicative quantum number, called R-parity, with values +1 for ordinary particles and −1 for SUSY particles. If R-parity is conserved, as it is assumed here, then sparticles are produced in pairs and the Lightest Supersymmetric Particle (LSP), to which all sparticles eventually decay, must be stable. In most models the LSP is the lightest neutralino χ_1^0, which is a stable, massive and weakly-interacting particle, and therefore an excellent candidate for the universe cold dark matter. As the neutrino, the lightest neutralino escapes experimental detection if produced e.g. in a collider experiment, and thus is expected to give rise to events with missing energy.

The most powerful SUSY searches have been performed so far at LEP and Tevatron, with complementary phenomenology and results.

At LEP, pairs of SUSY particles should be more or less democratically produced mainly through γ or Z exchange, e.g. $e^+e^- \to Z/\gamma \to \tilde{\ell}^+\tilde{\ell}^-$. Sparticles accessible at LEP must be relatively light, therefore in most cases are expected to decay directly to their SM partner and the LSP: $\tilde{p} \to p\chi_1^0$. This gives rise to simple final states, usually consisting of two acoplanar (i.e. not back-to-back) objects (for instance two leptons in the case of slepton pair production) accompanied by missing energy.

SUSY searches at LEP2 have been unsuccessful, and have excluded sparticle masses of up to ~ 100 GeV [17], i.e. up to the machine kinematic limit for the production of pairs of particles. This is true for all sparticles except two cases: gluinos, which are not expected to be copiously produced at LEP because they are not electroweakly-interacting particles; and the lightest neutralinos, because $\chi_1^0\chi_1^0$ production gives rise to invisible final states. In the latter case, however, the LEP experiments have been able to set an indirect mass lower limit by using the results from other searches (e.g. $\chi^+\chi^-$) and the fact that in minimal SUSY models the masses of the various sparticles are related. The limit found in this way is $m(\chi_1^0) > 45$ GeV at the 95% C.L. [17].

This limit can be compared to the results of direct searches for cold dark matter performed by experiments looking for neutralinos, or more generally for Weakly-Interacting Massive Particles (WIMP's), coming from the galactic halo and interacting with the detectors via WIMP-nuclei scattering. Figure 17 shows

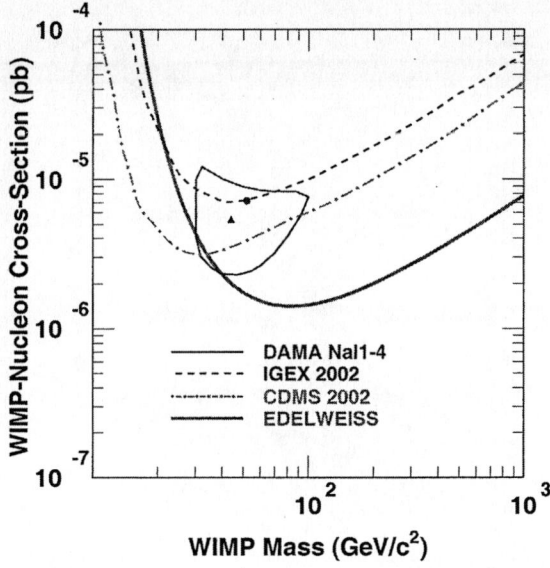

Fig. 17. The regions of the plane WIMP mass versus WIMP-nucleon spin-independent cross-section favoured at the 3σ level by the DAMA experiment (closed contour, [18]), or excluded at the 90% C.L. by the EDELWEISS experiment (solid line, [19]), by the CDMS experiment (dash-dotted line, [20]), and by the IGEX experiment (dashed line, [21]). From reference [19]

the region of the plane WIMP mass versus WIMP cross-section favoured by the DAMA experiment at Gran Sasso if the observed annual modulation in the rate of nuclear recoils [18] is attributed to galactic neutralinos. Also shown are the regions excluded by other experiments of similar scope, in particular a very recent result from the EDELWEISS experiment [19]. It can be seen that the LEP limit $m(\chi_1^0) > 45$ GeV is complementary to these results because it does not depend on the neutralino-nucleon cross-section, and therefore is able to exclude regions of the plane where direct searches have no sensitivity.

Unlike at LEP, at hadron colliders (Tevatron, LHC) the sensitivity to SUSY particles is not expected to be democratic. Squark and gluino production, which is due to strong processes like $q\bar{q} \to g \to \tilde{g}\tilde{g}$, has a much larger cross-section than the production of electroweak sparticles like charginos or sleptons. And since squarks and gluinos must be heavy ($m > 200\text{-}300$ GeV) given the present experimental limits, they are expected to decay in cascades with several intermediate steps (as depicted in Fig. 18), and hence to give rise to complicated final states with many jets, leptons and missing energy. Such spectacular signatures can be easily recognized from the SM backgrounds.

At the Tevatron Run 1 CDF and D0 have found no evidence for squark or gluino production, and have therefore been able to exclude masses below 200-300 GeV [22]. In Run 2 they should be able to discover these sparticles up to masses of $\sim 400\text{-}450$ GeV if they exist.

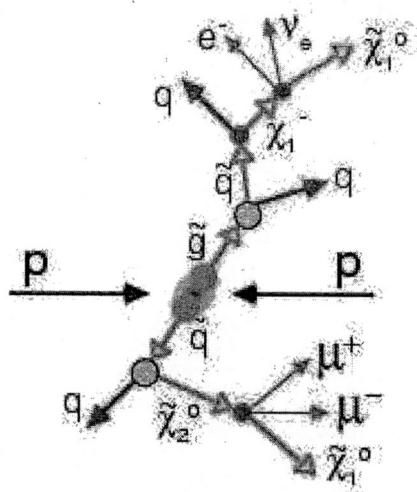

Fig. 18. Sketch of a pp collision producing a $\tilde{q}\tilde{g}$ pair, followed by cascade decays of both sparticles

In summary, Tevatron has a much higher sensitivity to squarks and gluinos than LEP, thanks to the higher centre-of-mass energy and to the strong production of these sparticles at hadron colliders. On the other hand LEP has set the best limits on the masses of electroweak sparticles, for which the signal-to-background ratio is small at hadron colliders. This is one example of complementarity between the two machines.

Finally, the LHC has a discovery potential for squarks and gluinos up to masses of 2.5-3 TeV, and therefore should be able to say the final word about low-energy SUSY, for the reasons mentioned above. If nothing is found at the LHC, Supersymmetry will most likely be ruled out.

5 Conclusions

Over the last ten years, high-energy physics experiments operating at various machines have performed precise measurements of the fundamental particles and their interactions, with accuracies of better than 0.1% in most cases. They have also looked for manifestations of new physics in a large variety of topologies.

A large amount of outstanding results have been produced, which are very challenging and constraining for any theoretical scenario. They demonstrate remarkable experimental achievements and innovative ideas in the field of accelerator and detector developments, data analysis techniques, etc. They have also triggered a huge theoretical effort to provide more and more accurate predictions in order to make sense of these beautiful measurements.

These results represent a big success for the Standard Model. All predicted particles have been discovered (except the Higgs boson). In addition, the theory structure, the predicted interactions and their phenomenological consequences, have been understood and confirmed at the level of higher-order radiative corrections up to energy scales of several hundred GeV, i.e. all the way back to $\sim 10^{-10}$ s after the Big Bang.

In spite of these successes, the Standard Model cannot the ultimate theory of the elementary particles and their interactions. Several problems with the Higgs mechanism, and more in general the major issue of the origin of the particle masses, as well as the many unexplained questions, call for new physics and strongly motivate future machines like the LHC.

Acknowledgements

I am grateful to Luigi Di Lella and Patrick Janot for their help and comments.

References

1. A. Pich: these Proceedings
2. G. Arnison et al.: Phys. Lett. **B122**, 103 (1983) M. Banner et al.: Phys. Lett. **B122**, 476 (1983) G. Arnison et al.: Phys. Lett. **B126**, 398 (1983) M. Banner et al.: Phys. Lett. **B129**, 130 (1983)

3. S. L. Glashow: Nucl. Phys. **22**, 579 (1961) S. Weinberg: Phys. Rev. Lett. **19**, 1264 (1967) A. Salam: in *Proceedings of the 8th Nobel Symposium*, 367 (1968)
4. http://www.slac.stanford.edu
5. The LEP Collaborations: 'A combination of preliminary electroweak measurements and constraints on the Standard Model', CERN-EP/2001-098, hep-ex/0112021 LEP Electroweak Working Group:
   http://lepewwg.web.cern.ch/LEPEWWG/plots/summer2002/
6. L. Arnaudon et al.: Phys. Lett. **B284**, 431 (1992)
7. F. Abe et al.: Phys. Rev. Lett. **73**, 225 (1994) S. Abachi et al.: Phys. Rev. Letters **74**, 2632 (1995)
8. F. Abe et al.: Phys. Rev. Lett. **80**, 2767 (1998)
9. G.P. Zeller et al.: Phys. Rev. Lett. **88**, 091802 (2002)
10. The LHC Study Group: 'The Large Hadron Collider conceptual design', 1995, CERN/AC/95-05
11. R. Barate et al.: Phys. Lett. **B495**, 1 (2000)
12. LEP Higgs Working Group: 'Search for the Standard Model Higgs Boson at LEP', LHWG Note/2002-01,
    https://lephiggs.web.cern.ch/LEPHIGGS/papers/index.html
13. M. Carena et al.: 'Report of the Higgs Working Group for Run 2 of the Tevatron', hep-ph/0010338
14. ATLAS Collaboration: 'Detector and physics performance Technical Design Report', 1999, CERN/LHCC/99-15
15. CMS Collaboration: 'The Compact Muon Solenoid Technical Proposal', 1994, CERN/LHCC/94-38
16. J. Ellis: these Proceedings
17. LEP SUSY Working Group: http://lepsusy.web.cern.ch/lepsusy/
18. R. Bernabei et al.: Phys. Lett. **B480**, 23 (2000)
19. A. Benoit et al.: 'Improved exclusion limits from the EDELWEISS WIMP Search', astro-ph/0206271 (2002)
20. D. Abrams et al.: 'Exclusion limits on the WIMP-nucleon cross-section from the Cryogenic Dark Matter Search', astro-ph/0203500 (2002)
21. A. Morales et al.: Phys. Lett. **B532**, 8 (2002)
22. T. Affolder et al.: Phys. Rev. Lett. **88**, 041801 (2002) S. Abachi et al.: Phys. Rev. Lett. **75**, 618 (1995)

Astroparticle Physics with the L3+C Detector

Pierre Le Coultre[1], on behalf of the L3 collaboration

[1] ETH-Zürich, Switzerland

Abstract. The L3+C detector, which was located at the CERN electron-positron collider, LEP, has been used to collect a total of $12 \cdot 10^9$ cosmic ray triggers during the years 1999 and 2000. Thanks to the unique properties of this detector, research topics relevant to various current problems in cosmic ray muon and astroparticle physics can be studied. The status of the data analysis on different subjects is presented.

1 Introduction

We give an overview of the physics topics which are presently being studied while analysing the cosmic ray data collected by the L3+C experiment. The L3 detector was installed at LEP,CERN, as a huge microscope to measure e^+ e^- collisions. The aim of this presentation is to demonstrate that large particle physics detectors can also be used as telescopes for astroparticle physics research.

2 The L3+C Detector

The L3+C detector [1] consists of two parts:
1.) A huge solenoidal magnet with a volume of 1000 m^3 and a field of 0.5 Tesla. Inside high precision drift chambers are installed [2]. Calorimeters and vertex detectors mounted in a support tube are not used for the purpose of cosmic ray muon studies. On top of the magnet 207 m^2 of plastic scintillators are installed for timing purposes. The detector is located 30 m under ground at an altitude of 450 m above sea level. Its geometrical acceptance amounts to 200 m^2sr. The muon momentum threshold is only 15 GeV/c; the momentum resolution equals 7.1 % at 100 GeV/c. L3+C has a unique opportunity compared to other cosmic ray spectrometers: Installed at an electron-positron collider it can record also muons from Z-boson decays and therefore check its momentum measurement and its detection efficiency. The angular resolution (limited by the multiple scattering in the rock overburden) amounts to less than 3.5 mrad above 100 GeV/c. Trigger and data acquisition are independent of the L3 experiment. A GPS clock is used for the timing of the experiment.
2.) The second part of the L3+C detector system consists of an air shower scintillator array installed on the roof of the surface building in order to estimate the size of the shower associated with the muon measured in the L3 spectrometer. It consists of 47 plastic scintillators viewed by photomultipliers through wavelength shifting fibers. The array area amounts to 30 x 54 m^2. A trigger rate of

typically 1.7 Hz is observed (energy threshold around 10 TeV), and 30% of the showers are accompanied by muons detected inside of L3. 25 showers above 10^{15} eV were recorded each day, out of which 2.5 have the core inside of the array.

L3+C started data taking in 1999 and stopped in Nov.2000. Some $12 \cdot 10^9$ muon triggers were collected, corresponding to an effective lifetime of 312 days.

3   Physics Topics

3.1   Differential Muon Momentum Spectrum

A compilation of existing measurements of the vertical momentum spectrum [3] between 10 and 1000 GeV shows that systematical errors are causing uncertainties of the absolute flux values up to 25 %. The main motivation to measure the muon spectrum precisely is the possibility to "normalize" the calculated muon neutrino spectrum, since there is a "one-to-one" relation between the two spectra (same parent particles). For discussing "atmospheric" neutrino oscillation the knowledge of the flux is fundamental. L3+C has presented preliminary results at last year's International Cosmic Ray Conference (ICRC) [4] for vertical muons between 40 and 1000 GeV/c. The systematical errors still being 7.7 %, while the goal is to reach less than 3%. Of same interest is the charge ratio as a function of the momentum, where the systematic effects are less severe. At this year's conferences we expect to present improved results on the vertical spectrum, the charge ratio, and the angular dependence (new). This will sharpen the conditions on possible primary composition and parameters of interaction models at high energies, as well as constrain further the atmospheric muon neutrino flux.

3.2   Primary Composition

The primary composition of cosmic rays above 10^{14} eV is badly known. L3+C has the opportunity to analyze the composition with a new technique: Its spectrometer associated with a surface air shower scintillator array allows to record the muon spectrum for different shower sizes. The preliminary measurements [5] show that the spectra get steeper, which indicates that the primary composition gets heavier towards the knee region. The energy is distributed among more particles and therefore one observes less high energy muons. - More recent analyses made on the muon multiplicity as a function of the shower size, the shower age and the zenith angle dependence also confirm this trend, which is in agreement with the most recent conclusions from the XXVIIth ICRC (2001).

3.3   Solar Flare Signals

The Baksan collaboration running an underground muon detector, has claimed to have observed 500 GeV muons related to a solar flare occuring on the 29 Sep. 1989 (a 5.5 σ effect) [6]. The question whether protons can be accelerated to very high energies by the sun in such occasions is therefore of interest. Since the L3+C detector was taking data during the recent solar maximum, the L3 collaboration had the opportunity to collect data also on the 14th of July 2000 when a particularly strong solar flare occured. A preliminary analysis has not revealed any strong signal, neither at the time of the flare, nor up to 12 hours before or after. An estimation of the primary proton or neutron flux limit above 20 GeV is presently being calculated with the help of simulation programs.

3.4   Primary Anti-Proton to Proton Ratio at 1 TeV

Existing data on the ratio of primary anti-protons to protons are available up to 50 GeV/c [7] and are in agreement with secondary production models. L3+C should be able to give a very reliable upper limit around 1 TeV by recording the "shadow of the Moon": primary nuclei, mainly protons in the range of energy of interest here, are deviated by the earth magnetic field to the east. High energy muons produced in air showers keep nearly the original direction of the primary. Protons hitting the moon will be absorbed and therefore a lack of muons from the west side of the moon will be observed ("shadow"). The extent of the shift relative to the Moon's position, as well as the shape of the shadow depend on the momentum of the primary, the size and the direction of the field along its path. L3+C gets clear shadow signals for different muon energy ranges on the west side of the moon. Anti-protons would be deviated in the opposite direction and produce an "anti-shadow" on the east side of the Moon. From the absence of such a signal an upper limit on the anti-proton to proton ratio will be extracted.

3.5   Bursts from Point Sources and Gamma Ray Bursts

L3+C has some good features to look for point sources: a full sky survey can be made 24 hours a day, in a practically unexplored energy range (γ energies above 100 GeV) due to the low muon energy threshold. The possibility to choose the energy threshold off-line allows for an optimisation of the signal to background ratio. The background is continuously monitored and well measured, and the sources can be followed along their trajectory across the sky. The difficulties while detecting γ induced showers are of course the possible interaction of the γ on its long way to the earth with the IR or the 2.7^0K radiation, AND the fact, that gamma induced showers produce not many muons. The event rate estimation [8] shows that L3+C, with its geometrical acceptance of 200 m^2, has no chance to detect a dc-signal, but bursts could be. Some candidate events are presently under investigation.

There exist experimental hints from other experiments, that very high energy γs are also present in gamma ray bursts (GRB); according to the HEGRA collaboration possibly up to 16 TeV [9]. For such cases we may have a chance to detect a few muons. Several models also predict these numbers and assuming that the γ ray spectrum gets flatter at high energies, L3+C may observe up to 5 muons in one second against a very low background.

3.6   Correlations over Large Distances

Two groups have claimed to have observed coincident events over large distances (order 100 km) [10]. Meant is the simultaneous detection of particles in far distant detectors. A possible explanation could be the interaction of a nuclearite (strangelet) with interstellar dust. The escaping particles would enter the atmosphere and produce simultaneous showers... – The collaboration is analysing data collected simultaneously during 6 months by the Cosmo-Aleph group and L3+C for time coincident events. The distance between the two detectors amounts to 6 km.

3.7   Exotic Events

In some underground experiments unexplained events have been observed: two or three charged tracks under large angles originating from a common vertex inside or outside of the detector   [11]. This is also one of L3+C's present research.

4   Conclusions

The presented list of physics topics shows the potential multipurpose aspect of present and future large detectors at accelerators. Selected experiments in the field of cosmic rays, astrophysics, and elementary particle physics could be performed. In particular if the necessary tools have already been foreseen at an early stage of the designing phase of the detector the results could be superior to many presently dedicated experiments on cosmic rays.

References

1. L3+C detector: O. Adriani et al., to appear in *Nucl.Instr. Meth.*, 2002
2. L3 detector: B.Adeva et al., *Nucl.Instr. Meth.* A323 (1992) 109
3. V.A. Naumov, hep-ph/0201310, to appear in the Proc. of the 2nd Workshop on Methodical Aspects of Underwater/Ice Neutrino Telescopes, Hamburg, Germany, 15-16 Aug. 2001;
   G. Fiorentini, V.A. Naumov and F.L. Villante, *Phys. Lett.* B510 (2001) 173
4. P. Le Coultre, the L3 collab., Proc. of the XXVIIth ICRC, Hamburg, 2001, HE 2.01
5. H. Wilkens, the L3 collab., Proc. of the XXVIIth ICRC, Hamburg, 2001, HE 133
6. BUST: S.N. Karpov, et al., *Il Nuovo Cimento* 21 (1998) 551
7. M. Boezio et al., the CAPRICE collab., *ApJ* 561 (2001) 787
8. F. Halzen, T. Stanev, G.B. Yodh; *Phys. Rev.* D55 (1997) 4475
9. L. Padilla et al., the HEGRA collab., *Astron. Astrophys.* 337 (1998) 43
10. O. Carrel, M. Martin; *Phys. Lett.* B325 (1994) 526
11. H.S. Chen et al., in CCAST-WL workshop series: Vol. 53, Dec.1995, Beijing

Beyond the Standard Model

John Ellis

Theoretical Physics Division, CERN, Geneva, Switzerland

Abstract. Some possible directions for physics beyond the Standard Model are reviewed, with particular emphasis on topics relevant to astrophysics and cosmology, including supersymmetry at the TeV scale and neutrino physics. Constraints on the minimal supersymmetric extension of the Standard Model with universal soft supersymmetry-breaking terms (CMSSM) are discussed. The prospects for observing supersymmetry at accelerators and in non-accelerator experiments are reviewed using benchmark scenarios to focus the discussion. The status and prospects of experiments on neutrino masses and oscillations are also reviewed, as is the possibility of observing transitions between different charged leptons.

1 Introduction

Accelerator data provide no evidence for physics beyond the Standard Model. Nevertheless, particle physicists are convinced that there must be accessible physics beyond the Standard Model, because it leaves many fundamental questions unanswered. We seek a *Grand Unified Theory*, because the Standard Model has three independent gauge couplings and (potentially) a CP-violating phase in QCD. We seek a *Theory of Flavour*, because the Standard Model has six random-seeming quark masses, three disparate charged-lepton masses, three weak mixing angles and the CP-violating Kobayashi-Maskawa phase. Finally, we seek the *Origin of Mass*: is this due to a Higgs boson, and is it accompanied by supersymmetric particles? This electroweak symmetry-breaking sector has at least two free parameters, so the Standard Model has a total of 19 parameters, without even addressing the more fundamental questions of the origins of the particle quantum numbers.

Non-accelerator neutrino experiments [1,2] now provide us with the first direct evidence for physics beyond the Standard Model, convincing us that neutrinos oscillate and have different non-zero masses. To describe these, we need three neutrino mass parameters, three neutrino mixing angles and three CP-violating phases in the neutrino sector. Moreover, we should not forget about gravity, with at least two parameters to understand: Newton's constant $G_N \equiv m_P^{-2} \sim (10^{19}$ GeV$)^{-2}$ and the cosmological 'constant', which recent data suggest is non-zero [3], and may not even be constant. Talking of cosmology, we would need at least one extra parameter to produce an inflationary potential, and at least one other to generate the baryon asymmetry, which cannot be explained within the Standard Model.

Particle theorists focus on three main issues in physics beyond th Standard Model: *unification* – the quest for a single framework for all gauge interactions, *flavour* – the quest for explanations of the proliferation of quark and lepton types, their mixings and CP violating phases, and *mass* – the quest for the origin of particle masses and an explanation why they are so much smaller than the Planck mass $m_P \sim 10^{19}$ GeV. Beyond all these beyonds, other theorists seek a *Theory of Everything* that includes gravity, reconciles it with quantum mechanics, explains the origin of space-time and why we live in four dimensions (if we do so).

Physics beyond the Standard Model is therefore a very broad subject, so I am very selective in this talk. For reasons that I describe in Section 2, many theorists believe that supersymmetry is the inescapable framework for discussing physics at the TeV scale and beyond. In the rest of this talk, I first discuss the constraints imposed on (the simplest) supersymmetric models by the available experimental and cosmological constraints, then address the prospects for understanding $g_\mu - 2$ in supersymmetric models, the prospects for detecting sparticles directly at present and future colliders, and the prospects for non-collider experiments, including the searches for dark matter. Then I review the growing evidence for neutrino oscillations, which, combined with supersymmetry, suggest that flavour-violating decays of charged leptons such as $\mu \to e\gamma$ decay might be observable in forthcoming experiments.

2   The Electroweak Vacuum

The generation of particle masses requires the breaking of gauge symmetry in the vacuum:

$$m_{W,Z} \neq 0 \Leftrightarrow < 0|X_{I,I_3}|0 > \neq 0 \tag{1}$$

for some field X with isospin I and third component I_3. The measured ratio

$$\rho \equiv \frac{m_W^2}{m_Z^2 \cos^2 \theta_W} \simeq 1 \tag{2}$$

tells us that X mainly has $I = 1/2$ [4], which is also what is needed to generate fermion masses. The key question is the nature of the field X: is it elementary or composite? A fermion-antifermion condensate $v \equiv < 0|X|0 > = < 0|\bar{F}F|0 > \neq 0$ would be analogous to what we know from QCD, where $< 0|\bar{q}q|0 > \neq 0$, and conventional superconductivity, where $< 0|e^-e^-|0 > \neq 0$. However, analogous 'technicolour' models of electroweak symmetry breaking [5] fail to fit the values of the radiative corrections ϵ_i to ρ and other quantities extracted from the precision electroweak data provided by LEP and other experiments, as seen in Fig. 1 [6]. One cannot exclude the possibility that some calculable variant of technicolour might emerge that is consistent with the data, but for now we focus on elementary Higgs models.

Within this framework, the data favour a relatively light Higgs boson, with $m_H \simeq 115$ GeV, just above the exclusion unit provided by direct searches at LEP,

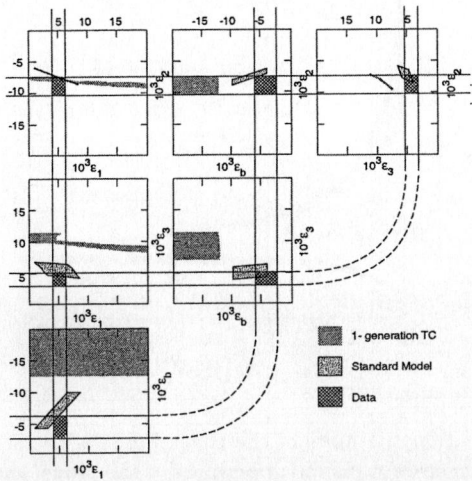

Fig. 1. Predictions for the radiative corrections ϵ_i in the Standard Model and a minimal one-generation model [7] are compared with the precision electroweak data [6]

being the 'most-probable' [8]. This is one reason why many theorists were excited by the possible sighting during the last days of LEP of a Higgs boson, with a preferred mass of 115.6 GeV [9]. If this were to be confirmed, it would suggest that the Standard Model breaks down at some relatively low energy $\lesssim 10^3$ TeV [10]. As seen in Fig. 2, above this scale the effective Higgs potential of the Standard Model becomes unstable as the quartic Higgs self-coupling is driven negative by radiative corrections due to the relatively heavy top quark [11]. This is not necessarily a disaster, and it is possible that the present electroweak vacuum might be metastable, provided that its lifetime is longer than the age of the Universe [12]. However, we would surely feel more secure if such instability could be avoided.

This may be done by introducing new bosons ϕ coupled to the Higgs field [10]:

$$\lambda_{22}|H|^2\,|\phi|^2 \quad : \quad M_0^2 \equiv \lambda_{22}v^2 \tag{3}$$

As seen in Fig. 3a, the effective potential is very sensitive to the coupling parameter M_0: for $M_0 \leq 70.9$ GeV in this example, the potential still collapses, whereas for $M_0 \geq 71.0$ GeV the potential blows up instead. Thus the bosonic coupling (3) must be finely tuned [10]. This occurs naturally in supersymmetry, in which the Higgs bosons are accompanied by fermionic partners $\tilde{H}$. As seen in Fig. 3b, again the Higgs coupling blows up in the absence of the $\tilde{H}$, whereas it is well behaved in the minimal supersymmetric extension of the Standard Model (MSSM).

The avoidance of fine tuning has long been the primary motivation for supersymmetry at the TeV scale [13]. This issue is normally formulated in connection with the hierarchy problem: why/how is $m_W \ll m_P$, or equivalently why is $G_F \sim 1/m_W^2 \gg G_N = 1/m_P^2$, or equivalently why does the Coulomb potential in an atom dominate over the Newton potential, $e^2 \gg G_N m_p m_e \sim (m/m_P)^2$,

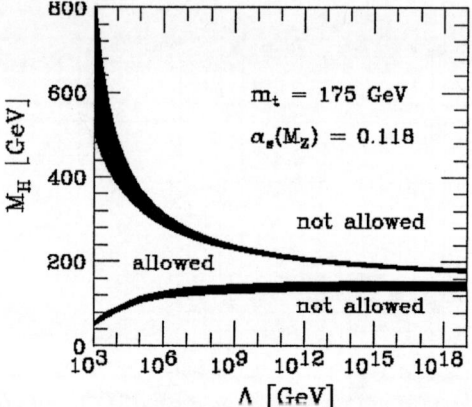

Fig. 2. The range allowed for the mass of the Higgs boson if the Standard Model is to remain valid up to a given scale Λ. In the upper part of the plane, the effective potential blows up, whereas in the lower part the present electroweak vacuum is unstable [11]

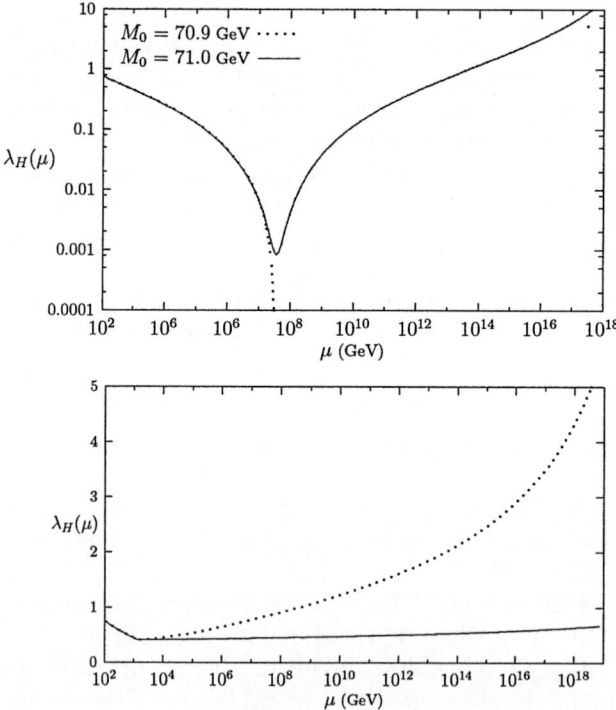

Fig. 3. (a) If the quartic coupling M_0 (3) is too large, the effective potential blows up (solid line), whereas it is unstable if M_0 is too small (dotted line), indicating a need for fine tuning. (b) This occurs naturally in a supersymmetric model (solid line) but not if the $\tilde{H}$ are omitted (dotted line) [10]

where $m_{p,e}$ are the proton and electron masses? One might think naively that it would be sufficient to set $m_W \ll m_P$ by hand. However, radiative corrections tend to destroy this hierarchy. For example, one-loop diagrams generate

$$\delta m_W^2 = \mathcal{O}\left(\frac{\alpha}{\pi}\right) \Lambda^2 \gg m_W^2 \tag{4}$$

where Λ is a cut-off representing the appearance of new physics, and the inequality in (4) applies if $\Lambda \sim 10^3$ TeV, and even more so if $\Lambda \sim m_{GUT} \sim 10^{16}$ GeV or $\sim m_P \sim 10^{19}$ GeV. If the radiative corrections to a physical quantity are much larger than its measured values, obtaining the latter requires strong cancellations, which in general require fine tuning of the bare input parameters. However, the necessary cancellations are natural in supersymmetry, where one has equal numbers of bosons B and fermions F with equal couplings, so that (4) is replaced by

$$\delta m_W^2 = \mathcal{O}\left(\frac{\alpha}{\pi}\right) |m_B^2 - m_F^2| . \tag{5}$$

The residual radiative correction is naturally small if

$$|m_B^2 - m_F^2| \lesssim 1 \text{ TeV}^2 \tag{6}$$

Note that this argument is logically distinct from that in the previous paragraph. There supersymmetry was motivated by the control of logarithmic divergences, and here by the absence of quadratic divergences.

3   The MSSM

The MSSM has the same gauge interactions as the Standard Model, and similar Yukawa couplings. A key difference is the necessity of two Higgs doublets, in order to give masses to all the quarks and leptons, and to cancel triangle anomalies. This duplication is important for phenomenology: it means that there are five physical Higgs bosons, two charged $H^{\pm}$ and three neutral h, H, A. Their quartic self-interactions are determined by the gauge interactions, solving the vacuum instability problem mentioned above and limiting the possible mass of the lightest neutral Higgs boson. However, the doubling of the Higgs multiplets introduces two new parameters: $\tan\beta$, the ratio of Higgs vacuum expectation values and μ, a parameter mixing the two Higgs doublets.

There are two key experimental hints in favour of supersymmetry. One is provided by the LEP measurements of the gauge couplings, that are in very good agreement with supersymmetric GUTs [14] if sparticles weigh ~ 1 TeV. This agreement appears completely fortuitous in composite Higgs models [5], and is difficult (though not impossible [15]) to reproduce accurately in models with large extra dimensions [16]. The other experimental hint is provided by the preference of the precision electroweak data for a relatively light Higgs boson [8]. In the MSSM, one predicts $m_h \lesssim 130$ GeV [17], right in the preferred range, whereas composite Higgs model generally predict heavier effective Higgs masses.

The gauge symmetries of the MSSM would permit the inclusion of interactions that violate baryon number and/or lepton number [18]:

$$\lambda L L E^c + \lambda' Q D^c L + \lambda'' U^c D^c D^c \tag{7}$$

where the $L(Q)$ are left-handed lepton (quark) doublets and the $E^c(D^c, U^d)$ are conjugates of the right-handed lepton (quark) singlets. Their possible appearance is ignored in this talk, in which case the lightest supersymmetric particle is stable, and hence a candidate for dark matter [19]. In the following this is assumed to be a neutralino, i.e., a mixture of the $\tilde{\gamma}, \tilde{H}$ and $\tilde{Z}$.

The final ingredient in the MSSM is the soft supersymmetry breaking, in the form of scalar masses m_0, gaugino masses $m_{1/2}$ and trilinear couplings A [20]. These are presumed to be inputs from physics at some high-energy scale, e.g., from some supergravity or superstring theory, which then evolve down to lower energy scale according to well-known renormalization-group equations. In the case of the Higgs multiplets, this renormalization can drive the effective mass-squared negative, triggering electroweak symmetry weaking [21]. In this talk, it is assumed that the m_0 are universal at the input scale [1], as are the $m_{1/2}$ and A parameters. In this case the free parameters are

$$m_0, m_{1/2}, A \quad \text{and} \quad \tan \beta , \tag{8}$$

with μ being determined by the electroweak vacuum conditions, up to a sign.

This constrained MSSM (CMSSM) serves as the basis for the subsequent discussion. It has the merit of being sufficiently specific that the different phenomenological constraints can be combined meaningfully. On the other hand, it is just one of the phenomenological possibilities offered by supersymmetry [23].

4   Constraints on the CMSSM

Important constraints on the CMSSM parameter space are provided by direct searches at LEP and the Tevatron collider, as seen in Fig. 4. One of these is the limit $m_{\chi^\pm} \gtrsim 103$ GeV provided by chargino searches at LEP, where the third significant figure depends on other CMSSM parameters. LEP has also provided lower limits on slepton masses, of which the strongest is $m_{\tilde{e}} \gtrsim 99$ GeV, again depending only sightly on the other CMSSM parameters, as long as $m_{\tilde{e}} - m_\chi \gtrsim 10$ GeV. The most important constraints on the u, d, s, c, b squarks and gluinos are provided by the Tevatron collider: for equal masses $m_{\tilde{q}} = m_{\tilde{g}} \gtrsim 300$ GeV. In the case of the $\tilde{t}$, LEP provides the most stringent limit when $m_{\tilde{t}} - m_\chi$ is small, and the Tevatron for larger $m_{\tilde{t}} - m_\chi$. Their effect is almost to exclude the range of parameter space where electroweak baryogenesis is possible [24].

Another important constraint is provided by the LEP limit on the Higgs mass: $m_H > 114.1$ GeV. This holds in the Standard Model, for the lightest

---

[1] Universality between the squarks and sleptons of different generations is motivated by upper limits on flavour-changing neutral interactions [22], but universality between the soft masses of the L, E^c, Q^c, D^c and U^c is not so well motivated.

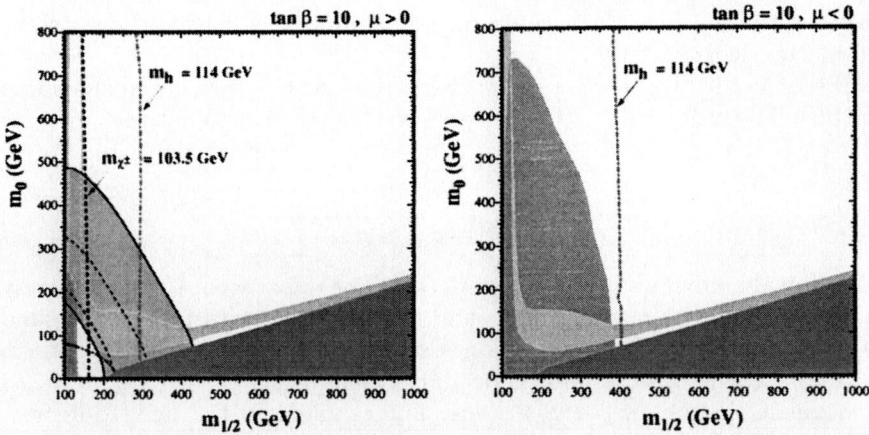

Fig. 4. Compilations of phenomenological constraints on the CMSSM for $\tan\beta = 10$ and (a) $\mu > 0$, (b) $\mu < 0$. Representative contours of the selectron, chargino and Higgs masses are indicated, as is the likely physics reach of Run II of the Tevatron Collider in (a). The dark shaded regions are excluded because the LSP is charged, whereas a neutralino LSP has acceptable relic density (10) in the light-shaded regions [25]. The medium-shaded region in (b) is excluded by $b \to s\gamma$ [26]. Panel (a) also includes a shaded region favoured by the most recent measurement of $g_\mu - 2$

Higgs boson h in the general MSSM for $\tan\beta \lesssim 5$, and in the CMSSM for all $\tan\beta$, at least as long as CP is conserved [2]. Since m_h is sensitive to sparticle masses, particularly $m_{\tilde{t}}$, via loop corrections:

$$\delta m_h^2 \propto \frac{m_t^4}{m_W^2} \ln\left(\frac{m_{\tilde{t}}^2}{m_t^2}\right) + \ldots \tag{9}$$

the Higgs limit also imposes important constraints on the CMSSM parameters, principally $m_{1/2}$ as seen in Fig. 4.

Also shown in Fig. 4 is the constraint imposed by measurements of $b \to s\gamma$ [26]. These agree with the Standard Model, and therefore provide bounds on chargino and charged Higgs masses, for example. For moderate $\tan\beta$, the $b \to s\gamma$ constraint is more important for $\mu < 0$, as seen in Fig. 4b, but it is also significant for $\mu > 0$ when $\tan\beta$ is large.

Figure 4 also displays the regions where the supersymmetric relic density $\rho_\chi = \Omega_\chi \rho_{critical}$ falls within the preferred range

$$0.1 < \Omega_\chi h^2 < 0.3 \tag{10}$$

The upper limit is rigorous, since astrophysics and cosmology tell us that the total matter density $\Omega_m \lesssim 0.4$, and the Hubble expansion rate $h \sim 1/\sqrt{2}$ to within about 10% (in units of km/s/Mpc). On the other hand, the lower limit

---

[2] The lower bound on the lightest MSSM Higgs boson may be relaxed significantly if CP violation feeds into the MSSM Higgs sector [27].

in (10) is optional, since there could be other important contributions to the overall matter density.

As is seen in Fig. 4, there are generic regions of the CMSSM parameter space where the relic density falls within the preferred range (10). What goes into the calculation of the relic density? It is controlled by the annihilation rate [19]:

$$\rho_\chi = m_\chi n_\chi : n_\chi \sim \frac{1}{\sigma_{ann}(\chi\chi \to \ldots)} \tag{11}$$

and the typical annihilation rate $\sigma_{ann} \sim 1/m_\chi^2$. For this reason, the relic density typically increases with the relic mass, and this combined with the upper bound in (10) then leads to the common expectation that $m_\chi \lesssim 1$ TeV. However, there are various ways in which the generic upper bound on m_χ can be increased along filaments in the $(m_{1/2}, m_0)$ plane. For example, if the next-to-lightest sparticle (NLSP) is not much heavier than χ: $\Delta m/m_\chi \lesssim 0.1$, the relic density may be suppressed by coannihilation: $\sigma(\chi+\text{NLSP}\to \ldots)$ [28]. In this way, the allowed CMSSM region may acquire a 'tail' extending to large m_χ, as in the case where the NLSP is the lighter stau: $\tilde\tau_1$ and $m_{\tilde\tau_1} \sim m_\chi$ as seen in Fig. 5 [29]. Another mechanism for extending the allowed CMSSM region to large m_χ is rapid annihilation via a direct-channel pole when $m_\chi \sim \frac{1}{2}m_{Higgs,Z}$ [30,31]. This may yield a 'funnel' extending to large $m_{1/2}$ and m_0 at large $\tan\beta$, as seen in Fig. 6 [31]. Another allowed region at large $m_{1/2}$ and m_0 is the 'focus-point' region [32], which is adjacent to the boundary of the region where electroweak symmetry breaking is possible, as seen in Fig. 7. However, in this region m_χ is not particularly large.

These filaments extending the preferred CMSSM parameter space are clearly exceptional, in some sense, so it is important to understand the sensitivity of the relic density to input parameters, unknown higher-order effects, etc. One

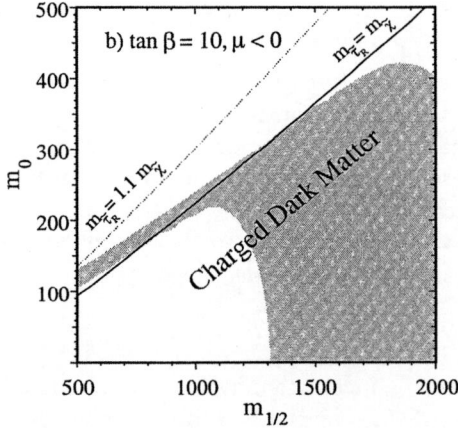

Fig. 5. The large-$m_{1/2}$ 'tail' of the $\chi - \tilde\tau_1$ coannihilation region for $\tan\beta = 10$ and $\mu < 0$ [29]

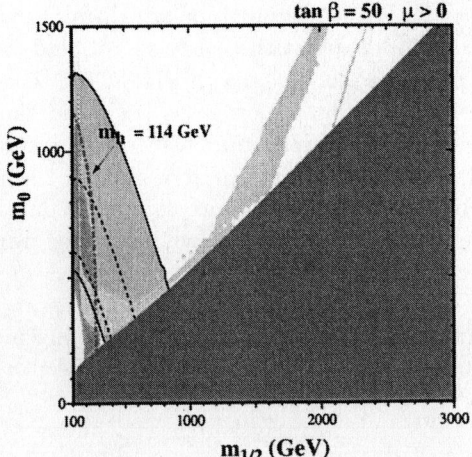

Fig. 6. The region where the cosmological relic density is in the preferred range (10) for $\tan\beta = 50$ and $\mu > 0$. Note the rapid-annihilation 'funnel' at intermediate $m_0/m_{1/2}$ [31]

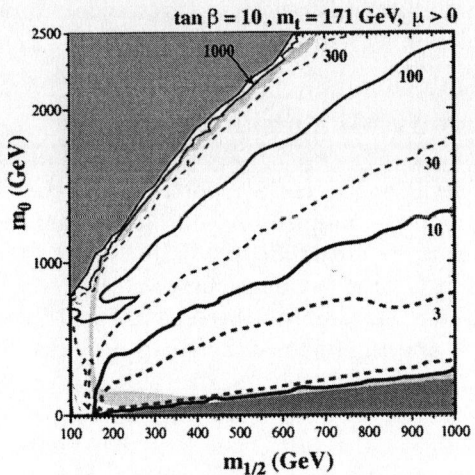

Fig. 7. The $m_{1/2}, m_0$ plane for $\tan\beta = 10$ and $\mu > 0$, including the 'focus-point' region [32] at large m_0, close to the boundary of the shaded region where electroweak symmetry breaking occurs, and exhibiting contours of the cosmological sensitivity (12) [33]

proposal is the relic-density fine-tuning measure [33]

$$\Delta^{\Omega} \equiv \sqrt{\sum_i \frac{\partial \ln(\Omega_\chi h^2)}{\partial \ln a_i}} \qquad (12)$$

where the sum runs over the input parameters, which might include (relatively) poorly-known Standard Model quantities such as m_t and m_b, as well as the CMSSM parameters $m_0, m_{1/2}$, etc. As seen in Fig. 7, the sensitivity Δ^Ω (12) is relatively small in the 'bulk' region at low $m_{1/2}, m_0$, and $\tan\beta$. However, it is somewhat higher in the $\chi - \tilde{\tau}_1$ coannihilation 'tail', and at large $\tan\beta$ in general. The sensitivity measure Δ^Ω (12) is particularly high in the rapid-annihilation 'funnel' and in the 'focus-point' region. This explains why published relic-density calculations may differ in these regions [34], whereas they agree well when Δ^Ω is small: differences may arise because of small differences in the treatments of the inputs.

It is important to note that the relic-density fine-tuning measure (12) is distinct from the traditional measure of the fine-tuning of the electroweak scale [35]:

$$\Delta_i \equiv \frac{\partial \ln m_W}{\partial \ln a_i} \qquad (13)$$

This electroweak fine-tuning is a completely different issue, and values of the Δ_i are not necessarily related to values of Δ^Ω. Electroweak fine-tuning is sometimes used as a criterion for restricting the CMSSM parameters. However, the interpretation of the Δ_i (13) is unclear. How large a value of Δ_i is tolerable? Different physicists may well have different pain thresholds. Moreover, correlations between input parameters may reduce its value in specific models.

5   Muon Anomalous Magnetic Moment

The BNL E821 experiment has recently reported an updated measurement of the muon anomalous magnetic moment $g_\mu - 2$ [36], which seems to indicate a 3.0-σ deviation from the Standard Model prediction based on e^+e^- data [37]. If this discrepancy is real, it is strong evidence for new physics at the TeV scale. As many authors have pointed out [38], this discrepancy could well be explained by supersymmetry if $\mu > 0$ and $\tan\beta$ is not too small, as exemplified in Figs. 4(a) and 6. Good consistency with all the experimental and cosmological constraints on the CMSSM is found for $\tan\beta \lesssim 10$ and $m_\chi \simeq 150$ to 350 GeV. Already before the measurement [36], the LHC was thought to have a good chance of discovering supersymmetry [39]. If the BNL result were to be confirmed, this would be almost guaranteed, as we now discuss.

6   Prospects for Observing Supersymmetry at Accelerators

As an aid to the assessment of the prospects for detecting sparticles at different accelerators, benchmark sets of supersymmetric parameters have often been found useful [40], since they provide a focus for concentrated discussion. A set of post-LEP benchmark scenarios in the CMSSM has recently been proposed [41], and are illustrated schematically in Fig. 8. They take into account the direct searches for sparticles and Higgs bosons, $b \to s\gamma$ and the preferred cosmological density range (10). About a half of the proposed benchmark points are consistent with the BNL $g_\mu - 2$ value at the $2 - \sigma$ level, but this was not imposed as an absolute requirement.

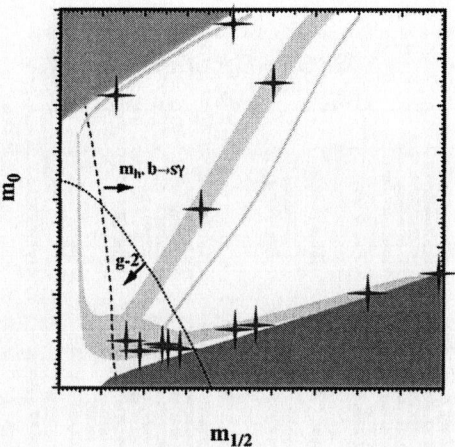

Fig. 8. Schematic overview of the benchmark points proposed in [41]. They were chosen to be compatible with the indicated experimental constraints, as well as have a relic density in the preferred range (10). The points are intended to illustrate the range of available possibilities

The proposed points were chosen not to provide an 'unbiased' statistical sampling of the CMSSM parameter space, whatever that means in the absence of a plausible *a priori* measure, but rather are intended to illustrate the different possibilities that are still allowed by the present constraints [41]. Five of the chosen points are in the 'bulk' region at small $m_{1/2}$ and m_0, four are spread along the coannihilation 'tail' at larger $m_{1/2}$ for various values of $\tan\beta$, two are in the 'focus-point' region at large m_0, and two are in rapid-annihilation 'funnels' at large $m_{1/2}$ and m_0. The proposed points range over the allowed values of $\tan\beta$ between 5 and 50. Most of them have $\mu > 0$, as favoured by $g_\mu - 2$, but there are two points with $\mu < 0$.

Various derived quantities in these supersymmetric benchmark scenarios, including the relic density, $g_\mu - 2, b \to s\gamma$, electroweak fine-tuning Δ and the relic-density sensitivity Δ^Ω, are given in [41]. These enable the reader to see at a glance which models would be excluded by which refinement of the experimental value of $g_\mu - 2$. Likewise, if you find some amount of fine-tuning uncomfortably large, then you are free to discard the corresponding models.

The LHC collaborations have analyzed their reach for sparticle detection in both generic studies and specific benchmark scenarios proposed previously [39]. Based on these studies, Fig. 9 displays estimates how many different sparticles may be seen at the LHC in each of the newly-proposed benchmark scenarios [41]. The lightest Higgs boson is always found, and squarks and gluinos are usually found, though there are some scenarios where no sparticles are found at the LHC. The LHC often misses heavier weakly-interacting sparticles such as charginos, neutralinos, sleptons and the other Higgs bosons.

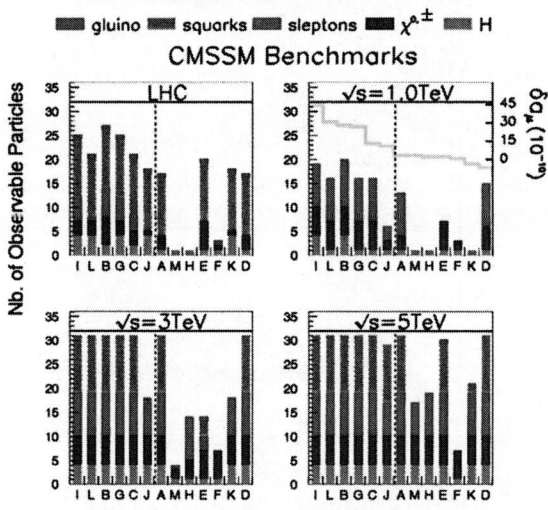

Fig. 9. Estimates of the numbers of different types of CMSSM particles that may be detectable [41] at (a) the LHC, (b) a 1-TeV linear e^+e^- collider [42], and (c,d) a 3(5)-TeV e^+e^- [43] or $\mu^+\mu^-$ collider [44]. Note the complementarity between the sparticles detectable at the LHC and at a 1-TeV linear e^+e^- collider

The physics capabilities of linear e^+e^- colliders are amply documented in various design studies [42]. Not only is the lightest MSSM Higgs boson observed, but its major decay modes can be measured with high accuracy, as seen in Fig. 10. Moreover, if sparticles are light enough to be produced, their masses and other properties can be measured very precisely, enabling models of supersymmetry breaking to be tested [45].

As seen in Fig. 9, the sparticles visible at an e^+e^- collider largely complement those visible at the LHC [41]. In most of benchmark scenarios proposed, a 1-TeV linear collider would be able to discover and measure precisely several weakly-interacting sparticles that are invisible or difficult to detect at the LHC. However, there are some benchmark scenarios where the linear collider (as well as the LHC) fails to discover supersymmetry. Only a linear collider with a higher centre-of-mass energy appears sure to cover all the allowed CMSSM parameter space, as seen in the lower panels of Fig. 12, which illustrate the physics reach of a higher-energy lepton collider, such as CLIC [43] or a multi-TeV muon collider [44].

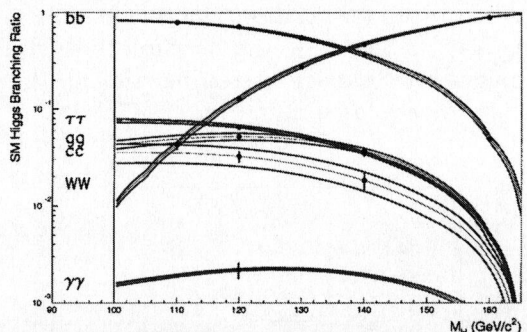

Fig. 10. Analysis of the accuracy with which Higgs decay branching ratios may be measured with a linear e^+e^- collider [46]

7   Search for Supersymmetric Dark Matter

Searches among cosmic rays with energies up to about 1 TeV provide several promising signatures for supersymmetric dark matter particles, that may enable this community to 'scoop' the LHC.

One possibility is to look for energetic *gamma rays* that may be emitted by LSP annihilations in the core of the Milky Way. The benchmark models indicate that these might have typical energies $\sim$ 10 GeV, and detectors such as GLAST with a threshold as low as $\sim$ 1 GeV might have a better chance, as seen in Fig. 11 [47]. The prospects for these searches depend on the degree to which the dark-matter particle density may be enhanced in the core of the Milky Way, which is uncertain by orders of magnitude. In Fig. 11, a middle-of-the-road enhancement by a factor 200 has been assumed.

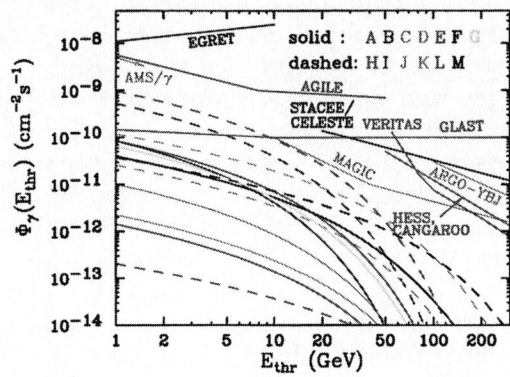

Fig. 11. Observations of γ rays from the galactic centre by GLAST and ground-based experiments may be able to test certain supersymmetric benchmark scenarios [47]

Another possibility is to look for *positrons* emitted by LSP annihilations in the halo of the Milky Way. In this case, the benchmark models indicate that energies ~ 100 GeV might be the most interesting, though the signal may be less promising than in the γ case, as seen in Fig. 12 [47].

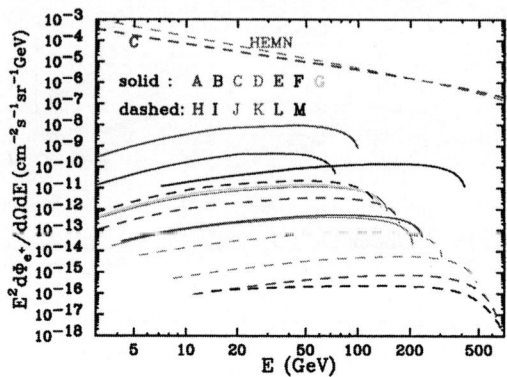

Fig. 12. Comparison between the cosmic-ray positron background and the fluxes from the annihilations of relic particles, calculated [47] in supersymmetric benchmark scenarios

One of the most promising signatures is *energetic muons* produced in the Earth by energetic neutrinos emitted by LSP annihilations in the centre of the Sun or Earth. In this case, as seen in Fig. 13, all energies up to ~ 1 TeV might be important. According to our calculations [47], the prospects for detecting relic annihilations in the core of the Sun appear more promising than those in the centre of the Earth. Some upper limits on the energetic solar muon flux have already been produced, most recently by the AMANDA Collaboration [48], which already begin to exclude some more extreme supersymmetric models.

The most convincing evidence for supersymmetric dark matter might eventually come from direct searches for the *scattering of relic particles* on nuclei in the laboratory [49]. Here the best chances seem to be offered by spin-independent scattering on relatively heavy nuclei. There has been a claim by the DAMA Collaboration [50] to have observed an annual modulation effect due to the scattering of dark matter particles, but this interpretation of their data has been largely excluded by the CDMS [51], EDELWEISS [52] and UKDMC experiments. In any case, reproducing the DAMA data would have required a scattering cross section much larger than predicted in the simple supersymmetric models studied in [47] and [53]. As illustrated in Fig. 14, future large cryogenic detectors, such as that proposed by the Heidelberg group, would good chances in many supersymmetric scenarios.

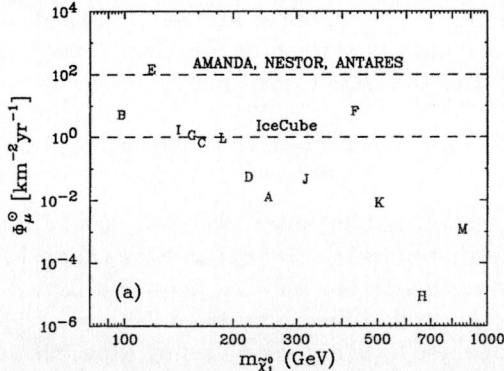

Fig. 13. Searches in IceCube and other km^2 detectors [48] for energetic muons originating from the interactions of high-energy neutrinos produced by the annihilations of supersymmetric relic particles captured inside the Sun may probe some supersymmetric benchmark scenarios [47]

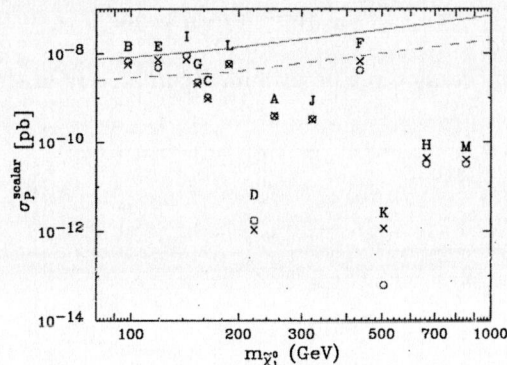

Fig. 14. Direct searches for the scattering of superysmmetric relic particles in underground detectors may probe some supersymmetric benchmark scenarios [47]

8   Neutrino Oscillations and Masses

There is no fundamental theoretical reason why neutrinos should not have masses and mix with one another. It is generally thought that particles are massless if and only if they are associated with an exact gauge symmetry. Examples include the photon and gluon, which are thought to be massless because of $U(1)$ and $SU(3)$ gauge symmetries, respectively. There is no exact gauge symmetry associated with lepton number L, so it is expected that lepton number should be violated at some level. If L is violated by two units: $\Delta L = 2$, then neutrino masses may arise.

The minimal renormalizable model of neutrino masses requires the introduction of weak-singlet 'right-handed' neutrinos N. These will in general couple to the conventional weak-doublet left-handed neutrinos via Yukawa couplings Y_ν that yield Dirac masses $m_D \sim m_W$. In addition, these 'right-handed' neutrinos

N can couple to themselves via Majorana masses M that may be $\gg m_W$, since they do not require electroweak symmetry breaking. Combining the two types of mass term, one obtains the seesaw mass matrix [54]:

$$(\nu_L, N) \begin{pmatrix} 0 & M_D \\ M_D^T & M \end{pmatrix} \begin{pmatrix} \nu_L \\ N \end{pmatrix}, \tag{14}$$

where each of the entries should be understood as a matrix in generation space. Diagonalizing the neutrino mass matrix (14) and the charged-lepton masses introduces in general a mismatch between the mass and flavour eigenstates [55], leading to oscillations between different neutrino flavours.

Direct evidence for these have been provided by experiments on atmospheric and solar neutrinos. As seen in Fig. 15, the atmospheric neutrino data are described well by [1]

$$\Delta m^2 \sim 2.5 \times 10^{-3} \text{ eV}^2, \theta \sim 45^0, \tag{15}$$

and the solar neutrino data are described well by [2]

$$\Delta m^2 \sim 5 \times 10^{-5} \text{ eV}^2, \theta \sim 30^0, \tag{16}$$

though other solutions cannot yet be excluded completely in this case.

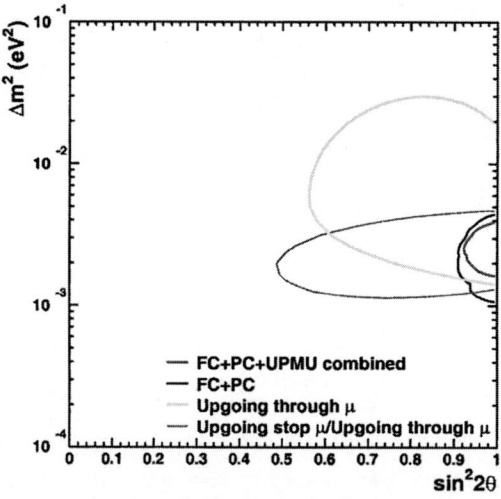

Fig. 15. A fit to the Super-Kamiokande data on atmospheric neutrinos [1] indicates near-maximal $\nu_\mu - \nu_\tau$ mixing with $\Delta m^2 \sim 2.5 \times 10^{-3}$ eV^2

Oscillation experiments do not, by themselves, determine the absolute values of the neutrino masses. Nevertheless, if we assume that the ranges (15, 16) characterize also the absolute magnitudes of the heaviest neutrino species, they would not provide a large amount of astrophysical dark matter, though future measurements of the cosmic microwave background radiation and large-scale structure might become sensitive to the atmospheric neutrino mass range (15) [56].

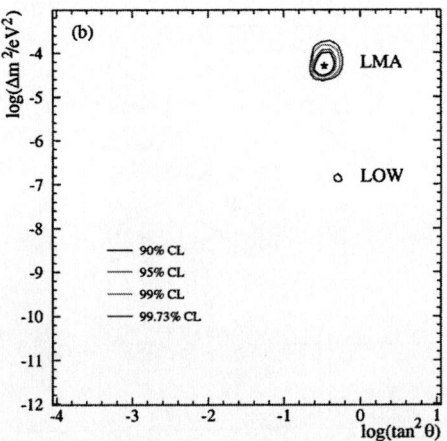

Fig. 16. A global fit to solar neutrino data, following the SNO measurements of the total neutral-current reaction rate, the energy spectrum and the day-night asymmetry, favours large mixing and $\Delta m^2 \sim 6 \times 10^{-5}$ eV^2 [2]

The BNL E821 report of a possible deviation from the Standard Model suggests that a non-trivial $\mu - \mu - \gamma$ vertex is generated at a scale $\lesssim 1$ TeV. Neutrino oscillations indicate that there are $\Delta L_\mu \neq 0$ processes, so it is natural to expect that there might also be a non-trivial $\mu - e - \gamma$ vertex. This is indeed the case in a generic supersymmetric seesaw model, where neutrino mixing induces slepton mixing [57]. Within this framework, the measurement of $g_\mu - 2$ fixes the sparticle scale, and $\Gamma(\mu \to e\gamma)$ may then be calculated within any given flavour texture, as illustrated in Fig. 17 [58].

The electric dipole moments of the electron and muon might also be measurable, and possibly a transverse asymmetry in $\mu \to 3e$ decay, which would demonstrate CP violation in the charged-lepton sector. This may provide another interesting interface with neutrino physics and cosmology [59]. The minimal supersymmetric seesaw model has six CP-violating phases: including one that may be measurable in neutrino oscillation experiments, two that could affect the neutrinoless double-β decay rate, and three other phases that may be responsible for our existence via leptogenesis in the early Universe [60].

9   Conclusions

As we have seen, future colliders such as the LHC and a TeV-scale linear e^+e^- collider have good prospects of discovering supersymmetry and making detailed measurements. In parallel, B and ν factories have good prospects of making inroads on the flavour and unification problems. Searches for dark matter and stopped-muon experiments also have interesting prospects.

Looking further beyond the Standard Model, how can one hope to test a Theory of Everything, including quantum gravity? This should be our long-term

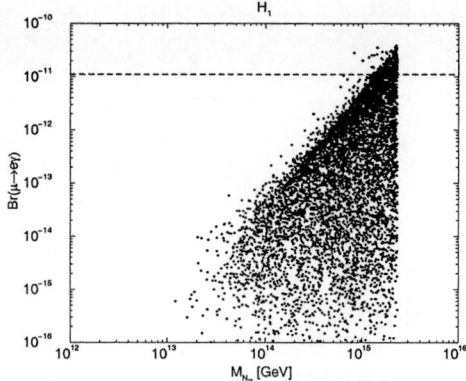

Fig. 17. Illustration that $\mu \rightarrow e\gamma$ decay may occur at a rate close to the present experimental upper limit, in the minimal supersymmetric seesaw model with parameters chosen to match the measured value of $g_\mu - 2$ [36] and neutrino oscillation experiments [1,2]

ambition, our analogue of the 'faint blue dot' towards which exoplanetary science is directed, and which motivates much of its funding. Testing a quantum theory of gravity will be relatively easy if there are large extra dimensions [61]. Much more challenging would be the search for observable effects if the gravitational scale turns out, after all, to be of the same order as the Planck mass $\sim 10^{19}$ GeV. Perhaps the only way to reconcile relativity with quantum mechanics is to modify one or the other, or both [62]? Testing the Theory of Everything may require thinking beyond the standard 'Beyond the Standard Model' box.

One thing seems certain: many of the fundamental problems in cosmology, such as the origin of matter and the nature of dark matter, can only be resolved by physics beyond the Standard Model. It is encouraging that many of the ideas for such physics can be tested by experiments in the near future.

References

1. Y. Fukuda *et al.* [Super-Kamiokande Collaboration], Phys. Rev. Lett. **81**, 1562 (1998) [arXiv:hep-ex/9807003].
2. Q. R. Ahmad *et al.* [SNO Collaboration], Phys. Rev. Lett. **89** (2002) 011301 [arXiv:nucl-ex/0204008]; Phys. Rev. Lett. **89** (2002) 011302 [arXiv:nucl-ex/0204009]; S. Fukuda *et al.* [Super-Kamiokande Collaboration], Phys. Lett. B **539** (2002) 179 [arXiv:hep-ex/0205075].
3. A. G. Riess *et al.* [Supernova Search Team Collaboration], Astron. J. **116** (1998) 1009 [arXiv:astro-ph/9805201]; S. Perlmutter *et al.* [Supernova Cosmology Project Collaboration], Astrophys. J. **517** (1999) 565 [arXiv:astro-ph/9812133]; N. A. Bahcall, J. P. Ostriker, S. Perlmutter and P. J. Steinhardt, Science **284** (1999) 1481 [arXiv:astro-ph/9906463].
4. D. A. Ross and M. J. Veltman, Nucl. Phys. B **95** (1975) 135.

5. For a classic review,see: E. Farhi and L. Susskind, Phys. Rept. **74** (1981) 277; for a recent review and references, see: K. D. Lane, *Technicolor 2000*, arXiv:hep-ph/0007304.

6. See, for example:
   G. Altarelli, arXiv:hep-ph/0011078, as updated in G. Altarelli, F. Caravaglios, G. F. Giudice, P. Gambino and G. Ridolfi, JHEP **0106** (2001) 018 [arXiv:hep-ph/0106029].

7. J. R. Ellis, G. L. Fogli and E. Lisi, Phys. Lett. B **343** (1995) 282.

8. See the LEP Collaborations, ALEPH, DELPHI, L3, OPAL, the LEP Electroweak Working Group and the SLD Heavy Flavour and Electroweak Groups, LEPEWWG/2001-01, available from http://lepewwg.web.cern.ch/LEPEWWG/Welcome.html.

9. LEP Higgs Working Group for Higgs boson searches, OPAL Collaboration, ALEPH Collaboration, DELPHI Collaboration and L3 Collaboration, *Search for the Standard Model Higgs Boson at LEP*, ALEPH-2001-066, DELPHI-2001-113, CERN-L3-NOTE-2699, OPAL-PN-479, LHWG-NOTE-2001-03, CERN-EP/2001-055, arXiv:hep-ex/0107029.

10. J. R. Ellis and D. Ross, Phys. Lett. B **506** (2001) 331 [arXiv:hep-ph/0012067].

11. For a review, see: T. Hambye and K. Riesselmann, arXiv:hep-ph/9708416.

12. G. Isidori, G. Ridolfi and A. Strumia, Nucl. Phys. B **609** (2001) 387 [arXiv:hep-ph/0104016].

13. L. Maiani, *Proceedings of the 1979 Gif-sur-Yvette Summer School On Particle Physics*, 1; G. 't Hooft, in *Recent Developments in Gauge Theories, Proceedings of the Nato Advanced Study Institute, Cargese, 1979*, eds. G. 't Hooft *et al.*, (Plenum Press, NY, 1980); E. Witten, Phys. Lett. B **105** (1981) 267.

14. J. Ellis, S. Kelley and D. V. Nanopoulos, Phys. Lett. B **260** (1991) 131; U. Amaldi, W. de Boer and H. Furstenau, Phys. Lett. B **260** (1991) 447; P. Langacker and M. x. Luo, Phys. Rev. D **44** (1991) 817; C. Giunti, C. W. Kim and U. W. Lee, Mod. Phys. Lett. A **6** (1991) 1745.

15. K. R. Dienes, E. Dudas and T. Gherghetta, Phys. Lett. B **436** (1998) 55 [arXiv:hep-ph/9803466] and Nucl. Phys. B **537** (1999) 47 [arXiv:hep-ph/9806292].

16. D. M. Ghilencea and G. G. Ross, Nucl. Phys. B **606** (2001) 101 [arXiv:hep-ph/0102306].

17. J. Ellis, G. Ridolfi and F. Zwirner, Phys. Lett. B **257** (1991) 83; M.S. Berger, Phys. Rev. D **41** (1990) 225; Y. Okada, M. Yamaguchi and T. Yanagida, Prog. Theor. Phys. **85** (1991) 1; Phys. Lett. B **262** (1991) 54; H.E. Haber and R. Hempfling, Phys. Rev. Lett. **66** (1991) 1815; R. Barbieri, M. Frigeni and F. Caravaglios, Phys. Lett. B **258** (1991) 167; P.H. Chankowski, S. Pokorski and J. Rosiek, Phys. Lett. B **274** (1992) 191; J.R. Espinosa and M. Quirós, Phys. Lett. B **266** (1991) 389; J.L. Lopez and D.V. Nanopoulos, Phys. Lett. B **266** (1991) 397; M. Carena, K. Sasaki and C.E.M. Wagner, Nucl. Phys. B **381** (1992) 66; P.H. Chankowski, S. Pokorski and J. Rosiek, Phys. Lett. B **281** (1992) 100; Nucl. Phys. B **423** (1994) 437; D.M. Pierce, A. Papadopoulos and S.B. Johnson, Phys. Rev. Lett. **68** (1992) 3678; A. Brignole, Phys. Lett. B **281** (1992) 284; M. Drees and M.M. Nojiri, Phys. Rev. D **45** (1992) 2482; V. Barger, M.S. Berger and P. Ohmann, Phys. Rev. D **49** (1994) 4908; G.L. Kane, C. Kolda, L. Roszkowski and J.D. Wells, Phys. Rev. D **49** (1994) 6173; R. Hempfling and A.H. Hoang, Phys. Lett. B **331** (1994) 99; P. Langacker and N. Polonsky, Phys. Rev. D **50** (1994) 2199; H.E. Haber, R. Hempfling and A.H. Hoang, Zeit. für Phys. C **75** (1997) 539; S. Heinemeyer, W. Hollik and G. Weiglein, Comput. Phys. Commun. **124**, 76 (2000) [hep-ph/9812320].

162   J. Ellis

18. For a review, see: H. Dreiner, arXiv:hep-ph/9707435.
19. J. Ellis, J.S. Hagelin, D.V. Nanopoulos, K.A. Olive and M. Srednicki, Nucl. Phys. B **238** (1984) 453; see also H. Goldberg, Phys. Rev. Lett. **50** (1983) 1419.
20. S. Dimopoulos and H. Georgi, Nucl. Phys. B **193** (1981) 150; N. Sakai, Z. Phys. C **11** (1981) 153.
21. K. Inoue, A. Kakuto, H. Komatsu and S. Takeshita, Prog. Theor. Phys. **68** (1982) 927 [Erratum-ibid. **70** (1982) 330]; L.E. Ibáñez and G.G. Ross, Phys. Lett. B **110** (1982) 215; L.E. Ibáñez, Phys. Lett. B **118** (1982) 73; J. Ellis, D.V. Nanopoulos and K. Tamvakis, Phys. Lett. B **121** (1983) 123; J. Ellis, J. Hagelin, D.V. Nanopoulos and K. Tamvakis, Phys. Lett. B **125** (1983) 275; L. Alvarez-Gaumé, J. Polchinski, and M. Wise, Nucl. Phys. B **221** (1983) 495.
22. J. R. Ellis and D. V. Nanopoulos, Phys. Lett. B **110** (1982) 44; R. Barbieri and R. Gatto, Phys. Lett. B **110** (1982) 211.
23. M. Dine and A. E. Nelson, Phys. Rev. D **48**, 1277 (1993) [hep-ph/9303230]; M. Dine, A. E. Nelson and Y. Shirman, Phys. Rev. D **51**, 1362 (1995) [hep-ph/9408384]; M. Dine, A. E. Nelson, Y. Nir and Y. Shirman, Phys. Rev. D **53**, 2658 (1996) [hep-ph/9507378]. D. E. Kaplan, G. D. Kribs and M. Schmaltz, Phys. Rev. D **62**, 035010 (2000) [hep-ph/9911293]; Z. Chacko, M. A. Luty, A. E. Nelson and E. Ponton, JHEP **0001**, 003 (2000) [hep-ph/9911323]. L. Randall and R. Sundrum, Nucl. Phys. B **557**, 79 (1999) [hep-th/9810155]; G. F. Giudice, M. A. Luty, H. Murayama and R. Rattazzi, JHEP **9812**, 027 (1998) [hep-ph/9810442].
24. M. Fukugita and T. Yanagida, Phys. Lett. B **174** (1986) 45; for a recent review, see: W. Buchmuller, arXiv:hep-ph/0107153.
25. J. R. Ellis, G. Ganis, D. V. Nanopoulos and K. A. Olive, Phys. Lett. B **502** (2001) 171 [arXiv:hep-ph/0009355].
26. M.S. Alam et al., [CLEO Collaboration], Phys. Rev. Lett. **74** (1995) 2885 as updated in S. Ahmed et al., CLEO CONF 99-10; BELLE Collaboration, BELLE-CONF-0003, contribution to the 30th International conference on High-Energy Physics, Osaka, 2000. See also K. Abe *et al.*, [Belle Collaboration], [arXiv:hep-ex/0107065]; L. Lista [BaBar Collaboration], [arXiv:hep-ex/0110010]; C. Degrassi, P. Gambino and G. F. Giudice, JHEP **0012** (2000) 009; M. Carena, D. Garcia, U. Nierste and C. E. Wagner, hep-ph/0010003; G. Isidori, talk at this conference.
27. M. Carena, J. R. Ellis, A. Pilaftsis and C. E. Wagner, Nucl. Phys. B **586** (2000) 92 [arXiv:hep-ph/0003180], Phys. Lett. B **495** (2000) 155 [arXiv:hep-ph/0009212]; and references therein.
28. S. Mizuta and M. Yamaguchi, Phys. Lett. B **298** (1993) 120 [arXiv:hep-ph/9208251]; J. Edsjo and P. Gondolo, Phys. Rev. D **56** (1997) 1879 [arXiv:hep-ph/9704361]. C. Boehm, A. Djouadi and M. Drees, Phys. Rev. D **62** (2000) 035012 [arXiv:hep-ph/9911496].
29. J. Ellis, T. Falk and K. A. Olive, Phys. Lett. B **444**, 367 (1998); J. Ellis, T. Falk, K. A. Olive and M. Srednicki, Astropart. Phys. **13** (2000) 181; M. E. Gómez, G. Lazarides and C. Pallis, Phys. Rev. D **61**, 123512 (2000) [hep-ph/9907261] and Phys. Lett. B **487**, 313 (2000) [hep-ph/0004028]; R. Arnowitt, B. Dutta and Y. Santoso, hep-ph/0102181.
30. M. Drees and M. M. Nojiri, Phys. Rev. D **47**, 376 (1993); H. Baer and M. Brhlik, Phys. Rev. D **53** (1996) 597 and Phys. Rev. D **57** (1998) 567; H. Baer, M. Brhlik, M. A. Diaz, J. Ferrandis, P. Mercadante, P. Quintana and X. Tata, Phys. Rev. D **63** (2001) 015007; A. B. Lahanas, D. V. Nanopoulos and V. C. Spanos, hep-ph/0009065.
31. J. R. Ellis, T. Falk, G. Ganis, K. A. Olive and M. Srednicki, Phys. Lett. B **510** (2001) 236 [arXiv:hep-ph/0102098].

32. J. L. Feng, K. T. Matchev and T. Moroi, Phys. Rev. Lett. **84**, 2322 (2000) [hep-ph/9908309]; J. L. Feng, K. T. Matchev and T. Moroi, Phys. Rev. D **61**, 075005 (2000) [hep-ph/9909334]; J. L. Feng, K. T. Matchev and F. Wilczek, Phys. Lett. B **482**, 388 (2000) [hep-ph/0004043].

33. J. R. Ellis and K. A. Olive, Phys. Lett. B **514** (2001) 114 [arXiv:hep-ph/0105004].

34. For other recent calculations, see, for example: A. B. Lahanas, D. V. Nano-poulos and V. C. Spanos, Phys. Lett. B **518** (2001) 94 [arXiv:hep-ph/0107151]; L. Roszkowski, R. Ruiz de Austri and T. Nihei, JHEP **0108**, 024 (2001) [arXiv:hep-ph/0106334].

35. J. Ellis, K. Enqvist, D. V. Nanopoulos and F. Zwirner, Mod. Phys. Lett. A **1**, 57 (1986); R. Barbieri and G. F. Giudice, Nucl. Phys. B **306** (1988) 63.

36. H. N. Brown *et al.* [Muon g-2 Collaboration], Phys. Rev. Lett. **86**, 2227 (2001) [hep-ex/0102017]; G. W. Bennett *et al.* [Muon g-2 Collaboration], Phys. Rev. Lett. **89** (2002) 101804 [Erratum-ibid. **89** (2002) 129903] [arXiv:hep-ex/0208001].

37. M. Davier, S. Eidelman, A. Hocker and Z. Zhang, arXiv:hep-ph/0208177; see also K. Hagiwara, A. D. Martin, D. Nomura and T. Teubner, arXiv:hep-ph/0209187; F. Jegerlehner, unpublished, as reported in M. Krawczyk, arXiv:hep-ph/0208076.

38. L. L. Everett, G. L. Kane, S. Rigolin and L. Wang, Phys. Rev. Lett. **86**, 3484 (2001) [arXiv:hep-ph/0102145]; J. L. Feng and K. T. Matchev, Phys. Rev. Lett. **86**, 3480 (2001) [arXiv:hep-ph/0102146]; E. A. Baltz and P. Gondolo, Phys. Rev. Lett. **86**, 5004 (2001) [arXiv:hep-ph/0102147]; U. Chattopadhyay and P. Nath, Phys. Rev. Lett. **86**, 5854 (2001) [arXiv:hep-ph/0102157]; S. Komine, T. Moroi and M. Yamaguchi, Phys. Lett. B **506**, 93 (2001) [arXiv:hep-ph/0102204]; S. P. Martin and J. D. Wells, Phys. Rev. D **64**, 035003 (2001) [arXiv:hep-ph/0103067]; H. Baer, C. Balazs, J. Ferrandis and X. Tata, Phys. Rev. D **64**, 035004 (2001) [arXiv:hep-ph/0103280]; J. Ellis, D. V. Nanopoulos and K. A. Olive, Phys. Lett. B **508** (2001) 65 [arXiv:hep-ph/0102331]; R. Arnowitt, B. Dutta, B. Hu and Y. Santoso, Phys. Lett. B **505** (2001) 177 [arXiv:hep-ph/0102344].

39. ATLAS Collaboration, *ATLAS detector and physics performance Technical Design Report*, CERN/LHCC 99-14/15 (1999); S. Abdullin *et al.* [CMS Collaboration], hep-ph/9806366; S. Abdullin and F. Charles, Nucl. Phys. B **547** (1999) 60; CMS Collaboration, Technical Proposal, CERN/LHCC 94-38 (1994).

40. See, for example: I. Hinchliffe, F. E. Paige, M. D. Shapiro, J. Soderqvist and W. Yao, Phys. Rev. D **55** (1997) 5520; TESLA Technical Design Report, DESY-01-011, Part III, *Physics at an e^+e^- Linear Collider* (March 2001).

41. M. Battaglia *et al.*, Eur. Phys. J. C **22** (2001) 535 [arXiv:hep-ph/0106204].

42. S. Matsumoto *et al.* [JLC Group], *JLC-1*, KEK Report 92-16 (1992); J. Bagger *et al.* [American Linear Collider Working Group], *The Case for a 500-GeV e^+e^- Linear Collider*, SLAC-PUB-8495, BNL-67545, FERMILAB-PUB-00-152, LBNL-46299, UCRL-ID-139524, LBL-46299, Jul 2000, hep-ex/0007022; T. Abe *et al.* [American Linear Collider Working Group Collaboration], *Linear Collider Physics Resource Book for Snowmass 2001*, SLAC-570, hep-ex/0106055, hep-ex/0106056, hep-ex/0106057 and hep-ex/0106058; TESLA Technical Design Report, DESY-01-011, Part III, *Physics at an e^+e^- Linear Collider* (March 2001).

43. R. W. Assmann *et al.* [CLIC Study Team], *A 3-TeV e^+e^- Linear Collider Based on CLIC Technology*, ed. G. Guignard, CERN 2000-08; CLIC Physics Study Group, http://clicphysics.web.cern.ch/CLICphysics/.

44. Neutrino Factory and Muon Collider Collaboration, http://www.cap.bnl.gov/mumu/mu_home_page.html; European Muon Working Groups, http://muonstoragerings.cern.ch/Welcome.html.

45. G. A. Blair, W. Porod and P. M. Zerwas, Phys. Rev. **D63** (2001) 017703 [hep-ph/0007107].
46. M. Battaglia and K. Desch, arXiv:hep-ph/0101165.
47. J. Ellis, J. L. Feng, A. Ferstl, K. T. Matchev and K. A. Olive, arXiv:astro-ph/0110225.
48. J. Ahrens *et al.* [AMANDA Collaboration], arXiv:astro-ph/0208006.
49. M. W. Goodman and E. Witten, Phys. Rev. D **31** (1985) 3059.
50. DAMA Collaboration, R. Bernabei *et al.*, Phys. Lett. B **436**, 379 (1998).
51. D. Abrams *et al.* [CDMS Collaboration], arXiv:astro-ph/0203500.
52. A. Benoit *et al.* [EDELWEISS Collaboration], Phys. Lett. B **513** (2001) 15 [arXiv:astro-ph/0106094].
53. J. R. Ellis, A. Ferstl and K. A. Olive, Phys. Lett. B **532** (2002) 318 [arXiv:hep-ph/0111064].
54. M. Gell-Mann, P. Ramond and R. Slansky, Proceedings of the Supergravity Stony Brook Workshop, New York, 1979, eds. P. Van Nieuwenhuizen and D. Freedman (North-Holland, Amsterdam); T. Yanagida, Proceedings of the Workshop on Unified Theories and Baryon Number in the Universe, Tsukuba, Japan 1979 (edited by A. Sawada and A. Sugamoto, KEK Report No. 79-18, Tsukuba); R. Mohapatra and G. Senjanovic, Phys. Rev. Lett. **44** (1980) 912.
55. Z. Maki, M. Nakagawa and S. Sakata, Prog. Theor. Phys. **28** (1962) 870.
56. S. Hannestad, arXiv:astro-ph/0211106.
57. J. Hisano, T. Moroi, K. Tobe and M. Yamaguchi, Phys. Rev. D **53** (1996) 2442; J. Hisano, D. Nomura and T. Yanagida, Phys. Lett. B **437** (1998) 351; J. Hisano and D. Nomura, Phys. Rev. D **59** (1999) 116005; S. F. King and M. Oliveira, Phys. Rev. D **60** (1999) 035003 [arXiv:hep-ph/9804283]. W. Buchmüller, D. Delepine and F. Vissani, Phys. Lett. B **459** (1999) 171; M. E. Gomez, G. K. Leontaris, S. Lola and J. D. Vergados, Phys. Rev. D **59** (1999) 116009; J. R. Ellis, M. E. Gomez, G. K. Leontaris, S. Lola and D. V. Nanopoulos, Eur. Phys. J. C **14** (2000) 319; W. Buchmüller, D. Delepine and L. T. Handoko, Nucl. Phys. B **576** (2000) 445; J. L. Feng, Y. Nir and Y. Shadmi, Phys. Rev. D **61** (2000) 113005; J. Sato and K. Tobe, Phys. Rev. D **63** (2001) 116010; T. Blazek and S. F. King, Phys. Lett. B **518** (2001) 109 [arXiv:hep-ph/0105005]. J. Hisano and K. Tobe, Phys. Lett. B **510** (2001) 197; S. Baek, T. Goto, Y. Okada and K. Okumura, hep-ph/0104146; S. Lavignac, I. Masina and C.A. Savoy, hep-ph/0106245.
58. J. R. Ellis, J. Hisano, S. Lola and M. Raidal, Nucl. Phys. B **621**, 208 (2002) [arXiv:hep-ph/0109125]; J. R. Ellis, J. Hisano, M. Raidal and Y. Shimizu, Phys. Lett. B **528**, 86 (2002) [arXiv:hep-ph/0111324]; arXiv:hep-ph/0206110.
59. J. R. Ellis and M. Raidal, Nucl. Phys. B **643** (2002) 229 [arXiv:hep-ph/0206174].
60. M. Fukugita and T. Yanagida, Phys. Lett. **B174**, 45 (1986).
61. I. Antoniadis and K. Benakli, Int. J. Mod. Phys. A **15** (2000) 4237 [arXiv:hep-ph/0007226].
62. J. R. Ellis, N. E. Mavromatos and D. V. Nanopoulos, arXiv:gr-qc/9909085.

Alpha: A Constant That Is Not a Constant?

G. Fiorentini and B. Ricci

Dipartimento di Fisica, Universitá di Ferrara and Istituto Nazionale di Fisica Nucleare, Sezione di Ferrara, I-44100 Ferrara, Italy

Abstract. We review the observational information on the constancy of the fine structure constant α. We find that small improvements on the measurement of ^{187}Re lifetime can provide significant progress in exploring the range of variability suggested by QSO data. We also discuss the effects of a time varying α on stellar structure and evolution. We find that radioactive dating of ancient stars can offer a new observational window.

1 Introduction

The possibility that some of the "fundamental constants" may depend on time was first discussed by Dirac [1]. He remarked that the huge ratio of electric to gravitational forces between a proton and an electron, about 10^{39}, was of the same order of magnitude as the age of the universe in units provided by the atomic constants, $e^2/m_e c^3$. If this coincidence is not casual, then one must have varying constants, their values changing as the age of the universe changes: *"This suggests that the above mentioned large numbers are to be regarded not as constants, but as simple functions of our present epoch, expressed in atomic units ... In this way we avoid the need of a theory to determine numbers of the order 10^{39}."* The approach of Dirac to what is now called the hierarchy problem opened a rich field of investigation. The variability of fundamental constants was analysed by Gamow, Dyson and others and then it was forgotten for a while.

Interest in this topic has been revived in the context of string theories, where all the coupling constants and parameters, except the string tension, are actually derived quantities, which are determined by the vacuum expectation values of the dilaton and moduli. Since all these fields evolve on cosmological scales the time variation of the constants of nature during the evolution of the universe arises as a natural possibility, see e.g. [2,3].

On the observational side, Webb et al. [4] have presented evidence for a cosmological evolution of the fine structure constant $\alpha = e^2/\hbar c$. The absorption spectra of diffuse clouds illuminated by quasars suggest that ten billion years ago α was slightly smaller, by about ten part per million. Of course this indication, if confirmed, would have enormous importance.

This short review attempts to provide an answer to some natural questions following the claim of ref.[4]:

i)What are the observational constraints on the variability of α and how do they compare with the result of ref. [4]?

ii)What are the prospects for improvements?

iii)What are the effects of a time varying α on stellar structure and evolution?

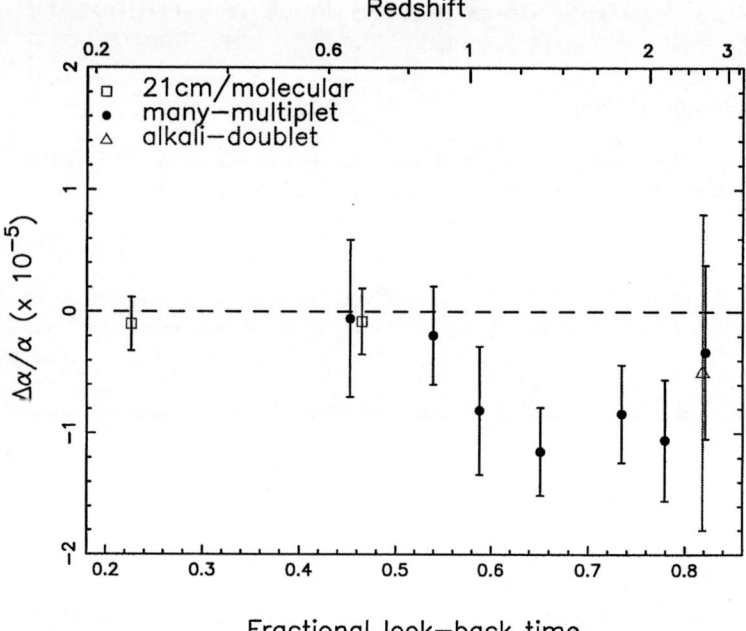

Fig. 1. $\Delta\alpha/\alpha$ vs. fractional look-back time to the Big Bang, from [4]

2   What Do Quasars Tell Us?

The measurement of the spectra of distant quasars as a mean to study possible variations of α was first suggested by Savedoff [5]. Narrow lines in quasar spectra are produced by absorption of radiation in intervening clouds of gases. Essentially one needs to identify two (sets of) lines, which depend differently on α, so as to extract the value of the redshift factor z together with the value of α at that epoch. The fine structure doublets of "alkali atoms" – a term used to denote atoms and atomic ions with just one electron in the outer shell – are well suited for this study.

This method has been used by several authors and it has been recently applied to a selection of high resolution observations, see [6]. No indication of a variable α has been found and the constraint $\Delta\alpha/\alpha = (-4.6\pm4.3[\text{stat}]\pm1.4[\text{sys}])10^{-5}$ has been obtained [6] on the possible deviation at $z = 2 \div 4$ from the present $(z = 0)$ value.

On the other hand, Webb et al. [4] have used a "many multiplet" method, where α is estimated from comparison of the lines of *different* atomic species, so as to obtain a sensitivity gain. The data are summarized in Fig. 1. In this way they claim to have found a deviation from the present α value over the redshift range $z = 1 \div 3$:

$$\Delta\alpha/\alpha = (-0.72 \pm 0.18) \cdot 10^{-5} \tag{1}$$

This result has been criticized in ref. [6] on the grounds that some systematic effect could mimic the variation of α. For example, the lines of the two atomic species considered in [4] are situated in different regions, so that calibration errors could simulate the effect of α variation. In contrast, the method based on the fine splitting of a line of the same species is not affected by these uncertainty sources.

Table 1. Summary on the variation of α

Source	$\Delta\alpha/\alpha$	Look back time (Gyr)	z^*	$\dot{\alpha}/\alpha$ (yr^{-1})	ref.
Laboratory	$\leq 1.6 \cdot 10^{-14}$	$4 \cdot 10^{-10}$	0	$\leq 4 \cdot 10^{-14}$	[7]
Oklo	$\leq 1 \cdot 10^{-7}$	1.8	$\simeq 0.1$	$\leq 6 \cdot 10^{-17}$	[11]
Meteorites	$\leq 4 \cdot 10^{-6}$	4.5	$\simeq 0.4$	$\leq 2 \cdot 10^{-15}$	–
^{12}C	$\leq 10^{-2}$	$\simeq 10$	$\simeq 1.5$	$\leq 10^{-12}$	–
stellar dating	$\leq 10^{-3}$	$\simeq 10$	$\simeq 1.5$	$\leq 10^{-13}$	–
QSO(doublet)	$\leq 10^{-4}$	$\simeq 11 - 13$	2–4	$\leq 10^{-14}$	[6]
QSO(multiplet)	$+1 \cdot 10^{-5}$	$\simeq 8 - 12$	1–3	$+1 \cdot 10^{-15}$	[4]
CMB	$\leq 5 \cdot 10^{-2}$	$\simeq 14$	$\simeq 10^3$	$\leq 3 \cdot 10^{-12}$	[19]
BBN	$\leq 1 \cdot 10^{-2}$	$\simeq 14$	$\simeq 10^9$	$\leq 7 \cdot 10^{-13}$	[19]

* The red shift – time connection is estimated for $H_o = 68$ Km/s/Mpc, $\Omega_M = 0.3$ and $\Omega_\Lambda = 0.7$ ($t_u \simeq 14$ Gyr).

3 Quasars and the Rest of the World

The available information on the variability of α is summarized in Table 1. Measurement in the laboratory are sensitive to extremely tiny variations $\Delta\alpha/\alpha \simeq 10^{-14}$, however on a time scale of just a few months. Essentially, one is comparing two clocks (a Hg$^+$ atomic clock and a Hydrogen maser), with frequencies which depend differently on α [7]. Experiments with cold atoms will provide a significant sensitivity gain. In fact, the ultimate limit for frequency measurement is observation time. Cold atoms in the laboratory fall due to gravity, whereas atoms in free fall do not fall at all, so let's go to space. This is the idea of an extremely interesting project on the International Space Station, which is expected to explore changes of α to the level $\Delta\alpha/\alpha \simeq 10^{-16}$ [8].

The physics of the fission reactor which nature operated at Oklo about two billion years ago provides a very important constraint. The footprints of natural fission arise from the abundances of rare earth isotopes at the Oklo site, which look similar to those produced by fission *today*. These isotopic abundances are related to large capture cross section of thermal neutrons, which correspond to nuclear resonances at about the thermal energies. The similarity of the abundances means thus that, in two billion years, nuclear energy levels has not varied by more than $kT \simeq 0.1$ eV, a very small range in comparison with the nuclear physics scale. The Coulomb contribution to the difference of nuclear energy levels. $E_{Cou} \simeq \alpha/r_{nuc}$, is thus strongly fixed, corresponding to $|\Delta\alpha/\alpha| \lesssim 10^{-7}$ (barring from accidental cancellation due to variations of other fundamental parameters). This was first pointed out in [9] and then discussed with much greater detail in [10,11].

The Oklo bound arises, essentially, from the fact that the Q value of a nuclear reaction contains an electromagnetic contribution which is sensitive to changes of α. The constancy of Q, within a small scale of order kT, follows from the observation that reaction rates are the same, now and at the Oklo time. Conceptually, one is again comparing two (nuclear) clocks operating at different times.

A similar argument can be applied to radioactive dating methods. The point is that the lifetimes τ of radioactive nuclei depend on the Q-value of the decay. In addition, α-decay rates have an exponential dependence on α, corresponding to the exponentially small tunnel probability. The most sensitive process is $^{187}Re \rightarrow ^{187}Os + e + \bar{\nu}$ due to a very small Q-value: $\Delta\tau/\tau \simeq 1.8 \cdot 10^4 \Delta\alpha/\alpha$ [12], see Table 2. The laboratory measurement $\tau_{1/2}(lab) = (42.3 \pm 0.7)$Gyr (68%C.L.) [15] can thus be compared [17] with the value inferred from Re/Os measurement in ancient meteorites $\tau_{1/2}(met) = (41.6 \pm 0.42)$ Gyr [16], dated by means of different radioactive methods (e.g. U/Th method, which is much less weakly affected by variation of α). The agreement within errors (again apart for accidental cancellations) provides a significant constraint, $\overline{\Delta\alpha/\alpha} = (1 \pm 1)10^{-6}$, where the bar denotes an average over the meteorite lifetime. It is not as strong as the Oklo bound, however it explores earlier times. Furthermore, two independent constraints (Oklo and meteorites) are important if one consider the possibility of simultaneous variations of several fundamental parameters.

Cosmic microwave background (CMB) yields information on the variability of α at even earlier times, since decoupling between radiation and matter occurs at the recombination epoch, the time when temperature is so low that atoms can be stable. This clearly depends on the atomic binding energy and thus on α [18]. A change of α also affects the Big-Bang Nucleosynthesis (BBN) mainly through the neutron-proton mass difference, which fixes the neutron density at the weak interactions freeze-out and consequently the primordial ^4He abundance. According to most recent analysis, both CMB and BBN are consistent with a a constant α and constrain $\Delta\alpha/\alpha$ to the per cent level, at $z \simeq 10^3$ and 10^9 respectively [19].

For comparing the different information one has to make assumptions about the time evolution of α. The simplest hypothesis is a linear time dependence, which is used in the fifth column of Table 1. In this case the QSO positive result only conflicts with the Oklo bound, which provides the most strict constraint.

It is interesting to observe that the meteorites give a bound close to the QSO signal. Improvements in the laboratory measurements of the ^{187}Re lifetime would thus be relevant. The planned atomic physics experiment on the ISS should be capable of exploring the region suggested by QSO.

Let us remark that the QSO-Oklo apparent conflict can be avoided if linearity does not hold. In addition α could depend on place. Also one has to remind that α is not alone; its evolution has to be accompanied by the evolution of other coupling constants, otherwise unification of interactions at the present epoch is just occasional. Actually, unification requires that a change of α is accompanied by a much stronger change in strong interaction parameters, see e.g. [20]. A change of α corresponds to a change of the QCD scale parameter $\Delta\Lambda_{QCD}/\Lambda_{QCD} \simeq 40\Delta\alpha/\alpha$. This has important consequences, since nucleon and pion masses scale as $M_n \propto \Lambda_{QCD}$ and $M_\pi \propto \sqrt{\Lambda_{QCD}}$. Nuclear radii, which depend on the range of the nuclear force, are thus also sensitive to a change of Λ_{QCD}.

In this situation, the analysis becomes much more complex, see e.g. [20–23] and a complete discussion has not yet been performed. Generally, one can remark the following points:

-Information on the change of α give also information on the couplings of other interactions.

-Several different sources of information are needed, for disentangling the contributions of different effects.

For these reasons, improvements of the various methods, which explore different space-time regions and receive contributions from different interactions, are important so as to confirm or constrain a possible variation of fundamental "constants".

Table 2. α–dependence of nuclear halflives, from [12]

Nucleus	Decay	$\tau_{1/2}$ [yr]	$d(\ln\tau)/d(\ln\alpha)$
^{238}U	α	$2 \cdot 10^9$	$\simeq -500$
^{40}K	EC	$1.3 \cdot 10^9$	$\simeq +30$
^{187}Re	β	$4 \cdot 10^{10}$	$\simeq +18000$

4   Stars and α

Stars are a useful laboratory for studying fundamental physics. In principle, at least, a change of α over very long times can affect stellar structure and evolution. Clearly, a change of α will affect nuclear fusion rates and opacity. For the former, one has an exponential effect in the tunnelling probability, whereas opacity scales with a small power of α (e.g. $\kappa \propto \alpha^2$ for Thomson scattering). In this spirit we shall briefly discuss a few relevant points:

4.1   The Sun

Helioseismology has provided us with a detailed knowledge of the present solar interior, well in agreement with Standard Solar Model (SSM) calculations which have been performed by assuming that α has been constant. Can a time dependent α spoil this agreement? We have constructed solar models with a linearly time dependent α such that the difference between solar formation and present is $\Delta\alpha/\alpha = 10^{-2}$. Note that this is a much larger variation than that implied by Oklo and/or meteorite constraints. The tiny difference in sound speed with respect to the SSM is well within the errors of helioseismic determination, see Fig. 2.

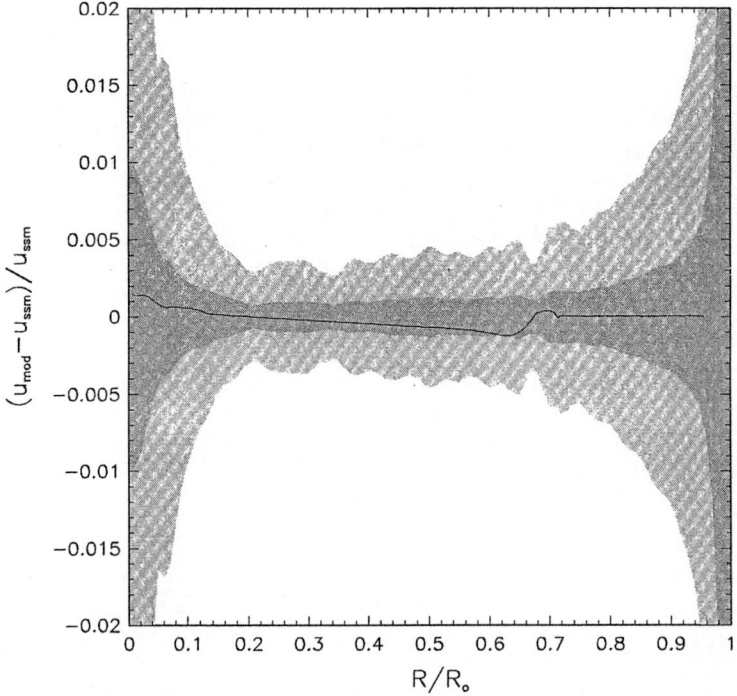

Fig. 2. Relative difference (model-SSM)/SSM of (isothermal) sound speed squared as a function of the radial coordinate for a time variation of α such that the difference between sun formation and present is $\Delta\alpha/\alpha = 10^{-2}$. The 1σ (3σ) helioseismic uncertainty corresponds to the dark (light) area, from [24]

4.2 Globular Clusters

As well known, the evolution of these systems provides a powerful method for determining the age of the Galaxy, see. e.g. [25]. Is this dating method affected by a time variation of α? Again the answer is negative, even for variation at the percent level over the Galaxy age, see Fig. 3, where we present isochrones at $t = 11$ Gyr calculated for the M68 cluster, as an illustrative example.

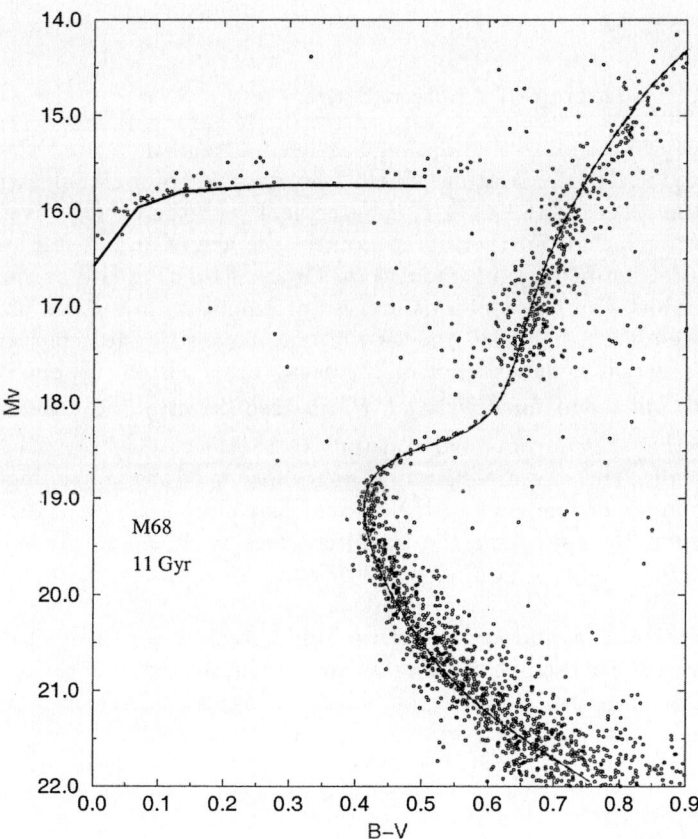

Fig. 3. Isochrones calculated with constant α (solid line) and with a time dependent α such that the difference between cluster formation and present is $\Delta\alpha/\alpha = 10^{-2}$ (red dashed line). The two curves look almost indistinguishable

4.3   Stellar Nucleosynthesis of ^{12}C

Our very existence relies on a nuclear accident, i.e. a suitably ^{12}C excited level which allows the carbon synthesis by means of $\alpha + \alpha + \alpha \rightarrow^{12} C^* \rightarrow^{12} C + \gamma$. Carbon synthesis occurs at $kT \simeq 10$KeV and the resonance position is measured at $(m_{12}^* - 3m_4)c^2 = 379.5$Kev. The observation of carbon in ancient stars implies that some 10 Gyr ago the resonance energy was the same, within kT. Thus the Coulomb contribution to the energy difference levels (about α/r_{nuc}) has not changed by more than kT, which implies $\Delta\alpha/\alpha \leq 10^{-2}$. Essentially, this is the same argument as for the Oklo reactor, however the bound is weaker since kT is larger.

4.4   Radioactive Dating of Ancient Stars

In the last few years, radioactive dating has been extended beyond the solar system, see e.g. [13]. Thorium dating of field halo stars and globular cluster stars yields ages on the order of (15 ± 4) Gyr, in agreement with the value derived from globular cluster evolution. Furthermore, recently the age of an old star ($\tau \simeq 12$ Gyr) has been determined by means of *both* Th and U dating [14] so that two clocks are available! The two methods are in agreement within errors of about 3 Gyr, under the assumption that nuclear lifetimes have remained constant. By exploiting the different α dependence of the decay rates, (from [12] one derives $\frac{d(ln\tau)}{d(ln\alpha)} = -450$ and -470 for ^{238}U and ^{232}Th respectively) the coherence of results implies that α has remained constant to the level $\Delta\alpha/\alpha \simeq 2.5 \cdot 10^{-2}$ on a 12 Gyr scale. There is a substantial cancellation of the α varying effect due to the similar α–dependence of the two nuclear clocks. One can achieve a stronger constraint by comparing the Uranium clock with the dating provided by globular cluster evolution (which is not affected by α changes). In this case one has $\Delta\alpha/\alpha \lesssim 10^{-3}$.

The measurement of stellar age from Uranium decay is at presently limited by incomplete knowledge of oscillator strengths and production rates of the elements produced in the r-process. However, significant progress can be expected, as theory and observation shall progress.

Acknowledgements

We are extremely grateful to C. Bonadiman, C. Chiosi, V. Castellani, S. Degl'Innocenti, F. Fusi-Pecci, H. Fritsch, G. Ottonello and F.L. Villante for useful discussions.

References

1. P.A.M. Dirac: Nature **139**, 323 (1937)
2. T. Damour and A.M. Polyakov, Nucl. Phys. **B 243**, 532 (1994).

3. E. Witten, in *Sources and Detection of Dark Matter and Dark Energy*, ed. D. Cline, (Springer 2000), pp. 27–36.

4. J. Webb et al. Phys. Rev. Lett. **87** 091301 (2001).

5. M. P. Savedoff, Nature **264** 340 (1956).

6. D.A. Varshalovich, A. Y. Potekhin and A.V. Ivanchik physics/000406, in *X-ray and Inner-Shell Processes*, ed by R.W. Dunford, D.S. Gemmel, E.P. Kanter, B. Kraessig, S.H. Southworth, L. Young, AIP Conf. Proc. (AIP, Melville, 2000) vol. 506, p. 503

7. T. Prestage et al., Phys. Rev. Lett. **74**, 3511 (1995).

8. www.cnes.fr/activities/connaissance/physique/aces/1sommaire_aces.htm

9. A. I. Shlyakhter Nature **264**, 340 (1976)

10. T. Damour and F. Dyson, Nuc. Phys. **B 480**, 37 (1996)

11. Y. Fujii et al., Nuc. Phys. **B 573**, 377 (2000)

12. F.F. Dyson, *The fundamental constant and their time variation*, in "Aspects of Quantum Theory", eds. A. Salam and E.P. Wigner (Cambridge University Press 1972), pp

13. J.W. Truran et al., astro-ph/0109526, in *Astrophysical Ages and Time Scales*, ASP Conference Series, Vol. TBD 2001, eds. T. von Hipped, N. Manset and C. Simpson.

14. R. Cayrel et al., Nature **409**, 691 (2001).

15. M. Lindner et al., Geoch. Cosmoch. Acta **53**, 1597 (1989)

16. M.I. Smoliar, R. J. Walker and J.W. Morgan, Science **271**, 1099 (1996).

17. M. Lindner et al., Nature **320**, 246 (1986); J.M. Luck and C.J. Allegre, Nature **302**, 130 (1983).

18. E.W. Kolb, M.J. Perry and T.P. Walker Phys. Rev. **D 33**, 869 (1986).

19. P.P. Avelino et al. astro-ph/0102144 (2001), Phys. Rev. **D 64**, 103503 (2001).

20. P. Langacker, G. Segre and M.J. Strassler, hep-ph/0112233, Phys. Lett. **B 528**, 121 (2002).

21. X. Calmet and H. Fritzsch, hep-ph/0112110 (2001); X. Calmet and H. Fritzsch, hep-ph/0204258 (2002).

22. K.A. Olive et al., hep-ph/0205269 (2002).

23. V.V. Flambaum and E.V. Shuryak, hep-ph/0201303 (2002)

24. S. Degl'Innocenti, W. A. Dziembowski, G. Fiorentini and B. Ricci, Astrop. Phys. **7**, 77 (1997).

25. S. Cassisi et al., Astron. Astrophy. Suppl. Ser.**134**, 103 (1999)

Gamma-Ray Bursts:
The Most Powerful Cosmic Explosions

E.P.J. van den Heuvel

Astronomical Institute "Anton Pannekoek"
and Center for High Energy Astrophysics, University of Amsterdam, The Netherlands

1 Introduction and Summary

The field of Gamma Ray Burst (GRB) research is one in which discovery by serendipity plays an important role. Serendipity in general means: one searches for something but finds something else, which often is more interesting. Generally in astrophysics this comes about because one has a new instrument that can measure some physical aspect at least an order of magnitude better than was possible before. For example, the new instrument has an order of magnitude better sensitivity, or spectral resolution or angular resolution. The discovery of the GRBs was itself a classical example of serendipity. They were discovered in 1967 with the US military Vela satellites, which had been built to monitor whether countries were keeping to the Nuclear Test Ban Treaty that had been signed earlier in the sixties. To this end the Vela satellites were built to be sensitive to the gamma ray flash of nuclear explosions in the Earth's atmosphere or in space. To check for possible radioactivity produced by explosions on the backside of the Moon, the Vela satellites had very wide orbits extending halfway to the Moon. There were always several of them orbiting the Earth at any given time. In 1967 they detected gamma ray flashes of much longer duration than expected from a nuclear explosion, and from the differences in arrival time of these flashes in the different Vela satellites the Los Alamos scientists could roughly determine the direction from which the flashes came. It turned out that they did not come from Earth but from the sky. The discoverers were so surprised by this result that they studied the bursts for a long time, until they were absolutely sure that this was a real phenomenon. In 1973 they presented their discovery to an astrophysical audience [33], which caused a sensation. Theorists produced dozens of theories about their possible origin, ranging from comets colliding with neutron stars to nuclear wars of extraterrestrial civilizations. For 30 years the places of origin of the GRBs remained a mystery, as none could be identified with a known object. It was not even known whether they came from nearby, in our own galaxy, or from very far away, at cosmological distances. Only in 1997, thanks to the Wide-Field imaging hard X-ray cameras aboard the Italian-Dutch BeppoSAX satellite it became possible to swiftly determine their positions on the sky with sufficient accuracy that identification with known objects could be made. It was found that GRBs exhibit afterglows at optical and soft X-ray wavelengths, which persist for a few days [57, 10]. This allowed localizing their places of origin in distant galaxies with redshifts between 0.3 and 4.5, corresponding to "distances" (look-back times) of between 4 and 11 billion lightyears. From this it became clear that GRBs are the most powerful photon-emitters in the Universe with fluxes that for a short time rival the total photon emission of the observable Universe. They are the most powerful explosions in the Universe since the Big Bang, sometimes reaching an intrinsic optical brightness a million times that of a supernova. It has since been found that GRBs occur outside the nuclei of their host galaxies, and that some GRBs coincide with peculiar and very energetic supernovae. This indicates that they are stellar phenomena, presumably related to the death of very massive stars. The host galaxies in most cases show evidence of a high formation rate of massive stars, which independently suggests that GRBs are related to the evolution of these short-lived massive stars.

It is interesting to note that the Wide Field hard X-ray cameras of BeppoSAX had not been designed to study GRBs. They were built by the Utrecht Space Research group with the purpose of studying transient and variable X-ray sources in the galactic bulge (see below). The fact that they turned out to be the ideal instruments to solve the problem of the places of origin of the GRBs was another case of serendipity, confirming again the rule that if one builds an instrument that can do things that no other instrument could do before (in this case: determine within a few hours the position of a hard X-ray source with arcminute accuracy), one will achieve new breakthroughs. This review is organized as follows. In section 2 a brief historical review is given of the study of GRBs, followed in section 3 with an overview of the properties of GRB afterglows. Section 4 reviews the "relativistic fireball" model for the origin of GRBs and their afterglows. Section 5 deals with the host galaxies of GRBs and with the relation between GRBs, supernovae and other possible progenitors. Finally, in section 6, we argue that GRBs are the ideal probes for the star-formation history of the Universe up to the earliest times. This potential is now within reach, as from next year GRB missions will enable us to detect GRBs and their afterglows out to redshifts of between 10 and 20, and possibly even beyond. For earlier reviews I refer to [16, 58, 11].

2   Some History

2.1   IPN and Compton-GRO BATSE

In the seventies and eighties many interplanetary spacecrafts were equipped with small GRB probes, such that by using the differences of arrival times of a burst in different spacecrafts, one could determine the direction from which the burst was emitted. This Inter Planetary Network (IPN) has remained important up till today (e.g. cf. [28, 9]). It can provide accurate positions of GRBs, but it generally takes several days before these are available. By that time in most cases the afterglow has faded which, in hindsight, must be the reason that despite the accurate positions, the IPN did not lead to the discovery of the places of origin of the GRBs. A great step forward in GRB studies occurred in March 1991 with the launch of NASA's Compton-Gamma Ray Observatory, a 17-ton

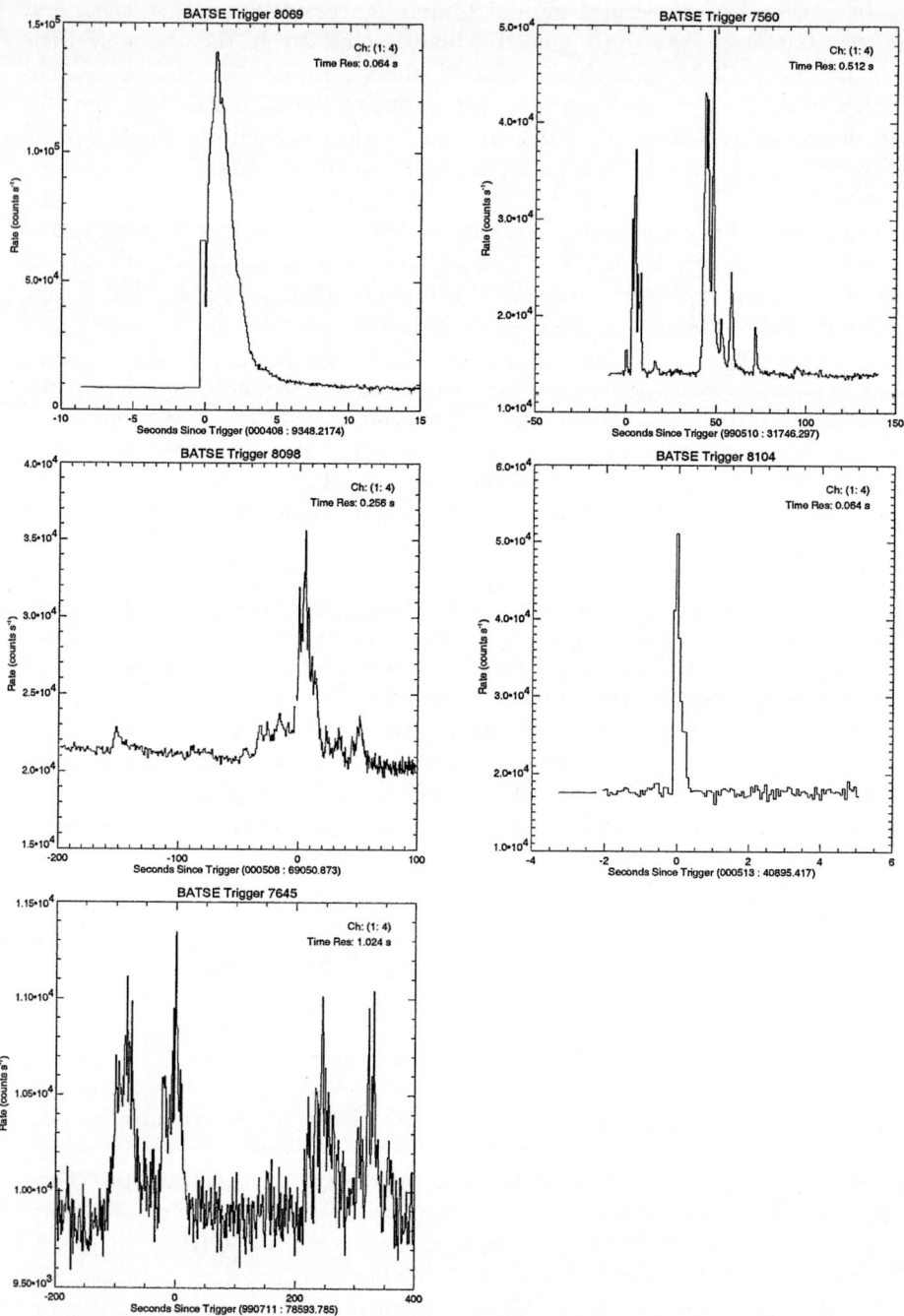

Fig. 1. Some characteristic time profiles of Gamma Ray Bursts (see Fishman and Meegan 1995)

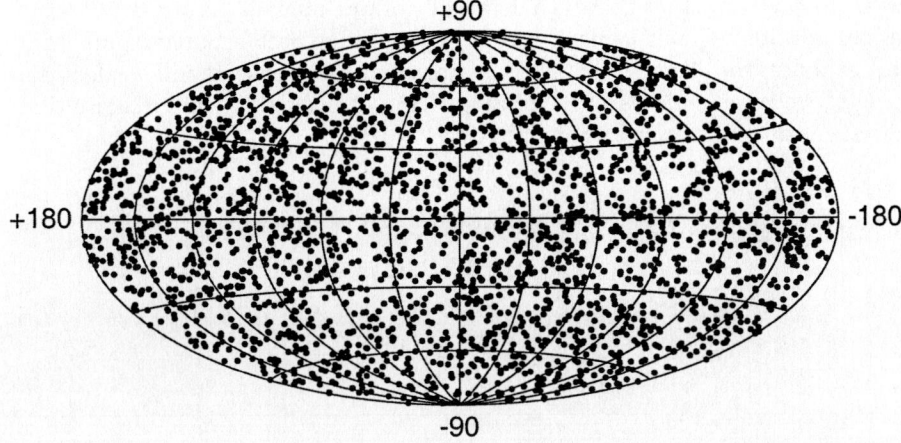

Fig. 2. 2704 BATSE gamma-ray bursts. The figure shows the celestial distribution of 2704 BATSE triggered bursts in galactic coordinates. If GRBs were of galactic origin a concentration towards the galactic plane (the horizontal axis in this figure) would be expected (courtesy NASA)

observatory mission in the same class as the Hubble Space Telescope. Compton-GRO was equipped with a variety of instruments, among them BATSE (Burst and Transient Source Experiment), which could observe GRBs occurring at any place in the sky not obscured by the Earth and localize them with an accuracy of about one degree (twice the diameter of the full Moon). Figure 1 shows some characteristic burst profiles, which range form very smooth to extremely spiky, showing, in some cases, spikes with millisecond durations. The durations of bursts range from a fraction of a second to several minutes. In the latter cases the bursts show many spikes occurring at apparently random times. BATSE discovered that GRBs occur about once per day. In the nine years of operation it observed close to 3000 GRBs and discovered that their sky distribution is completely smooth [40, 7], showing no relation whatsoever with our Galaxy (see Fig. 2). This in itself narrowed down very much the possible places of origin of the GRBs. Only two possibilities remained: either they arise in a spherical halo of our Galaxy, presumably made up of old neutron stars (as argued by Lamb [36]), or they originate in galaxies at cosmological distances (as argued by Paczyński [46]), as these are distributed completely smoothly over the sky. A strong argument for the latter possibility was put forward by Meegan et al. [40] from a plot of peak flux distributions, such as depicted in Fig. 3. If GRBs have the same intrinsic peak luminosity one would expect, if they have a smooth distribution throughout a static Euclidian space, that the peak fluxes P received on Earth show the typical number (N) distribution:

$$N(:)P^{-3/2} \tag{1}$$

However, Fig. 3 shows that there is a deficiency of weak bursts as compared to the Euclidian relation. This is typically what one would expect if bursts come from distant galaxies: the expansion of the Universe will then redshift and weaken the fluxes that we receive on Earth. The plot of Fig. 3 fits well with a distribution such that most GRBs arise at redshifts $z \geq 2$.

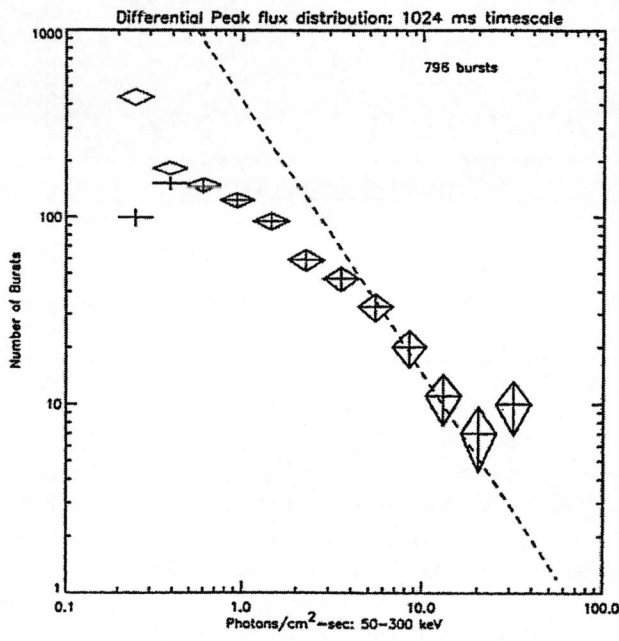

Fig. 3. The peak flux distribution of 796 Gamma Ray Bursts observed by BATSE (Pendleton et al. 1996). The flux is measured over the energy range 50–300 keV. The figure clearly shows a deficiency of faint bursts as compared to what is expected for a homogeneous source distribution in static 3-dimensional Euclidian space (the dashed line; explanation in the text)

Nevertheless, the still substantial positional uncertainties of the BATSE positions of GRBs, of order of one degree, made it impossible to find counterparts of GRBs at other wavelengths. The typical field of view of a large optical telescope is of order 10 arc minutes, and it would have taken many exposures with large telescopes to completely cover a BATSE error box and search among the millions of faint stars in this error box which star might have changed brightness. As this is very time-consuming, any short-lasting optical afterglow of a burst will have faded by the time the entire error box has been explored. As a result, unfortunately, BATSE did not solve the problem of the places of origin of the GRBs.

2.2   BeppoSAX's Wide Field Cameras and the Discovery of GRB Afterglows

For hard X-rays, like for Gamma Rays, there is no reflecting optics possible. Therefore, in order to reach high positional accuracy in hard X-rays, one has to use another technique: the shadowmask camera. This technique was successfully applied by the Netherlands Space Research Organization SRON with a camera aboard the Russian space station Mir. Using this experience, two such cameras, with a 40x40-degree field of view, were mounted in BeppoSAX, back to back, with the purpose of making long-term variability studies of the X-ray sources in the galactic bulge. At its front end the camera has a metal shadow mask, with a random pattern of small holes. Of any X-ray source in the field of view the mask casts a shadow on the detector. Mathematical deconvolution algorithms show that, if the random pattern of holes of the mask is known, the positions of all X-ray sources in the 40x40 degrees field of view of the camera can be retrieved from the observed shadow pattern. The bottleneck is the speed of the computation in the deconvolution. In the case of the BeppoSAX Wide Field Cameras (WFCs) a positional accuracy of a few arc minutes could be reached, at first, within eight hours, later within a few hours. In view of its large field of view the WFC can observe all X-ray sources in the galactic bulge at the same time. In the seventies and eighties it had become clear that GRBs emit a small fraction of their energy, of order a few percent, in the hard X-ray range, below 30 keV, where the BeppoSAX WFCs are sensitive. Therefore, F. Frontera and E. Costa proposed to use the anti-coincidence shield of BeppoSAX, which is sensitive to gamma rays, as a GRB monitor in combination with the WFCs (e.g. see [27]). If a GRB is observed from the direction at which the WFCs are pointed, one checks whether a signal is observed in the WFCs. If this is the case, one can determine the position of the GRB with arc-minute accuracy. Once this position is known, it is communicated to ground-based optical and radio observatories. At the same time, the satellite is re-oriented to point the imaging soft X-ray telescopes aboard BeppoSAX at the source position (these so-called "Narrow Field Instruments (NFIs)" have a field of view of order 30 arc minutes, and can determine a source position with 0.2 arc minute precision). BeppoSAX was launched on April 30, 1996, but the first time this observing strategy was successfully applied was with the GRB of 28 February 1997. Some ten hours after the occurrence of the GRB the satellite was re-oriented and the NFIs discovered a soft X-ray afterglow at the position given by the WFC. The afterglow faded in the course of about two days [10], as depicted in Fig. 4. Independently, van Paradijs and collaborators some 21 hours after the burst took an exposure of the WFC position with the 4.2m William Herschel Telescope on La Palma, and another one of the same field 8 days later with the 2.5m Isaac Newton Telescope. They discovered that a "star" of magnitude 20 on the 28 February exposure had disappeared on the exposure of March 8, 1997 [57], as depicted in Fig. 5. The BeppoSAX team had already earlier, with the GRB of 11 January 1997, discovered a fading X-ray afterglow, but this was the first time an optical afterglow was discovered and identification with a known object

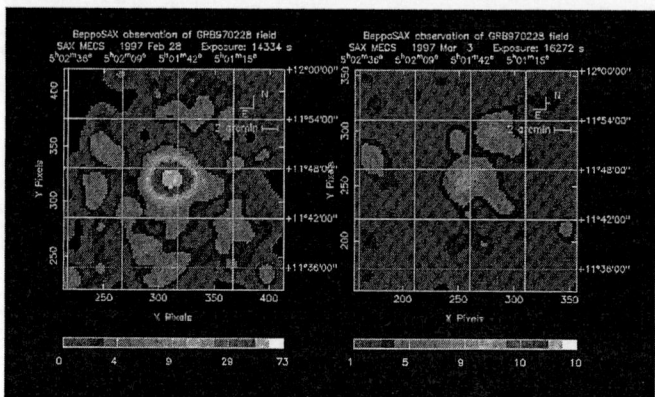

Fig. 4. Discovery images of the X-ray afterglow of GRB 970228 with the BeppoSAX Narrow-Field Instruments (Costa et al. 1997)

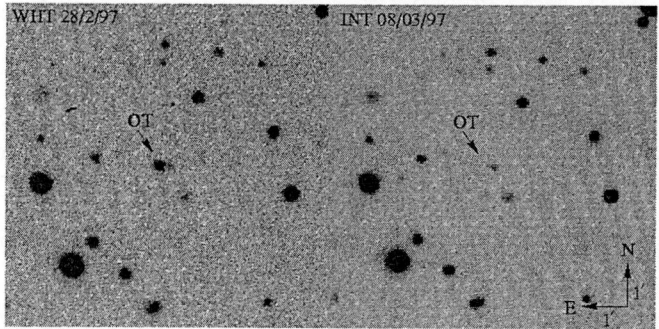

Fig. 5. Discovery images of the optical afterglow of GRB 970228 made on La Palma with the William Herschel Telescope on February 28, 1997 (left) and with the Isaac Newton Telescope on March 8, 1997 (right). The OT visible on 28 February has disappeared on March 8 (Van Paradijs et al.1997)

became possible. The afterglows decayed fast, according to a power law with a typical exponent between -1.1 and -1.3 (see for example Fig. 7). When the Hubble Space Telescope was pointed at the position of GRB970228 on March 26, 1997, the optical and infrared afterglow were still visible (I=24, V=26) and the afterglow was found to be asymmetrically located in a very faint galaxy [55]. (Only in 1999 with the Keck Telescope the redshift of this galaxy could be determined: z=0.69 [13]). As the identification of the afterglow of GRB970228 became known only after March 8, 1997, when the optical afterglow had faded, it was not possible to obtain the afterglow spectrum. This succeeded for the first time with the burst of May 8, 1997, which was rapidly identified by Bond

[6] using a Kitt Peak 1-m telescope, upon which the Keck Telescope took its spectrum, which was found to have a redshift of z=0.835 [44], corresponding to a distance of about 7 billion lightyears ("look back time"). This definitively established the cosmological nature of GRBs. Also the radio afterglow of this burst was discovered [17, 21] and could be followed for about 3 months. Later that year, on 14 December 1997, a GRB was discovered in a host galaxy with a redshift of 3.4 [50]. At present the largest redshift of a GRB afterglow recorded is z=4.5, for GRB000131, measured with ESO's VLT by the GRACE collaboration [2, 3].

3 Afterglows

So far, as of May 2002, from 52 localizations (12 by IPN), 40 afterglows were detected in X-rays (mostly by BeppoSAX, several also by RXTE and HETE-II), 32 at optical wavelengths and 22 at radio wavelengths. In about 20 cases the redshift could be measured and the host galaxy could be identified (see section 5). Figure 6 shows the isotropic energy output vs. redshift of the 17 bursts for which redshifts were known by January 2002). The figure shows that most redshifts cluster between 1 and 2, and the isotropic energy output ranges between 10^{52} and 3.10^{54} erg. The latter equals the rest-mass energy of about one solar mass. In all but one case the optical afterglows were not detected earlier than a few hours after the burst. The one exception was the burst of 23 January

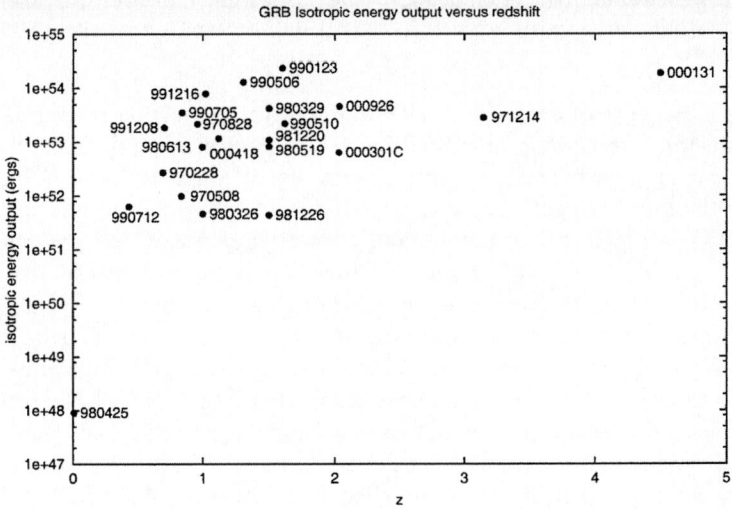

Fig. 6. Isotropic energy output versus redshift z of the 17 GRBs for which redshifts were known by January 2002 (after Vreeswijk 2002). If the energy emission is not isotropic (i.e.: "beamed") the total energy output of the bursts can be much lower (see text)

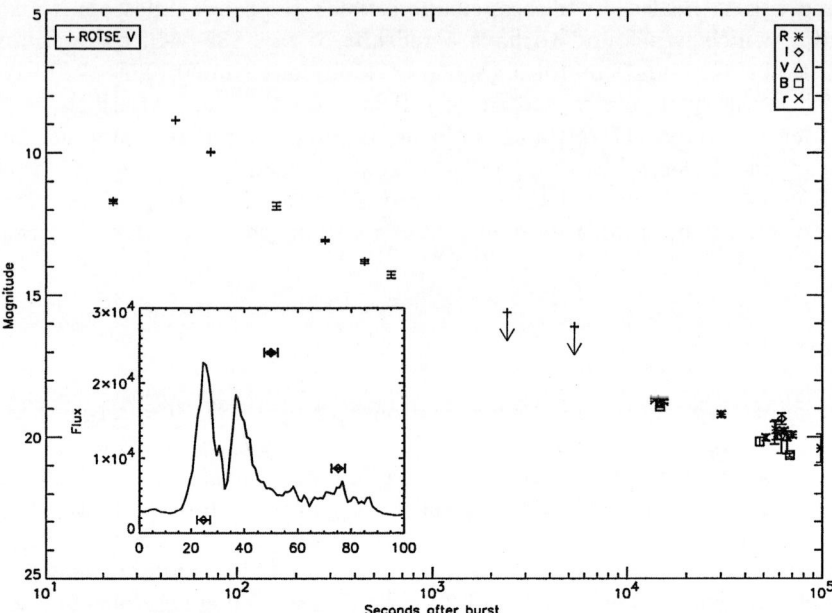

Fig. 7. R-band lightcurve of the afterglow of GRB 990123. The various symbols indicate observations by different observatories. The brightest six points are from the ROTSE robotic telescope (Akerlof et al. 1999), which show that this burst, at redshift 1.6, could have been seen with a small pair of binoculars. The dashed line indicates the power-law decay of the late lightcurve, with exponent -1.12 ± 0.03 (after Galama et al. 1999)

1999, for which an optical detection was made within 20 seconds after the start of the gamma-ray burst, with the ROTSE robotic telescope [1]. This showed that the optical afterglow reached a maximum visual magnitude V=8.86, 44.4 seconds after the burst trigger. At that magnitude the afterglow could have easily been seen with a simple pair of binoculars, even though this GRB and its host galaxy have a redshift of z=1.60! Figure 7 shows the lightcurve of the afterglow of this burst [23]. At its maximum optical brightness this afterglow was about one million times brighter than a supernova at the same distance. If it had been located in our galaxy, at the distance of the Orion Nebula, it would have been seen on Earth as bright as the Sun. One sees from this that with GRB-afterglows one will be able to probe the cosmos out to very much larger distances (and thus: till much earlier times) than with supernovae. This will become possible with the SWIFT mission, which is to be launched in September 2003 [25]. SWIFT will provide within 15 seconds a positional accuracy of 1 to 4 arc minutes and after 60 seconds: of 3 arc seconds. One therefore may expect to find in the near future many more cases like GRB990123. In principle one will be able to observe such afterglows out to redshift z=20, i.e.: well into the "Dark Ages", before the occurrence of the re-ionization of the Universe. As GRBs are stellar phenomena, connected with massive star death (see sections 5 and 6) one will thus be able with GRBs to probe star and galaxy formation up till the very earliest times. The radio afterglows provide important information on the evolution of the size of the emitting object. A nice example is the radio afterglow of GRB970508, which was followed with the VLA by Frail et al. [17] for almost 3 months. At the beginning the detected radio flux exhibited enormous fluctuations on short timescales, while later on the amplitude of the fluctuations decreased and the timescale of the fluctuations became longer. This is typically what one expects for a source located in the interstellar medium of a galaxy, which starts out with very small dimensions but rapidly increases in size. When it is still very small, then - like a scintillating star - the motion of the source relative to the density fluctuations in the interstellar medium of the host galaxy will cause the flux emitted into the direction of the Earth to violently fluctuate. However, when the source has grown in size, its area projected against the local interstellar medium will cover many density fluctuations, so the effects of the density fluctuations on the flux emitted towards the Earth will be averaged out and the amplitude of the observed variations will be diminished and also the timescales will become longer. From these observations it is clear that the source starts out roughly from stellar dimensions.

An important aspect is the question whether the energy of a GRB is emitted isotropically or whether it is beamed. With beamed we mean here, whether the electromagnetic radiation emitted by the GRB source is emitted in a narrow cone which happens to be directed into our direction. The required isotropic luminosities are enormous, so "beaming" is a way to reduce the energy requirements for GRB models. If the radiation is beamed in a cone with an opening half angle of 10 degrees, the energy requirements are reduced by a factor 100 with respect to the isotropic case, so they become typically of order 10^{50} to 10^{52} erg, i.e. roughly similar to typical supernova energies (which are of order 10^{51} erg). Collapse of the core of a massive star to a neutron star releases roughly $0.15 M_\odot c^2 = 10^{53}$ erg in gravitational binding energy, of which 99 per cent is emitted in the form of neutrinos, and about one per cent produces the supernova explosion and mass ejection. The matter which produces the GRB is ejected with high Lorentz factors, in the range 100 to 1000 (see below), and when this matter impacts onto the surrounding local circumstellar or interstellar medium, the afterglows at other wavelengths are expected to be produced (see below). (The relativistic motion of the emitting particles will cause the radiations emitted by these particles to be beamed into the forward direction, so: relativistic motion in the form of a "jet" will also lead to radiation emitted into a narrow cone). From a study of the spectral evolution of the afterglows at all wavelengths, one can in principle deduce the opening angle of the cone ("jet") into which the matter is ejected from the source (it would lead too far to describe this analysis technique here). From an analysis of the spectral evolution of a number of afterglows Frail et al. [18] found that the typical half opening angles of the jets are about 5 degrees, and that the real ejection energies of the matter in GRB sources is typically in the range 2.10^{50} to 2.10^{51} erg.

4   The Relativistic Fireball Model; Evidence for Jets

The fact the total flux of GRBs varies on a timescale of milliseconds indicates a source size $\leq$ 300 km. This, in combination with the source luminosity of order 10^{49} erg/sec or more implies a gigantic photon density in the source and since a sizeable fraction of the photons has energies $\geq$ 0.5MeV, one expects the source to be optically thick to pair creation by the two-photon process. However, the observed spectra of GRBs are typically those of an optically thin synchrotron source: at the lower energies a power law of index 1/3 and at the high energies a power law of index -p/2, where -p is the power-law index of the energy distribution of the relativistic electrons. This contradiction constitutes what is called "the compactness problem" of GRB sources. [This problem would still occur if the GRB sources would be in our own galaxy, at 10^{10} times lower luminosities.] Cavallo and Rees [8] pointed out that this contradiction can be resolved if the entire emitting region is expanding with highly relativistic velocities, i.e.: with a high Lorentz factor Γ.

This is due to the fact that in that case a time interval experienced as short for an observer at Earth, can be long at the emitting source, such that the source size in reality is much larger than it would be if the emitting surface were not moving. This can be shown as follows. If the source is moving towards us with a velocity v, then photons emitted at the source at times t and $t + dt$ will reach the Earth at times that differ on Earth by an interval dT given by:

$$dT = dt(1 - v/c) \tag{2}$$

For v very close to c, this leads to:

$$dT = dt(1 - \beta^2)/(1 + v/c) = dt/2\Gamma^2 \tag{3}$$

since $v/c \approx 1$, and $(1 - \beta^2) = 1/\Gamma^2$.

Thus, if the source expands with a Lorentz factor $\Gamma = 1000$, an interval of a millisecond observed at Earth is an interval of 2000 seconds at the source. In that time the source has moved over 10^9 km. At this source size, the compactness problem completely disappears. Also reasoning from another side, one comes to the conclusion that the source should expand at relativistic speeds. This is because if the source would indeed have a size of only 300 km and contains 10^{51} erg in photon energy, mixed with matter, the energy density in the source would be so gigantic that it would immediately propel the matter to expand with relativistic velocities. Thus, both reasoning from the side of the "compactness problem" and from the impossibility of keeping so much energy together in such a small volume, one comes to a relativistic expanding "fireball" as the "engine" of the GRB. This relativistic fireball model was put forward by Mészáros and Rees [41, 42, 43] and Rees and Mészáros [52].

The amount of matter entrained in the fireball needs not be large: if the Lorentz factor is $\Gamma = 1000$, and the GRB energy is 10^{51} erg, the total accelerated mass needs only to be 0.1 Earth mass (5.10^{-7} solar mass). This low value of the "baryon load" is also a requirement, because otherwise the baryons would cause the photons to degrade to low energies before they escape, and no GRB would be observed. The required energies and sizes of the "GRB Engine" are typically values expected for stellar collapse (supernovae). This makes it likely that GRBs are somehow connected with massive star deaths or massive star remnants.

As the GRB spectra are those of an optically thin synchrotron source, one appears to be dealing here with relativistic electrons moving in magnetic fields. In astrophysics the most general mechanism for accelerating particles and creating magnetic fields is: in shocks. For this reason, one postulates the occurrence of what are called "internal shocks" for producing the primary GRB that lasts for at most a few minutes [cf. 52, 48]. These internal shocks are thought to arise due to a series of short-lasting relativistic mass ejection events with different Lorentz factors produced by the "engine". Matter ejections from different events catch up with one another and collide, producing the shocks.

The afterglows at other wavelengths are then, according to these models, produced when the ejected matter collides with the surrounding interstellar or circumstellar matter. Circumstellar matter can, in the case of massive stars, be wind matter ejected by the star during its evolution prior to collapse. When the ejected matter thus collides with the surrounding matter, a so-called "external shock" is produced which sweeps up surrounding matter, causing the expansion to be gradually slowed down [cf. 41, 42, 43]. This external shock model for producing the afterglows has been successfully worked out further by Piran and collaborators (e.g. see the review by Piran [48]). This model also predicts the occurrence of a reverse shock, which at very early times in the GRB produces a short-lasting very high optical luminosity. The detection of the enormously bright optical afterglow of GRB990123 during the first minutes of the GRB appears to provide a beautiful confirmation of this prediction (see Fig. 7).

The spectral evolution of GRB afterglows, interpreted in terms of the "external shock" model, from gamma to radio wavelengths, provides important diagnostics about the physical parameters of the ambient medium and about the "beaming" of the ejecta (the opening angle of the jet of ejected matter); I refer for this to [61, 18]. As mentioned above, the opening half angle of the jets derived from the observed spectral evolution of the afterglows is typically of order 5 degrees [18].

5   Host Galaxies and Star Formation Rates

In all cases of an optical identification of a GRB afterglow, a galaxy was found at the position of the afterglow. In most cases the galaxy is so distant and faint and has such a small angular size (often not more than 0.5 arc second and in some cases even as small as 0.1 arc second) that the Hubble Space Telescope is needed to make the identification and to discern its structure. This implies, at the same time, that the astrometric position of the afterglow must be determined with an accuracy of order 0.1 arcsec or better if one wishes to find the localization of the GRB on the host. In general this is possible, and accuracies better than 0.02 arc sec have been reached in a number of cases (see [5, 31, 3] for detailed

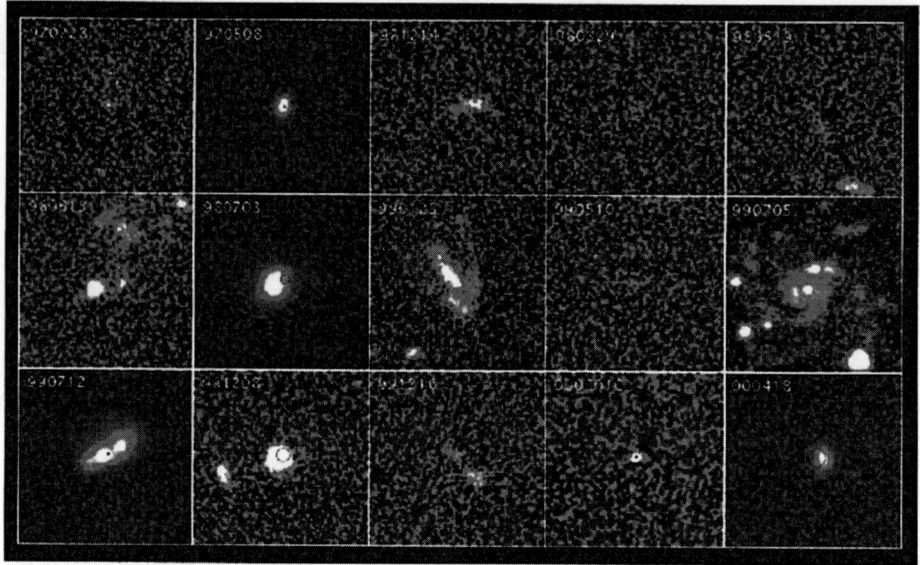

Fig. 8. Collection of 15 Hubble Space Telescope images of GRB host galaxies. The circles indicate the position where the GRB occurred (Vreeswijk 2002)

discussions). Figure 8 shows a compilation by Vreeswijk and Fruchter [20, 60] of 15 HST images of host galaxies, with the positions of the GRB afterglows indicated by the open circles. In most cases the hosts are blue dwarf galaxies. In the case of GRB990705 it is a grand-design spiral. In a number of cases the blue dwarf hosts are disturbed and elongated, resembling the blue compact galaxies in the Hubble Deep Field (North). Such disturbed small galaxies probably are the "building blocks" from which at a later epoch larger galaxies will be constituted. Plotted in the color-magnitude diagram of the Hubble Deep Field (North) galaxies, the GRB host galaxies fall in the area of the bluest galaxies in this diagram, indicating a large population of young massive stars, and thus: a high star formation rate [19]. Continuum radio detections in a number of cases show radio fluxes that confirm high star formation rates. The connection between high continuum radio emission and massive star formation comes about as follows. Stars with masses above 10 $M_\odot$ live very short, typically of order 10^7 yrs or shorter and terminate their lives with a supernova explosion. Therefore, when the massive star formation rate is high, also the supernova rate is high, and the many supernova remnants in such a galaxy produce a large synchrotron radio continuum emission; this method for estimating the star formation rate was first suggested by Vreeswijk [59], and successfully applied by him [60]. In other cases the strong H-alpha and O[III] 5007 Å emission indicates a - for these small galaxies -high formation rate of massive stars. For a small dwarf galaxy, a star formation rate of a few solar masses per year is already a very high rate, making them "starbursts". In several cases also star formation rates of over 100 solar masses per year have been detected from continuum radio and IR emission, but these are rare. All in all, it appears that the hosts have a relatively high level of star formation per unit luminosity [14, 31, 26, 3].

This all concerns the formation of massive stars, as all above described star formation indicators measure only the formation rate of the short-lived hot and luminous massive stars. The locations of the GRB afterglows on the images of their host galaxies in Fig. 8 indicate clearly that the GRBs occur at random locations in their host galaxies, i.e.: they are not related to the nuclei of galaxies (see also [5]). This, together with the facts that the dimensions and energies of GRB "engines" are typically those of stellar collapse (see above) and that the hosts have an elevated rate of massive-star formation, provides strong evidence indicating that GRBs are stellar phenomena and are somehow related to deaths of massive stars.

6   The Relation Between GRBs and Supernovae; Hypernovae and Other Possible Progenitors

6.1   Relation Between GRBs and Supernovae

A direct indication of a relation between GRBs and massive-star deaths came about with the discovery by Galama et al. [22] that GRB980425 which occurred on 25 April 1998, coincided in position and time with the explosion of a supernova (SN1998bw) in the relatively nearby spiral galaxy ESO 184-G82 (Fig. 9). This galaxy has a small redshift (z=0.008) corresponding to a distance of only 45 Mpc. As GRB980425 looks like a normal GRB at a cosmological distance, it was intrinsically very much fainter (by a factor of at least ten thousand) than "normal" GRBs. It turned out that SN1998bw with which it coincided was of a most unusual type: it was the most luminous radio supernova ever observed [35], and its synchrotron radio spectrum indicated that matter was ejected here with mildly relativistic velocities (Lorentz factor about 2). Its optical spectrum

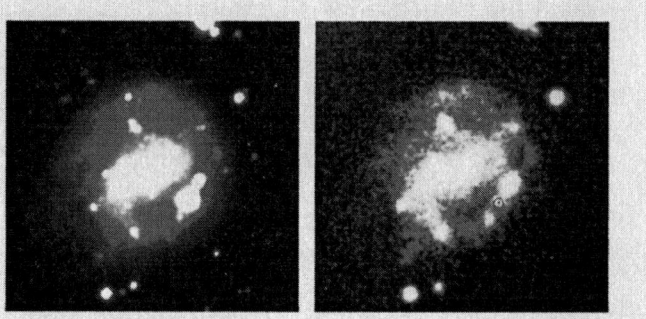

Fig. 9. Image of the Galaxy ESO 184-G82 with (left) and without (right) supernova 1998bw (Galama et al. 1998b). The right-hand picture was taken several years earlier than the left-hand one

showed that it was of the very rare Type Ic, which means that it had neither hydrogen nor helium in its spectrum. The widths of the emission lines in its spectrum indicated ejection velocities of 60,000 km/sec, an order of magnitude higher than normal in supernovae. From these velocities and estimates of the ejected amounts of mass it was derived that the energy of this supernova explosion was unusually large: some 2 to 3 times 10^{52} erg, which is more than an order of magnitude larger than usual supernova energies (cf. [29]). Theoretical modeling of the supernova lightcurve and spectrum showed that here at least 0.7 solar masses of ^{56}Ni was ejected (some ten times more than in normal supernovae) indicating that the exploding star was a carbon-oxygen star with a mass between 6 and 14 solar masses and a collapsing core mass of more than 3 $M_\odot$ [29, 39, 63]. As $3M_\odot$ is the absolute upper limit to the mass of a stable neutron star [32, 56], the conclusion is that here the core must have collapsed to a black hole. This implies that GRB980425/SN1998bw was the first formation event of a stellar black hole ever observed! Subsequently, it was discovered by Bloom et al. [4] that the late afterglow lightcurve of GRB980326 shows a brightening, peaking some 3 weeks after the GRB, which fits exactly an underlying supernova lightcurve similar to that of SN1998bw, at a redshift of 0.9, which presumably is the redshift of this GRB. And slightly later, Reichart [53] and Galama et al. [24] discovered a similar "supernova hump" in the late lightcurve of GRB970228; again the brightness of the maximum of this hump fits that of a supernova at the redshift of this GRB ($z=0.69$). With CHANDRA, Piro [49] discovered the redshifted iron emission complex of 6.7 keV in the afterglow of GRB991216 ($z=1.0$), at energy 3.5 keV. The ejection of Fe in this event again suggests an underlying supernova. So, in at least four cases we have now evidence for a relation between a GRB and a (very peculiar) supernova, and in one case we know that this supernova was the formation event of a stellar black hole.

6.2   The Hypernova Model

A relation between GRBs and black hole formation had in fact already been predicted by Woosley [62], who gave the very energetic birth event of a black hole the name "hypernova". Woosley and collaborators have shown [39, 63] that if one considers a rotating stellar core that collapses to a black hole, the angular momentum of the collapsing core prevents that all collapsing matter will immediately disappear into the black hole: in stead the black hole that forms will be surrounded by a disk or torus of nuclear matter in keplerian rotation that will only gradually - due to angular momentum loss by friction in this differentially rotating torus/disk - be accreted by the black hole. Numerical hydrodynamic calculations of the collapse process support this idea and show that the accretion of the torus/disk may take several minutes, and during this accretion so much energy is produced (in the form of neutrinos and electromagnetic radiation) that a small part of the disk matter is ejected in the form of highly relativistic jets, perpendicular to the disk [39]. These jets are so powerful that in the case of a C-O star they are able to pierce the entire star and appear almost undisturbed at the stellar surface at the rotational poles of the star and disappear into space.

The relativistic jets presumably produce the GRB and their impact onto the surrounding circumstellar matter produces the afterglows at other wavelengths (see above).

6.3 "Long" and "Short" Bursts; The Origin of Short Bursts

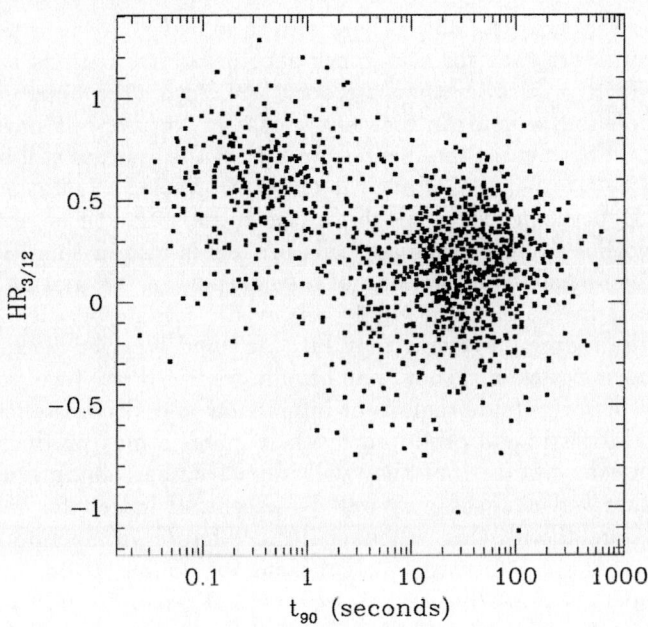

Fig. 10. Plot of hardness versus duration of BATSE bursts (cf. Kouveliotou et al. 1993) shows that there are two categories of bursts: a group with hard spectra and short durations (less than 2 seconds) and a group with softer spectra and longer durations (2 seconds to several minutes). Afterglows are known only for the second category

This "Hypernova" model for the production of GRBs appears to be able to at least qualitatively explain the production of the GRBs of which we presently know the afterglows and host galaxies. These bursts have relatively "long" durations, between a few seconds and several minutes. However, such bursts form only a fraction of about two third of all known BATSE GRBs. Figure 10 shows a plot of the spectral hardness ("gamma ray color") of BATSE bursts versus their durations. The figure shows that the bursts fall into two clear categories [34]: (1) a group with relatively "hard" spectra and short durations, of less than about 2 seconds, and (2) a group with "softer" spectra and durations of 2 seconds to a few hundred seconds. All afterglows detected so far are of GRBs of the second category. So far, never an afterglow has been detected of a "short" burst, and we therefore have no indication of their places of origin, nor of their nature. An attractive scenario for the origin of the short bursts is: the merger of double neutron star binaries. This model for the origin of GRBs was already proposed by Paczyński [45] and subsequently worked out in detail by Piran and colleagues [15]. The merger of a close double neutron star is a process that certainly occurs in Nature: the Hulse-Taylor binary pulsar PSR 1913+16 will merge about 300 million years from now due to losses of orbital energy and angular momentum in the form of gravitational waves. We know now at least 3 of such close binaries in our galaxy, and the merger rate of such systems in galaxies like ours is about one per 10^5 to 10^6 years (cf. [38]). There also is expected to be a population of double neutron stars that form at much shorter orbital periods and merge within a few million years after they are born [12]. Also close binaries consisting of a black hole and a neutron star must exist in Nature, but have not yet been discovered. In both cases the merger product of the systems will be a black hole surrounded by a torus/disk of nuclear matter in Keplerian rotation. Like in the case of the Hypernova discussed above, the torus/disk will be accreted by the black hole within a short time, and this process is also in this case expected to lead to the formation of ejection of a small part of the matter (of order of an Earth mass) in the form of highly relativistic jets perpendicular to the torus/disk. As the merging systems may be "old" and received kick velocities from two supernova explosions, they may have moved far away from their birth places, and may be far outside regions of interstellar and circumstellar matter. Therefore, the relativistic jets produced by their mergers may produce a GRB, but it is by no means certain that they will impact on surrounding matter, so afterglows at other wavelengths may well be absent. If indeed for most short bursts afterglows are absent, this will make it much harder, if not impossible, to identify their birthplaces. Numerical calculations for double neutron star mergers and black hole-neutron star mergers were carried out by Ruffert and Janka [54]. See also [30]. These authors found that indeed highly relativistic jets are produced by these mergers (typical Lorentz factors 100 to 1000), and that the durations of the GRBs produced by these jets will indeed be typically those of the short-duration bursts: 0.01 to 1 second. However, the energy produced by a double neutron star merger (of order 10^{49} ergs) was too low for a GRB, whereas that produced by a black hole-neutron star merger could reach 10^{51} erg, i.e. high enough for a GRB. The ejected amount of matter is typically less than a few Earth masses.

These calculations seem promising for our understanding of the origin of the short duration bursts. However, in order to really get more insight in the origin of these bursts, we also need to have observations of afterglows, which so far have been lacking. Hopefully, HETE-II and SWIFT may help to elucidate the places of origin of these bursts in the near future.

7 Concluding Remarks; Prospects for the Near Future

Since the discovery of the optical afterglows of Gamma Ray Bursts in 1997 we now know that:

- GRBs originate in galaxies at cosmological distances (redshifts ranging from 0.3 to 4.5) with an elevated rate of star formation.
- GRBs are beamed stellar phenomena with relativistic bulk motion of the source, with Lorentz factors 100 to 1000 and total energies of order 10^{51} erg.
- Early afterglows of GRBs can reach an intrinsic optical brightness (as observable on Earth) a million times that of a supernova. This will allow one to observe the optical afterglows of GRBs out to redshifts of at least z=20.

It was argued convincingly by Lamb and Reichart [37], Djorgovski et al.[14] and Andersen [3] that, since host galaxies of GRBs are all actively star forming galaxies, by searching for GRB host galaxies one obtains the most ideal unbiased sample of actively star forming galaxies in the Universe. Therefore, by studying GRBs out to redshifts z=20 one will be able to obtain an unbiased measurement of the evolution of the star-formation rate in the Universe until the earliest times, long before the epoch of re-ionization, which presumably is located between z= 7 to 10.

This will become feasible already very soon, with the SWIFT mission [25], which is due to be launched in September 2003. SWIFT, in combination with robotic telescopes for very rapid follow up, and the VLT and Keck for high-resolution spectroscopy, will form a very powerful combination for realizing the study of the complete star-formation history of the Universe. A most interesting aspect of this all is that, like in the case of the supernova-cosmology projects, this type of ultra-high redshift cosmology will make use of stars, not of galaxies or quasars, since GRBs are stellar phenomena. Indeed, stars are the basic building blocks of galaxies and history has shown time and again that it is almost always through the study of the physics of stars that great progress is made in astrophysics.

Acknowledgement

I thank my colleagues of the GRACE[1] collaboration for many enlightening discussions on GRB physics.

References

1. C. Akerlof, R. Balsano, S. Barthelmy, J. Bloch, P. Butterworth et al.: Nature 398, 400-402 (1999)
2. M.I. Andersen, J. Hjorth, H. Pedersen, et al. In: *Gamma Ray Bursts in the Afterglow Era*, eds. by E. Costa, F. Frontera and J. Hjorth (Springer Heidelberg 2001) pp. 133-135
3. M.I. Andersen: Ph.D. Thesis, University of Oulu, Finland (2002)

---

[1] GRACE = Gamma Ray burst Afterglow Collaboration at ESO. The GRACE collaboration, established in 1999 by the late Jan van Paradijs, consists of groups from Denmark, Germany, France, Italy, The Netherlands, Spain, The UK and the US.

4. J.S. Bloom, S.R. Kulkarni, S.G. Djorgovski, A.C. Eichelberger, P. Cote: Nature, 401, 453-456 (1999)
5. J.S. Bloom and S.R. Kulkarni. In:*Gamma Ray Bursts in the Afterglow Era,* eds. E. Costa, F. Frontera and J. Hjorth (Springer, Heidelberg 2001) pp. 209-211
6. H.E. Bond: I.A.U. Circ, 6654 (1997)
7. M.S. Briggs, W.S. Paciesas, G.N. Pendleton, C.A. Meegan, G.J. Fisherman et al. Ap.J. 451, 40-63 (1996)
8. G. Cavallo and M.J. Rees: M.N.R.A.S. 183,359-365 (1978)
9. T.L. Cline, K.C. Hurley, S. Barthelmy et al. In: *Gamma Ray Bursts in the Afterglow Era,* eds. E. Costa, F. Frontera and J. Hjorth (Springer, Heidelberg 2001) pp. 375-377
10. E. Costa, F. Frontera, J. Heise, M. Feroci, J. In 't Zand et al.: Nature 387, 783-785 (1997)
11. E. Costa, F. Frontera and J. Hjorth (eds.): *Gamma Ray Bursts in the Afterglow Era,* (Springer, Heidelberg 2001) pp.459
12. J. Dewi, O. Pols: M.N.R.A.S. (in press) (2002)
13. S.G. Djorgovski, S.R. Kulkarni, J.S. Bloom, D.A. Frail: GCN 289 (1999)
14. S.G. Djorgovski, S.R. Kulkarni, J.S. Bloom, D.A. Frail et al. In: *Gamma Ray Bursts in the Afterglow Era,* eds. E. Costa, F. Frontera, J. Hjorth (Springer, Heidelberg 2001) pp. 218-225
15. D. Eichler, M. Livio, T. Piran, D.N. Schramm: Nature 340, 126-128 (1989)
16. G.J. Fishman, C.A. Meegan: Ann. Rev. Astron. Ap. 33, 415-458 (1995)
17. D.A. Frail, S.R. Kulkarni, L. Nicastro, M. Feroci, G.B. Taylor: Nature 389, 261-263 (1997)
18. D.A. Frail, S.R. Kulkarni, R. Sari et al.: Ap. J. 562, L55 (2001)
19. A. Fruchter: Oral presentation at conference *Gamma Ray Bursts in the Afterglow Era,* Rome (unpublished) (1999)
20. A. Fruchter: Private Communication (2002)
21. T.J. Galama, R.A.M.J. Wijers, M. Bremer et al.: Ap. J. 500, L101-105 (1998a)
22. T.J. Galama, P.M. Vreeswijk, J. van Paradijs et al.: Nature 395, 670-672 (1998b)
23. T.J. Galama, M.S. Briggs, R.A.M.J. Wijers, P.M. Vreeswijk et al.: Nature 398, 394-396 (1999)
24. T.J. Galama, N. Tanvir, P.M. Vreeswijk, R.A.M.J. Wijers et al.: Ap. J. 536, 185-194 (2000)
25. N. Gehrels: *Gamma Ray Bursts in the Afterglow Era,* eds. E. Costa, F. Frontera and J. Hjorth (Springer, Heidelberg 2001) pp 357-360
26. J. Hjorth, B. Thomsen, S.R. Nielsen, M.I. Andersen et al.: Ap. J. 576, 113-119 (2002)
27. K. Hurley: *Gamma Ray Bursts and Neutrion Star Physics,* eds. E.P. Liang, V. Petrosian (Am. Inst. of Phys., New York 1986) 141:1
28. K. Hurley, M. Feroci, M. Cinti, E. Costa et al.: Ap. J. 534, 258-264 (2000)
29. K. Iwamoto, P.A. Mazzali, K. Nomoto et al.: Nature 395, 672-674 (1998)
30. H.T. Janka, T. Eberl, M. Ruffert, C.L. Fryer: Ap. J. 527, L39 (1999)
31. A.O. Jaunsen, J. Hjorth, G. Björnsson, M.I. Andersen et al.: Ap. J. 546, 127-133 (2001)
32. V. Kalogera, G. Baym: Ap. J. 470, L61-L64 (1996)
33. R.W. Klebesadel, I.B. Strong, R.A. Olson: Ap. J. 182, L85-88 (1973)
34. C. Kouveliotou, C.A. Meegan, G.J. Fishman, M.P. Bhat, M.S. Briggs: Ap. J. 413, L101-104 (1993)
35. S.R. Kulkarni, D.A. Frail, M.H. Wieringa et al.: Nature 395, 663-669 (1998)

36. D.Q. Lamb: Pub. Astr. Soc. Pacific 107, 1152 (1995)
37. D.Q. Lamb, D.E. Reichart. In: *Gamma Ray Bursts in the Afterglow Era,* eds. E. Costa, F. Frontera and J. Hjorth (Springer, Heidelberg 2001) pp 226-232
38. D.R. Lorimer, E.P.J. Van den Heuvel: M.N.R.A.S. 283, L37-40 (1996)
39. A.I. MacFayden, S.E. Woosley: Ap. J. 524, 262-289 (1999)
40. C.A. Meegan, G.J. Fishman, R.B. Wilson, W.S. Paciesas, G.N. Pendleton et al.: Nature 355, 143-145 (1992)
41. P. Mészáros, M.J. Rees: Ap. J. 397, 570 (1992)
42. P. Mészáros, M.J. Rees: Ap. J. 405, 278 (1993)
43. P. Mészáros, M.J. Rees: Ap. J. 476, 232 (1997)
44. M. Metzger, S.G. Djorgovski, S.R. Kulkarni, C. Steidel et al.: Nature 387, 878 (1997)
45. B. Paczyński: Ap. J. 308, L43-46 (1986)
46. B. Paczyński: Pub. Astr. Soc. Pacific 107, 1167 (1995)
47. G. Pendleton, R. Mallozzi, W. Paciesas, M. Briggs et al.: Ap. J. 464, 606 (1996)
48. T. Piran: Phys. Reports 314, 575-667 (1999)
49. L. Piro. In: *Gamma Ray Bursts in the Afterglow Era,* eds. E. Costa, F. Frontera and J. Hjorth (Springer, Heidelberg 2001) pp 97-105
50. A.N. Ramaprakash, S.R. Kulkarni, D.A. Frail et al.: Nature 393, 43-46 (1998)
51. M.J. Rees, P. Mészáros: M.N.R.A.S. 258, L41-43 (1992)
52. M.J. Rees, P. Mészáros: Ap. J. 430, L93 (1994)
53. D.E. Reichart: Ap. J. 521, L111-115 (1999)
54. M. Ruffert, H.T. Janka: Astron. Astrophys. 344, 573-606 (1999)
55. K.C. Sahu, M. Livio, L. Petro et al.: Nature 387, 476-478 (1997)
56. G. Srinivasan. In: *Black Holes in Binaries and Galactic Nuclei,* eds. L. Kaper, E.P.J. van den Heuvel, P. Woudt (Springer, Heidelberg 2001) pp 45-52
57. J. van Paradijs, P. Groot, T.J. Galama, C. Kouveliotou et al.: Nature 386, 686-689 (1997)
58. J. van Paradijs, C. Kouveliotou, R.A.M.J. Wijers: Ann. Rev. Astr. Ap. 379-425 (2000)
59. P.M. Vreeswijk: Oral contrib. at meeting *Gamma Ray Bursts in the Afterglow Era,* Rome (1999)
60. P.M. Vreeswijk: Ph. D. Thesis, University of Amsterdam (2002)
61. R.A.M.J. Wijers, T.J. Galama: Ap. J. 523, 177-186 (1999)
62. S.E. Woosley: Ap. J. 405, 273-277 (1993)
63. S.E. Woosley: *Gamma Ray Bursts in the Afterglow Era,* eds. E. Costa, F. Frontera and J. Hjorth (Springer, Heidelberg 2001) pp 257-262

Astrophysics with High Energy Gamma Rays

Heinrich J. Völk

Max-Planck-Institut für Kernphysik, Saupfercheckweg 1,
D-69117 Heidelberg, Germany

Abstract. Recent results, the present status and the perspectives of high energy gamma-ray astronomy are described. Since the satellite observations by the Compton Gamma Ray Observatory and its precursor missions have been reviewed extensively, emphasis is on the results from the ground-based gamma-ray telescopes. They concern the physics of Pulsar Nebulae, Supernova Remnants in their assumed role as the Galactic sources of Cosmic Rays, Jets from Active Galactic Nuclei, and the Extragalactic Background radiation field due to stars and dust in galaxies. Since the gamma-ray emission is nonthermal, this kind of astronomy deals with the pervasive high-energy nonequilibrium states in the Universe. The present build-up of larger and more sensitive instruments, both on the ground and in space, gives fascinating prospects also for observational cosmology and astroparticle physics. Through realistically possible further observational developments at high mountain altitudes a rapid extension of the field is to be expected.

1 Introduction

There is a general consensus that the main energy sources for high energy γ-ray emission are extreme objects in the Universe with high energy turnover. This belief stems from the observation of young Pulsars, Supernova Remnants (SNRs) and Active Galactic Nuclei (AGNs), and we shall discuss some of these detections here.

We know that high energy γ-rays are abundantly produced in collisions of charged particles that have been accelerated in collective processes. They involve ionized systems (i.e. collisionless plasmas), large scale mass motions, and electromagnetic fields. The nonthermal processes are important because they are part of the major energy dissipation processes like shock waves that arise in explosive events, at the breaking of supersonic flows in the form of winds and jets from galaxies, and during mergers of galaxies and clusters. Another important source of particle acceleration should be the dissipative angular momentum transport in magnetized accretion flows near compact objects. Therefore we can plausible assume that a major fraction of the random kinetic energy in the Universe is nonthermal, with many of the particles at ultrarelativistic energies. And we can see them in gamma rays.

High energy γ-rays might also be due to rare decays of ultra-heavy particles or arise from annihilations of the lightest supersymmetric particles that are widely believed to make up the nonbaryonic Dark Matter. Such indirect identifications with the aid of γ-ray observations are goals of astroparticle physics, even though no effects have been found until now. All γ-ray detectors include nevertheless the fundamental issue of Dark Matter search as part of their observation programs.

Physically speaking, the most interesting γ-ray features are *localized* emissions. They should immediately portray the generating energetic processes and thus give us new insights into the astrophysics of the sources. Localized emissions from deep gravitational potential wells like the center of our Galaxy also appear as the most promising indicators for accumulations of weakly interacting massive cold Dark Matter particles (WIMPs).

From the point of view of Cosmic Ray (CR) physics γ-ray astronomy is an indirect form of energetic particle detection. But it is a crucial one since even ultrarelativistic charged particles are strongly deflected from straight line orbits by the interstellar and intergalactic magnetic fields. The direct detection of the charged particles will therefore not help us to find their sources even if they happen to reach the Earth. A possible exception are the CRs with the highest energies $\sim 10^{20}$ eV (see the paper by A.A. Watson in these Proceedings). At the much lower energies $\leq 10^{15}$ eV, where energetic particle distributions typically contain almost all their energy *density*, only neutral secondary collision products like γ-rays or neutrinos point back to the sources. Due to their different production modes and their vastly different interaction strength with matter they should give complementary results. Intensive R&D efforts are presently made to develop neutrino astronomy, using large volumes of polar ice and ocean water, as discussed by F. Halzen in these Proceedings. I shall concentrate here on high energy γ-ray astronomy.

Since space born γ-ray astronomy has been reviewed extensively in the past (e.g. [56]), I will put more emphasis the recent results from ground-based telescopes and their impact on major physics questions. The astronomical objects that have been successfully studied are as diverse as Pulsar Nebulae, Supernova Remnants, AGN jets, and the diffuse Extragalactic background radiation field in the Optical/Infrared wavelength range. At the end I will indicate the perspectives of the overall field for the future.

2   High Energy γ-Ray Detectors

High energy γ-rays are measured with pair production detectors. They have been used on satellites up to energies of the order of 10 GeV and on the ground above a threshold of about 200 GeV, with special detector arrangements reaching energies as low as 50 GeV.

A typical space instrument is shown in Fig. 1. To reject the dominant flux of charged CR particles, the detector is covered by an anti-coincidence shield. Following the NASA telescope SAS-II, ESA's Cos B was a remarkably long-lived mission, to be finally succeeded by the EGRET-instrument on CGRO. These detectors had a large field of view ~ 1 sr and a small effective area < 1 m^2, for energies 30 MeV $\leq E \leq$ 30 GeV (e.g. [56]). Since the termination of CGRO no satellite experiment is operating in this energy range.

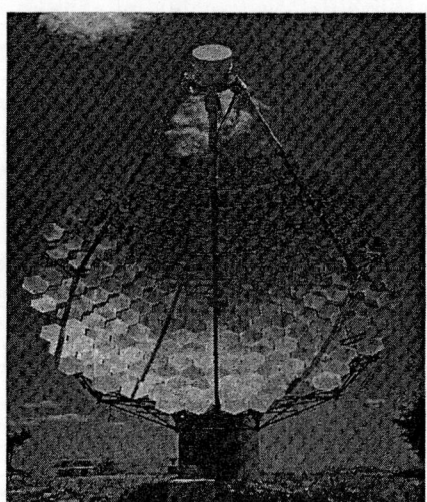

Fig. 1. (Left) The EGRET instrument on the Compton Gamma Ray Observatory (CGRO). Left: overall instrument with anti-coincidence shield. Right: schematic of conversion of the incoming primary γ-ray into an e^+e^- pair that is tracked in a spark chamber, with energy measurement in the calorimeter on the bottom

Fig. 2. (Right) The 10 m Whipple telescope in Arizona, of the now VERITAS collaboration. The Alt/Azimuth reflector consists of tessellated glass mirrors that focus the atmospheric Cherenkov light onto a camera whose pixels are fast phototubes

At γ-ray energies above about 5 GeV, the atmospheric shower containing many pairs can be used on the ground to detect the associated Cherenkov light. Today the standard instruments are imaging optical telescopes (Fig. 2). Thus the atmosphere itself is part of the detector. The dominant background of showers from nuclear CRs can be suppressed by analysis of the shower images, separating the broad hadronic showers from the concentrated electromagnetic γ-ray-showers. Actually this is best achieved with a stereoscopic array of Cherenkov telescopes, as pioneered by the HEGRA telescope system on La Palma[1]. Presently, four major instruments of this type are operating: Whipple/VERITAS (Arizona)[2], CANGAROO (Australia)[3], CAT (French Pyrenees)[4], and HEGRA.

In contrast to the satellite detectors they have a small field of view of a few degrees and a low duty cycle $\sim 10\%$ due to the restriction to clear and moonless nights. However the effective telescope area $\sim$ few $\times 10^4 \text{m}^2$ is extremely large, the energy resolution achieved is $\sim 15\%$, and the angular resolution with $\sim 0.1°$ per event is an order of magnitude better than the satellite telescopes flown up to now. Also non-imaging detector systems, using large mirror areas from solar

[1] http://www.mpi-hd.mpg.de/hfm/CT/CT.html.

[2] http://veritas.sao.arizona.edu/veritas/technical-details.shtml.

[3] http://icrhp9.icrr.u-tokyo.ac.jp/.

[4] http://lpnp90.in2p3.fr/ cat/index.html.

power plants, are being developed to observe at lower thresholds ~ 20 GeV, like CELESTE [5], STACEE [6], or GRAAL [7], but we do not have space to discuss them here.

Space and/or Ground?

As a result of these technological developments the two types of instruments are not only complementary in their energy range but also in their instrumental capabilities. It is therefore no surprise that next generation instruments are developed both for the ground as well as for space. The costs differ by more than an order of magnitude.

3   Gamma-Ray Astrophysics Results at High Energies

Pulsar Nebulae of Rotating Neutron Stars and their Magnetic Fields

Pulsed GeV γ-rays have been detected from eight young Pulsars by EGRET. This emission is usually attributed to radiation from electron - photon cascades in the Pulsar magnetosphere (e.g. [55]), even though recently an alternative radiation mechanism has been proposed due to magnetic reconnection in the winds of Pulsars with streams of alternating magnetic polarity [38].

Unpulsed soft X-ray synchrotron emission from quite a number of extended nebulae around Pulsars has been detected with the ROSAT and ASCA telescopes (e.g [15]), [35]) at rather high luminosities: $L_X \sim 10^{-3} L_{\rm spin-down}$. Except for the Crab Nebula and PSR B1706 - 44 the implied Inverse Compton (IC) TeV γ-ray emission by the same electrons in the Cosmic Microwave Background (CMB) is only marginally measurable from most sources at present. Yet the coming generation of γ-ray telescopes should be able to detect a significant number of these objects – and therefore the average B-field – because the radiant luminosities should be comparable, $L(E_{\rm IC}/L(E_{\rm syn}) = U_{\rm ph}/U_B \leq 10^{-1}$ for interstellar-type magnetic fields $B \sim 10^{-5}{\rm G} = B_{-5}$ which might be typical for many sources [4]; here $U_{\rm ph}$ and U_B denote the energy densities of the photon field and the magnetic field, respectively. For the Crab Pulsar, and therefore presumably for other Pulsars as well, the Nebula is thought to arise through an ultrarelativistic wind of e^+e^- pairs, dissipating at a termination shock and accelerating particles that populate a slowly expanding hot and partially nonthermal bubble [52], [36], [13]. The spectrum of the emitted radiation has been modeled phenomenologically by e.g. [27] and [14]. For the Crab Nebula almost the entire energy loss of the rotating neutron star is emitted by synchrotron radiation from the Nebula, that means, in a nonthermal form (Fig. 3)[8]. This picture is of course quite simplified

---

[5] http://wwwcenbg/extra/Astroparticule/celeste/index.html.

[6] http://www.astro.ucla.edu/ stacee/.

[7] http://rplaga.tripod.com/almeria/.

[8] There is in fact no sign for synchrotron emission from a 'thermal' shocked wind component with a characteristic energy scale expected in UV/soft X-rays, in possible contrast to the Vela Pulsar (S. Bogovalov, private communication). Using an argu-

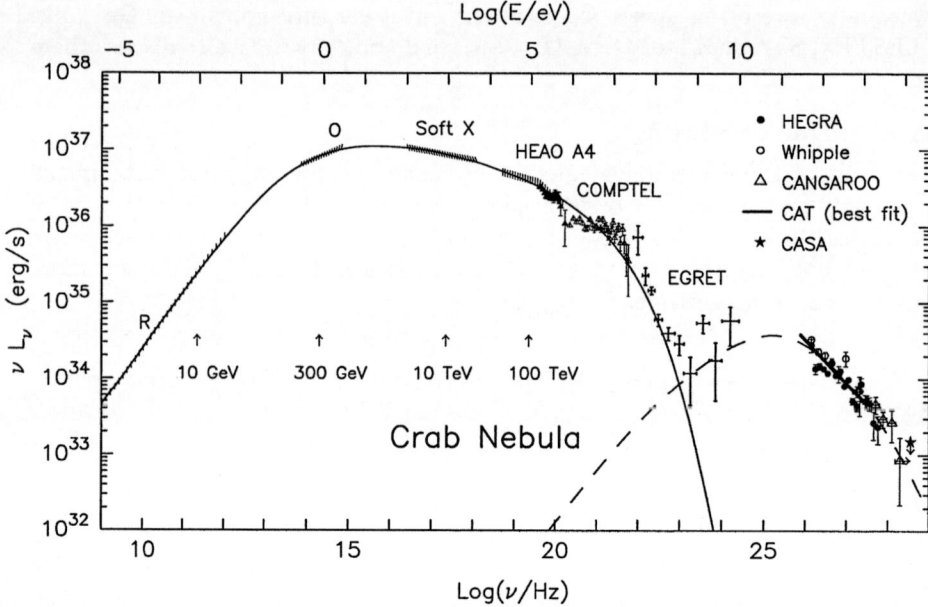

Fig. 3. Spectral energy density νL_ν of the Crab Nebula's emission. The high intensity synchrotron emission reaches into the γ-ray range as observed by several satellite instruments. For this specific young object it exceeds by far the flux in the second hump, presumably due to inverse Compton radiation, as observed with several Cherenkov telescopes (after [14])

as the recent highly resolved X-ray observations of the Crab with the Chandra telescope (http://chandra.harvard.edu/photo/0052/index.html) show. But the basic physics ingredients of the Rees & Gunn model seem remarkably robust.

In the very young Crab Nebula $L(E_{IC}/L(E_{syn}) \sim 10^{-3}$ which implies a high average magnetic field strength $B \simeq 3 \times 10^{-4}$ Gauss. Thus, despite the fact that it is the strongest steady source of nonthermal radiation in the sky – in particular also in TeV γ-rays – it is a comparatively inefficient IC emitter due to its enormous intrinsic field $B \gg B_{-5}$. Even though magnetic field strengths of this magnitude are expected from equipartition arguments for this object, and had been inferred from the spectral steepening of the synchrotron emission in the far infrared wavelength range [44], the independent measurements of the synchrotron as well as the IC emission, put such inferences on a solid experimental basis.

For the second case, PSR B1706 - 44, the steady IC flux is about 1/5 of the Crab flux at TeV energies. However, this does not mean a huge synchrotron

---

ment from diffusive shock acceleration theory we speculate that in the extremely luminous Crab Nebula the nonlinear backreaction of the accelerated particles on the flow has smoothed the shock completely, so that the 'thermal' component is only adiabatically compressed wind, remaining essentially 'cold' (e.g. [30]).

luminosity by analogy. Rather we have $L(E_{IC})/L(E_{syn}) \sim 10$, according to [37], [26]. Thus, either the B-field in the Nebula is on average extremely small, as in an old extended bubble, or the particles IC scatter outside the Nebula where the field is probably indeed very low. Large scale morphological information may be needed to interpret the measurement [4].

What can we expect from future γ-ray measurements of a large sample for an improved understanding of Pulsar Nebulae?

If PSR B1706 - 44 turned out to be the more typical case than the Crab Nebula, then we might have a modified picture: In analogy to the lack of equilibrium of nonrelativistic wind bubbles around massive stars (e.g. [22]), the Pulsar Nebula would likely be dynamically unstable allowing the bubble to cool, not in the form of synchrotron cooling but rather by escape of the relativistic particle component into the surrounding diffuse Supernova Remnant. The spatially integrated *total* nonthermal emission should still be comparable to the spin-down luminosity, with the changing magnetic field now playing primarily the role of a wind and particle accelerator and not that of a cooling agent.

The possible appearance of a shocked thermal component of the nebula is expected to exhibit characteristic synchrotron and IC signatures that would directly allow a measurement of the Lorentz factor of the Pulsar Wind. The relative strength of the thermal component may in addition be a indicator for the degree of nonlinearity of the acceleration process.

Shell Type Supernova Remnants and the Origin of Galactic Cosmic Rays

If we disregard the possible compact remnant, the physics interest in diffuse, shell-type Supernova Remnants (SNRs) is readily enumerated: as an ensemble the SNRs lead to the largest mechanical energy input into the Interstellar Medium of galaxies. The strong blast wave from the explosion, sweeping up the circumstellar medium, suggests efficient diffusive shock acceleration of charged nuclei to a power law source spectrum roughly $\propto E^{-2}$. Maximum energies should possibly reach 10^{15} eV and the elemental ratios should very roughly correspond to cosmic abundances. In short, SNRs are suspected to be the sources of the Galactic CRs. Essentially by default.

Even SNRs fulfill the enormous energy requirement for the replenishment of the Cosmic Rays in the Galaxy of $\sim 10^{40}$ erg/yr not by a large margin; at least 10% of the entire mechanical energy released in the event must on average be converted into nonthermal energy of relativistic nuclei.

The only experimental test of this widespread belief consists in direct multiwavelength observations of SNRs. Due to the inferred hard source spectra, with about equal energy per decade, the best test uses very high energy γ-rays and/or neutrinos since all background radiations fall off more steeply with particle energy. Up to now the required sensitivity can only be approached by γ-ray experiments and this prospect has been one of the driving forces behind the development of high energy γ-ray astronomy. Notwithstanding these goals, all γ-ray detectors operating up to now have been only marginally sensitive in the face of existing flux estimates [31], [47].

Also CR electrons are detected on the top of the atmosphere, at a 1 percent flux level in comparison with CR nuclei. And they are equally assumed to be accelerated in SNRs.

Although electrons contribute only to a negligible degree to the CR energy density, a high energy electrons emits synchrotron radiation as well as IC γ-rays very efficiently compared with the production rate of π^0-decay γ-rays per CR nucleus and the overall IC emission can be of the same order of magnitude as the hadronic γ-ray emission from SNRs [45]. In fact, nonthermal electron synchrotron emission has been inferred for a number of SNRs not only at radio wavelengths but also in hard X-rays, and in recent years several such sources have also been reported in TeV γ-rays. It was therefore as simple as it was tempting to interpret them in terms of IC emission alone.

Theoretical Estimates of γ-ray Emission

From the point of view of CR nucleon origin, the γ-ray emission from electrons in SNRs appears as a curse rather than a blessing, since it requires a separation of hadronic from leptonic γ-rays. On the other hand we may as well turn this fact into an advantage. Estimates of the γ-ray production in SNRs are based on diffusive shock acceleration theory. Although one of the best developed theories in astrophysics, its application to the time-dependent situation of an evolving point explosion in a large scale magnetic field is difficult. In addition, the overall evolution is a highly nonlinear dynamical problem due to the high acceleration efficiency and the backreaction of the accelerated particles on the structure of the shock (e.g. [42]). Thus several sub-processes, like the strength of injection of suprathermal particles into the acceleration process, which cannot be calculated very accurately require observational input. Here the synchrotron channel is important since the proton and electron spectra have the same form for relativistic energies (apart from the trivial radiation losses). In this way also the unknown magnetic field strength is constrained.

The estimated flux values also depend on the character of the progenitor star's evolution such as the mass loss for massive stars, and the type of Supernova explosion, and they depend on the overall Supernova energetics. Purely astronomical parameters, like source distance and ambient gas density are obviously important as well. They can only be determined by comprehensive multi-wavelength observations. And although certain parameter combinations are fixed by the given thermal X-ray luminosity and overall SNR dynamics, the combined uncertainty in the estimated γ-ray flux may reach an order of magnitude. Given the marginal instrumental sensitivity, it is not surprising that there were few detections and many non-detections in the past five years. The history of these observational efforts is quite interesting.

Early Observations: Upper Limits

Early observations ($\sim$ 10 hours) of radio-bright SNRs, associated with a number of so-called unidentified EGRET sources, were unsuccessful. This concerned especially the well-known core collapse SNRs γ-Cygni and IC-443. Both the Whipple [23] and HEGRA [63] found only upper limits above a few 100 GeV. Extrapolations from the γ-ray energies > 100 MeV of the EGRET observations to 1 TeV are not necessarily appropriate since the EGRET fluxes might be rather contaminated by Pulsar emissions in that detector's very large field of view, cutting off beyond about 20 GeV. The a priori flux estimates for the π^0-decay γ-rays just about bracket the experimental upper limits.

Deep Observations of Historical Supernovae

After significantly deeper observations ($\sim$ 100 hours) detections were reported by the CANGAROO collaboration [59], [60] for the Type Ia Supernova SN 1006 from the year 1006 AD, the X-ray brightest SNR in the Southern Hemisphere [40] and presumably the result of the deflagration of an accreting White Dwarf in a low density environment, as well as for the X-ray-detected southern SNR RX J1713.7-3946 [46], possibly a core collapse SN in the neighborhood of some dense gas clouds [57]. The HEGRA collaboration [7] announced the detection of Cassiopeia A (Cas A), the youngest Galactic SN from around 1680 AD, and presumably the result of the core collapse of a massive Wolf-Rayet progenitor. Cas A is also the brightest nonthermal radio source in the sky. HEGRA [8] also reported a very low upper limit – at 3 percent of the flux from the Crab Nebula – for Tycho's SNR from 1572 AD, a Type Ia Supernova seen by Tycho Brahe (on whose planetary orbit observations Johannes Kepler based his laws of planetary motions). The total observation time for Cas A was 230 h, the deepest γ-ray measurement at TeV energies made up to know. All four objects had been been detected before in the radio continuum and in hard X-rays, where also the latter do presumably contain a significant synchrotron contribution.

SN 1006 has been phenomenologically modeled by many authors as an IC source due to the X-ray synchrotron electrons in the Cosmic Microwave Background (see e.g. [59]). This presupposes a rather small interstellar magnetic field of about 4 μG in order to explain the high γ-ray flux, given the synchrotron flux, and then no π^0-decay γ-ray flux is needed. In the opposite case of a significantly larger magnetic field, a significant hadronic component is required, with a shell-type morphology because of the gas compression at the shock [2].

Based on new data, the CANGAROO collaboration has recently reconsidered its initial IC interpretation of the γ-ray emission from SNR RX J1713.7-3946 [32]. From a comparison of the inferred IC spectrum and expected forms of a hadronic spectrum with the measured γ-ray spectrum it was argued that the γ-ray nature of the object should rather be hadronic. This claim did not remain uncontested, based on multi-wavelength arguments [53], [25]. In the absence of an overall theoretical model for this poorly understood source, for which is not

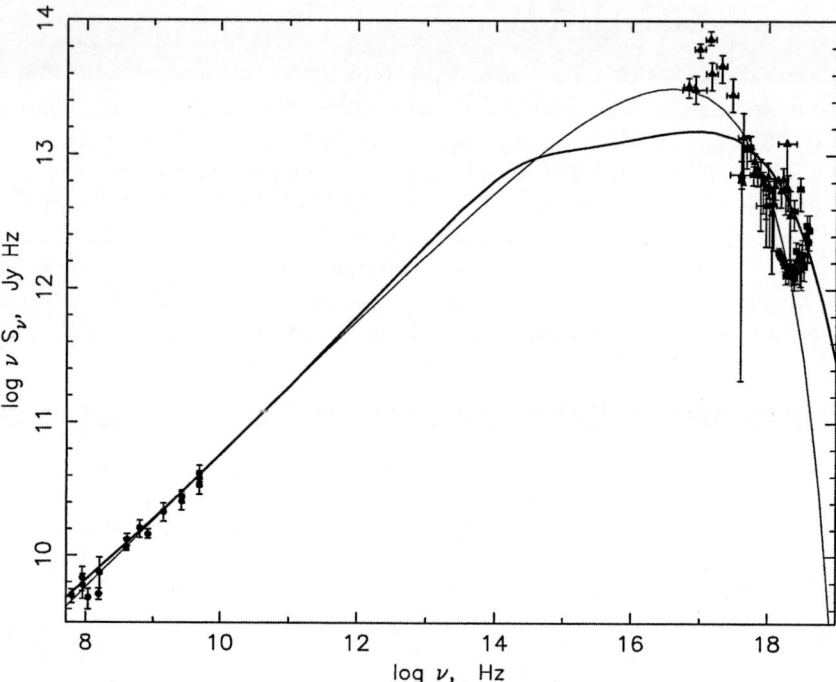

Fig. 4. Theoretical synchrotron spectral energy density for SN 1006 compared with radio and X-ray observations. The best fit is for efficient proton acceleration, an effective upstream magnetic field of 20μ Gauss and an electron to proton ratio $K_{ep} = 1.5 \times 10^{-3}$ (*solid line*). The (physically not plausible) case of inefficient proton acceleration and low field strength of 4μ G (*dashed line*) gives a harder radio spectrum, high nonthermal X-ray emission and $K_{ep} = 4 \times 10^{-2}$ (from [18])

clear whether it is due to a Type Ia Supernova or due to a core collapse, it is difficult to see how such controversies can be resolved.

A self-consistent estimate for SN 1006, based on time-dependent nonlinear acceleration theory, calculates the space and time evolution of the overall SNR dynamics together with the electron synchrotron, π^0-decay and IC spectra, with the CMB as primary photon target [16], [17], [18]. It shows (Fig. 4) that not only a high effective magnetic field strength of $\approx 20\mu$G upstream of the SNR shock is required to describe the overall synchrotron spectrum, from the radio to X-rays [54]; [34]; [12], but that one also needs a strong nonlinear shock modification due to efficient acceleration of nuclei in order to explain the steep radio spectrum. This is consistent with ion injection theory. The theoretical γ-ray spectrum is then dominated by π^0-decay, even though only by a moderate factor of order 5 (Fig. 5), with the π^0-decay γ-ray spectrum extending up to almost 100 TeV, and with a dipolar γ-ray emission morphology along the external magnetic field direction. This predicted γ-ray spectrum agrees reasonably well with the EGRET upper limits [48] and the latest TeV results [60]. An artificially assumed low

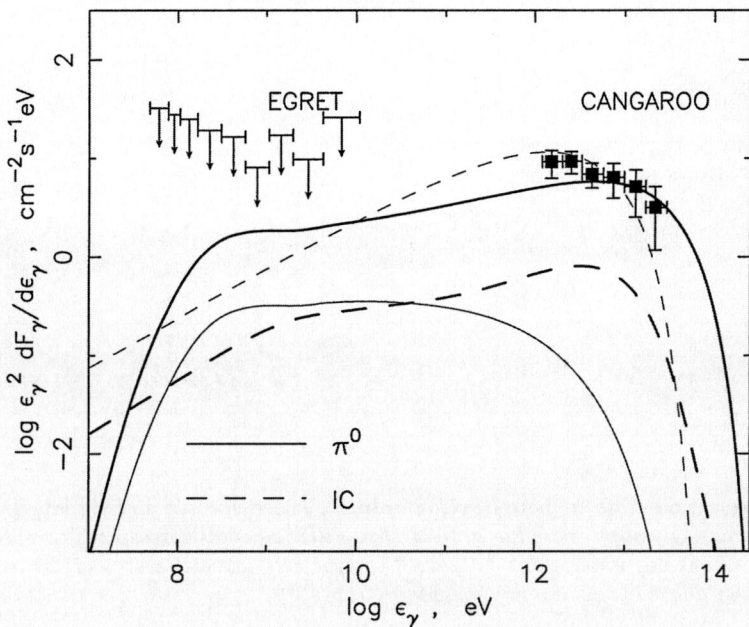

Fig. 5. Theoretical IC (*dashed lines*) and π^0-decay (*solid lines*) γ-ray spectral energy densities for SN 1006. Efficient proton acceleration is indicated by the thick curves, inefficient proton acceleration by the thin curves. The recent high energy γ-ray flux data and the EGRET upper limits [48] are also shown (from [18])

magnetic field, combined with a physically implausible low proton acceleration efficiency, describes the synchrotron observations considerably poorer and the IC γ-ray spectrum reaches less than about 10 TeV. As a consequence, good spatial as well as spectral coverage from a minimum of 100 GeV and preferably even from about 100 MeV up to the highest measurable γ-ray energies is needed to ultimately resolve this issue for the high energy nuclear particles of interest.

It is also important that an analogous conclusion can be drawn from the radio/X-ray spectra of Tycho's SNR which is calculated to lie just below the present γ-ray detection limit [64]. Also for Cas A – a more difficult object through its complex mass loss history before the explosion – the theory suggests a hadronic γ-ray emission [19]. Recent work strongly supports this interpretation [20]. It will be one of the important tasks for the coming Northern Hemisphere Cherenkov telescopes to obtain good γ-ray spectra from these sources.

HEGRA Galactic Plane Survey

A limited survey of the Galactic Plane (from longitude 85° to − 3°) with the HEGRA system using on average 2.8 hours of integration time per pointing (Fig. 6) yielded only upper limits for suspected individual sources. The scan also included 19 known SNRs. Source stacking of the SNR candidate sources

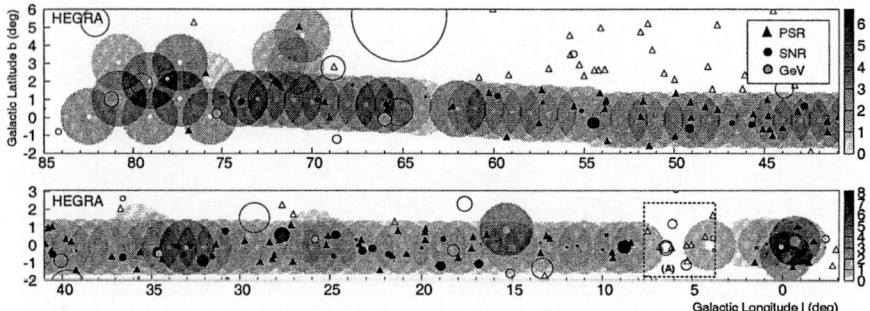

Fig. 6. Observation time in hours (*right ordinate*) used for the individual Galactic plane scan points given by the large *gray circles*. *Filled symbols* correspond to potential sources for which an upper limit is given. *Symbol size* gives the size of the source. Objects in the *dashed box* were not included (from [9])

gives a combined upper limit about a factor 2 above the expected π^0-decay flux [9]. This shows that the nondetections are consistent with a dominant hadronic γ-ray emission.

Conclusions Regarding a SNR Origin for the Galactic Cosmic Rays

The search – under the SNR lamp post – for Galactic CR origin, one of the problems of the century, has made remarkable progress through high energy γ-ray astronomy during the last years. The few detections and the many nondetections are consistent with a SNR origin of the dominant nuclear CR component. We can expect that the new ground-based arrays, coming on line these years, will decide this question.

Blazar Jet Emission

Active Galactic Nuclei (AGNs) are presumably accretion-powered supermassive Black Holes in the center of galaxies. The associated jet-like outflows are often characterized by apparent superluminal motion. Amongst the different AGNs there is a class of objects, the Blazars, which in the optical show a dominant nonthermal continuum together with broad lines, while being highly variable in time. The continuum may be attributed to acceleration processes in the jet (Fig. 7).

Rather unexpectedly about 70 such objects have been found in the GeV region by the EGRET instrument and this was one of the major discoveries of CGRO. The high γ-ray luminosity that dominates the overall spectral energy density distribution from the AGN suggests that the γ-ray emission is strongly Doppler–boosted by coming from a relativistic jet ([21]) whose bulk motion is essentially directed towards the observer. In fact, the apparent luminosity $L_{\rm app}$ is connected with the intrinsic source luminosity $L_{\rm int}$ through $L_{\rm app} = \delta^4 L_{\rm int}$, where

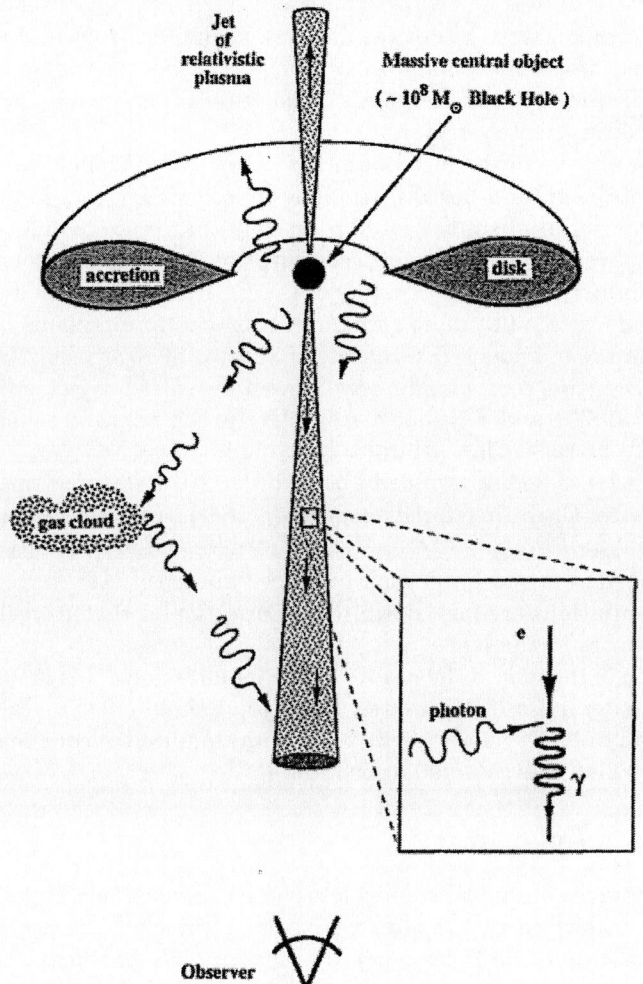

Fig. 7. Schematic of a Blazar. An IC mechanism for the radiation from the bulk relativistic AGN jet is assumed for definiteness, where the ambient photon field may come from various sources, including internal synchrotron photons produced in the jet itself (SSC). For Blazars the jet is pointing close to the line of sight

$\delta = \Gamma^{-1}(1 - \beta \cos\Theta)^{-1} > 1$ is the so-called Doppler factor for bulk motion with Lorentz factor Γ at an angle Θ relative to the line of sight. In the limiting case of Θ going to zero $\delta = 2\Gamma$; δ might well be ~ 10.

Due to its small detection area the EGRET instrument had long integration times which prevent the analysis of correlations with the fluxes in other wavelength ranges on short time scales. Yet these are especially interesting since the radiative electron cooling times at very high energies are short enough so that their radiation amplitude may follow the dynamical time variations of the system.

In a subclass of Blazars the optical lines are negligible or even absent. These BL Lac objects, named after the prototype galaxy, show the maximum of their synchrotron emission at X-ray energies. The corresponding γ-ray emission is then expected to peak in the TeV region.

Several such objects have indeed been found in the TeV range. And since the effective areas of the ground-based Cherenkov telescopes are very large, they can follow rapid time variations much more effectively than a space detector limited by photon statistics. Due to intergalactic absorption TeV Blazars must be rather nearby at redshifts $z \ll 1$, whereas many of the EGRET AGNs are distant, very luminous Quasars with a flat radio spectrum. It is therefore perhaps not surprising that the number of known TeV Blazars is a factor of 10 smaller. At distances of the order of 150 Mpc, corresponding to the well-measured objects Mkn 421 and Mkn 501 at $z = 0.031$ and $z = 0.034$, respectively, the nonthermal efficiency of such sources can be rather low. In order to avoid intrinsic TeV γ-ray absorption in the radiating jet, δ should typically be of order 10. Requiring the flux to be comparable to the Crab flux for detectability, the intrinsic luminosity has to be roughly equal to $L_{\gamma,\mathrm{intr}}^{\mathrm{source}} \sim (\delta/10)^{-4} \times 10^{40}\,\mathrm{erg/sec} \sim 10^{-5} \times L_{\mathrm{Edd}}(10^7\,M_\odot)$, where $L_{\mathrm{Edd}}(10^7\,M_\odot)$ is the Eddington luminosity of a $10^7 M_\odot$ accreting object. The observed variations are fast, down to sub-hour scales that translate to sub-parsec spatial scales of the jet.

The question is then as to the nature of the jet emission. This is first of all an interesting question in itself. Ultimately however, it should also reveal the origin and composition of AGN jets. Presently the γ-ray studies concern the nature of the jet emission and intergalactic absorption.

Nature of the Jet Emission

The two main sources of radiation from jets can be energetic electrons producing IC emission from low energy photon fields, or extremely high energy protons that generate photo-pion and photo-pair cascades or directly radiate synchrotron emission.

At the comparatively low luminosities of BL Lac sources, the most plausible IC target fields for leptonic jets are the radio synchrotron photons from the same population of energetic electrons. The double peaked spectrum typically inferred from quasi-simultaneous X-ray and high energy γ-ray observations (e.g. [61]; Fig. 8) then suggests electron cooling to be about equally distributed between synchrotron and IC losses and fast enough to explain the good temporal correlation observed between the γ-ray with the X-ray fluxes. In such a synchrotron self-Compton (SSC) interpretation the γ-ray flux should increase quadratically with the synchrotron flux. This is about what is observed (see also [41]).

Of course this interpretation is not unique. Alternative hadronic models (e.g [43]) require protons of extremely high energies $\leq 10^{19}$ eV in the jet. They produce pions on the abundant low frequency photon fields longward of far infrared wavelengths, i.e. ultimately a gamma signal and a extremely high energy neutrino signal. Such protons could possibly also be the sources of the ultra-high

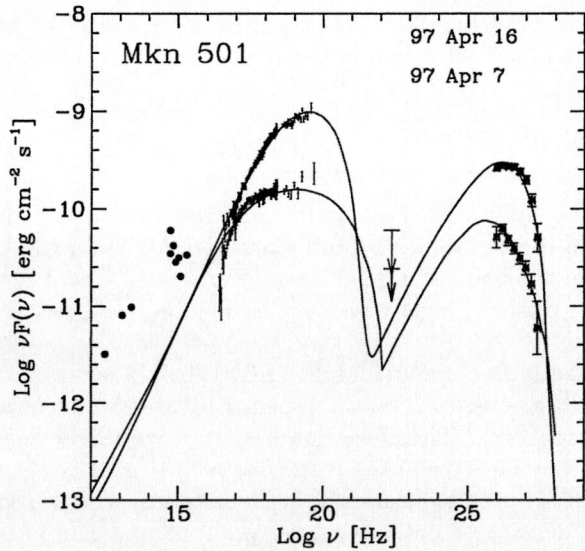

Fig. 8. Multi-wavelength SSC modeling of the X-ray and TeV γ-ray energy fluxes from Mkn 501, observed at different days with the *BeppoSAX* satellite and the CAT telescope, respectively. Upper limits in the 100 MeV region are from EGRET. The Inverse Compton peak lies in the Klein-Nishina domain, with Intergalactic absorption being neglected (from [61])

energy CRs in our Milky Way. And they may possibly power the high energy emission of flat spectrum Quasars that appear optically thick for TeV emission. Unfortunately the cooling of protons on photons is a rather slow process. And for TeV γ-ray sources also a jet optical depth smaller than unity for pair creation is required at TeV energies in order to allow the escape of the photons produced in the interior. This limits the photon density available for proton cooling.

Pion production has to compete with *proton* synchrotron emission which becomes important at such high energies [1]. It is a faster process for BL Lac objects as long as the magnetic field strength in the γ-ray emission region is of order 100 G. Then also essentially all energy goes into TeV γ-rays. As a consequence, at least in the TeV range, hadronic jets favor the proton synchrotron channel. Such very large magnetic fields O(100) G are actually required for the necessary fast acceleration, given the observed fast time variations. Such high B-fields presumably require a massive cold hadronic jet component to ensure dynamical equilibrium.

What are the jets made of?

It is quite surprising that such an extreme alternative has not been resolved until now. Detailed *time-dependent* modeling of simultaneous multi-wavelength observations will be necessary to clearly distinguish between leptonic and hadronic jets in BL Lac objects. Obviously this is at the same time one of the prerequisites for an understanding of the jet origin in the first place.

Intergalactic γ-ray Absorption
on the Extragalactic Background Light

The spectrum of the diffuse Extragalactic radiation field has a double peak structure due to the direct radiation from stars and AGNs in the UV, optical and near infrared on the one hand, and the mid and far infrared reradiation of absorbed starlight by dust at longer wavelengths, all integrated over the evolution of the Universe.

In this Extragalactic Background Light (EBL) TeV γ-rays are absorbed by pair production with a cross section that peaks at about one quarter of the Thompson cross section σ_T. Therefore one can approximately relate the two photon energies in the form $(E/1\text{TeV}) \approx (h\nu/1\text{eV})^{-1}$ and write the optical depth in the form $\tau(E) = \xi(\sigma_T/4) h\nu \, n_{ph}(h\nu) \times$ distance, where $n_{ph}(h\nu)$ is the differential number density of the low energy photons and ξ is of order unity. Thus, for a constant spectral energy density of the EBL, $\tau(E)$ increases linearly with E and the TeV spectra from Extragalactic sources will have an imprint in the form of characteristic absorption features with a high energy cutoff. These absorption features should give information on the spectrum of the EBL in an elegant way, an information that direct observations of the EBL can only yield through a difficult and uncertain subtraction of dominant foreground radiations such as the Zodiacal Light or the so-called Galactic Cirrus; this subtraction is especially problematic in the mid infrared region.

The uncertainty in this method lies in the unknown primary γ-ray source spectrum. Thus for example Mkn 501 shows an exponential cutoff proportional to $E^{-1.9} \exp^{-E/E_0}$, with $E_0 = 6.2$ TeV which is roughly consistent with observational estimates of the EBL spectrum [6]. However, on a 3σ level, the source Mkn 421 (at about the same redshift) appears to have a significantly lower cutoff energy of $E = 3.6$ TeV [11], precluding the possibility of the cutoff being only an absorption feature of the EBL.

Therefore we need to measure γ-ray sources at different, in fact higher distances. Fortunately, the BL Lac object H 1426 + 428 at the fourfold redshift $z = 0.129$ could recently be detected and its TeV spectrum measured by HEGRA is within the errors *consistent with the characteristic absorption features* expected [11]. The source spectrum was assumed to be $\propto E^{-1.92}$, consistent with X-ray synchrotron observations (Fig. 10). The absorption feature consists in a strong hardening of the observed γ-ray spectrum between about 2 and 5 TeV (Fig. 9).

One can also fit a power law to the HEGRA data alone – with no further justification than the simplicity of a 2-parameter straight line – obtaining a somewhat lower overall statistical significance. However, this power law is quite flat and it deviates therefore strongly from a steep absorbed spectrum at energies below 1 TeV. In fact, the Wipple [50] and CAT [29] telescopes have recently confirmed the expected steep spectrum below 1 TeV.

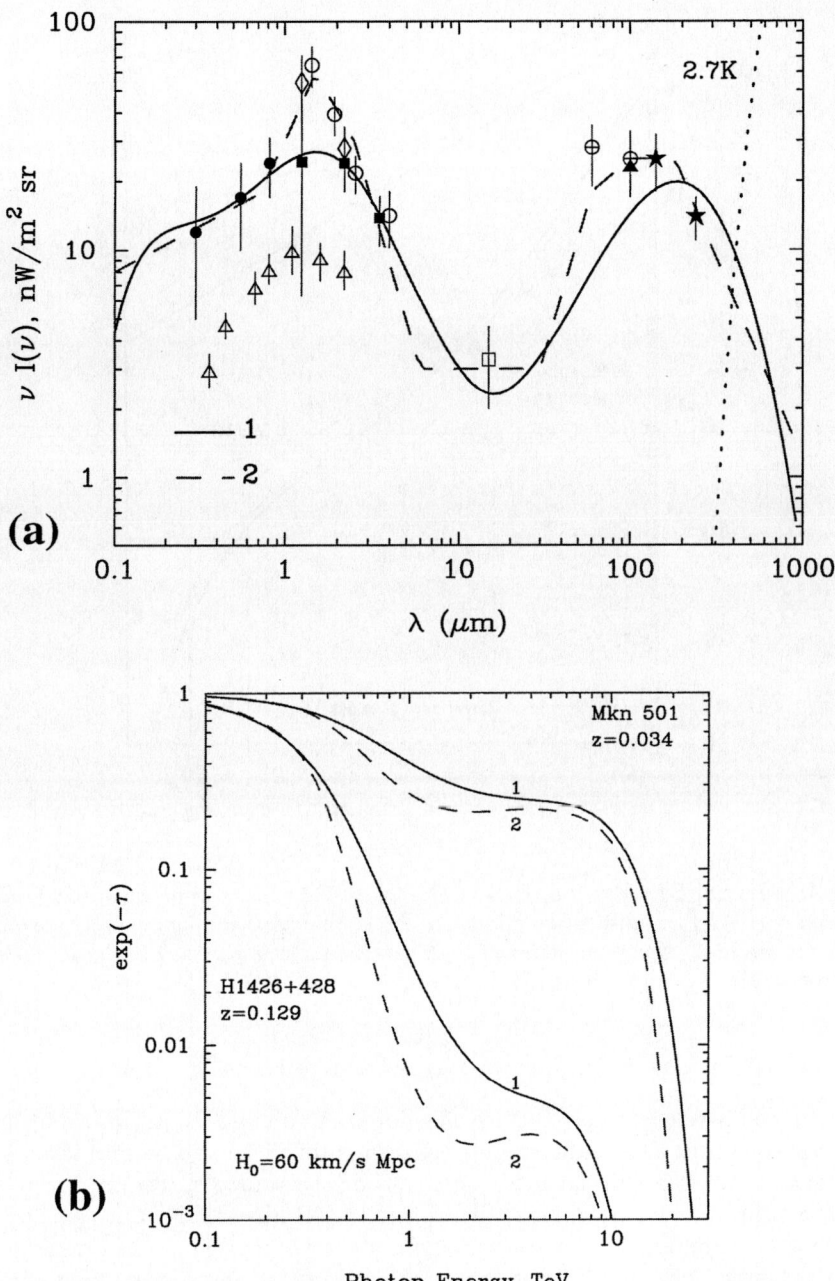

Fig. 9. Different empirical approximations to the direct measurements of the spectral energy density of the EBL (**a**), and their energy-dependent absorption effect on TeV photons from Mkn 501 and H 1426 + 428, respectively (**b**). The optical depth is denoted by τ (from [11])

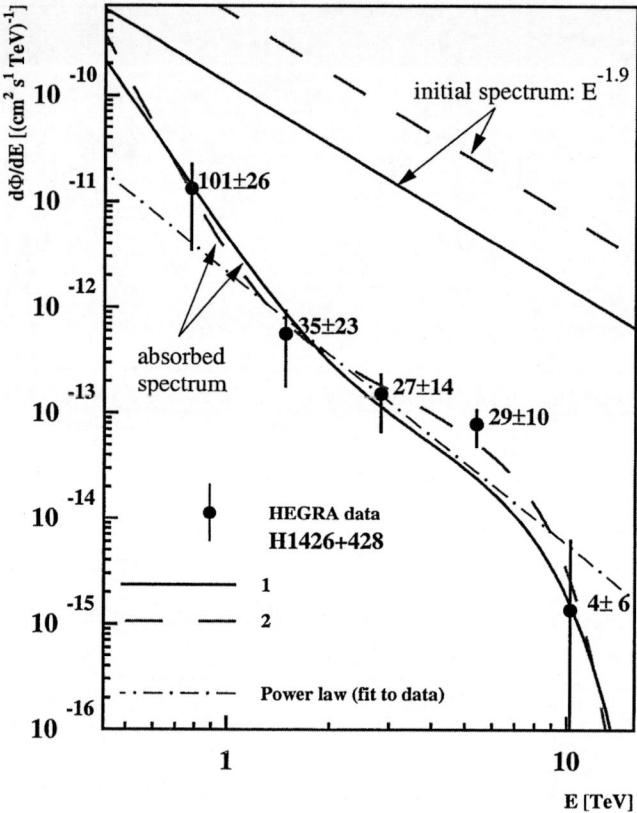

Fig. 10. Differential HEGRA spectrum of H 1426 + 428 and its approximation (*solid* and *dashed* curve) by the absorption effect, cf. Fig. 9, on initial (primary source) spectra $\propto E^{-1.9}$ suggested by X-ray measurements. A power law fit is given by the *dash-dotted* curve (from [11])

Perspectives for the EBL from γ-ray measurements

These results show clearly that TeV cutoffs alone contain insufficient information, in contrast to earlier expectations [58], because cutoffs can also be mimicked by several effects, foremost by an intrinsic cutoff of the source spectrum due to a finite maximum particle energy, or by the Klein-Nishina effect. On the other hand the characteristic wavelength variation of the absorption characteristics of the γ-ray spectra, measured at different redshifts, offers the prospect of making accurate and convincing γ-ray determinations of the EBL in the near and mid infrared in the near future.

4   Future Perspectives of High Energy γ-ray Astronomy

Physics Questions

With the next generation of instruments coming on line a much larger number of sources will be detected. Such an increase by an order of magnitude gives good reasons to expect that several of the major physics problems which we have discussed here, will be solved. This should become especially true, in one way or the other, for the origin of the Galactic Cosmic Rays from SNRs. Another area of research, not mentioned above at all, will be the 3-dimensional nonthermal structure of the Galaxy, together with its halo in the form of the Galactic Wind (e.g. [51]). It should find its complement in investigations at low radio frequencies with the proposed Square Kilometer Array.

Besides these developments γ-ray astronomy will increasingly move to Extragalactic sources and to observational cosmology and, at least in a serendipitous form, to Astroparticle Physics.

Beyond intrinsic AGN physics, the instrumental sensitivity increase will allow studies of nearby starburst galaxies and through them the expected formation of a strong nonthermal component *throughout the Universe* can be studied. Complementary studies aim at the nonthermal component of galaxy clusters. By its large size and expected turbulent agitation, the Intracluster Medium should not only confine the visible thermal matter and the Dark Matter, but also the relativistic hadronic component since its formation. This means that clusters of galaxies are closed systems, preserving not only the chemical but also the nonthermal history and entropy production since structure formation started [62].

Strong emitters of very high energy γ-rays like the jets from flat spectrum Quasars are expected to be surrounded by a halo of $e^+ e^-$ pairs due to the absorption of very high energy γ-rays with $E_\gamma \sim 100$ TeV in the EBL and subsequent magnetic isotropization in an intergalactic field $> 10^{-12}$ G. The Compton upscattering of photons from the Cosmic Microwave Background initiates a cascade that becomes observable at lower γ-ray energies when the space between us and the source becomes ultimately transparent [3]. The halos would be visible even if the jet points in an arbitrary direction relative to the observer due to the magnetic isotropization. Measurements of the angular size $\sim 1°$ and γ-ray energy spectrum of such halos should allow the determination of the Hubble constant, i.e. the *absolute distance* of these objects, and a determination of the *local* (in redshift z) EBL. Even though such measurements promise to be difficult, and although there exists a substantial confusion problem at larger redshifts, the possible rewards are correspondingly high.

It is also worth to emphasize the perspectives for Astroparticle Physics, for instance by γ-ray observations of the Galactic Center region. One set of simulations of the mass density $\rho(r)$ of Cold Dark Matter particles in the gravitational potential well of the Galaxy suggests a rise $\varrho(r) \propto r^{-1}$ with decreasing radius in the innermost region [49]. However, other simulations come to more extreme results (see e.g. [39] for a recent convergence study): $\varrho(r) \propto r^{-3/2}$ for very small r, while agreeing for larger radii with the r^{-1}-dependence. Depending on whether there is a strong density cusp or not, the annihilation rate of e.g. Dark Matter neutralinos $\propto \varrho^2$ in the very Galactic Center could therefore be quite high or rather small (e.g. [24]) and γ-ray observations of the Galactic Center will not only be interesting from an astronomical point of view. Calculations of the neutralino annihilation flux (e.g. [33]) suggest the appearance of a line, possibly at energies between 100 GeV and 1 TeV, besides a continuum that is strongly falling off with γ-ray energy. Fluxes may be at the percent level of the Crab Nebula. Observations with the coming generation of Cherenkov telescopes could thus provide a test of different halo models of our Galaxy and/or put meaningful constraints on SUSY parameter space.

Next Generation Instruments

The next space projects in high energy γ-ray astronomy will be NASA's Gamma-ray Large Astronomical Space Telescope GLAST[9] with an expected launch in 2007, and its small brother, the Italian precursor mission Astrorivelatore Gamma ad Immagini LEggero AGILE[10], whose launch is presently foreseen for 2004. Both detectors are based on silicon strip technology. Comparable in sensitivity to EGRET, AGILE will have a much larger FoV of 3 sr and thus be very good for surveys. Similarly, it is largely its survey capability which will distinguish GLAST from ground based Cherenkov telescopes. However, GLAST will also be more than an order of magnitude step beyond EGRET in sensitivity, angular resolution, and spectral coverage. The energy range will extend to hundreds of GeV, even though for reasons of statistics its de facto energy range will usually be limited to some tens of GeV. GLAST will, first of all, be used to investigate the large number of unidentified EGRET sources, left over from the CGRO mission. Beyond, it is expected to find hundreds of AGNs, to localize a fair number of SNRs, and to search for extended Extragalactic objects.

On the ground, in the complementary energy region above $\sim$ 50 GeV, the future has already begun in Australia with the first 10 m Cherenkov telescope of the 2x2 stereoscopic array CANGAROO III[11] operating since more than one year. The next telescopes are expected to start observations soon. In Namibia, the first 12 m - telescope of the 2x2 array of the Phase I of the High Energy Stereoscopic System H.E.S.S.[12] became operational this June 2002 (Fig. 11), to be followed by the other three components in time steps of 6 months. The 17 m single MAGIC telescope[13] is due to be commissioned in La Palma still in 2002. And in a few years the 7-telescope 10 m array VERITAS[14] will follow in Arizona.

Typically these instruments will have an energy threshold around 100 GeV, and an order of magnitude increase in sensitivity at 1 TeV compared to the pre-

---

[9] http://glast.gsfc.nasa.gov/.

[10] http://agile.mi.iasf.cnr.it/Homepage/.

[11] http://icrhp9.icrr.u-tokyo.ac.jp/c-ii.html.

[12] http://www.mpi-hd.mpg.de/hfm/HESS/HESS.html.

[13] http://hegra1.mppmu.mpg.de/MAGICWeb/.

[14] http://veritas.sao.arizona.edu/veritas/index/shtml.

Fig. 11. The first of the four 12 m telescopes of H.E.S.S. Phase I in Namibia. The tessellated mirror consists of 380 aluminized glass mirrors of 60 cm diameter. The focal plane detector ('camera') has 960 ultrafast photomultiplier pixels, covering an area of 1.4 m that corresponds to a field of view of 5°. The energy threshold is about 100 GeV and the sensitivity is about 10^{-12} erg/(cm²s) above 100 GeV and about 10^{-13} erg/(cm²s) above 1 TeV for 50 h of observation. (Photograph F. Toussenel, June 2002)

vious generation instruments with their excellent angular resolution of 0.1°. The energy resolution $\Delta E/E$ is about 15 %. Several of the ground based instruments will have made detailed observations already years before GLAST. Nevertheless the superior survey capability and the lower energy range will still reserve GLAST important and unique goals.

The attraction of γ-ray astronomy at high energies is that it is a young field. Whereas satellite instruments appear limited in their capabilities simply by the required sizes and masses, this is not really true for ground-based Cherenkov telescopes. Putting them on high mountain altitude, like ESO's ALMA site at 5000 m a.s.l., a future large extension in threshold down to about 5 GeV is possible with a 2x2 array of 20 m telescopes, while basically retaining the enormous effective area in the 10^5 m^2 range [5]. As a consequence, close-by γ-ray bright objects like the Vela Pulsar could be detected in seconds with such an array.

Acknowledgements

I thank F.A. Aharonian, E.G. Berezhko, D. Breitschwerdt, L. Costamante, W. Hofmann, D. Horns, and M. Panter for valuable discussions, and J. Suppanz-Pirsch for expert help with the manuscript.

References

1. F.A. Aharonian: New Astronomy **5**, 377 (2000)
2. F.A. Aharonian, & A.M. Atoyan: A&A **351**, 330 (1999)
3. F.A. Aharonian, P.S. Coppi, H.J. Völk: ApJ **423**, L5 (1994)
4. F.A. Aharonian, A.M. Atoyan, T. Kifune: MNRAS **291**, 162 (1997)
5. F.A. Aharonian, A.K. Konopelko, H.J. Völk, H. Quintana: Astropart. Phys. **15**, 335 (2001)
6. F.A. Aharonian, et al. (HEGRA): A&A **349**, 11 (1999)
7. F.A. Aharonian, et al. (HEGRA): A&A **370**, 112 (2001)
8. F.A. Aharonian, et al. (HEGRA): A&A **373**, 292 (2001)
9. F.A. Aharonian, et al. (HEGRA): to be published in A&A
10. F.A. Aharonian et al. (HEGRA): A&A, in press; astro-ph/0205499 (2002)
11. F.A. Aharonian, et al. (HEGRA): A&A **384**, L23a (2002c)
12. Allen, G.E., Gotthelf, E.V., Petre, R.: Evidence of 10 – 100 TeV Electrons in Supernova Remnants. Proc. 26th ICRC (Salt Lake City), **3**, 480 (1999)
13. J. Arons: Space Sci Rev. **75**, 235 (1996)
14. A. M. Atoyan & F.A. Aharonian: MNRAS **278**, 525 (1996)
15. W. Becker, K.T. Brazier, J. Trümper: A&A **298**, 528 (1995)
16. E.G. Berezhko, L.T. Ksenofontov, S.I. Pethukov: "Radio-, X-ray and Gamma-ray Emission Produced in SN 1006 by Accelerated Cosmic Rays". In: *Proc. 26th ICRC (Salt Lake City) 1999*, **3**, 431
17. J.H. Berezhko, L.T. Ksenofontov, H.J. Völk: "Emission of SN 1006 produced by accelerated cosmic rays". In: *Proc. 27th ICRC (Hamburg) 2001* **6**, 2465
18. E. G. Berezhko, L.T. Ksenofontov, H.J. Völk: Submitted to A&A (2002)
19. E.G. Berezhko, G. Pühlhofer, H.J. Völk: "Gamma-ray emission from Cassiopeia A produced by accelerated cosmic rays". In: *Proc. 27th ICRC (Hamburg)(2001)* **6**, 24
20. E. G. Berezhko, G. Pühlhofer, H.J. Völk: A&A, in preparation
21. R. D. Blandford, M.J. Rees (1978): Proc. Pittsburg Conf. on BL Lac Objects, 341
22. D. Breitschwerdt, F.D. Kahn: MNRAS **235**, 1011 (1988)
23. J.H. Buckley et al. (Whipple): A&A **329**, 639 (1998)
24. J.H. Buckley, G. Jungmann (1998): "Gamma-Rays from Neutralino Annihilation: Prospects for ACT Detection". In: *Towards a Major Atmospheric Cherenkov Detector – IV, 1998*,ed. by M. Cresti (Padova, 1998) 26
25. Y.M. Butt, D.F. Torres, G.E. Romero, T. Dame, J.A. Combi: Nature **418**, 499 (2002)

26. P.M. Chadwick et al.: Astropart. Phys. **9**, 131 (1998)
27. O.C. De Jager, A.K. Harding: ApJ **396**,161 (1992)
28. A. Djannati-Atai et al. (CAT): A&A **350**, 17 (1999)
29. A. Djannati-Atai et al. (CAT): astro-ph/0207618, to appear in A&A (2002)
30. L. O'C. Drury, H.J. Völk: ApJ **248**, 344 (1981)
31. L. O'C. Drury, F.A. Aharonian, H.J. Völk: A&A **287**, 959 (1994)
32. R. Enomoto et al. (CANGAROO): Nature **416**, 823 (2002)
33. P. Gondolo, J. Edsjö, L. Bergström, P. Ullio, E.A. Baltz: astro-ph/0012234 (2000)
34. A.J.S. Hamilton, C.L. Sarazin, A.E. Szymkowiak: ApJ **300**, 698 (1986)
35. N. Kawai, K. Tamura, S. Shibata: "New Detection of X-Ray Pulsar Nebulae by ASCA". In: *Proc. IAU Colloq. 160* (1996) *The Hot Universe*, ed. by K. Koyama et al.(Kluwer, Dordrecht 1996), 265
36. C. F. Kennel, F. V. Coroniti: ApJ **283**, 694 (1984)
37. T. Kifune et al. (CANGAROO): ApJ **438**, L91 (1995)
38. J. G. Kirk, O. Skjaeraasen, Y.A. Gallant: A&A **388**, L29 (2002)
39. A. Klypin, A. V. Kravtsov, J. S. Bullock, J. R. Primack: ApJ **554**, 915 (2001)
40. K. Koyama: Nature **378**, 255 (1995)
41. H. Krawczynski, P.S. Coppi T. Maccarone, F.A. Aharonian: A&A **353**, 97 (2000)
42. M.A. Malkov, L.O'C Drury: Rep. Prog. Phys. **64**, 429 (2001)
43. K. Mannheim: A&A **269**, 67 (1993)
44. P.L. Marsden, F.C. Gillet, R.E. Jennings, J.P. Emerson, T. de Jong: ApJ **278**, L29 (1984)
45. A. Mastichiadis: A&A **305**, L53 (1996)
46. H. Muraishi et al. (CANGAROO): A&A **354**, L57 (2000)
47. T. Naito & F. Takahara: J. Phys. G: Nucl. Part. Phys. **20**, 477 (1994)
48. T. Naito T. Yoshida, M. Mori, T. Tanimori: Astron. Nachr. **320**, 205 (1999)
49. J.F. Navarro, C.S. Frenk, S.D.M. White: ApJ **462**, 563 (1996)
50. H. Petry et al. (VERITAS): astro-ph/0207506, to appear in ApJ. (2002)
51. V.S. Ptuskin, H.J. Völk, V.N. Zirakashvili, D. Breitschwerdt: A&A **321**, 434 (1997)
52. M.J. Rees, J.E. Gunn: MNRAS **167**, 1 (1974)
53. O. Reimer, M. Pohl: A&A **390**, L43 (2002)
54. S.P. Reynolds: ApJ **459**, L13 (1996)
55. M.A. Ruderman, P.G. Sutherland: ApJ **196**, 51 (1975)
56. V. Schönfelder (Ed.): *The Universe in Gamma Rays* (Springer, Berlin, Heidelberg, 2001)
57. P. Slane, B.M. Gaensler, T.M. Dame, J.P. Hughes, P.P. Plucinsky, A. Green: ApJ **525**, 357 (1999)
58. F.W. Stecker, O.C. de Jager, M.H. Salamon: ApJ **390**, L49 (1992)
59. T. Tanimori(CANGAROO): ApJ **479**, L25 (1998)
60. T. Tanimori et al. (CANGAROO) (1998): "Study of the TeV gamma-ray spectrum of SN1006 around the NE Rim". In: *Proc. 27th ICRC Hamburg, 1998* **6**, 2465
61. F. Tavecchio, L. Maraschi, E. Pian, L. Chiappetti, A. Celotti, G. Fossati, G. Ghisellini, E. Palazzi, C.M. Raiteri, R.M. Sambruna, A. Treves, C.M. Urry, M. Villata, A. Djannati-Atai: ApJ **554**, 725 (2001)
62. H.J. Völk, F.A. Aharonian, D. Breitschwerdt: Space Sci. Rev. **75**, 279 (1996)
63. H.J. Völk (HEGRA): "Particle Acceleration and Gamma-Ray Production in Shell Remnants". In: *Proc. Towards a Major Atmospheric Cherenkov Detector-V*, ed. by O.C. de Jager, Space Research Unit 1997, Westprint-Potchefstrom, South Africa 1997, ISBN 1-86822-295-0, 87–106
64. H.J. Völk, E.G. Berezhko, L.T. Ksenofontov, G.P. Rowell: To be published in A&A (2002)

Cosmic Rays at the Highest Energies

A.A. Watson

Department of Physics and Astronomy, University of Leeds,
Leeds LS2 9JT, UK

Abstract. Reasons for the current interest in cosmic rays above 10^{19} eV are described. The latest results on the energy spectrum, arrival direction distribution and mass composition of cosmic rays are reviewed. The enigma set by the existence of ultra high-energy cosmic rays remains and ideas proposed to explain it are discussed. Progress with the construction of the Pierre Auger Observatory and plans for the EUSO space instrument are outlined.

1 Introduction

For the purposes of this review I define ultra high-energy cosmic rays (UHECRs) as those cosmic rays having energies above 10^{19} eV. There is currently great interest in them, partly because we have little idea as to how Nature creates particles or photons of these energies. Also we know enough about their energy spectrum and arrival direction distribution to believe that we have an additional problem: their sources must be reasonably nearby (within 100 Mpc) but there is no evidence of the anisotropies anticipated if the galactic and inter-galactic magnetic fields are as weak as astronomers tell us.

The distance limit comes from a combination of well-understood particle physics and the universality of the 2.7 K radiation. Interactions of protons and heavier nuclei with this, and other, radiation fields degrade the energy of particles rather rapidly. In the case of protons, the reaction is photopion production, while heavier nuclei are photodisintegrated by the 2.7 K radiation and the diffuse infrared background. These effects were first recognised by Greisen, and by Zatsepin and Kuzmin, and lead to the expectation that the energy spectrum of cosmic rays should terminate rather sharply above 4×10^{19} eV (the GZK cut-off). Above 4×10^{19} eV about 50% of particles must come from within 130 Mpc, while at 10^{20} eV the corresponding distance is 20 Mpc.

It is possible that some or all of the UHECRs are photons but, if so, the sources must be even closer. Photons of these energies are strongly attenuated by pair production and at 10^{20} eV the relevant electromagnetic fields are diffuse radio photons in the $1 - 10$ MHz band. The flux of such photons is poorly known but the mean free path for pair production seems unlikely to be more than a few Mpcs.

The most recent data suggest that particles do exist with energies beyond the GZK cut-off and that the arrival direction distribution is isotropic. The mass of the cosmic rays above 10^{19} eV is not known, although there are recent experimental limits on the fraction of photons that constrain a class of models proposed to resolve the enigma.

2   Measurement of UHECR

The properties of UHECRs are obtained by studying the cascades, or extensive air showers (EAS), they create in the atmosphere. Many methods of observing these cascades have been explored but currently two approaches seem to be most effective. In one, the density pattern of particles striking an array of detectors laid out on the ground is used to infer the primary energy. At 10^{19} eV the footprint of the EAS on the ground is several square kilometres so detectors can be spaced many hundreds of metres apart. Alternatively, on clear moonless nights, the fluorescence light emitted when shower particles excite nitrogen molecules in the atmosphere can be observed by large photomultiplier cameras. This technique, uniquely, allows the rise and fall of the cascade in the atmosphere to be inferred.

The primary energy of the initiating particle or photon is deduced in different ways. For the detector arrays, Monte Carlo calculations have shown that the particle density at distances from $400 - 1200$ m is closely proportional to the primary energy. Such a density can be measured accurately (usually to around 20%) and the primary energy inferred from conversion relations, that are mass independent at the 10% level, found by calculation. The estimate of the energy depends on the realism of the representation of features of particle interactions within the Monte Carlo model, at energies well above accelerator energies. The currently favoured model (QGSJET) is based on QCD and is matched to accelerator measurements. Although this model appears to describe a variety of data from TeV energies up to 10^{20} eV [24], one cannot be certain of the systematic error in the energy estimates.

For the fluorescence detectors, the primary energy is found by integrating the number of electrons in the cascade curve and assuming that their rate of energy loss is close to that at the minimum of the dE/dx curve for electrons, ~ 2.2 MeV per g cm^{-2} in the case of air. A small, model-dependent, correction must be made to account for the energy carried by muons and neutrinos into the ground. Ideally, one wants to compare estimates of the primary energy made in the same shower by the two techniques operating simultaneously, but this has yet to be done at these energies. So far all that has been possible is to compare estimates of the fluxes at nominally the same energy.

3   The Energy Spectrum, Arrival Direction Distribution and Mass of UHECRs

The most important parameters to measure are the energy spectrum arrival direction and mass distribution of the incoming UHECRs.

3.1   Energy Spectrum

Until relatively recently, data on the energy spectrum from a number of experiments had seemed to be in good accord [25]. The rates of events at 10^{19} eV reported by different experiments were in agreement at the $10 - 15\%$ level. In addition, preliminary data from the Utah-based HiRes group, reported at the International Cosmic Ray Conference in 1999, contained 7 events above 10^{20} eV, in good agreement with the number anticipated from the flux seen by the Japanese AGASA array.

The situation has now changed dramatically. At the international meeting in Hamburg (August 2001), the AGASA group [30] reported additional data, quite consistent with their earlier work, and described 17 events above 10^{20} eV. The HiRes group reported on monocular data obtained with one of their cameras from an exposure slightly greater than that of the AGASA group [19]. Assuming a spectrum similar to that reported by the AGASA group, the HiRes team had expected to see about 20 events above 10^{20} eV, but observed only 2. This unexpected discrepancy is not yet understood.

The data from Fly's Eye (the earliest fluorescence experiment), Haverah Park (a ground array that used water-Cherenkov detectors), HiRes and AGASA are shown in Fig. 1. There are several points to note. The Haverah Park energy estimates have been re-assessed [1] using the QGSJET model. In the range 3×10^{17} to 3×10^{18} eV there is very good agreement between the Fly's Eye, Haverah Park and HiRes results. A recent Haverah Park analysis [2] suggests that protons and iron are in the ratio 35:65 in this energy range. With this mixture agreement between the spectra is even better. This implies that the QGSJET model provides an adequate description of important features of showers up to 10^{18} eV. However, the AGASA energies have been estimated with the QGSJET model under the assumption that the primaries are protons at energies above 3×10^{18} eV, the lowest AGASA energy plotted. There is no evidence as to what mass species is dominant at the highest energies but the methods used would lead to an estimate lower by about 20% if iron nuclei were assumed. This change would be insufficient to reconcile the AGASA-HiRes differences, particularly with regard to the point at which the spectrum slope flattens above 10^{18} eV. However a combination of a change in the QGSJET model and iron primaries (for which there is no evidence) might go some way to aligning the different results at the highest energies as would a systematic change in the aperture of acceptance near to the AGASA threshold.

There are also unanswered questions about the HiRes data. The 'disappearance' of the events, reported as being above 10^{20} eV in 1999, is attributed to a better understanding of the atmosphere which is now claimed to be clearer than had previously been supposed. The Hamburg results [19] were prepared using an 'average atmosphere' so presumably subsequently some events will be assigned larger energies and some smaller ones. Two further issues need resolving. Firstly, an accelerator-based calibration of the fluorescence yield [20] led to the claim 'that the fluorescence yield of air between 300 and 400 nm is proportional to the electron dE/dx.' This claim is not consistent with information tabulated in the

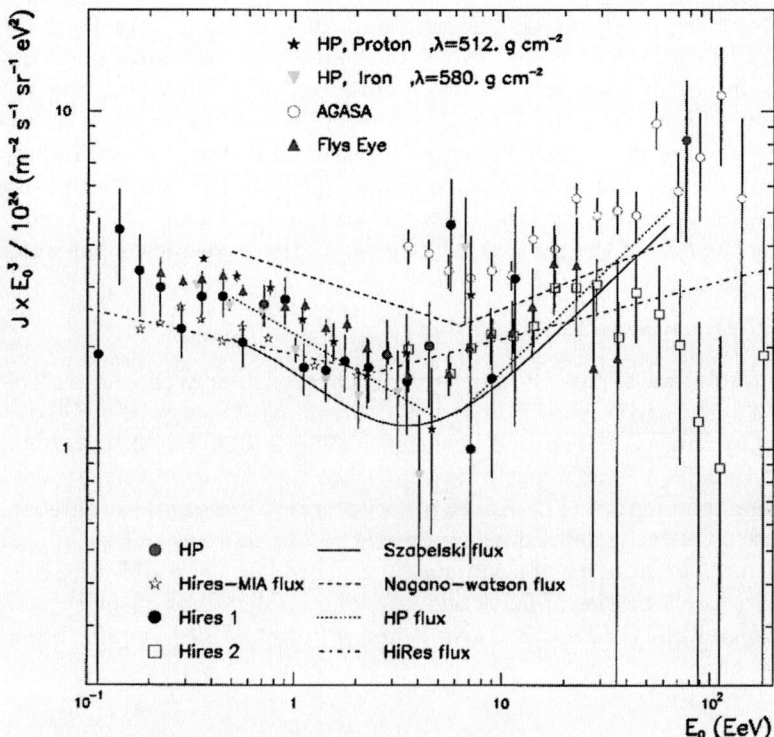

Fig. 1. A composite energy spectrum from AGASA, Fly's Eye, Haverah Park and HiRes. This plot was prepared with the help of Maximo Ave. The Agasa and HiRes spectra were reported at the Conference in Hamburg 2001[30,19]

paper, where it is shown that the yield from 50 keV electrons is similar to that from 1.4 MeV electrons. Also the dE/dx curve plotted there, normalised to the 1.4 MeV measurements, does not fit the accelerator data for 300, 650 and 1000 MeV electrons. The latter discrepancy is about 15 – 20% and in such a direction as would increase the HiRes energies. Secondly, Nagano et al. [26] has described a new measurement of the yield in air from 1.4 MeV electrons. In what seems to be a very careful study, they find that the earlier results [20] gave a higher yield at 356.3 nm and 391.9 nm than is found now. Nagano attributes the absence of background corrections as being responsible for at least some of the discrepancies [27]. The longer wavelengths become increasingly important, because of Rayleigh scattering, when showers are observed at the large distances common at the highest energies. The magnitude of the adjustments that need to be made to the HiRes data are presently unclear and further fluorescence yield measurements are certainly required. Some of these might usefully be made at CERN by an Auger-EUSO collaboration. It is worth noting that measurements in 1970 [18] of fluorescence from nitrogen at 391.4 nm support Nagano's measurement.

At the Hamburg meeting, the HiRes group also reported data from their stereo system. See also [22]. With 20% of the monocular exposure, they found 1 event with an energy estimated as being close to 3×10^{20} eV, the energy of the largest event found with the Fly's Eye detector [10]. My opinion is that the spectra from AGASA and HiRes will come together as further understanding is gained of the models and of the atmosphere. Knowledge of the mass composition will also help considerably. For now it seems certain that trans-GZK events do exist but that the flux of them is less certain than appeared a few years ago.

3.2   Arrival Direction Distribution

The angular resolution of current shower arrays and of fluorescence detectors is typically $2 - 3°$. The arrival direction of the 59 events with energy above 4×10^{19} eV registered by the AGASA group is shown in Fig. 2 [32]. The distribution is isotropic and there is no preference for events to come from close to the galactic or the super-galactic planes. The AGASA group draw attention to a number of clusters, where a cluster is defined as a grouping of 2 or more events within $2.5°$. It is claimed that the number of doublets (5) and triplets (1) could have arisen by chance, with probabilities of 0.1% and 1%. The implications of such clusters would be profound but the case for them is not yet proven. The angular bin was

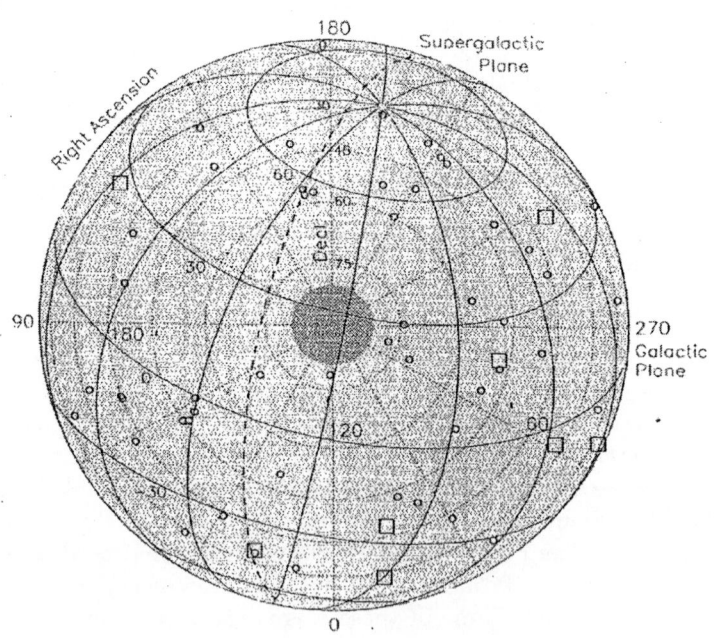

Fig. 2. AGASA arrival direction distribution for 59 events above 4×10^{19} eV. The most energetic events ($> 10^{20}$ eV) are shown by squares [32]

not defined a priori and the data set used to make the initial claim for clusters is also being used in the 'hypothesis testing' phase. Furthermore, I note that the directions of the 7 most energetic events observed by Fly's Eye, Haverah Park, Yakutsk and Volcano Ranch do not line up with any of the 6 cluster directions.

It is hard to understand the isotropy observed at 10^{20}eV if the local extragalactic magnetic field is really just 10^{-9} gauss. A proton of 10^{20} eV would be deflected by only about 2° over a distance of 20 Mpc if the field has a 1 correlation length of about 1Mpc[21]. If the fields were much higher, as has been suggested [15], then the lack of anisotropy might be understood, but more energy is then stored in the magnetic field and this may create other difficulties. Similarly, if the charge of the particles initiating the showers was much higher than Z=1, the isotropy could be explained. Hence measurement of the mass composition is of crucial importance.

3.3   Mass Composition

Interpretation of the data on UHECRs is hampered by our lack of knowledge of the mass of the incoming particles. Data from several experiments can be interpreted as indicating a change from a dominantly iron beam near 3×10^{17} eV to a dominantly proton beam at 10^{19} eV (see [2,4] for recent discussions). But the situation is unclear and quite open at higher energies. The data are just too limited and the interpretations are ambiguous as both the fluorescence detectors and ground arrays rely on shower models to deduce composition information.

It is unlikely that the majority of the events claimed to be near 10^{20} eV have photons as parents as some of the showers have the normal numbers of muons (the tracers of primaries that are nuclei) and the profile of the most energetic fluorescence event is inconsistent with that of a photon primary [16]. Furthermore, there is now evidence that less than 40% of the events at 10^{19} eV are photon-initiated. This limit has been set in two ways. Taking the energy spectrum as measured by Fly's Eye as being independent of the mass of the incoming particles, the rate of showers coming at large angles to the vertical can be calculated. Using Haverah Park data, it has been found that the observed rate of inclined showers is much higher than would be expected if the primary particles were mainly photons [3]. A more traditional attack on the problem by the AGASA group, searching for showers which have significantly fewer muons than normal, has given the same upper limit [31].

It is unlikely that many events are created by neutrinos as the distribution of zenith angles would be different from that observed. Indeed, in all aspects so far measured, events of 10^{20} eV look like events of 10^{19} eV, but ten times larger, and this can be reiterated as we go to lower and lower energies were nuclei seem certain to be the progenitors of showers.

4   Theoretical Interpretations

The UHECR enigma is attracting significant theoretical attention. Some ideas suppose a form of electromagnetic acceleration while others invoke new physics.

Currently it is popularly believed that cosmic rays with energies up to about 10^{15} eV are energised by a process known as 'diffusive shock acceleration'. Supernovae explosions are identified as the likely sites, although so far there is no direct evidence for acceleration of nuclei by supernova remnants at any energy. The diffusive shock process, which has its roots in some early ideas of Fermi, has been extensively studied since its conception in the late 1970s. In [12] it is shown that the maximum energy attainable is given by $E = kZeBR\beta c$, where B is the magnetic field in the region of the shock, R is the size of the shock region and k is a constant less than 1. The same result has been obtained by a number of people, e.g. [17], and most authors agree upon it. However, some claim that the diffusive shock acceleration process can be modified to give much higher energies than indicated by the equation and that radio galaxy lobes, in particular, are probable acceleration sites. It is difficult to see how an energy of 3×10^{20} eV can be accounted for if the size of the shock region is 10 kpc and the magnetic field is 10 μG (values thought typical of lobes of radio galaxies), as even the optimum estimate of the energy reachable in such an environment is lower by a factor of 3 than the observational upper limit. It could be that the magnetic fields are stronger than is usually supposed, a line of argument that also comes from the arrival direction work mentioned above.

Proposals have been made which dispense with the need for electromagnetic acceleration. Attention has usually been focused on the highest energy events ($> 10^{20}$ eV). However, it is my view that proposers of some of the more exotic mechanisms often overlook one or more important points. Any mechanism able to explain the highest energy events must also explain those above about 3×10^{18} eV, where the galactic component probably disappears. The spectrum above this point is possibly too smooth to imagine that there are two or more radically different components – although this might be seen as an almost philosophical argument, particularly in the light of Fig. 1! In addition, the solutions proposed must produce particles at the top of the atmosphere that can generate showers of the type we see and now understand rather well. Finally, source energetics cannot be ignored: there seems little point in inventing a mechanism to 'solve' the GZK cut-off problem that requires a source region that is unrealistically energetic.

An overview of the various non-electromagnetic processes, the so-called 'top down mechanisms', proposed can be found in [25] and I will only discuss one of these here. It has been suggested that UHECR arise from the decay of super-heavy relic particles. In this picture, the cold dark matter is supposed to contain a small admixture of long-lived super-heavy particles with a mass $> 10^{12}$ GeV and a lifetime greater than the age of the Universe [8]. It is argued that such particles can be produced during reheating following inflation, or through the decay of hybrid topological defects such as monopoles connected by strings. I find it hard to judge how realistic these ideas are but the decay cascade from a particular candidate [5] has been studied in some detail [11] and [29]. A feature of the decay cascade is that an accompanying flux of photons and neutrinos is predicted which may be detectable with a large enough installation. In particular photons are expected to be between 2 and 10 times as numerous as protons above 10^{19} eV. The anisotropy question has been examined and specific predictions have been made for the anisotropy that would be seen by a Southern Hemisphere observatory [9,13,6,23]. Observation of the predicted anisotropy, plus the identification of appropriate numbers of neutrinos and photons, would be suggestive of a super-heavy relic origin. However, the experimental results on the photon/proton ratio at 10^{19} eV described above clearly do not support it, or topological defect models that also predict large photon fluxes [3,16,31].

5   Detectors of the Future

It must be clear from what has been said above that more data on UHECRs are badly needed. The AGASA array of 100 km^2 is, inevitably, drawing to the end of its useful life. The HiRes instrument is taking data but does not have sufficient aperture to resolve the questions now being posed in a reasonable time, particularly if the flux above 10^{20} eV does turn out to be as low as is implied by their preliminary results [19]. Therefore, two new instruments are being developed with the aim of increasing the number of events above 10^{20} eV by a very large factor. These instruments, the Pierre Auger Observatory [28] and the EUSO space instrument [14], will be briefly described.

5.1   The Pierre Auger Observatory

The Pierre Auger Observatory was conceived to measure the properties of the highest energy cosmic rays with unprecedented statistical precision. When completed, it will consist of two instruments, constructed in the Northern and Southern Hemispheres, each covering an area of 3000 km^2. Two instruments are necessary for essentially the same reasons as optical telescopes are built in both hemispheres. The design calls for a hybrid detector system with 1600 particle detector elements and three or four fluorescence detectors at each of the sites. The particle detectors will be 1.2 m deep water-Cherenkov tanks arranged on a 1.5 km hexagonal grid. Cherenkov detectors have been selected because water acts as a very effective absorber of the multitude of low energy electrons and photons found at distances of about 1 km from the shower axis. In addition the tanks respond well to muons, nearly all of which traverse the whole of the detector.

At the Southern site (see Fig. 3) fluorescence detectors will be set up at four locations. Possibly one will be near the centre of the particle array with the others on small promontories at the array edge: the site is close to the town of Malargue in Mendoza Province, Argentina, some five hours drive from Mendoza City. During clear moonless nights, signals will be recorded in both the fluorescence detectors and the particle detectors, while for roughly 90% of the time only particle detector information will be available. Data from the water-tanks, which are powered with solar panels, are sent to the office building using a purpose built radio link and a commercial microwave system. Each tank runs

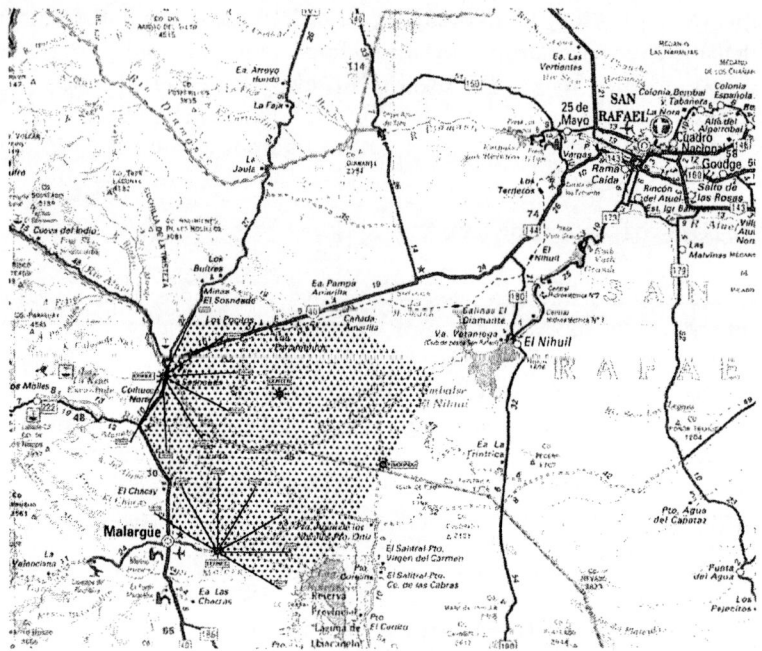

Fig. 3. Plan of the Pierre Auger Observatory near Malargue, Mendoza Province, Argentina. Most of the water tanks will be located on the Pampa to the north east of the town of Malargue, which is about 200 km south of the city of San Rafael. Each dot within the area to the left of route 40 marks the planned position of a water tank. They are separated by 1.5 km. Fluorescence detectors will be established at the sites marked Leones, Morades, Center and Coihueco

autonomously sending low-level triggers to the centre at 20 Hz. When an appropriate grouping in space and time is identified by the data-logging computer, data from that group of detectors, and nearby neighbours, are requested and transmitted. The microwave towers are located at the sites of the fluorescence detectors, which run from a conventional electrical supply. Relative arrival times at detectors are measured using the GPS network.

Construction of the central laboratory in Malargue began in March 1999 and this, and an office building, provide excellent infra-structural facilities for the project. An engineering array, containing 40 water tanks and a section of a fluorescence detector was completed in September 2001 and the design of all of the sub-systems of the Observatory has now been demonstrated. Many fluorescence events have been recorded since the first were registered in May 2001 and a large number of 'between tank' coincidences have been seen since the first were obtained some two months later. The first 'hybrid' events were recorded in December 2001 and 75 had been obtained when the prototype fluorescence detector was dismantled so that construction of the final instrument could begin. The hybrid events promise to yield valuable details about many parameters, including the lateral distribution of the signals seen by the water tanks and the accuracy with which a single eye can be used to obtain the shower core position.

Preliminary analysis of the events from the engineering array is underway and there is great confidence that the observatory will work as designed. By early 2003 it is planned that a further 100 tanks will have been installed and instrumented, together with two complete fluorescence detectors at Los Leones and Coihueco. With this level of instrumentation, serious science can begin as approximately 150 km^2 of area will be monitored. There are obvious point source targets to be examined, such as Cen A and the galactic centre, but the earliest science may come from searches for photons made using various techniques that are now being developed. It is expected that the full Observatory in the southern hemisphere will be completed during 2005, with the four fluorescence detectors becoming operational about a year earlier. When the Auger Observatory at Malargue has operated for 10 years, it is expected that over 300 events above 10^{20} eV will have been recorded.

5.2   EUSO (and OWL)

Achieving an exposure greater than that targeted by the Auger Observatory is a formidable challenge. A promising line is the development of an idea due to Linsley [7]. The concept is to observe fluorescence light produced by showers from space with satellite-borne equipment. It is proposed to monitor $\sim 10^5$ km^2 sr (after allowing for an estimated 8% on-time). Preliminary design studies have been carried out in Italy, Japan and the USA. An Italian-led collaboration has proposed a design that is under Phase A study for flight on the International Space Station. This is known as EUSO (the Extreme Universe Space Observatory), and has the potential to detect neutrinos in large numbers, as well as UHECRs. Observations are scheduled to start in 2008: the twin satellite OWL project, which is being developed at NASA, will follow sometime later. These projects require considerable technological development but may be the only way to push to energies beyond whatever energy limits are found with the planned Auger instruments.

6   Conclusions

There are many reasons to be intrigued by the highest energy cosmic rays, not least the fact that we have no idea where, or how, they are created. Nor do we know what the highest energy will turn out to be as it seems likely that the present limit is set only by the observation time with the instruments used so far. Large increases in data are expected in the coming decade from the Pierre Auger Observatory and EUSO. These will very likely need fresh particle physics input and more astronomical data for their full interpretation, so the future looks extremely exciting.

Acknowledgements

I am grateful to the organisers for inviting me to the ESO-CERN-ESA symposium. On-going support of PPARC to work on ultra high-energy cosmic rays at the University of Leeds is gratefully acknowledged. I also thank my many colleagues in the Pierre Auger project for helping to make a 10-year-old dream become a reality. The work in [18] was drawn to my attention by P. Privitera.

References

1. Ave, M., et al., 2002, Astroparticle Physics (in press); astro-ph/0112253
2. Ave, M., et al., 2002, Astroparticle Physics (in press); astro-ph/0203150
3. Ave, M., et al., 2000, Phys Rev Letters **85** 2244
4. Ave, M., et al., 2002, Astroparticle Physics (in press); astro-ph/0112071
5. Benakli, K., Ellis, J. and Nanopolous, D.V., 1999, Phys Rev D **59** 047301
6. Benson, A., Smialkowshi, A. and Wolfendale, A.W., 1999, Astroparticle Physics **10** 313
7. Benson, R. and Linsley, J., 1981, Proc 17th Int Conf on Cosmic Rays (Paris) **8** 145
8. Berezinsky, V., Kachelreiss, M. and Vilenkin, A., 1997, Phys Rev Lett **22** 4302
9. Berezinsky, V. and Mikhailov, A.A., 1998, astro-ph/9810277
10. Bird, D., et al., 1995, Astrophys J **441** 144
11. Birkel, M. and Sarkar, S., 1998, Astroparticle Physics **9** 297
12. Drury, L. O'C., 1994, Contemporary Physics **35** 232
13. Dubovsky, S.L. and Tinyakóv, P.G., 1998, hep-ph/9808446
14. EUSO project: www.ifcai.pa.cnr.it/~EUSO/
15. Farrar, G.R. and Piran, T., 2000, Phys Rev Letters **84**, 3527
16. Halzen, F., et al., 1995, Astroparticle Physics **3** 151
17. Hillas, A. M., 1984, Ann. Rev. Astronomy & Astrophysics **22**, 425
18. Hirsh, M.N., Poss, E., and Eisner, P.N., 1970, Phys Rev A **1** 1615
19. Jui, C.H. et al., 2001, Proc 27th Int Conf on Cosmic Rays (Hamburg) **1** 354
20. Kakimoto, F., et al., 1996, Nucl Inst and Methods **A372** 527
21. Kronberg, P.P., 1994, Rep Prog Phys **57** 325
22. Loh, E., 2002, Talk at the NEEDS workshop, Karlsruhe, April 2002: http://www.iklanl.fzk.de/ needs/
23. Medina Tanco, G.A. and Watson, A.A., 1999, Astroparticle Physics **12** 25
24. Nagano, M. et al., 2000, Astroparticle Physics **13** 277
25. Nagano, M. and A. A. Watson, 2000, Rev Mod Phys **27** 689
26. Nagano, M., et al., 2001, Proc 27th Int Conf on Cosmic Rays (Hamburg) **2** 675
27. Nagano, M., private communication, September 2001
28. Pierre Auger Observatory: www.auger.org/
29. Rubin, N. A., 1999, M Phil Thesis, University of Cambridge
30. Sakaki, N., et al., 2001, Proc 27th Int. Conf. on Cosmic Rays (Hamburg)1 333
31. Shinosaki, K, et al.,2001, Proc 27th Int. Conf. on Cosmic Rays (Hamburg) **1** 346
32. Takeda, M, et al., 2001, Proc 27th Int. Conf. on Cosmic Rays (Hamburg) **1** 341

High-Energy Neutrinos from Cosmic Rays

Francis Halzen

Department of Physics, University of Wisconsin,
1150 University Avenue, Madison, WI 53706

Abstract. We introduce neutrino astronomy from the observational fact that Nature accelerates protons and photons to energies in excess of 10^{20} and 10^{13} eV, respectively. Although the discovery of cosmic rays dates back close to a century, we do not know how and where they are accelerated. We review the facts as well as the speculations about the sources. Among these gamma ray bursts and active galaxies represent well-motivated speculations because these are also the sources of the highest energy gamma rays, with emission observed up to 20 TeV, possibly higher.

We discuss why cosmic accelerators are also expected to be cosmic beam dumps producing high-energy neutrino beams associated with the highest energy cosmic rays. Cosmic ray sources may produce neutrinos from MeV to EeV energy by a variety of mechanisms. The important conclusion is that, independently of the specific blueprint of the source, it takes a kilometer-scale neutrino observatory to detect the neutrino beam associated with the highest energy cosmic rays and gamma rays. The technology for commissioning such instruments exists.

1 The Highest Energy Particles: Cosmic Rays, Photons and Neutrinos

1.1 The New Astronomy

While conventional astronomy spans 60 octaves in photon frequency, from 10^4 cm radio-waves to 10^{-14} cm gamma rays of GeV energy, successful efforts are underway to probe the Universe at yet smaller wavelengths and larger photon energies; see Fig. 1. Gamma rays, gravitational waves, neutrinos and very high-energy protons are explored as astronomical messengers. As exemplified time and again, the development of novel ways of looking into space invariably results in the discovery of unanticipated phenomena. As is the case with new accelerators, observing only the predicted will be slightly disappointing.

Why pursue high-energy astronomy with neutrinos or protons despite the considerable instrumental challenges? A mundane reason is that the Universe is not transparent to photons of TeV energy and above (in ascending factors of 10^3, units are: GeV/TeV/PeV/EeV/ZeV). For instance, a PeV energy photon cannot deliver information from a source at the edge of our own galaxy because it will annihilate into an electron pair in an encounter with a 2.7 Kelvin microwave photon before reaching our telescope. Only neutrinos can reach us without attenuation from the edge of the Universe at all energies.

At EeV energies, proton astronomy may be possible. Above 50 EeV the arrival directions of electrically charged cosmic rays are no longer scrambled by the

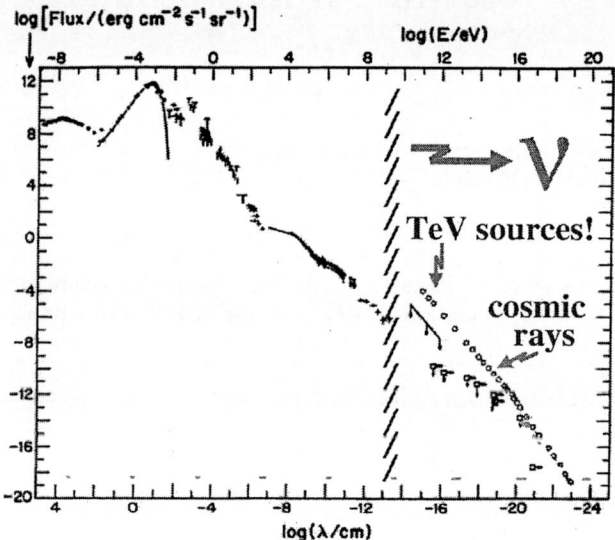

Fig. 1. The diffuse flux of photons in the Universe, from radio waves to GeV-photons. Above tens of GeV, only limits are reported although individual sources emitting TeV gamma rays have been identified. Above GeV energy, cosmic rays dominate the spectrum

ambient magnetic field of our own galaxy. They point back to their sources with an accuracy determined by their gyroradius in the intergalactic magnetic field B:

$$\frac{\theta}{0.1^\circ} \cong \frac{\left(\frac{d}{1 \text{ Mpc}}\right)\left(\frac{B}{10^{-9}\text{ G}}\right)}{\left(\frac{E}{3\times10^{20}\text{ eV}}\right)}, \tag{1}$$

where d is the distance to the source. Speculations on the strength of the intergalactic magnetic field range from 10^{-7} to 10^{-12} Gauss in the local cluster. For a distance of 100 Mpc, the resolution may therefore be anywhere from sub-degree to nonexistent. Proton astronomy should be possible at the very highest energies; it may also provide indirect information on intergalactic magnetic fields. Determining the strength of intergalactic magnetic fields by conventional astronomical means has been challenging.

1.2   The Highest Energy Cosmic Rays: Facts

In October 1991, the Fly's Eye cosmic ray detector recorded an event of energy $3.0^{+0.36}_{-0.54}\times10^{20}$ eV [1]. This event, together with an event recorded by the Yakutsk air shower array in May 1989 [2], of estimated energy $\sim 2\times10^{20}$ eV, constituted at the time the two highest energy cosmic rays ever recorded. Their energy corresponds to a center of mass energy of the order of 700 TeV or ~ 50 Joules, almost 50 times the energy of the Large Hadron Collider (LHC). In fact, all

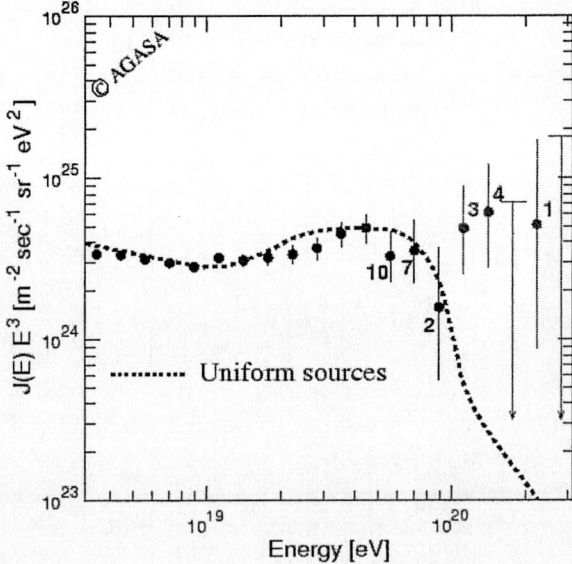

Fig. 2. The cosmic ray spectrum peaks in the vicinity of 1 GeV and has features near 10^{15} and 10^{19} eV referred to as the "knee" and "ankle" in the spectrum, respectively. Shown is the flux of the highest energy cosmic rays near and beyond the ankle measured by the AGASA experiment. Note that the flux is multiplied by E^3

active experiments [3] have detected cosmic rays in the vicinity of 100 EeV since their initial discovery by the Haverah Park air shower array [4]. The AGASA air shower array in Japan[5] has now accumulated an impressive 10 events with energy in excess of 10^{20} eV [6].

The accuracy of the energy resolution of these experiments is a critical issue. With a particle flux of order 1 event per km^2 per century, these events are studied by using the earth's atmosphere as a particle detector. The experimental signature of an extremely high-energy cosmic particle is a shower initiated by the particle. The primary particle creates an electromagnetic and hadronic cascade. The electromagnetic shower grows to a shower maximum, and is subsequently absorbed by the atmosphere. The shower can be observed by: i) sampling the electromagnetic and hadronic components when they reach the ground with an array of particle detectors such as scintillators, ii) detecting the fluorescent light emitted by atmospheric nitrogen excited by the passage of the shower particles, iii) detecting the Cerenkov light emitted by the large number of particles at shower maximum, and iv) detecting muons and neutrinos underground.

The bottom line on energy measurement is that, at this time, several experiments using the first two techniques agree on the energy of EeV-showers within a typical resolution of 25%. Additionally, there is a systematic error of order 10% associated with the modeling of the showers. All techniques are indeed subject to the ambiguity of particle simulations that involve physics beyond the LHC. If the final outcome turns out to be an erroneous inference of the energy of the

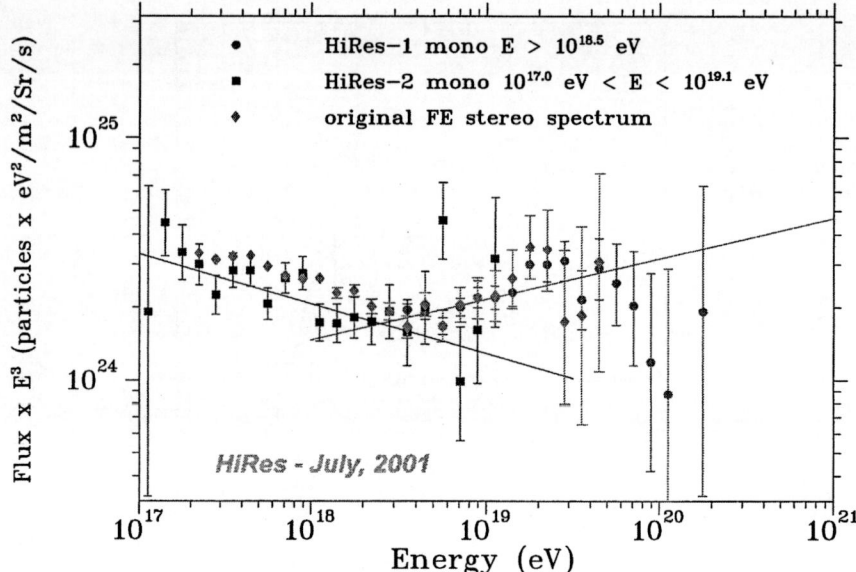

Fig. 3. As in Fig. 2, but as measured by the HiRes experiment

shower because of new physics associated with particle interactions at the Λ_{QCD} scale, we will be happy to contemplate this discovery instead.

The premier experiments, HiRes and AGASA, agree that cosmic rays with energy in excess of 10 EeV are not galactic in origin and that their spectrum extends beyond 100 EeV. They disagree on almost everything else. The AGASA experiment claims evidence that the highest energy cosmic rays come from point sources, and that they are mostly heavy nuclei. The HiRes data does not support this. Because of low statistics, interpreting the measured fluxes as a function of energy is like reading tea leaves; one cannot help however reading different messages in the spectra (see Fig. 2 and Fig. 3).

1.3   The Highest Energy Cosmic Rays: Fancy

Acceleration to > 100 EeV? It is sensible to assume that, in order to accelerate a proton to energy E in a magnetic field B, the size R of the accelerator must be larger than the gyroradius of the particle:

$$R > R_{\mathrm{gyro}} = \frac{E}{B}. \tag{2}$$

That is, the accelerating magnetic field must contain the particle orbit. This condition yields a maximum energy

$$E \sim \gamma B R \tag{3}$$

by dimensional analysis and nothing more. The γ-factor has been included to allow for the possibility that we may not be at rest in the frame of the cosmic

Table 1. Requirements to generate the highest energy cosmic rays in astrophysical sources.

Conditions with $E \sim 10$ EeV		
• Quasars	$\gamma \cong 1$	$B \cong 10^3$ G $M \cong 10^9 M_{sun}$
• Blazars	$\gamma \gtrsim 10$	$B \cong 10^3$ G $M \cong 10^9 M_{sun}$
• Neutron Stars	$\gamma \cong 1$	$B \cong 10^{12}$ G $M \cong M_{sun}$
Black Holes		
$\vdots$		
• GRB	$\gamma \gtrsim 10^2$	$B \cong 10^{12}$ G $M \cong M_{sun}$

accelerator. The result would be the observation of boosted particle energies. Theorists' imagination regarding the accelerators has been limited to dense regions where exceptional gravitational forces create relativistic particle flows: the dense cores of exploding stars, inflows on supermassive black holes at the centers of active galaxies, annihilating black holes or neutron stars. All speculations involve collapsed objects and we can therefore replace R by the Schwarzschild radius

$$R \sim GM/c^2 \tag{4}$$

to obtain

$$E \propto \gamma BM . \tag{5}$$

Given the microgauss magnetic field of our galaxy, no structures are large or massive enough to reach the energies of the highest energy cosmic rays. Dimensional analysis therefore limits their sources to extragalactic objects; a few common speculations are listed in Table 1.

Nearby active galactic nuclei, distant by ~ 100 Mpc and powered by a billion solar mass black holes, are candidates. With kilogauss fields, we reach 100 EeV. The jets (blazars) emitted by the central black hole could reach similar energies in accelerating substructures (blobs) boosted in our direction by Lorentz factors of 10 or possibly higher. The neutron star or black hole remnant of a collapsing supermassive star could support magnetic fields of 10^{12} Gauss, possibly larger. Highly relativistic shocks with $\gamma > 10^2$ emanating from the collapsed black hole could be the origin of gamma ray bursts and, possibly, the source of the highest energy cosmic rays.

The above speculations are reinforced by the fact that the sources listed are also the sources of the highest energy gamma rays observed. At this point, however, a reality check is in order. The above dimensional analysis applies to the Fermilab accelerator: 10 kilogauss fields over several kilometers corresponds to 1 TeV. The argument holds because, with optimized design and perfect alignment of magnets, the accelerator reaches efficiencies matching the dimensional limit. It is highly questionable that nature can achieve this feat. Theorists can imagine acceleration in shocks with an efficiency of perhaps 10%.

The astrophysics of accelerating particles to Joule energies is so daunting that many believe that cosmic rays are not the beams of cosmic accelerators but the decay products of remnants from the early Universe, such as topological defects associated with a Grand Unified Theory (GUT) phase transition.

Are Cosmic Rays Really Protons: the GZK Cutoff? All experimental signatures agree on the particle nature of the cosmic rays – they look like protons or, possibly, nuclei. We mentioned at the beginning of this article that the Universe is opaque to photons with energy in excess of tens of TeV because they annihilate into electron pairs in interactions with the cosmic microwave background. Protons also interact with background light, predominantly by photoproduction of the Δ-resonance, i.e. $p + \gamma_{CMB} \to \Delta \to \pi + p$ above a threshold energy E_p of about 50 EeV given by:

$$2E_p\epsilon > \left(m_\Delta^2 - m_p^2\right) . \tag{6}$$

The major source of proton energy loss is photoproduction of pions on a target of cosmic microwave photons of energy ϵ. The Universe is, therefore, also opaque to the highest energy cosmic rays, with an absorption length of

$$\lambda_{\gamma p} = (n_{\text{CMB}} \, \sigma_{p+\gamma_{\text{CMB}}})^{-1} \tag{7}$$

$$\cong 10 \text{Mpc}, \tag{8}$$

when their energy exceeds 50 EeV. This so-called GZK cutoff establishes a universal upper limit on the energy of the cosmic rays. The cutoff is robust, depending only on two known numbers: $n_{\text{CMB}} = 400 \, \text{cm}^{-3}$ and $\sigma_{p+\gamma_{\text{CMB}}} = 10^{-28} \, \text{cm}^2$ [7–10].

Cosmic rays do reach us with energies exceeding 100 EeV. This presents us with three options: i) the protons are accelerated in nearby sources, ii) they do reach us from distant sources which accelerate them to even higher energies than we observe, thus exacerbating the acceleration problem, or iii) the highest energy cosmic rays are not protons.

The first possibility raises the challenge of finding an appropriate accelerator by confining these already unimaginable sources to our local galactic cluster. It is not impossible that all cosmic rays are produced by the active galaxy M87, or by a nearby gamma ray burst which exploded a few hundred years ago.

Stecker [11] has speculated that the highest energy cosmic rays are Fe nuclei with a delayed GZK cutoff. The details are complicated but the relevant quantity in the problem is $\gamma = E/AM$, where A is the atomic number and M the nucleon mass. For a fixed observed energy, the smallest boost towards GZK threshold is associated with the largest atomic mass, i.e. Fe.

Could Cosmic Rays be Photons or Neutrinos? Above question naturally emerges in the context of models where the highest energy cosmic rays are the decay products of remnants or topological structures created in the early universe with typical energy scale of order 10^{24} eV. In these scenarios the highest energy cosmic rays are predominantly photons. A topological defect will suffer a chain decay into Grand Unified Theory (GUT) particles X and Y, that subsequently decay to familiar weak bosons, leptons and quark or gluon jets. Cosmic rays are,

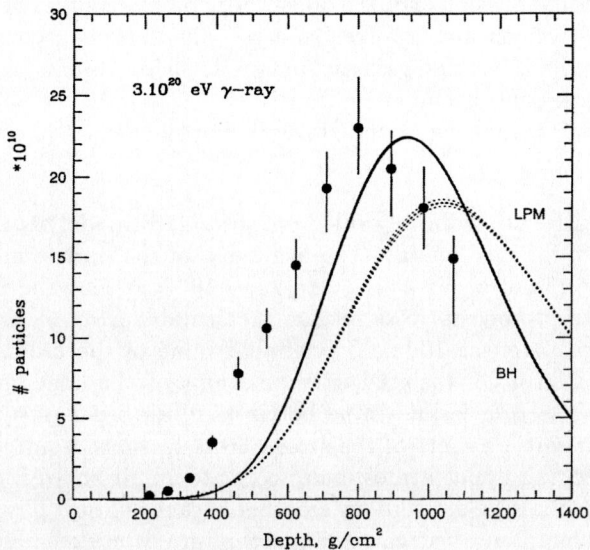

Fig. 4. The composite atmospheric shower profile of a 3×10^{20} eV gamma ray shower calculated with Landau-Pomeranchuk-Migdal (dashed) and Bethe-Heitler (solid) electromagnetic cross sections. The central line shows the average shower profile and the upper and lower lines show 1 σ deviations – not visible for the BH case, where lines overlap. The experimental shower profile is shown with the data points. It does not fit the profile of a photon shower

therefore, predominately the fragmentation products of these jets. We know from accelerator studies that, among the fragmentation products of jets, neutral pions (decaying into photons) dominate, in number, protons by close to two orders of magnitude. Therefore, if the decay of topological defects is the source of the highest energy cosmic rays, they must be photons. This is a problem because there is compelling evidence that the highest energy cosmic rays are not photons:

1. The highest energy event observed by Fly's Eye is not likely to be a photon [12]. A photon of 300 EeV will interact with the magnetic field of the earth far above the atmosphere and disintegrate into lower energy cascades – roughly ten at this particular energy. The detector subsequently collects light produced by the fluorescence of atmospheric nitrogen along the path of the high-energy showers traversing the atmosphere. The atmospheric shower profile of a 300 EeV photon after fragmentation in the earths magnetic field, is shown in Fig. 4. It disagrees with the data. The observed shower profile does fit that of a primary proton, or, possibly, that of a nucleus. The shower profile information is sufficient, however, to conclude that the event is unlikely to be of photon origin.

2. The same conclusion is reached for the Yakutsk event that is characterized by a huge number of secondary muons, inconsistent with a pure electromagnetic cascade initiated by a gamma ray.

3. The AGASA collaboration claims evidence for "point" sources above 10 EeV. The arrival directions are however smeared out in a way consistent with primaries deflected by the galactic magnetic field. Again, this indicates charged primaries and excludes photons.
4. Finally, a recent reanalysis of the Haverah Park disfavors photon origin of the primaries [4].

Neutrino primaries are definitely ruled out. Standard model neutrino physics is understood, even for EeV energy. The average x of the parton mediating the neutrino interaction is of order $x \sim \sqrt{M_W^2/s} \sim 10^{-6}$ so that the perturbative result for the neutrino-nucleus cross section is calculable from measured HERA structure functions. Even at 100 EeV a reliable value of the cross section can be obtained based on QCD-inspired extrapolations of the structure function. The neutrino cross section is known to better than an order of magnitude. It falls 5 orders of magnitude short of the strong cross sections required to make a neutrino interact in the upper atmosphere to create an air shower.

Could EeV neutrinos be strongly interacting because of new physics? In theories with TeV-scale gravity, one can imagine that graviton exchange dominates all interactions and thus erases the difference between quarks and neutrinos at the energies under consideration. The actual models performing this feat require a fast turn-on of the cross section with energy that violates S-wave unitarity [13-21].

We have exhausted the possibilities. Neutrons, muons and other candidate primaries one may think of are unstable. EeV neutrons barely live long enough to reach us from sources at the edge of our galaxy.

2   A Three Prong Assault on the Cosmic Ray Puzzle

We conclude that, where the highest energy cosmic rays are concerned, both the accelerator mechanism and the particle physics are enigmatic. The mystery has inspired a worldwide effort to tackle the problem with novel experimentation including air shower arrays covering an area of several times 10^3 square kilometers[22] and arrays of multiple air Cerenkov telescopes[23]. We here discuss kilometer-scale neutrino observatories. While these have additional missions such as the search for dark matter[24], their observations are likely to have an impact on cosmic ray physics.

Why we anticipate that secondary photons and neutrinos are associated with the highest energy cosmic rays is sketched in Fig. 5. The cartoon draws our attention to the fact that cosmic accelerators are also cosmic beam dumps that produce secondary photon and neutrino beams. Accelerating particles to TeV energy and above requires relativistic, massive bulk flows. These are likely to originate from the exceptional gravitational forces associated with black holes or neutron stars. Accelerated particles therefore pass through intense radiation fields or dense clouds of gas surrounding the black hole leading to the production of secondary pions. These subsequently decay into photons and neutrinos that

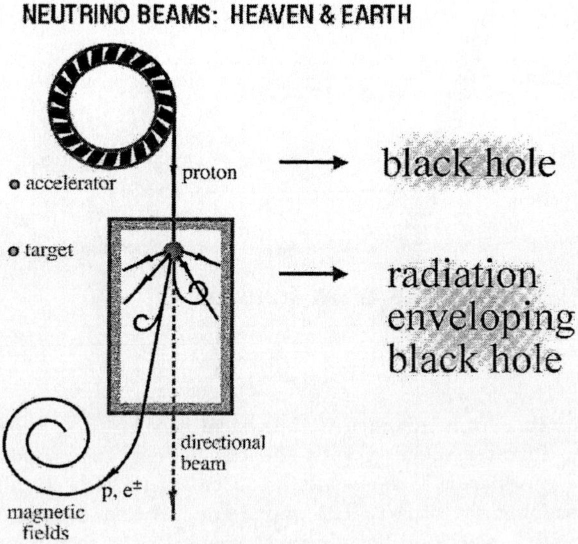

NEUTRINO BEAMS: HEAVEN & EARTH

Fig. 5. Diagram of cosmic accelerator and beam dump. See text for discussion

accompany the primary cosmic ray beam. Example of beam dumps include the external photon clouds or the UV radiation field that surrounds the central black hole of active galaxies, or the matter falling into the collapsed core of a dying supermassive star producing a gamma ray burst. The target material, whether a gas of particles or of photons, is likely to be sufficiently tenuous for the primary proton beam and the secondary photon beam to be only partially attenuated. However, shrouded sources from which only neutrinos can emerge, as in terrestrial beam dumps at CERN and Fermilab, are also a possibility.

How many neutrinos are produced in association with the cosmic ray beam? The answer to this question, among many others[25,26], provides the rational for building kilometer-scale neutrino detectors.

Let's first consider the question for the accelerator beam producing neutrino beams at an accelerator laboratory. Here the target absorbs all parent protons as well as the muons, electrons and gamma rays (from $\pi^0 \to \gamma + \gamma$) produced. A pure neutrino beam exits the dump. If nature constructed such a "hidden source" in the heavens, conventional astronomy has not revealed it. It cannot be the source of the cosmic rays, however, for which the dump must be partially transparent to protons.

In the other extreme, the accelerated proton interacts once, thus producing the observed high-energy gamma rays [37]. It subsequently escapes the dump. We refer to this as a transparent source without absorption. Particle physics directly relates the number of neutrinos to the number of observed cosmic rays and gamma rays[27]. Every observed cosmic ray interacts once, and only once, to produce a neutrino beam determined only by particle physics. The neutrino flux

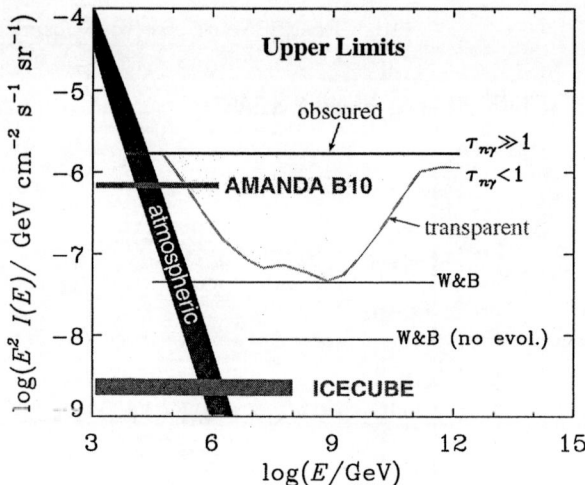

Fig. 6. The neutrino flux from compact astrophysical accelerators. Shown is the range of possible neutrino fluxes associated with the the highest energy cosmic rays. The lower line, labeled "transparent", represents a source where each cosmic ray interacts only once before escaping the object. The upper line, labeled "obscured", represents an ideal neutrino source where all cosmic rays escape in the form of neutrons. Also shown is the ability of AMANDA and IceCube to test these models

for such a transparent cosmic ray source is referred to as the Waxman-Bahcall flux [28–31] and is shown as the horizontal lines labeled "W&B" in Fig. 6. The calculation is valid for $E \simeq 100$ PeV. If the flux is evaluated at both lower and higher cosmic ray energies, however, larger values are found. This is shown as the non-flat line labeled "transparent" in Fig. 6. On the lower side, the neutrino flux is higher because it is normalized to a larger cosmic ray flux. On the higher side, there are more cosmic rays in the dump to produce neutrinos because the observed flux at Earth has been reduced by absorption on microwave photons, the GZK-effect. The increased values of the neutrino flux are also shown in Fig. 6. The gamma ray flux of π^0 origin associated with a transparent source is qualitatively at the level of observed flux of non-thermal TeV gamma rays from individual sources[27].

Nothing prevents us, however, from imagining heavenly beam dumps with target densities somewhere between those of hidden and transparent sources. When increasing the target photon density, the proton beam is absorbed in the dump and the number of neutrino-producing protons is enhanced relative to those escaping the source as cosmic rays. For the extreme source of this type, the observed cosmic rays are all decay products of neutrons with larger mean-free paths in the dump. The flux for such a source is shown as the upper horizontal line in Fig. 6.

The above limits are derived from the fact that theorized neutrino sources do not overproduce cosmic rays. Similarly, observed gamma ray fluxes constrain potential neutrino sources because for every parent charged pion ($\pi^{\pm} \to l^{\pm} + \nu$), a neutral pion and two gamma rays ($\pi^0 \rightarrow \gamma + \gamma$) are produced. The electromagnetic energy associated with the decay of neutral pions should not exceed observed astronomical fluxes. These calculations must take into account cascading of the electromagnetic flux in the background photon and magnetic fields. A simple argument relating high-energy photons and neutrinos produced by secondary pions can still be derived by relating their total energy and allowing for a steeper photon flux as a result of cascading. Identifying the photon fluxes with those of non-thermal TeV photons emitted by supernova remnants and blazers, we predict neutrino fluxes at the same level as the Waxman-Bahcall flux[32]. It is important to realize however that there is no evidence that these are the decay products of π^0's. The sources of the cosmic rays have not been revealed by photon or proton astronomy [33–36]; see however reference [37].

For neutrino detectors to succeed they must be sensitive to the range of fluxes covered in Fig. 6. The AMANDA detector has already entered the region of sensitivity and is eliminating specific models which predict the largest neutrino fluxes within the range of values allowed by general arguments. The IceCube detector, now under construction, is sensitive to the full range of beam dump models, whether generic as or modeled as active galaxies or gamma ray bursts. IceCube will reveal the sources of the cosmic rays or derive an upper limit that will qualitatively raise the bar for solving the cosmic ray puzzle. The situation could be nothing but desperate with the escape to top-down models being cut off by the accumulating evidence that the highest energy cosmic rays are not photons. In top-down models, decay products eventually materialize as quarks and gluons that fragment into jets of neutrinos and photons and very few protons.

3   High Energy Neutrino Telescopes

Although neutrino telescopes have multiple interdisciplinary science missions, the search for the sources of the highest-energy cosmic rays stands out because it clearly identifies the size of the detector required to do the science[38].

Whereas the science is compelling, the real challenge has been to develop a reliable, expandable and affordable detector technology. Suggestions to use a large volume of deep ocean water for high-energy neutrino astronomy were made as early as the 1960s. In the case of the muon neutrino, for instance, the neutrino (ν_μ) interacts with a hydrogen or oxygen nucleus in the water and produces a muon travelling in nearly the same direction as the neutrino. The blue Cerenkov light emitted along the muon's ~kilometer-long trajectory is detected by strings of photomultiplier tubes deployed deep below the surface. With the first observation of neutrinos in the Lake Baikal and the (under-ice) South Pole neutrino telescopes, there is optimism that the technological challenges to build neutrino telescopes can hopefully be met.

The first generation of neutrino telescopes, launched by the bold decision of the DUMAND collaboration to construct such an instrument, are designed to reach a large telescope area and detection volume for a neutrino threshold of order 10 GeV. The optical requirements of the detector medium are severe.

A large absorption length is required because it determines the spacings of the optical sensors and, to a significant extent, the cost of the detector. A long scattering length is needed to preserve the geometry of the Cerenkov pattern. Nature has been kind and offered ice and water as adequate natural Cerenkov media. Their optical properties are, in fact, complementary. Water and ice have similar attenuation length, with the role of scattering and absorption reversed. Optics seems, at present, to drive the evolution of ice and water detectors in predictable directions: towards very large telescope area in ice exploiting the long absorption length, and towards lower threshold and good muon track reconstruction in water exploiting the long scattering length.

DUMAND, the pioneering project located off the coast of Hawaii, demonstrated that muons could be detected by this technique[39], but the planned detector was never realized. A detector composed of 96 photomultiplier tubes located deep in Lake Baikal was the first to demonstrate the detection of neutrino-induced muons in natural water[40,41]. In the following years, *NT-200* will be operated as a neutrino telescope with an effective area between $10^3 \sim 5 \times 10^3 \, \mathrm{m}^2$, depending on energy. Presumably too small to detect neutrinos from extraterrestrial sources, *NT-200* will serve as the prototype for a larger telescope. For instance, with 2000 OMs, a threshold of 10~20 GeV and an effective area of $5 \times 10^4 \sim 10^5 \, \mathrm{m}^2$, an expanded Baikal telescope would fill the gap between present detectors and planned high-threshold detectors of cubic kilometer size. Its key advantage would be low threshold.

The Baikal experiment represents a proof of concept for deep ocean projects. These do however have the advantage of larger depth and optically superior water. Their challenge is to find reliable and affordable solutions to a variety of technological challenges for deploying a deep underwater detector. The European collaborations ANTARES[42–44] and NESTOR[45–47] plan to deploy large-area detectors in the Mediterranean Sea within the next year. The NEMO Collaboration is conducting a site study for a future kilometer-scale detector in the Mediterranean[48].

The AMANDA collaboration, situated at the U.S. Amundsen-Scott South Pole Station, has demonstrated the merits of natural ice as a Cerenkov detector medium[49]. In 1996, AMANDA was able to observe atmospheric neutrino candidates using only 80 eight-inch photomultiplier tubes[49].

With 302 optical modules instrumenting approximately 6000 tons of ice, AMANDA extracted several hundred atmospheric neutrino events from its first 130 days of data. AMANDA was thus the first first-generation neutrino telescope with an effective area in excess of 10,000 square meters for TeV muons[50]. In rate and all characteristics the events are consistent with atmospheric neutrino origin. Their energies are in the 0.1–1 TeV range. The shape of the zenith angle distribution is compared to a simulation of the atmospheric neutrino signal in Fig. 7. The variation of the measured rate with zenith angle is reproduced by the simulation to within the statistical uncertainty. Note that the tall geometry of the detector strongly influences the dependence on zenith angle in favor of more vertical muons.

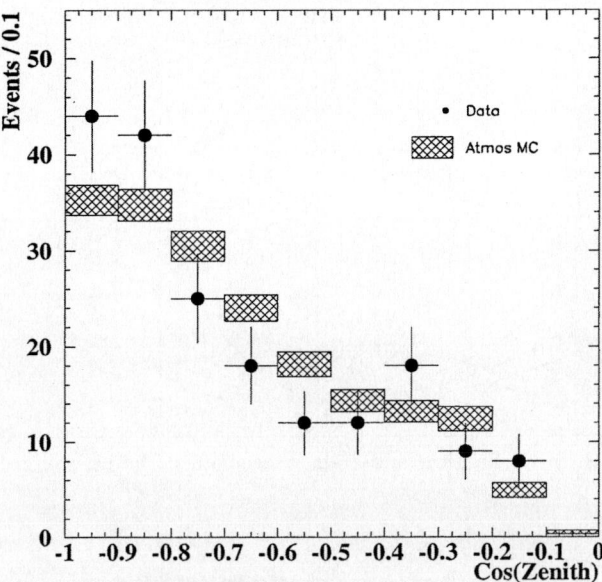

Fig. 7. Reconstructed zenith angle distribution. The points mark the data and the shaded boxes a simulation of atmospheric neutrino events, the widths of the boxes indicating the error bars

The arrival directions of the neutrinos are shown in Fig. 8. A statistical analysis indicates no evidence for point sources in this sample. An estimate of the energies of the up-going muons (based on simulations of the number of reporting optical modules) indicates that all events have energies consistent with an atmospheric neutrino origin. This enables AMANDA to reach a level of sensitivity to a diffuse flux of high energy extra-terrestrial neutrinos of order[50] $dN/dE_\nu = 10^{-6}E_\nu^{-2}\,\mathrm{cm}^{-2}\,\mathrm{s}^{-1}\,\mathrm{sr}^{-1}\,\mathrm{GeV}^{-1}$, assuming an E^{-2} spectrum. At this level they exclude a variety of theoretical models which assume the hadronic origin of TeV photons from active galaxies and blazars[11]. Searches for neutrinos from gamma-ray bursts, for magnetic monopoles, and for a cold dark matter signal from the center of the Earth are also in progress and, with only 138 days of data, yield limits comparable to or better than those from smaller underground neutrino detectors that have operated for a much longer period.

In January 2000, AMANDA-II was completed. It consists of 19 strings with a total of 677 OMs arranged in concentric circles, with the ten strings from AMANDA forming the central core of the new detector. First data with the expanded detector indicate an atmospheric neutrino rate increased by a factor of three, to 4–5 events per day. AMANDA-II has met the key challenge of neutrino astronomy: it has developed a reliable, expandable, and affordable technology for deploying a kilometer-scale neutrino detector named IceCube.

IceCube is an instrument optimised to detect and characterize sub-TeV to multi-PeV neutrinos of all flavors (see Fig. 9) from extraterrestrial sources. It will consist of 80 strings, each with 60 10-inch photomultipliers spaced 17 m apart.

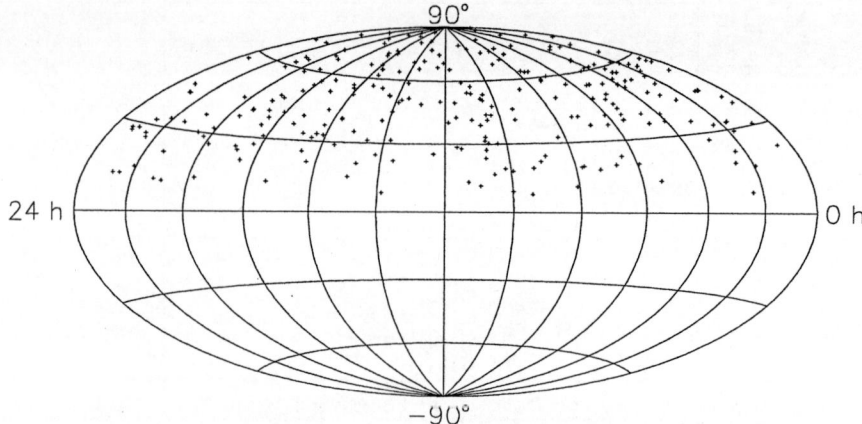

Fig. 8. Distribution in declination and right ascension of the up-going events on the sky

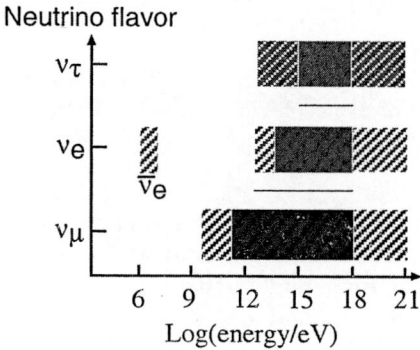

Fig. 9. Although IceCube detects neutrinos of any flavor above a threshold of ~ 0.1 TeV, it can identify their flavor and measure their energy in the ranges shown

The deepest module is 2.4 km below the surface. The strings are arranged at the apexes of equilateral triangles 125 m on a side. The effective detector volume is about a cubic kilometer, its precise value depending on the characteristics of the signal. IceCube will offer great advantages over AMANDA II beyond its larger size: it will have a much higher efficiency to reconstruct tracks, map showers from electron- and tau-neutrinos (events where both the production and decay of a τ produced by a ν_τ can be identified) and, most importantly, measure neutrino energy. Simulations indicate that the direction of muons can be determined with sub-degree accuracy and their energy measured to better than 30% in the logarithm of the energy. Even the direction of showers can be reconstructed to better than 10° in both θ, ϕ above 10 TeV. Simulations predict a linear response in energy of better than 20%. This has to be contrasted with the logarithmic energy resolution of first-generation detectors. Energy resolution is critical because, once one establishes that the energy exceeds 100 TeV, there is no atmospheric neutrino background in a kilometer-square detector.

At this point in time, several of the new instruments, such as the partially deployed Auger array and HiRes to Magic to Milagro and AMANDA II, are less than one year from delivering results. With rapidly growing observational capabilities, one can express the realistic hope that the cosmic ray puzzle will be solved soon. The solution will almost certainly reveal unexpected astrophysics, if not particle physics.

For a recent review of neutrino astronomy and its relationship to cosmic rays, see Ref. [51].

Acknowledgements

This work was supported in part by DOE grant No. DE-FG02-95ER40896 and in part by the Wisconsin Alumni Research Foundation.

References

1. D.J. Bird *et al.*: Phys. Rev. Lett. **71**, 3401 (1993)
2. N.N. Efimov *et al.*: *ICRR Symposium on Astrophysical Aspects of the Most Energetic Cosmic Rays*, ed. M. Nagano and F. Takahara (World Scientific, 1991)
3. http://www.hep.net/experiments/all_sites.html, provides information on experiments discussed in this review. For a few exceptions, we will give separate references to articles or websites.
4. M. Ave *et al.*: Phys. Rev. Lett. **85**, 2244 (2000)
5. http://www-akeno.icrr.u-tokyo.ac.jp/AGASA/
6. *Proceedings of the International Cosmic Ray Conference*, Hamburg, Germany, August 2001
7. K. Greisen: Ann. Rev. Nucl. Science **10**, 63 (1960)
8. F. Reines: Ann. Rev. Nucl. Science **10**, 1 (1960)
9. M.A. Markov, I.M. Zheleznykh: Nucl. Phys. **27**, 385 (1961)
10. M. A. Markov in *Proceedings of the 1960 Annual International Conference on High-energy Physics at Rochester*, ed. by E.C.G. Sudarshan, J.H. Tinlot, A.C. Melissinos (1960)
11. F.W. Stecker, M.H. Salamon: Astrophys. J. **512**, 521 (1992), astro-ph/9808110
12. R.A. Vazquez *et al.*: Astroparticle Physics **3**, 151 (1995)
13. J. Alvarez-Muniz, F. Halzen, T. Han, D. Hooper: Phys. Rev. Lett. **88**, 021301 (2002), hep-ph/0107057.
14. R. Emparan, M. Masip, R. Rattazzi: Phys. Rev. D **65**, 064023 (2002), hep-ph/0109287.
15. P. Jain, D.W. McKay, S. Panda, J.P. Ralston: Phys. Lett. B **484**, 267 (2000), hep-ph/0001031.
16. A. Jain, P. Jain, D.W. McKay, J.P. Ralston: hep-ph/0011310.
17. C. Tyler, A.V. Olinto, G. Sigl: Phys. Rev. D **63**, 055001 (2001), hep-ph/0002257.
18. S. Nussinov, R. Shrock; Phys. Rev. D **59**, 105002 (1999), hep-ph/9811323.
19. S. Nussinov, R. Shrock: Phys. Rev. D **64**, 047702 (2001), hep-ph/0103043.

20. G. Domokos, S. Kovesi-Domokos: Phys. Rev. Lett. **82**, 1366 (1999), hep-ph/9812260.
21. G. Domokos, S. Kovesi-Domokos, P.T. Mikulski: hep-ph/0006328.
22. A. Watson: these proceedings
23. H.J. Volk: these proceedings
24. C. Tao: these proceedings
25. F. Halzen: 'The case for a kilometer-scale neutrino detector', in *Nuclear and Particle Astrophysics and Cosmology*, Proceedings of Snowmass 94, ed. by R. Kolb, R. Peccei
26. F. Halzen: 'The Case for a Kilometer-Scale Neutrino Detector: 1996', in *Proc. of the Sixth International Symposium on Neutrino Telescopes*, ed. by M. Baldo-Ceolin, (Venice,1996)
27. F. Halzen, E. Zas: Astrophys. J. **488**, 669 (1997), astro-ph/9702193
28. J.N. Bahcall, E. Waxman: Phys. Rev. D **64** (2001), hep-ph/9902383
29. E. Waxman, J.N. Bahcall: Phys. Rev. D **59** (1999), hep-ph/9807282
30. K. Mannheim, R.J. Protheroe, J.P. Rachen: Phys. Rev. D **63**, 023003 (2001), astro-ph/9812398
31. J.P. Rachen, R.J. Protheroe, K. Mannheim: presented at *The 19th Texas Symposium on Relativistic Astrophysics: Texas in Paris*, Paris, France, 14–18 Dec 1998, astro-ph/9908031
32. J. Alvarez-Muniz, F. Halzen: UW-Madison report MADPH-00-1167 (2002)
33. T.K. Gaisser, R.J. Protheroe, T. Stanev: Ap.J. **492**, (1998) 219
34. L. O'C. Drury, F.A. Aharonian, H.J. Völk: A & A **287**, (1994) 959
35. J.A Esposito, S.D. Hunter, G. Kanbach, P. Sreekumar: Ap.J. **461**, (1996) 820
36. W. Bednarek, R. J. Protheroe: Phys. Rev. Lett. **79**, 2616 (1997)
37. The Cangoroo collaboration: Nature **416**, 797 (2002)
38. T. K. Gaisser, F. Halzen, T. Stanev: Phys. Rept. **258**, 173 (1995) [Erratum ibid. **271**, 355 (1995)], hep-ph/9410384
    J.G. Learned, K. Mannheim,: Ann. Rev. Nucl. Part. Science **50**, 679 (2000)
39. 'Cosmic Rays in the Deep Ocean', the *DUMAND Collaboration* (J.!Babson et al.). ICR-205-89-22, Dec 1989, 24pp, Published in Phys. Rev. D **42**, 3613 (1990)
40. I.A. Belolaptikov et al.: Astroparticle Physics **7**, 263 (1997)
41. V.A. Balkanov et al.: Astro. Part. Phys. **14**, 61 (2000)
42. E. Aslanides et al, astro-ph/9907432 (1999)
43. F. Feinstein [ANTARES Collaboration]: Nucl. Phys. Proc. Suppl. **70**, 445 (1999)
44. T. Montaruli [ANTARES Collaboration]: *Proceedings of TAUP 2001: Topics in Astroparticle and Underground Physics*, Assergi, Italy, 8–12 Sep 2001, hep-ex/0201009
45. L. Trascatti, in *Procs. of the 5th International Workshop on Topics in Astroparticle and Underground Physics (TAUP 97)*, Gran Sasso, Italy, 1997, ed. by A. Bottino, A. di Credico, P. Monacelli: Nucl. Phys. **B70** (Proc. Suppl.), 442 (1998)
46. P.K. Grieder [NESTOR Collaboration]: Nuovo Cim. **24C**, 771 (2001)
47. L. Trasatti [NESTOR Collaboration]: Nucl. Phys. Proc. Suppl. **70**, 442 (1999)
48. Talk given at the *International Workshop on Next Generation Nucleon Decay and Neutrino Detector (NNN 99)*, Stony Brook, 1999, Proceedings to be published by AIP.
49. The AMANDA collaboration: Astroparticle Physics **13**, 1 (2000)
50. E. Andres et al.: Nature **410**, 441 (2001)
51. F. Halzen, D. Hooper: Repts. Prog. Phys., in press, astro-ph/0204527.

Unravelling the X-Ray Background

Günther Hasinger

Max-Planck-Institut für extraterrestrische Physik,
D-85741 Garching, Germany

1 Introduction

Deep X-ray surveys indicate that the cosmic X-ray background (XRB) is largely due to accretion onto supermassive black holes, integrated over cosmic time. In the soft (0.5-2 keV) band more than 90% of the XRB flux has been resolved using 1.4 Msec observations with ROSAT [17] and recently 1-2 Msec Chandra observations [29,4] and 100 ksec observations with XMM-Newton [18]. In the harder (2-10 keV) band a similar fraction of the background has been resolved with the above Chandra and XMM-Newton surveys, reaching source densities of about 4000 deg^{-2}. Surveys in the very hard (5-10 keV) band have been pioneered using BeppoSAX, which resolved about 30% of the XRB [9]. XMM-Newton and Chandra have now also resolved the majority (60-70%) of the very hard X-ray background. Optical follow-up programs with 8-10m telescopes have been completed for the ROSAT deep surveys and find predominantly Active Galactic Nuclei (AGN) as counterparts of the faint X-ray source population [31,23] mainly X-ray and optically unobscured AGN (type-1 Seyferts and QSOs) and a smaller fraction of obscured AGN (type-2 Seyferts). The X-ray observations have so far been about consistent with population synthesis models based on unified AGN schemes [6,13], which explain the hard spectrum of the X-ray background by a mixture of absorbed and unabsorbed AGN, folded with the corresponding luminosity function and its cosmological evolution. According to these models, most AGN spectra are heavily absorbed and about 80% of the light produced by accretion will be absorbed by gas and dust [7]. However, these models are far from unique and contain a number of hidden assumptions, so that their predictive power remains limited until complete samples of spectroscopically classified hard X-ray sources are available. In particular they require a substantial contribution of high-luminosity obscured X-ray sources (type-2 QSOs), which so far have only scarcely been detected. The cosmic history of obscuration and its potential dependence on intrinsic source luminosity remain completely unknown. Gilli et al. e.g. assumed strong evolution of the obscuration fraction (ratio of type-2/type-1 AGN) from 4:1 in the local universe to much larger covering fractions (10:1) at high redshifts (see also [7]). The gas to dust ratio in high-redshift, high-luminosity AGN could be completely different from the usually assumed galactic value due to sputtering of the dust particles in the strong radiation field [15]. This might provide objects which are heavily absorbed at X-rays and unobscured at optical wavelengths.

After having understood the basic contributions to the X-ray background, the general interest is now focussing on understanding the physical nature of these sources, the cosmological evolution of their properties, and their role in models of galaxy evolution. We know that basically every galaxy with a spheroidal component in the local universe has a supermassive black hole in its centre [11]. The luminosity function of X-ray selected AGN shows strong cosmological density evolution at redshifts up to 2, which goes hand in hand with the cosmic star formation history [26]. While the comoving space density of optically and radio-selected QSO has been shown to decline significantly beyond a redshift of 2.5 [30,33], the statistical quality of X-ray selected AGN high-redshift samples still needs to be improved [26]. The new Chandra and XMM-Newton surveys are now providing strong additional constraints here. Optical identifications for the deepest Chandra and XMM-Newton fields are still in progress, however a mixture of obscured and unobscured AGN with an increasing fraction of obscuration at lower flux levels seems to be the dominant population in these samples too [3,29,35,36] (see below). Interestingly, first examples of the long-sought class of high-redshift, high-luminosity, heavily obscured active galactic nuclei (type-2 QSO) have been detected in deep Chandra fields [28,35] and in the XMM-Newton deep survey in the Lockman Hole field [24]. In this paper we give an update on the optical identification work in the Chandra Deep Field South, which thanks to the efficiency of the VLT has progressed furthest among the deepest X-ray surveys.

2   The Chandra Deep Field South (CDFS)

The Chandra X-ray Observatory has performed deep X-ray surveys in a number of fields with ever increasing exposure times [27,20,12] and has completed a 1 Msec exposure in the Chandra Deep Field South (CDFS [29]) and a 2 Msec exposure in the Hubble Deep Field North (HDF-N [4]. The Megasecond dataset of the CDFS is the result of the coaddition of 11 individual Chandra ACIS-I exposures with aimpoints only a few arcsec from each other. The nominal centre of the CDFS is α=3:32:28.0, δ=-27:48:30 (J2000). This field was selected in a patch of the southern sky characterized by a low galactic neutral hydrogen column density $N_H = 8 \times 10^{19}$ cm^{-2} and a lack of bright stars [29]. In Fig. 1, we show the colour composite Chandra image of the CDFS. This was constructed by combining images (smoothed with a Gaussian with $\sigma = 1''$ in three bands (0.3-1 keV, 1-3 keV, 3-7 keV), which contain approximately equal numbers of photons from detected sources. Blue sources are those undetected in the soft (0.5-2 keV) band, most likely due to intrinsic absorption from neutral hydrogen with column densities $N_H > 10^{22}$ cm^{-2}. Very soft sources appear red. A few extended low surface brightness sources are also readily visible in the image. The CDFS was also observed with XMM-Newton for a total of $\sim$ 500 ksec in July 2001 and January 2002 in guaranteed observation time (PI: J. Bergeron). Due to high background conditions some data were lost and a total of $\sim$ 370 ksec has finally been accumulated [19].

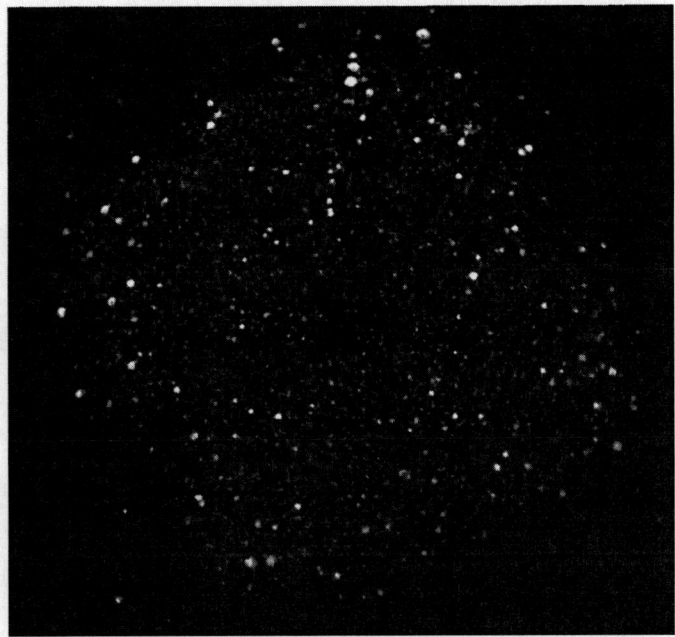

Fig. 1. Color composite image of the Chandra Deep Field South of 940 ks (pixel size=0.984″, smoothed with a (=1″ Gaussian). The image was obtained combining three energy bands: 0.3-1 keV, 1-3 keV, 3-7 keV (respectively red, green and blue)

3   Optical Identifications in the CDFS

Our primary optical imaging was obtained using the FORS1 camera on the ANTU (UT-1 at VLT) telescope. The R band mosaics from this data cover 13.6′ × 13.6′ to depths between 26 and 26.7 (Vega magnitudes). These data do not cover the full CDFS area and must be supplemented with other observations. The ESO Imaging Survey (EIS) has covered this field to moderate depths in several bands [2,37]. The EIS data have been obtained using the Wide Field Imager (WFI) on the ESO-MPG 2.2 meter telescope at La Silla. Figure 2 shows Chandra X-ray contours in a selected area of the CDFS superposed on a deep BRK multicolour image. The positioning is better than 0.5″ and we readily identify likely optical counterparts in 85% of the cases (78% for the shallower WFI data). Note the very red object in the lower right, which is only detected at K. Figure 3 shows the classical correlation between optical (R-band) magnitude and X-ray flux of the CDFS-objects. Generally the 0.5-2 keV flux is given, however, for Chandra sources not detected in the soft band, the 2-10 keV flux is given. Sources are marked according to their optical classification (see below). The Chandra data extend the previous ROSAT range by a factor of 40 in flux and to substantially fainter optical magnitudes. While the bulk of the type-1 AGN population still follows the general correlation along a constant f_X/f_{opt} line, the type-2 AGN cluster at higher X-ray to optical flux ratios. There is also a

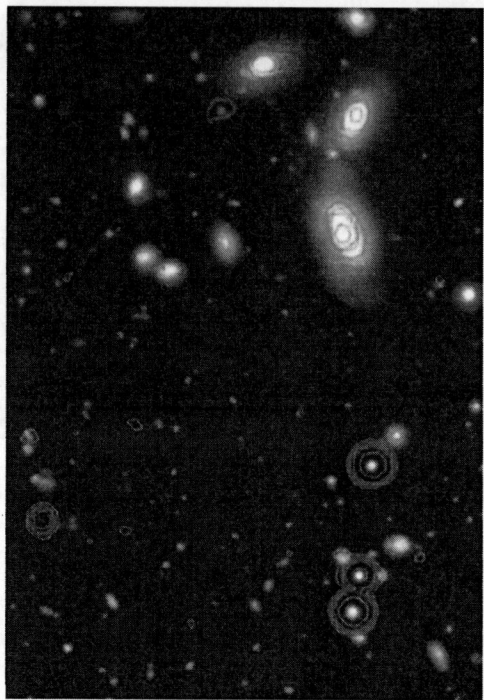

Fig. 2. Cutout of a part of the CDFS. A deep FORS R-image has been combined with the EIS WFI B-image and the GOODS ISAAC K-image. X-ray contours are overplotted on the optical/NIR data. The image shows diffuse X-ray emission for the bright galaxies. The very red counterpart in the lower right is only visible in the deep GOODS K-band image (from [19])

new population of normal galaxies showing up at significantly brighter optical magnitudes.

Optical spectroscopy has been carried out in 11 nights with the ESO Very Large Telescope (VLT) in the time frame April 2000 - December 2001, using deep optical imaging and low resolution multiobject spectroscopy with the FORS instruments with individual exposure times ranging from 1-5 hours. Some preliminary results including the VLT optical spectroscopy have already been presented [28,29]. The complete optical spectroscopy will be published in [36]. Redshifts could be obtained so far for 169 of the 346 sources in the CDFS, of which 123 are very reliable (high quality spectra with 2 or more spectral features), while the remaining optical spectra contain only a single emission line, or are of lower S/N. For objects fainter than R=24 reliable redshifts can be obtained if the spectra contain strong emission lines. For the remaining optically faint objects we have to resort to photometric redshift techniques. Nevertheless, for a subsection of the sample at off-axis angles smaller than 8 arcmin we obtain a spectroscopic completeness of about 60%.

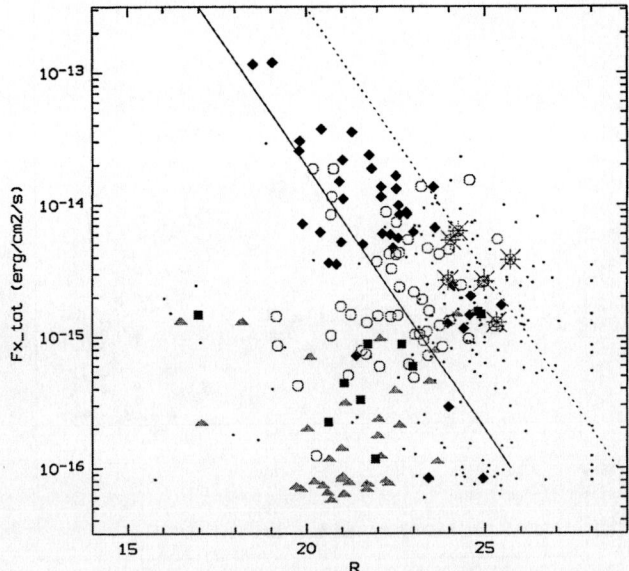

Fig. 3. X-ray flux versus R-band magnitude for the CDFS-sources. Objects are marked according to their X-ray/optical classification: filled diamonds correspond to type-1 AGN, open hexagons to type-2 AGN, triangles to galaxies and squares to extended X-ray sources. The large asterisks indicates type-2 QSOs (see text). Small dots refer to spectroscopically unidentified CDFS sources, the brighter ones of which have photometric redshifts. Optically empty error circles have been plotted at R=28. The solid line corresponds to an X-ray to optical flux ratio of 1, the dashed line is at an optical limit 3 magnitudes fainter

Type-1 AGN (Seyfert-1 and QSOs) can be often readily identified by the broad permitted emission lines in their optical spectra. Luminous Seyfert-2 galaxies show strong forbidden emission lines and high-excitation lines indicating photoionization by a hard continuum source. However, already in the spectroscopic identifications of the ROSAT Deep Surveys it became apparent, that an increasing fraction of faint X-ray selected AGN shows a significant, sometimes dominant contribution of stellar light from the host galaxy in their optical spectra, depending on the ratio of optical luminosity between nuclear and galaxy light [23]. If an AGN is much fainter than its host galaxy it is not possible to detect it optically. Many of the counterparts of the faint X-ray sources detected by Chandra and XMM-Newton show optical spectra dominated by their host galaxy and only a minority have clear indications of an AGN nature (see also [3]. In these cases, the X-ray emission could still be dominated by the active galactic nucleus, while a contribution from stellar and thermal processes (hot gas from supernova remnants, starbursts and thermal halos, or a population of X-ray binaries) can be important as well. Therefore X-ray diagnostics in addition to the optical spectroscopy can be crucial to classify the source of the

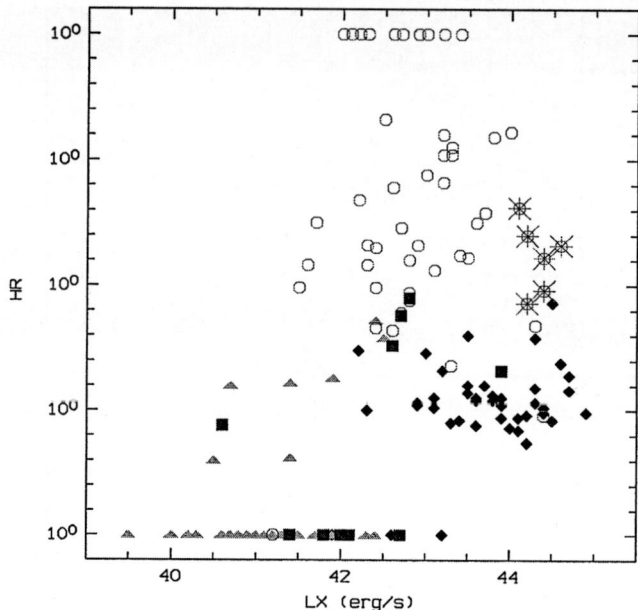

Fig. 4. Hardness ratio versus rest frame luminosity in the total 0.5-10 keV band. Symbols as in Fig. 3. A critical density universe with $H_0 = 50\ km\ s^{-1}\ Mpc^{-1}$ has been adopted. Luminosities are not corrected for possible intrinsic absorption

X-ray emission. AGN have typically X-ray luminosities above $10^{42}\ erg\ s^{-1}$ and power law spectra, often with significant intrinsic absorption [25]. Local, well-studied starburst galaxies have X-ray luminosities typically below $10^{42}\ erg\ s^{-1}$ and very soft X-ray spectra. Thermal haloes of galaxies and the intergalactic gas in groups can have higher X-ray luminosities, but have soft spectra as well. Following [29], we show in Fig. 6 the hardness ratio as a function of the luminosity in the 0.5-10 keV band for 165 sources for which we have optical spectra and rather secure classification [36]. The hardness ratio is defined as HR = (H-S)/(H+S) where H and S are the net count rates in the hard (2-7 keV) and the soft band (0.5-2 keV), respectively. The X-ray luminosities are not corrected for internal absorption and are computed in a critical density universe with H_0=50 km s^{-1} Mpc^{-1}. Different source types are clearly segregated in this plane. Type-1 AGNs (black diamonds) have luminosities typically above $10^{42}\ erg\ s^{-1}$, with hardness ratios in a narrow range around HR$\approx$ -0.5. This corresponds to an effective $\gamma = 1.8$, commonly found in type-1 AGN. Type-2 AGN are skewed towards significantly higher hardness ratios (HR>0), with (absorbed) luminosities in the range $10^{41-44}\ erg\ s^{-1}$. Direct spectral fits of the XMM-Newton and (some) Chandra spectra clearly indicate that these harder spectra are due to neutral gas absorption and not due to a flatter intrinsic slope [25]. Therefore the unabsorbed, intrinsic luminosities of type-2 AGN would fall in the same range as those of type-1's. In Fig. 4, we also indicate the type-2 QSOs (asterisks), the first one of which was discovered in the CDFS [28]. In the meantime, more examples have been found in the CDFS and elsewhere (e.g. [35]). It is interesting to note that no high-luminosity, very hard sources exist in this diagram. This is a selection effect of the pencil beam surveys: due to the small solid angle, the rare high luminosity sources are only sampled at high redshifts, where the absorption cut-off of type-2 AGN is redshifted to softer X- ray energies. Indeed, the type-2 QSOs in this sample are the objects at $L_X > 10^{44}\ erg\ s^{-1}$ and HR>-0.2. The type-1 QSO in this region of the diagram is a BAL QSO with significant intrinsic absorption. About 10% of the objects have optical spectra of normal galaxies (marked with triangles), luminosities below 10^{42} erg s^{-1} and very soft X-ray spectra (several with HR=-1), as expected in the case of starbursts or thermal halos. Those at $L_X < 10^{41.5}\ erg\ s^{-1}$ and HR larger than -0.7 are at particularly low redshifts. However, a separate subset has harder spectra (HR>-0.5), and luminosities above $10^{41}\ erg\ s^{-1}$. In these galaxies the X-ray emission is likely due to a mixture of low level AGN activity and a population of low mass X-ray binaries (see also [3]). Therefore the deep Chandra and XMM-Newton surveys detect for the first time the population of normal starburst galaxies out to intermediate redshifts [27,12,24]. These galaxies might become an important means to study the star formation history in the universe completely independently from optical/UV, sub-mm or radio observations.

4   The Redshift Distribution

Figure 5 shows the optical magnitudes of the spectroscopically identified CDFS sources as a function of redshift. There is a segregation between type-1 and type-2 AGN at high redshifts, most likely because the optical light from type-1 AGN contains a significant non-thermal contribution in addition to the host galaxy. Reliable redshifts can be obtained at the VLT typically for objects with R<25.5, however, some incompleteness already sets in around R=23. The CDFS has a spectroscopic completeness of about 60%, which is mainly caused by the fact that about 40% of the counterparts are optically too faint to obtain reliable spectra. Photometric redshift estimates of the remainder of the sources indicate a redshift distribution similar to the spectroscopic one. The completeness of 60% therefore allows us to compare the redshift distribution with predictions from X-ray background population synthesis models [13], based on the AGN X-ray luminosity function and its evolution as determined from the ROSAT surveys [26], which predict a maximum at redshifts around z=1.5. It is interesting to note, that contrary to these expectations, the bulk of the CDFS objects are found at redshifts below 1. The redshift distribution peaks at z~0.7, even if the normal star forming galaxies in the sample are removed. This clearly demonstrates that the population synthesis models will have to be modified to incorporate different luminosity functions and evolutionary scenarios for intermediate- redshift, low-luminosity AGN. In Fig. 5 there is an interesting accumulation of redshifts in the range z=0.6-0.8. We have discovered two large-scale structures at redshifts z=0.66 and z=0.73, respectively, which are made up of type-1 and type-2 AGN

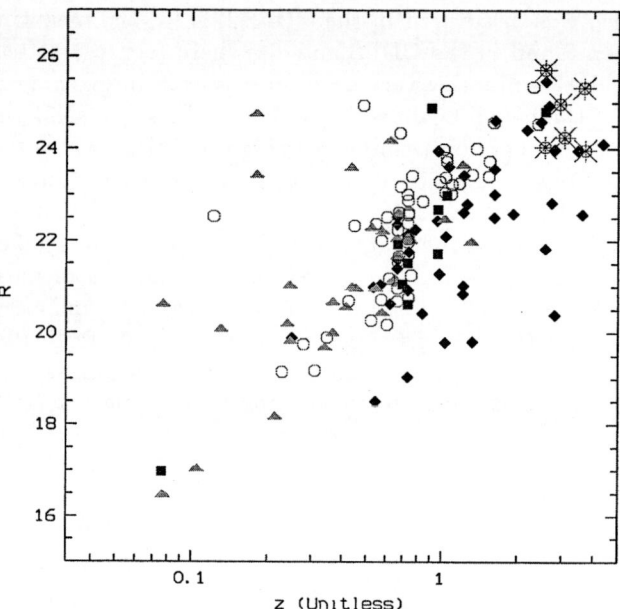

Fig. 5. Optical magnitudes as a function of redshift for the CDFS objects. Symbols are as in Fig. 4. An accumulation of objects in two redshift bins around z=0.7 is due to a large scale structure in the CDFS [14]

as well as normal galaxies in roughly the same proportion as observed in the field. The objects in these redshift spikes are distributed across a large fraction of the field, so that they are probably sheet-like structures [14]. Both of them are also seen in the K-band galaxy survey selected in the same field [5] and also correspond to several X-ray clusters in the field. It will be interesting to study the correlation of active galaxies to field galaxies in these sheets and to try to determine the role that galaxy mergers play in the triggering of the AGN activity. Finally, there may be a relation between the surprisingly low redshift of the bulk of the Chandra sources, the existence of the sheets at the same redshift and the strongly evolving population of dusty starburst galaxies inferred from the ISO mid-infrared surveys [10]. By no means does the CDFS redshift distribution confirm the prediction by Haiman & Loeb [16], that a large number (∼100) QSO at redshifts larger than 5 should be expected in any ultra deep Chandra survey. The highest redshift in the CDFS thus far is 3.7, while there are two objects at z=4.4 and z=5.2, respectively, in the HDF-N [4] and one QSO at z=4.5 in the Lockman Hole [32]. This suggests a cut-off of the X-ray selected QSO space density at high redshift.

5   Summary and Outlook

Deep X-ray surveys have shown that the cosmic X-ray background (XRB) is largely due to accretion onto supermassive black holes, integrated over cosmic time. The findings are consistent with the notion that most larger galaxies contain black holes which have been active in the past. However, the characteristic hard spectrum of the XRB can only be explained if most AGN spectra are heavily absorbed [6]. Thus about 80-90% of the light produced by accretion must be absorbed by gas and dust clouds, which may reside in nuclear starburst regions that feed the AGN [7]. The star formation history has been determined in the last years based on optical and UV measurements [34]. However, deep submillimeter surveys with SCUBA have revealed the existence of a large population of hitherto undetected dust-enshrouded galaxies [21], which may provide the dominant contribution to the star formation rate at higher redshifts. The spectral shape of the X-ray background may be related to the dust obscuration of the far-infrared sources, which are believed to be the high-z equivalents of the ultra luminous IR galaxies (ULIRGs). For some ULIRGs the presence of heavily obscured AGN has been inferred by BeppoSAX (e.g. [38]). Therefore a relation between the faint X-ray and far-infrared source populations is expected. Indeed, a large fraction of the faint hard Chandra and XMM-Newton sources have infrared counterparts in deep ISOCAM images [8,1] and the redshift distribution of faint X-ray sources and Mid-IR sources is similar (see above). The so-far deepest X-ray and SCUBA observations [20] did pick up only very few common objects. Even deeper X-ray images in conjunction with deep surveys at the peak wavelength of the far-infrared background e.g. with SIRTF, are therefore required. The Chandra Deep Field South has been selected as one of the deep fields in the SIRTF legacy programme "Great Observatories Origins Deep Survey" (GOODS), which will produce the deepest observations with the SIRTF IRAC instrument at $3.6-8\mu$ and with the MIPS instrument at 24μ and together with the Chandra data provide the necessary depth and statistics to finally establish the FIR/X-ray relation. In addition to the data described here, a large number of supporting observations across a wide range of the electromagnetic spectrum are being carried out or planned. We have proposed to complement the already existing Chandra Megasecond observations with two 500 ksec ACIS-I pointings to homogenize and increase the exposure in the GOODS area to 2 Msec. Three small regions in the CDFS had already been observed with HST, which provides excellent morphology of the AGN host galaxies and photometry for the faintest optical counterparts [22]. The whole CDFS has already been covered by an extensive set of pointings with the new Advanced Camera for Surveys (ACS) in BVIz to "near HDF" depth. Following up the deep EIS survey in the CDFS, ESO has started a large program to image the GOODS area with the VLT to obtain deep JHKs images in some 32 ISAAC fields. The first imaging data covering the central 50 arcmin^2 have recently been made public. Optical spectroscopy across the whole field will be obtained with very high efficiency using VIRMOS on the VLT. The multiwavelength coverage of the field will be complemented by deep radio data from the VLA at 6 cm (already obtained) and

ATCA at 20 cm. The CDFS/GOODS will therefore ultimately be one of the patches in the sky providing a combination of the widest and deepest coverage at all wavelengths and thus a legacy for the future.

References

1. Alexander D.M., Aussel H., Bauer F.E., et al., 2002, ApJ 568, L85
2. Arnouts S., Vandame B., Benoist C., et al., 2001, A&A. 379, 740
3. Barger, A. J., Cowie, L. L., Mushotzky, R. F., Richards, E. A., 2001, AJ 121, 662
4. Brandt W.N., Alexander D.M., Bauer, F.E., Hornschemeier A.E., 2002, astro-ph/0202311
5. Cimatti, A., Daddi E., Mignoli M., et al., 2002, A&A 381, L.68
6. Comastri, A.; Setti, G.; Zamorani, G.; Hasinger, G., 1995, A&A 296, 1
7. Fabian A.C., Barcons X., Almaini O., Iwasawa K., 1998, MNRAS 297, L11
8. Fadda D., Flores H., Hasinger G., 2002, A&A 383, 838
9. Fiore F., La Franca F., Giommi P., et al., 1999, MNRAS 306, 55
10. Franceschini A., Fadda D., Cesarsky C., et al., 2002, ApJ 568, 470
11. Gebhardt K., Bender R., Bower G., et al., 2000, ApJ 539, 13
12. Giacconi, R., Rosati P., Tozzi P., et al., 2001, ApJ 551, 624
13. Gilli, R., Salvati, M., Hasinger, G., 2001, A&A 366, 407
14. Gilli, R., Cimatti, A., Daddi, E., Hasinger G., Rosati, P., Szokoly G., et al., 2003, ApJ submitted
15. Granato G.L., Danese L., Francheschini A., 1997, ApJ 486, 147
16. Haiman, Z. & Loeb A., 1999, ApJ 519, 479
17. Hasinger, G., Burg, R., Giacconi, R., et al., 1998, A&A 329, 482
18. Hasinger, G., Altieri, B., Arnaud, M., et al., 2001, A&A 365, 45
19. Hasinger, G., Bergeron, J., Mainieri, V., Rosati, P., Szokoly, G. & CDFS Team, 2002, ESO Messenger 108, 11
20. Hornschemeier, A.E., Brandt, W.N., Garmire, G.P., et al., 2000, ApJ 541
21. Hughes D.H., Serjeant S., Dunlop J., et al., 1998, Nature 394, 241
22. Koekemoer A.M., Grogin N.A., Schreier E.J., 2002, ApJ 567, 657
23. Lehmann, I., Hasinger, G., Schmidt, M., et al., 2001, A&A 371, 833
24. Lehmann I., Hasinger G., Murray S.S, Schmidt M., 2002, astro-ph/0109172
25. Mainieri V., Bergeron J., Rosati P., et al., 2002, A&A 393, 425
26. Miyaji, T., Hasinger, G., Schmidt, M., 2000, A&A 353, 25
27. Mushotzky, R.F., Cowie L.L., Barger, A.J., Arnaud, K.A., 2000, Nature 404, 459
28. Norman C., Hasinger G., Giacconi R., et al. 2002, ApJ 571, 218
29. Rosati P., Tozzi P., Giacconi R., et al., 2002, ApJ 566, 667
30. Schmidt, M., Schneider, D.P. & Gunn J.E., 1995, AJ 114, 36
31. Schmidt, M., Hasinger, G., Gunn, J.E., et al., 1998, A&A 329, 495
32. Schneider, D.P., Schmidt, M., Hasinger, G., et al., 1998, AJ 115, 1230
33. Shaver P.A. et al., 1996, Nature 384, 439
34. Steidel C.C., Adelberger K.L., Giavalisco M., Dickinson M., Pettini M., 1999, ApJ 519, 1
35. Stern D., Moran E.C., Coil A.L., et al., 2002, ApJ 568, 71
36. Szokoly, G., Hasinger G., Rosati, P. et al., 2003 (in prep.)
37. Vandame et al. 2001, astro-ph/0102300
38. Vignati P., Molendi S., Matt G., et al., 1999, A&A 349, L57

What Do We *Really* Know About Neutron Stars?

Lodewijk Woltjer

Observatoire de Haute Provence, F-04780 Saint-Michel l'Observatoire, and
Osservatorio Astrofisico di Arcetri, Largo Enrico Fermi 5, I-50125 Firenze

Abstract. The masses and cooling histories of neutron stars appear to be compatible with conventional models and do not provide evidence for stars composed of strange matter. Pulsar ages and magnetic fields are discussed. When radio quiet X-ray emitting neutron stars are included, the supernova rate and the formation rate of neutron stars appear to agree rather well. The velocities of neutron stars are discussed and much of the evidence for large "kicks" at birth seems to be ambiguous.

1 Introduction

The neutron was discovered in 1932 by J. Chadwick. Two years later W. Baade and F. Zwicky [3] proposed that a supernova is associated with the formation of a neutron star and that in this process cosmic-rays are accelerated. Today this scenario is still accepted, but the site of cosmic-ray acceleration has moved outwards to the shock waves generated in the interstellar gas by the supernova explosion. Neutron stars were further considered by Landau [20] and detailed models were presented in 1939 by Oppenheimer and Volkoff [24]. Assuming the neutron gas to be composed of free, cold neutrons and taking into account relativistic effects, they obtained the mass-radius relation and showed that, as in the case of white dwarfs, there is a maximum mass for which they found a value of 0.7 solar masses.

Subsequently it was found that the equation of state of the neutron gas is far from that of an ideal gas because of the strong interactions. This has led to an extensive literature in which many body interactions and the creation of particles play a role. For a typical "modern" equation of state it is found that over a range of masses the radius remains about 10 km and that the maximum mass is likely to be about 2 solar masses or perhaps slightly higher (e.g. [27]). However, softer equations of state have been proposed that lead to a lower maximum mass. Configurations rotating at maximum angular velocity would have a mass limit about 15% larger than non-rotating ones. However, to-date no such fast rotating neutron stars with periods below 1 ms have been found, though extensive searches have been made. Some authors have considered also the possibility that a stable state of "strange quark" matter could exist (e.g. [1]), which would change the equation of state.

The masses of some close double neutron stars may be determined from relativistic orbital effects with precisions at the 1% level. They are found to be scattered in a narrow range around 1.35 solar masses with a dispersion of 0.04

$M_\odot$ [32]. In wider binaries with one neutron star the uncertainties are much larger, but the data are generally consistent with these values. However, in Vela X-1 a very accurate set of radial velocity measurements has led to a neutron star mass of 1.8 ± 0.15 $M_\odot$ which is still to be multiplied with $(\sin i)^{-1}$, with i the unknown angle of inclination of the orbit [5]. These results suggest that the clustering of neutron stars around 1.35 $M_\odot$ is unrelated to the mass limit, but rather a consequence of stellar evolution. In fact, recent calculations by Timmes et al. [30] suggest that rather massive stars end their evolution by the collapse of an iron core with a Chandrasekhar like mass and that in some cases masses near 1.7–1.8 $M_\odot$ may result. With a small amount of silicon burning material falling back onto the neutron star after the explosion the observed mass distribution could be reproduced.

2   Cooling

Neutron stars are born hot, but lose much energy in copious neutrinos processes during their formation. Following this the interior cools more slowly mainly by the modified URCA process $n+N \rightarrow p+N+e+v_e$ and its inverse (e.g. [34]). In case pion or kaon condensates form more rapid cooling processes $n \rightarrow p+e+v_e$ become possible and the star cools very fast.

The interior may be heated by friction. When a magnetic neutron star slows down, the surface will rotate more slowly than the interior and heat will be liberated, which for medium age objects may be significant.

The temperature of the surface may be kept relatively high by conduction from the interior, which may be non uniform if magnetic fields are present. The surface then may also be heated by energetic particles accelerated in the magnetosphere creating "hot spots". Accretion may also heat the surface especially in binaries if matter is transferred to the neutron star. Though much less effective, accretion of interstellar matter onto slow moving cool neutron stars may also cause measurable radiation. Cooling of the surface occurs by radiation of photons.

In first approximation the temperature may be measured on the assumption that the surface radiates like a black body. However, recent studies have shown that atmospheres a few cm thick of hydrogen or helium may shift the spectral distribution to shorter wavelengths and lead to an overestimate of the effective temperature (e.g. [37]). Hot spots and a possible pulsar component in the radiation further complicate the picture, while interstellar absorption also modifies the spectrum. ROSAT has provided soft X-ray data on a dozen of neutron stars. The conclusion is that the predictions of cooling by the modified URCA process, with possibly some frictional heating in the older (10^7 y) objects added, can fully account for the observations [6]. More rapid cooling processes are incompatible with these results.

The conclusion of this brief review of the basic neutron star properties is that M(R), M_{limit} and T(t) may be fitted by a standard scenario. There are no problems, but many solutions!

3   Rotation Powered and Other Pulsars

Following the discovery of pulsars in 1968, neutron stars have become more fully integrated into general astrophysics. It became possible to determine rotation periods, magnetic fields ages and velocities, while it was found that the rotational energy of pulsars was coupled to various energetic processes. The age of a pulsar may be determined from the period and its time derivative as follows:

$$\tau = \frac{P}{(n-1)\dot{P}} \left[1 - \left(\frac{P_0}{P} \right)^{n-1} \right]$$

where n the breaking index $P\ddot{P}/\dot{P}^2$ is assumed to be a constant, which need not always be a valid assumption. If the pulsar spins down by magnetic dipole radiation $n = 3$ and if the period P is much longer than the period at birth P_0, we have for the characteristic age $\tau_c = \frac{1}{2}P/\dot{P}$. In four young pulsars it has been possible to measure n with as results $n = 2.5; \ 2.2; \ 2.8$ and 1.4 [22]. In very young pulsars higher multipoles may still play a rather large role, while the rapid decline of the rotational speeds leads to stresses and ruptures in the neutron star crust causing "glitches" in the period and further affecting the determination of n. In older pulsars the assumption $n = 3$ may well be appropriate, but for the moment this has not been proven. Most pulsars associated with supernova remnants seem to have ages τ_c compatible with the ages of these remnants. Since the latter also have uncertainties, this does not prove that in most cases $\tau = \tau_c$ to better than a factor of two.

A particular illustration of the problems is presented by the supernova remnant G 11.2-0.3. With a radius of only 2.5 pc it must be a young object, no more than at most a few thousand years old. In fact, it has been tentatively identified with a supernova observed in China in A.D. 383. However, the 65 ms X-ray pulsar has $\tau_c = 24000$ y [33]. It is, of course, very well possible that P_0 was around 62 ms and that this explains the discrepancy between τ and τ_c.

For spin down by magnetic dipole radiation we have

$$B^2 \propto I \, P\dot{P}$$

with B the magnetic field and I the moment of inertia of the neutron star ($\approx 10^{45}$ g cm^2). With masses and radii in a rather narrow range the latter should not be too different from one pulsar to another. In view of this relation it seems interesting to place the observed pulsars in a diagram with as coordinates P and $(P\dot{P})^{1/2}$ (Figure 1). In such a diagram we may also draw lines of constant $\tau_c = P^2/(2\,P\dot{P})$ and of constant rotational energy loss

$$\dot{E}_{\rm rot} = 4\pi^2 \, I \, \dot{P}/P^3 \, .$$

While the diagram may be looked at as simply a plot of observed points, the translation of the coordinates into B and $\dot{E}_{\rm rot}$ requires the assumption of a dipolar magnetic field and a moment of inertia derived from models. We note in

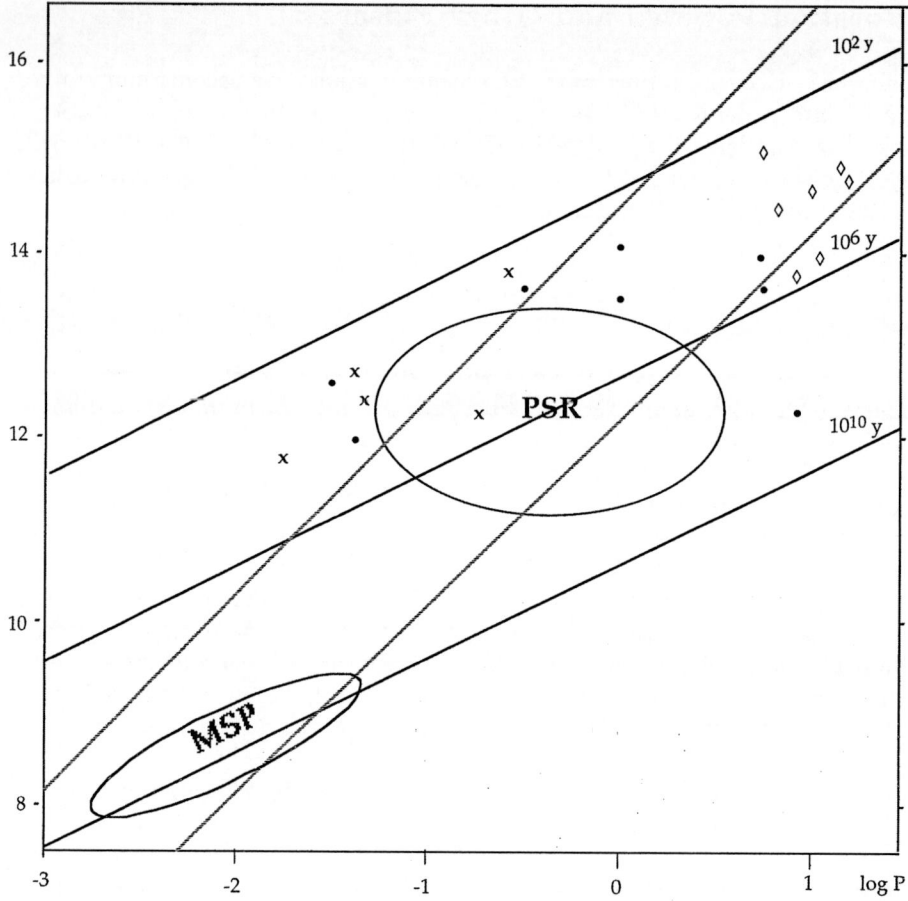

Fig. 1. The different types of pulsars. The horizontal axis gives the log of the period in seconds, the vertical axis $19.5 + 1/2 \log{(P\dot{P})}$, which in case of a dipole field at the surface of a neutron star with $I = 10^{45}$ gr cm^2 is equal to $\log B$. The full drawn lines represent equal values of τ_c the characteristic ages, the more inclined lines correspond to $\dot{E}_{\rm rot}$ equal to 10^{32} respectively 10^{36} erg s^{-1}. The main areas where radio pulsars (PSR) and millisecond pulsars (MSP) are found are indicated. Dots represent radio pulsars outside the PSR area, crosses X-ray pulsars without observed radio emission and lozenges "anomalous" X-ray pulsars and soft gamma-ray repeaters in which the main energy source for the emitted radiation cannot be the rotational energy. Note that the vertical axis of the diagram is not $\log P$ as in the usual presentation, in which lines of equal B add another set of skew lines

passing that even if the axis of the dipole were aligned to the rotation axis, the rotational energy loss would not be very different.

In the diagram there are two concentrations of points. The typical radio pulsars have fields of $10^{10} - 10^{13}$ G and ages of the order of a million years – a consequence of selection due to the low energy loss rate of slowly rotating pulsars and the difficulty of very slow pulsars to generate the voltages necessary for the pair production responsible for the radio emission. There is no clear evidence for magnetic field decay in pulsars. Since the interior is likely to be superconducting, this is perhaps not too surprising. Thus, the pulsars will evolve along a horizontal line at first rapidly and then slowly, becoming finally unobservable. So they accumulate in the PSR area in the diagram.

A second concentration of points is composed of the millisecond pulsars, generally pulsars in a double star system. When the other component evolves, it expands and may transfer mass to the neutron star. This leads to a transfer of orbital angular momentum to the neutron star and thereby to a rapid spin up to short periods – the record period being 1.56 ms. The magnetic fields of the MSP are rather weak – typically $10^8 - 10^9$ G. This may be the consequence of the accretion of mass which deforms or hides the stronger field of the neutron star.

In the upper left and middle of the diagram are the young pulsars, some with radio and X-ray emission, others with only X-ray emission. Several of the young pulsars are associated with supernova remnants.

Of particular interest are the X-ray pulsars in the upper right. With periods around 10 sec and inferred magnetic fields $\geq 10^{14}$ G, they appear to be generally young. Actually, in several cases they are associated with supernova remnants. However, their spindown energy loss is relatively low, well below their actual luminosity. So they cannot be powered by rotation. In fact, their spectra are much softer than short period X-ray pulsars, but their luminosities are well above those of simple cooling neutron stars. The absence of radio emission in strong field pulsars is sometimes ascribed to gamma-ray splitting, which would prevent the shower of positrons and electrons responsible for the radio waves from forming [4].

With their strong magnetic fields, these objects have been baptized "magnetars". It was proposed long ago that such objects could extract energy from the magnetic field by reconnection and dissipation in the same way as happens in solar flares [38]. In recent discussions detailed models have been constructed [13] in which the magnetic field in the interior is evacuated by ambipolar diffusion on time scales of $10^4 - 10^5$ y. If this happens enough energy could be liberated to account for the observed luminosity. However, in this case the magnetic fields inferred from $P\dot{P}$ are likely to be an overestimate [16].

Very similar objects are the Soft Gamma-ray Repeaters. In quiescent condition they appear to be almost indistinguishable from the preceding class. However, from time to time they produce a gamma-ray burst. Four such objects are known in our Galaxy and one in the Large Magellanic Clouds. Because their association with supernova remnants is not always evident, they have sometimes been taken to be somewhat older than other "magnetars", or to have higher velocities. Whether this is true remains to be seen. The gamma-ray bursts are not very different from the "cosmological" bursts, except for their somewhat softer spectra and lower energies. The same "fireball" scenario might apply.

So we now have several classes of pulsars:

Radio pulsars with sometimes optical, X- and gamma-ray emission.

X-ray pulsars without radio emission.

"Anomalous" X-ray pulsars with long periods and very strong fields.

Soft gamma-ray repeaters.

It is interesting to see a certain continuity between radio pulsars and "anomalous" X-ray pulsars. Recently a radio pulsar has been found with a period of 8.4 sec and a pulse width of only 1°. So maybe the radio pulses narrow with increasing fields and periods to finally disappear altogether.

There is an interesting relation between the magnetic field strength and the radius of various types of stars. Ordinary stars have radii of 10^6 km and fields up to 10^5 G. White dwarfs have radii of 10^4 km and recently one with a field of 5×10^8 G has been found [9]. Neutron stars have radii of 10 km and fields up to 10^{15} G. Thus, the maximum fields scale as R^{-2}. The most simple minded explanation would be that the fields are compressed while conserving flux [38]. Alternatively, the fields in the neutron stars might be generated in a convective dynamo during the supernova process (e.g. [31]).

4   Neutron Stars and Supernova Remnants

One of the problems encountered early on was that few supernova remnants appeared to have clearly associated radio pulsars. While this could perhaps be explained by the magnetic fields being buried under accreting matter and the pulsar only appearing after the supernova remnant had disappeared (1-2×10^4 y) or in some cases by high velocities of pulsars, the situation has been largely resolved by the X-ray observations. These have led to the discovery of the various classes of X-ray pulsars mentioned before. In addition, in some remnants non pulsating soft X-ray sources have been found which in some cases are variable on (year?) long time scales. Taking all classes together, neutron stars have now been found associated with 60% of the supernova remnants closer than 5.5 kpc from the sun. A recent addition has been the object at the center for Cas A which had been long looked for [28], although there is still some uncertainty if this is a neutron star or a black hole [26]. Already in 1970 Cavaliere and Pacini [11] suggested that the lack of a pulsar might be due to a rapid spin down as a result of a very strong magnetic field.

In a new radio pulsar sample Manchester [23] recently looked for surrounding supernova remnants with high sensitivity. Around eight pulsars with $\tau_c < 21$ ky, five probable remnants were found, while five pulsars with $\tau_c > 33$ ky had none. Thus, there seems to be a rather good correlation both ways between pulsars and supernova remnants. The absence of a perfect correlation is not surprising: the observability of gaseous remnants depends very much on the density of the surrounding interstellar medium, while some supernovae may leave black holes with uncertain characteristics. Also interstellar absorption affects the observability of soft sources.

Another case that illustrates the difficulty in finding a perfect correlation is the remnant G 29.6+0.1 which was discovered around an X-ray pulsar. However, when the remnant was found, the pulsar had almost disappeared and was too faint to detect pulsations [14].

Some time ago I tried [39] to determine the formation rates of massive stars, supernovae, pulsars and X-ray sources within 4 kpc from the sun with the following (updated) results:

Stars $8-40$ $M_\odot$	3×10^{-5} kpc^{-2} y^{-1}
"Historical" SN	$(4 \pm 2) \times 10^{-5}$
SN rate extrapolated from other galaxies	2×10^{-5}
Radio pulsars	$(0.5 \pm 0.2) \times 10^{-5}$
All X-ray neutron stars	$(2.1 \pm 0.7) \times 10^{-5}$

In view of the uncertainties, the agreement between the rate of formation of stars believed to generate neutron stars, of supernovae and of X- and radio-emitting neutron stars appears to be satisfactory.

The "historical" supernovae are particularly uncertain. The supernova remnant 3 C 58 has generally been identified with a "new star" seen in China in AD 1181. However, recent observations of the expansion of the radio source [7] suggest that 3 C 58 is 5000 years old. This is confirmed by the discovery of an X-ray pulsar at its center with p = 65 ms and τ_c = 5400 y [36]. On the other hand, Green and Stephenson [15] have strengthened the identification of 3 C 58 with the event of 1181. Aschenbach [2] has discovered another nearby remnant, which must be less than 1000 years old, since gamma-rays from the decay of ^{44}Ti ($t_{1/2} \approx 90$ y) have been observed [18]. Thus, the number of nearby "historical" supernovae has not changed, but the story shows the pitfalls and the incompleteness in identifying the "new stars".

5   Velocities of Neutron Stars

The distribution of neutron star velocities gives information on the "kicks" they receive in the supernova process and on possible acceleration thereafter. Frequently cited in the literature is the study of Lorimer et al. [21] who concluded that the mean three dimensional velocity of radio pulsars is about 500 km s^{-1}. On the basis of supposed associations between pulsars and supernova remnant velocities of 2000 km s^{-1} and more have been claimed. Other authors have found much lower mean velocities [17], [8], [39] of more typically 200 km s^{-1}. Cordes and Chernoff [12] concluded that two populations of pulsars exist: 86% belonging to a population with mean speed of 175 km s^{-1} and the remainder to a population with <v> = 700 km s^{-1}. Since the total sample contained only 49 pulsars, these results remain quite provisional.

In a careful analysis van den Heuvel and van Paradijs [35] found that kicks at birth ("a few" 100 km s^{-1}) are needed to prevent too many double neutron star binaries from forming and that in five X-ray binaries with a Be star as one of the components kicks ranging from $30-200$ km s^{-1} were required to explain the orbital characteristics. The most telling evidence for large kicks comes from the binary Cir X-1 which was found to have a radial velocity of 425 km s^{-1}. From simple celestial mechanics it was found that to account for this a kick in the range $500-1100$ km s^{-1} would be required [29]. However, Pal'shin and Tsygan [25] have described what they call a "photon rocket". In an X-ray binary with mass transfer after the supernova explosion the rotation axis of the neutron star will align rapidly with the orbital axis. If also the magnetic axis is aligned accretion will occur at the poles along the open field lines. Generally in stellar magnetic fields there is no very precise symmetry between the two poles, and so one could expect that also the accretion will be asymmetric and, as a consequence, the luminosity. They then show that an asymmetry factor of 0.2 in the accretion of 0.1 M$_\odot$ onto a neutron star in a binary of total mass of 2.5 M$_\odot$ leads to an acceleration of the system to 480 km s^{-1}, thus obviating the need for a large kick at birth. Janka and Müller [19] have tried to model kicks caused by neutrino processes and stochastic convection in the supernova process and had already difficulty arriving at kicks of 500 km s^{-1}.

It would be important to make careful measurements of parallaxes and proper motions with radio VLBI techniques of the pulsars with the largest claimed velocities. One interesting case – the "duck nebula" – has already been found where the association between the supernova remnant and the pulsar seemed clear and indicated a velocity of nearly 2000 km s^{-1}. However, direct measurements detected no motion at all with an upper limit of 500 km s^{-1}.

One final piece of evidence shows that numerous pulsars must be born with low velocities. In the globular cluster 47 Tucanae some 20 millisecond pulsars have been found and from the selection effects $200-1000$ inferred [10]. These pulsars are now found in peculiar binaries which originated by exchange collisions between the neutron stars and binaries existing in the cluster which contains some 10^6 stars in all. But the escape velocity from the cluster is less than 60 km s^{-1} and had the birth kicks been large they would all have left the cluster long ago.

References

1. Alcock, C., Olinto, A.V. 1988, Ann. Rev. Nucl. Part. Sci. 38, 161
2. Aschenbach, B. 1998, Nature 396, 141
3. Baade, W., Zwicky, F. 1934, Proc. Nat. Ac. Sci. Wash. 20, 254
4. Baring, M.G., Harding, A.K. 1998, ApJ 507, L55
5. Barziv, O., Kaper, L., van Kerkwijk, M.H., Telting, J.H., van Paradijs, J. 2001, A&A 377, 925
6. Becker, W. 1995, Ann. New York Ac. Sci. 759, 250 (17th Texas Symp.)
7. Bietenholz, M.F., Kassim, N.E., Weiler, K.W. 2001, ApJ 560, 772
8. Blaauw, A., Ramachandran, R. 1998, J. Astroph. Astron. 19, 19
9. Burleigh, M.R., Jordan, S., Schweizer, W. 1999, ApJ 510, L37
10. Camilo, F., Lorimer, D.R., Freire, P., Lyne, A.G., Manchester, R.N. 2000, ApJ 535, 975

11. Cavaliere, A., Pacini, F. 1970, ApJ 159, L21
12. Cordes, J.M., Chernoff, D.F. 1998, ApJ 505, 315
13. Duncan, R.C., Thompson, C. 1992, ApJ 392, L9
14. Gotthelf, E.V., Vasisht, G., Gaensler, B., Torii, K. 2000, ASP Conf. Ser. 202, 707
15. Green, D.A., Stephenson, F.R. 1999, Astronomy & Geophysics 40, 27
16. Harding, A.K., Contopoulos, I., Kazanas, D. 1999, ApJ 525, L125
17. Hartman, J.W. 1997, A&A 322, 127
18. Iyudin, A.F. et al. 1998, Nature 396, 142
19. Janka, H.T., Müller, E. 1995, Ann. New York Ac. Sci. 759, 269 (17th Texas Symp.)
20. Landau, L.D. 1938, Nature 141, 333
21. Lorimer, D.R., Bailes, M., Harrison, P.A. 1997, MNRAS 289, 592
22. Lyne, A.G., Pritchard, R.S., Graham-Smith, F., Camilo, F. 1996, Nature 381, 497
23. Manchester, R.N. 2001 in Young Supernova Remnants (ed. S.S. Holt, U. Hwang, AIP Conf. Proc. 565), 305
24. Oppenheimer, J.R., Volkoff, G. 1939, Phys. Rev. 55, 374
25. Pal'shin, V.D., Tsygan, A.I. 2000, ASP Conf. Ser. 202, 653
26. Pavlov, G.G., Zavlin, V.E., Aschenbach, B., Trümper, J., Sanwal, D. 2000, ApJ 531, L53
27. Srinivasan, G. 2002, A&A Rev. 11, 66
28. Tananbaum, B. et al. 1999, IAU Circular 7246
29. Tauris, T.M. et al. 1999, MNRAS 310, 1165
30. Timmes, F.X., Woosley, S.E., Weaver, T.A. 1996, ApJ 475, 834
31. Thompson, C., Duncan, R.C. 1996, ApJ 473, 322
32. Thorsett, S.E., Chakrabarty, C. 1999, ApJ 512, 288
33. Torii, K., Tsunemi, H., Dotani, T., Mitsuda, K., Kawai, N., Kinugasa, K., Saito, Y., Shibata, S. 1999, ApJ 523, L69
34. Tsuruta, S. 1998, Phys. Rep. 292, 1
35. van den Heuvel, E.P., van Paradijs, J. 1997, ApJ 483, 399
36. Weiskopf, M.C. et al. 2002, ApJ in press
37. Werner, K., Deetjen, J. 2000, ASP Conf. Ser. 202, 623
38. Woltjer, L. 1968, ApJ 152, 179
39. Woltjer, L. 1998, Mem. Soc. Astron. It. 69, 1079

Supermassive Black Holes in Galaxy Centers

Ralf Bender[1] and John Kormendy[2]

[1] University Observatory of the LMU, Scheinerstr. 1, D-81679 Munich, Germany, and
MPI for Extraterrestrial Physics, Giessenbachstr., D-85748 Garching, Germany
[2] Dept. of Astronomy, RLM 15.308, University of Texas, Austin, TX 78712, U.S.A.

In the past two years, many reviews (e. g., Kormendy and Gebhardt 2001) and conferences (e. g., Ho 2003) have been devoted to the subject of supermassive black holes (SMBHs) in galaxy centers. Therefore, this paper provides only a brief summary of key results and references.

After successful but slow ground-based work on a few strong cases (Kormendy and Richstone 1995), the search for SMBHs is now undergoing a major breakthrough. With the Hubble Space Telescope (HST), efficient SMBH searches are possible in relatively unbiased samples of galaxies. As a result, the number of SMBH cases has grown by a factor of ~ 5 in the past three years. From this work, we have learned that virtually all galaxies with bulge components (that is, elliptical galaxies and S0 – Sbc galaxies) contain supermassive black holes, or more conservatively, central dark objects (Magorrian et nuk. 1998; Richstone et nuk. 1998; Kormendy and Gebhardt 2001).

Are these black holes? In the Milky Way and in NGC 4258, astrophysically plausible alternatives to SMBHs can almost certainly be ruled out (Miyoshi et al. 1995; Maoz 1998; Schödel et al. 2002). On time scales much shorter than the Hubble time, dark clusters made of brown dwarf stars become luminous as the stars collide, merge, and become massive enough for nuclear energy generation. Clusters of stellar remnants (white dwarf stars, neutron stars, or stellar-mass black holes) evaporate as a result of two-body relaxation. HST STIS observations now make M 31 the third galaxy in which SMBH alternatives are very unlikely (Bender et al. 2003, Kormendy et al. 2003).

In ~ 40 other galaxies, evidence for central dark objects has been reported (for references, see Kormendy and Gebhardt 2001). In these, the constraints on the sizes and densities of the central dark objects are not strong enough to exclude dark clusters on the grounds of short relaxation times. However, in some cases, the clusters would be too tiny to have contained the remnants' progenitor stars without an embarrassingly large number of stellar collisions. Also, the combination of dynamical results and indirect evidence from active galactic nuclei (AGNs) makes a strong case for SMBHs. Particularly compelling is the observation of superluminal expansion of radio jets (Bridle and Perley 1984), which makes it clear that AGN engines are relativistically compact. Also, x-ray observations of Fe $K\alpha$ emission with line widths of $\sim 1/3$ of the speed of light point to SMBH engines (see Reynolds and Nowak 2002 for a review). SMBHs appear to be standard equipment in galaxy bulges (Richstone et nuk. 1998).

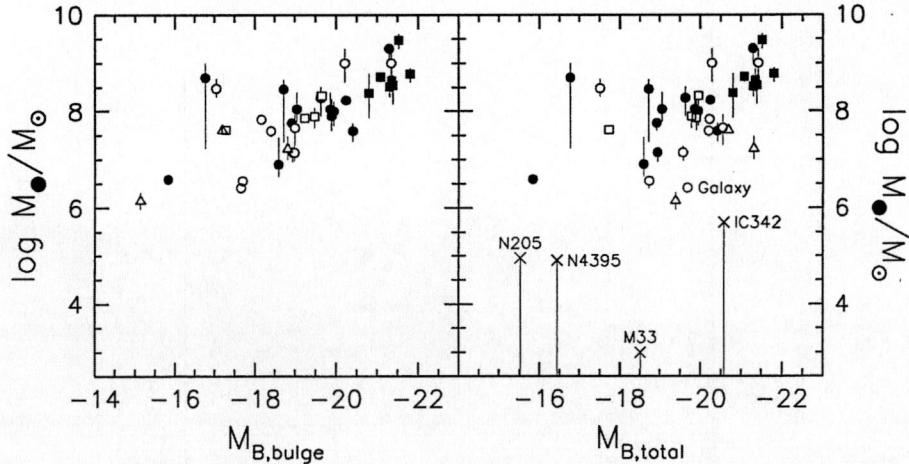

Fig. 1. Black hole mass *versus* (left) bulge and (right) total galaxy luminosity. Note: luminosities are measured in "absolute magnitudes", $M_{B,bulge} = -2.5 \log L_{B,bulge} +$ constant; these are not to be confused with masses M_{bulge} in the equations below. Filled symbols are for elliptical galaxies, open symbols are for bulges, and crosses are for bulgeless disks. Circles indicate masses derived from stellar dynamics, squares are based on ionized gas dynamics, and triangles are based on maser gas dynamics. This figure is from Kormendy and Gebhardt (2001)

If we assume that the dynamically detected dark objects are SMBHs, then we find that their masses correlate well with the bulge masses of their host galaxies (Kormendy 1993b; Kormendy and Richstone 1995; Magorrian et nuk. 1998). The most reliable observations (Kormendy and Gebhardt 2001, see Fig. 1) imply:

$$M_\bullet \sim 0.0013 \cdot M_{bulge} \ .$$

It is important to note that SMBH mass does not correlate with galaxy disks. Figure 1 shows how the correlation of $M_\bullet$ with bulge luminosity (left) is destroyed for disk galaxies (right: open symbols and crosses) when the disk luminosity is included. Especially compelling is the case of the bulgeless spiral M 33 (Gebhardt et nuk. 2001; Merritt et al. 2001). On the other hand, some pure disks have Seyfert nuclei and so presumably do contain SMBHs (Filippenko and Ho 2002). However, their masses appear to be *much* smaller than the canonical 0.13 % of the bulge mass implied by the left panel of Fig. 1. It will be crucial to improve the statistics on small SMBHs in disk galaxies.

There also is a tight correlation (Fig. 2) between SMBH mass and the velocity dispersion σ of the bulge at radii much larger than the dynamical sphere of influence of the SMBH (Gebhardt et nuk. 2000a, Ferrarese and Merrit 2000). The scatter is almost as small as the measurement errors (typically a factor of

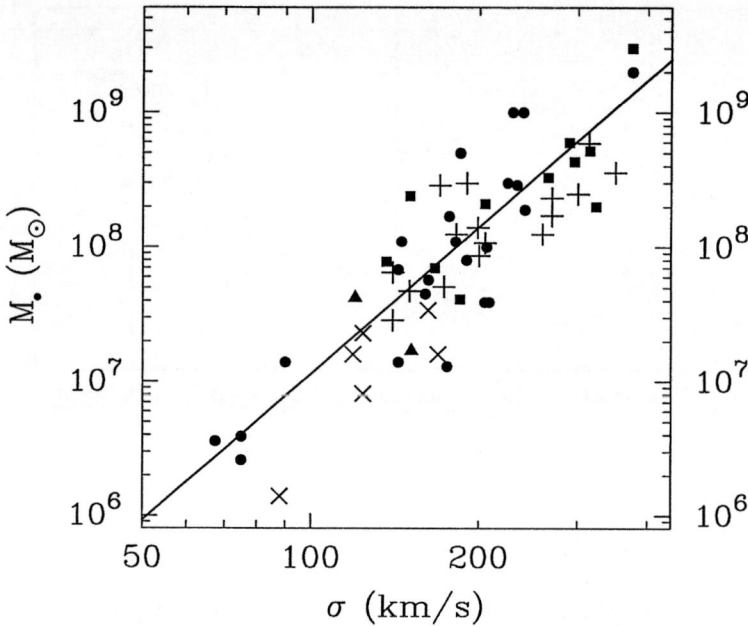

Fig. 2. The relation between black hole mass and bulge velocity dispersion (Kormendy and Gebhardt 2001). Different symbols indicate different methods of estimating the black hole mass and galaxy velocity dispersion. Filled symbols represent dynamical analyses of stars (circles), gas (squares), or masers (triangles). Crosses are based on reverberation mapping, and plus signs are based on ionization models

~ 2 in $M_\bullet$). Figure 2 gives:

$$\frac{M_\bullet}{M_\odot} \sim 0.1 \left(\frac{\sigma}{\mathrm{km\ s^{-1}}}\right)^4$$

(Tremaine et al. 2002). Together, Figs. 1 and 2 imply that, at a given bulge mass, more massive black holes live in more compact bulges (Kormendy et nuk. 2003). This connection between $M_\bullet$ and the bulge formation process is part of the growing evidence that SMBHs and bulges formed together.

Pseudobulges (Kormendy 1993a), which likely formed secularly from disk material, follow the same $M_\bullet - M_{B,\mathrm{bulge}}$ and $M_\bullet - \sigma$ relations as classical bulges and ellipticals (Kormendy and Gebhardt 2001). Again, the formation of bulges and the growth of SMBHs appears to have proceeded in lockstep.

Supermassive black holes have become critical to our understanding of why galaxies have cores, i. e., central regions in which the stellar density gradient is shallow compared to the rest of the galaxy. Galaxy mergers should destroy cores: smaller galaxies have higher densities and, in the absence of supermassive black holes, these high densities are preserved in a merger (e. g., Faber et nuk. 1997; Holley-Bockelmann and Richstone 1999, 2000). If the merger progenitors contain gas, then this gas is expected to fall to the center and further build up the stellar density (Mihos and Hernquist 1994). How, then, can the highest-luminosity galaxies – which surely are merger remnants – have the fluffiest centers? Many papers have explored the attractive idea that SMBH binaries formed in galaxy mergers decay by flinging stars away, thereby excavating a low-density region near the galaxy center (e. g., Begelman, Blandford, and Rees 1980; Ebisuzaki, Makino, and Okamura 1991; Makino and Ebisuzaki 1996; Quinlan 1996; Quinlan and Hernquist 1997; Faber et nuk. 1997; Milosavljević & Merritt 2001, Lauer et nuk. 2002). This has become a very hot topic. It will be important to check whether its predictions – e. g., high tangential velocity dispersions near galaxy centers – are convincingly seen.

SMBH masses derived via different techniques (stellar dynamics, ionized gas dynamics, maser emission from molecular gas, and reverberation mapping) agree very well. They imply indistinguishable $M_\bullet - M_{B,bulge}$ and $M_\bullet - \sigma$ correlations (e. g., Laor 1998, Gebhardt et nuk. 2000b). This is convincing evidence for the reliability of black hole mass determinations.

The $M_\bullet - M_{bulge}$ relation implies that the volume mass density of SMBHs in the local universe is consistent with the expected mass density of quasar remnants (Richstone et nuk. 1998; Yu and Tremaine 2002; Ferrarese 2002). Almost four decades after the seminal papers that launched the AGN paradigm (Salpeter 1964; Zel'dovich 1964; Lynden-Bell 1969, 1978), namely, the energy arguments that quasars shine by gas accretion onto SMBHs, the waste mass left behind after the quasar era appears to have been found.

Acknowledgements

It is a pleasure to thank the Nuker team for many years of productive collaboration. The Nuker team members are D. Richstone (PI), R. Bender, G. Bower, A. Dressler, S. M. Faber, A. Filippenko, K. Gebhardt, R. Green, C. Grillmair, L. C. Ho, J. Kormendy, T. R. Lauer, J. Magorrian, J. Pinkney, C. Siopis, and S. Tremaine.

References

1. Begelman, M. C., Blandford, R. D., and Rees, M. J,: Nature, **287**, 307 (1980)
2. Bender et al.: in preparation (2003)
3. Bridle, A. H., Perley, R. A.: ARA&A, **22**, 319 (1984)
4. Ebisuzaki, T., Makino, J., Okamura, S. K.: Nature, **354**, 212 (1991)
5. Faber et nuk.: AJ, **114**, 1771 (1997)
6. Ferrarese, L.: astro-ph/0203047 (2002)
7. Ferrarese, L., Merritt, D.: ApJ, **539**, L9 (2000)
8. Filippenko, A. V., Ho, L.: preprint (2002)
9. Gebhardt et nuk.: ApJ, **539**, L13 (2000a)
10. Gebhardt et nuk.: ApJ, **543**, L5 (2000b)
11. Gebhardt et nuk..: AJ, **122**, 2469 (2001)

12. Ho, L. C.: *Carnegie Observatories Centennial Symposium on Coevolution of Black Holes and Galaxies*, Cambridge: Cambridge University Press (2003)
13. Holley-Bockelmann, K., Richstone, D.: ApJ, **517**, 92 (1999)
14. Holley-Bockelmann, K., Richstone, D.: ApJ, **531**, 232 (2000)
15. Kormendy, J.: in *IAU Symposium 153, Galactic Bulges*, eds. H. Dejonghe, H. Habing, Dordrecht: Kluwer (1993a)
16. Kormendy, J.: in *The Nearest Active Galaxies*, eds. J. Beckman, L. Colina, H. Netzer, Madrid: Consejo Superior de Investigaciones Científicas, 197 (1993b)
17. Kormendy et nuk.: ApJ, submitted (2003)
18. Kormendy et al.: in preparation (2003)
19. Kormendy, J., Gebhardt, K.,: in *20th Texas Symposium on Relativistic Astrophysics*, eds. H. Martel, J. C. Wheeler, New York: AIP, 363 (2001)
20. Kormendy, J., Richstone, D.: ARA&A, **33**, 581 (1995)
21. Laor, A.: ApJ, **505**, L83 (1998)
22. Lauer, T. R. et nuk..: AJ, **124**, 1975 (2002)
23. Lynden-Bell, D.: Nature, **223**, 690 (1969)
24. Lynden-Bell, D.: Physica Scripta, **17**, 185 (1978)
25. Magorrian, J. et nuk.: AJ **115**, 2285 (1998)
26. Makino, J., Ebisuzaki, T.: ApJ, **465**, 527 (1996)
27. Maoz, E.: ApJ, **494**, L181 (1998)
28. Merritt, D., Ferrarese, L., Joseph, C. L.: Science, **293**, 1116 (2001)
29. Mihos, J. C., Hernquist, L.: ApJ, **437**, L47 (1994)
30. Milosavljević, M., Merrit, D.: ApJ **563**, 34 (2001)
31. Miyoshi, M. et al.: Nature, **373**, 127 (1995)
32. Quinlan, G. D.: NewA, **1**, 35 (1996)
33. Quinlan, G. D., Hernquist, L.: NewA, **2**, 533 (1997)
34. Reynolds, C. S., Nowak, M. A.: Physics Reports, in press, astro-ph/0212065 (2002)
35. Richstone, D. et nuk.: Nature, **395A**, 14 (1998)
36. Salpeter, E. E.: ApJ, **140**, 796 (1964)
37. Schödel et al.: Nature, **419**, 694 (2002)
38. Tremaine, S. et nuk.: ApJ, **574**, 740 (2002)
39. Yu, Q., Tremaine, S.: MNRAS, **335**, 965 (2002)
40. Zel'dovich, Ya. B.: Soviet Physics – Doklady, **9**, 195 (1964)

The Center of the Milky Way

Andreas Eckart

Universität zu Köln, Zülpicher Str.77, 50937 Köln, Germany

Abstract. The stellar proper motion data obtained with the MPE SHARP camera at the ESO NTT since 1992, now allow us to determine orbital accelerations for some of the most central individual stars of our Galaxy. This enables us for the first time to investigate the central mass distribution at the smallest currently accessible separations from Sgr A*. The analysis of stellar orbits clearly supports the presence of a supermassive nuclear black hole at the center of the Milky Way.

1 A Massive Black Hole at the Galactic Center

Near-infrared diffraction limited imaging using the MPE speckle camera SHARP at the 3.5 m New Technology Telescope (NTT) of the European Southern Observatory (ESO) since 1992 as well as high angular resolution spectroscopy allow us to determine the amount and concentration of mass at the center of the Milky Way (Eckart & Genzel 1996, Genzel et al. 2000, Eckart et al. 2002). Proper motions as well as the detection of acceleration of stars in the vicinity of the compact radio source Sgr A* (Ghez et al. 2000, Eckart et al. 2001a) indicate that this source is associated with a central black hole of about 3×10^6 M$_\odot$. Proper motions of up to 1400 km/s in the central arcsecond around Sgr A* (Eckart & Genzel 1996, 1997) as well as stellar orbital curvatures of the stars S1, S2, and S8 obtained with SHARP are in excellent agreement with measurements by Ghez et al. (1998, 2000) using NIRC on the Keck telescope.

Speckle spectroscopy with SHARP at the NTT (Genzel et al. 1997) as well as slit spectroscopy at the VLT with ISAAC (Eckart, Ott, Genzel 1999) and at the Keck telescope (Figer et al. 2000) suggest that several of the high velocity stars are of early type. Equipartition arguments that include limits on the proper motions of the radio source Sgr A* (<16km/s, Backer 1996, Reid et al. 1999, Genzel et al. 2000) and the estimated mass and known proper motion of the inner fast moving stars (Eckart, Genzel 1997, Genzel et al. 1997) result in a lower limit of the order of 10^5 M$_\odot$ that has to be associated with Sgr A*.

In general the motions of the stars in the central cluster do not deviate strongly from isotropy and are consistent with a spherical isothermal stellar cluster (Genzel et al. 2000). Estimates of the dark mass are usually obtained via the virial theorem, via the projected mass estimator as given by Bahcall & Tremaine (1981), or via a Jeans analysis (Genzel et al. 2000). These mass estimates are all based on the analysis of small samples of stars in annuli centered on the position of SgrA*. The observed projected stellar accelerations allow us only to derive lower limits to the enclosed mass from single stars. By inferring an estimate of the true physical separation of the stars from the center these lower limits can be statistically corrected for projection effects and compared to previous estimates of the amount and compactness of the enclosed dark mass. Using stars S1 and S2 we then find a value of $M_{acc} = (5 \pm 3) \times 10^6 M_\odot$. This estimate is obtained from *individual* stars as close to the massive black hole at the center of the Milky Way as currently possible. Within the uncertainties it is fully consistent with all previous results obtained at larger radii.

The VLBI maser nucleus of NGC 4258 (Greenhill et al. 1995, Myoshi et al. 1995) and the dark mass at the center of the Milky Way are currently the best and most compelling cases for the existence of super-massive nuclear black holes (Maoz 1998). The presence of a mass concentration accreting at radii as close as ∼20 Schwarzschild radii of a $3 \times 10^6 M_\odot$ massive black hole is also supported by the recent CHANDRA results (Baganoff et al. 2001). The current conclusion from the observational facts and a comparison to external nuclei is that this mass is most likely a single massive black hole.

2   Alternative Models

Boson stars (Kaup 1968) are supposed to be supported by the Heisenberg uncertainty principle. Ruffini & Bonazzola (1969) showed that - e.g. for a boson mass of 1GeV - a stable object of total mass of $10^{-19} M_\odot$ and 1 fm diameter could be formed. If a hypothetical weak repulsive force between bosons is introduced (Colpi, Shapiro & Wasserman 1986) then objects with total masses as large as they are found in galactic nuclei can be formed. For a large range of hypothetical boson masses they can have sizes of only several times there Schwarzschild radii. This makes it difficult to clearly distinguish observationally between boson stars and black holes as candidates for super-massive objects at the nuclei of galaxies (see also Torres, Capozziello & Lambiase 2000, Mielke & Schunck 2000).

Fermion ball scenarii as an attempt to explain large compact nuclear masses have been introduced by Viollier, Leimgruber & Trautmann (1992). They are stabilized by the degeneracy pressure of the corresponding fermion candidates, e.g. neutrinos. However, to explain all nuclear dark masses ranging between 3×10^6 (G.C.) and 3×10^9 $M_\odot$(M87) the putative, as yet unidentified neutrinos must have a mass in the range of 17keV. In the case of the Galactic Center the radius of the neutrino ball must then be of the order of 15 mpc (Munyaneza, Tsiklauri, & Viollier 1999, De Paolis et al. 2001). The low upper limits on the proper motion of SgrA* (e.g. Reid et al. 1999) as well as its compactness at radio wavelengths are not explained by such a model. Current estimates of the enclosed mass as a function of radius (Fig. 1) have the tendency to lie above the expected enclosed mass curve for a ∼15 mpc radius extended neutrino ball.

A resolved mass - and therefore gravitational potential that decreases towards smaller separations from the center position - allows to account for a decreasing radiative efficiency towards the middle. This may help to explain the low luminosity of SgrA*. That scenario, however, does not explain what hap-

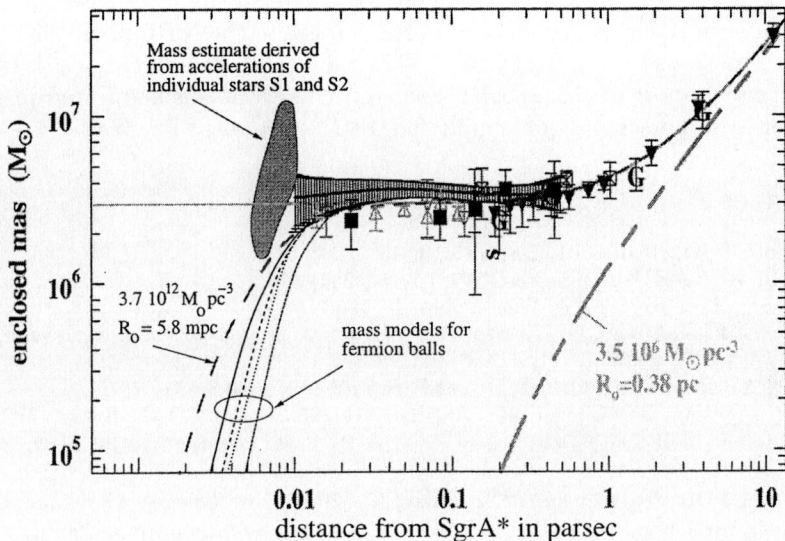

Fig. 1. The shaded area to the left indicates the corrected mass estimate obtained from the accelerations of stars S1 and S2 (Eckart et al. 2002). We also show the mass estimates obtained via Jeans modeling and other estimations based on proper motions and Doppler velocities (Fig. 17 by Genzel et al. 2000). For comparison we show enclosed mass curves expected for the overall stellar cluster (to the right), the most compact extended mass (plummer like density law $\rho(r) = \rho_o[1 + (r/R_o)^2]^{-\alpha/2}$ with α=5) consistent with the Jeans analysis (to the left; Genzel et al. 2000), and those for fermion balls with total masses ranging between 2.4 and $2.8 \times 10^6 \, M_\odot$ and neutrino masses between 14.35 and 17.5 keV (as given by Munyaneza, Tsiklauri & Viollier 1999)

pens to the permanently in-falling (baryonic) matter. It appears plausible that it will be trapped and condense to the bottom of the potential well. A scenario which defeats the purpose of having a ball of degenerated matter (esp. neutrinos) (Melia & Falke 2001). Future improved measurements of stellar trajectories will allow to determine amount and size of any possible extended mass contribution (Rubilar & Eckart 2000).

3   Summary and Conclusions

The analysis of stellar orbits as close to the center of the Milky Way as currently possible (Ghez et al. 2000, Genzel et al. 2000, Eckart et al. 2002) clearly supports the presence of a compact dark mass at the position of SgrA* which is likely in the form of of a massive black hole. Future sensitive measurements with interferometers will allow us to determine the enclosed mass at even smaller radii. Alternative models that describe the central object as a fermion ball

(Munyaneza & Viollier, 2002, Munyaneza, Tsiklauri, Viollier, 1998) predict an extended mass distribution which will become evident from the shapes of stellar orbits measured well within the central arcsecond.

Acknowledgements: The SHARP team is grateful to the ESO Director General and his staff to let us bring the SHARP camera to the NTT since 1999. The SHARP team is also thankful to the NTT and La Silla staff for their interest and technical support of the SHARP camera. This work was supported in part by the Deutsche Forschungsgemeinschaft (DFG) via grant SFB 494.

References

1. Baganoff, F.K., et al. 2001, Nature 413, 45
2. Bahcall, J.N. and Tremaine, S.C. 1981, ApJ 244, 805
3. Colpi, M., Shapiro, S.L., & Wasserman, I., 1986, Phys. Rev. D, 57, 2485
4. De Paolis, F., Ingrosso, G., Nucita, A.A., Orlando, D., Capozziello, S., Iovane, G., 2001, A&A 376, 853
5. Eckart, A., Genzel, R., Ott, T, Schödel, R., 2002, MNRAS 331, 917
6. Eckart, A., Ott, T., Genzel, R., 2001a, IAU 205, Proc. of IAU Symp. NO.205 on "Galaxies and Their Constituents at the Highest Angular Resolutions", p.41., Dordrecht:Reidel.
7. Eckart, A., Ott, T., Genzel, R., Rubilar, G., 2001b, in "Science with the LBT", Proc. of a workshop held at Ringberg Castle, Tegernsee / Germany 24-29 July 2000, Tom Herbst (ed.), Neumann Druck, Heidelberg, Germany, ISBN 3-00-008071-6
8. Eckart, A., Ott, T., Genzel, R., 1999, A&A, 352, 22
9. Eckart, A., Genzel, R., 1997, MNRAS 284, 576
10. Eckart, A., Genzel, R., Hofmann, R., Drapatz, S., Katterloher, R., Quirrenbach, A., Tacconi-Garman, L., 1997, Science with the VLT Interferometer, Proc. of the ESO workshop, held at Garching, Germany, 18-21 June 1996, p.259
11. Eckart, A., Genzel, R., 1996, Nature 383, 415
12. Figer, D., et al. 2000, ApJ 533, 49
13. Genzel, R., Pichon, C., Eckart, Gerhard, O.E., Ott, T., 2000, MNRAS 317, 348
14. Genzel, R., Eckart, A., Ott, T., Eisenhauer, F., 1997, MNRAS 291, 219
15. Genzel, R., Thatte, N., Krabbe, A., Kroker, H. and Tacconi-Garman, L.E. 1996, ApJ 472, 153
16. Ghez, A.M., Klein, B.L., Morris, M., Becklin, E.E., 1998, ApJ 509, 678
17. Ghez, A., et al., 2000, Nat. 407, 349
18. Greenhill, L. et al. 1995, ApJ 440, 619
19. Kaup, D.J., 1968, Phys. Rev. 172, 1331
20. Maoz E., 1998, ApJ, 494, L13
21. Melia, F., & Falke, H., 2001, Ann.Rev.Astr.Ap. 39, 309
22. Mielke, E.W., & Schunk, F.E., 2000, Nucl. Phys. B, 564, 1985
23. Munyaneza, F., Tsiklauri, D., & Viollier, R.D., 1999, ApJ 526, 744
24. Munyaneza, F., Tsiklauri, D., Viollier, R.D., 1998, ApJ 509, L105
25. Munyaneza, F., Viollier, R.D., 2002, ApJ 564, 274
26. Myoshi M., et al., 1995, Nature, 373, 127
27. Reid, M.J., et al., 1999, ApJ 524, 816
28. Rubilar, G.F. & Eckart, A., 2001, A&A 374, 95
29. Ruffini, R., & Bonazzola, S., 1969, 187, 1767
30. Torres, D.F., Capozziello, S., & Lambiase, G., 2000, Phys. Rev. D, 62, 104012
31. Viollier, R.D., Leimgruber, F.R., & Trautmann, D., 1992, Phys. Lett. B, 297, 132

Sources of Gravitational Waves

Bernard F. Schutz

Max Planck Institute for Gravitational Physics
(Albert Einstein Institute)
D-14476 Golm, Germany

Abstract. Gravitational wave detectors on the ground will begin taking quality data next year (2003), and the space-based LISA detector is planned for launch in 2011. I review here the physics of gravitational waves and survey the expected astronomical sources, with a focus on LISA. Observations by LISA are particularly interesting for cosmology: they have the potential to measure the acceleration history of the universe back to redshifts of 4 or 5.

1 Introduction to Gravitational Radiation

Gravitational radiation is the last major prediction of Einstein's general relativity that has not been directly verified. We have excellent quantitative but indirect evidence for the predictions of general relativity from observations of the Hulse-Taylor binary pulsar [1], but so far detectors have not been sensitive enough to make a direct detection. The weakness of their coupling to our detectors makes them attractive as probes of the universe: they interact with everything so weakly that they are not significantly scattered or absorbed as they travel to us. The "optical" depth of the universe to gravitational radiation is much smaller than one right back to the big bang. Despite this weakness, gravitational waves can carry prodigious energies, and their emission can have catastrophic consequences for their sources. This makes them even more interesting to detect.

The present time is a good one for reviewing what we know, or think we know, about gravitational wave sources, and therefore what kinds of information observations of gravitational wave are likely to bring:

- Giant interferometric detectors on the ground are nearing completion, and in 2003 they should begin observations at the sensitivity level of 10^{-21} that represents the threshold where it is not implausible that detections will be made in the first year or so; but neither is it certain that detections will happen so soon.
- The space-based detector LISA has recently been agreed as a joint project between ESA and NASA, for launch in 2011.
- Astronomers have become more and more confident that many – perhaps most – galaxies harbour supermassive black holes in their centres, some of which could be in binary pairs, whose mergers are LISA's most dramatic target.
- The discovery of the acceleration of the universe [2,3] raises the possibility that the "dark energy" responsible for this acceleration is time-dependent,

and that by measuring its history we might gain clues about fundamental physics theories. Because LISA will measure distances to merging black holes, it may offer a way to measure the acceleration history of the universe to high redshifts.

These topics form the focus of this review. We begin with a short introduction to the physics of gravitational waves and their sources. Our intention here is to develop back-of-the-envelope formulas that are helpful in understanding the main characteristics of sources and hence our expectations of what we can learn from observing them. The approximations are rough, factor-of-two accuracy. Behind these approximations lie a great number of detailed studies of sources, some of which we will touch on. While there is no room here for a full discussion of detailed calculations, it is possible to understand their results by using the approximations we introduce here.

1.1   The Physics of Gravitational Radiation

The early controversies and confusions about the existence and physical nature of gravitational waves were resolved in the period 1950-1980 through the work of many physicists, and the observations of the Hulse-Taylor system confirm that the theory is now well-understood. Waves are carriers of *tidal gravitational forces*, which is to say that what can be measured is only the *difference* between the time-dependent gravitational accelerations produced by the wave at different points. The action of the wave therefore is proportional, to lowest order, to the distance L between the locations being measured. The wave amplitude h is the dimensionless proportional factor: the wave produces a distance change between two free particles given by

$$\delta L = hL. \tag{1}$$

Waves are well-defined if h is small, so that the wave looks like a perturbation of a smooth background space. In this regime, which covers the propagation and detection of waves but not necessarily their generation, waves have a well-defined energy and momentum flux, localised to regions the size of a wavelength and time-spans of the order of one period.

The nonlinearity of general relativity makes realistic computations difficult. Even the two-body problem, which in Newtonian gravity is a classic undergraduate exercise, is not exactly solvable in general relativity. It is being solved by a combination of painstaking approximation methods and large-scale numerical simulations.

When two massive bodies orbit one another with a large enough separation that their orbital speeds are much smaller than c, then the post-Newtonian approximation can provide a very accurate representation of their motion. This is an asymptotic expansion in the orbital velocity v and in the weak gravitational field. Currently the approximations can be used up to the third post-Newtonian order (v^6 beyond Newtonian) [4]. The lowest order of this approximation for the emitted gravitational waves is called the *quadrupole approximation*. There is a convenient upper bound on the amplitude h of the radiation produced in the quadrupole approximation by a system with total mass M and size R at a distance r from the observer [5]:

$$h \leq \left(\frac{GM}{rc^2}\right)\left(\frac{GM}{Rc^2}\right). \tag{2}$$

This is simply the product of two dimensionless Newtonian potentials, the first the potential of the source at the observer, and the second the internal potential. The second term accounts for the driving force of self-gravity, which is assumed to govern the motions that emit the radiation. This term is always smaller than 1, so the amplitude of the waves is always less than the Newtonian potential of the source. The formula gives only an upper bound, because spherical motions produce no radiation at all. This bound is attained, however, by maximally non-symmetrical systems, like binary stars of equal mass.

In the same spirit there is an upper bound on the energy luminosity in gravitational waves:

$$L \leq \frac{3}{32}\frac{c^5}{G}\left(\frac{GM}{Rc^2}\right)^5. \tag{3}$$

The dimensionless Newtonian potential term shows how sensitive the luminosity is to the compactness of the source: a source that is only twice as large as another of the same mass will radiate only 3% of the luminosity, all other things being equal. The dimensions come from the constant $c^5/G = 3.6 \times 10^{59} \mathrm{erg\, s^{-1}}$, which the formula shows is an upper bound on *all* gravitational luminosities.

Like electromagnetic waves, gravitational waves are *transverse*: they act in the plane perpendicular to their propagation direction. There are two independent polarisations, conventionally called "+" and "×". These are illustrated in Fig. 1.

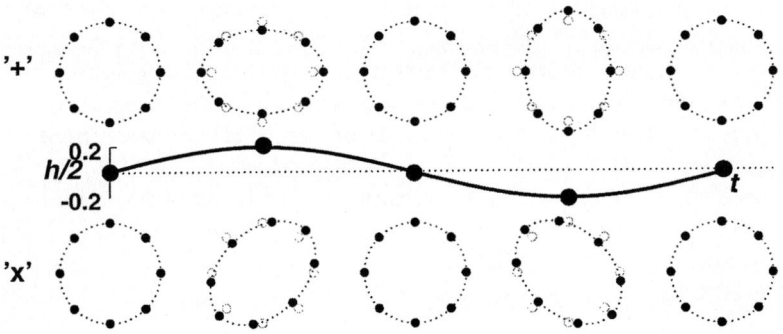

Fig. 1. The two independent, orthogonal polarisations of a gravitational wave. The dots illustrate positions of free particles in a plane transverse to the wave. The successive arrangements of the dots show the way their proper distances change as the wave passes

All gravitational wave detectors are polarised, so they receive only one polarisation component. This is one reason why many observations require more than one detector to interpret them.

A key to using polarisation information to interpret observations is to understand that the polarisation of a wave follows the source motion. Although Fig. 1 illustrates the action of a wave on free particles, it could just as well depict the motions of source masses that lead to the given polarisation. The polarisation matches as closely as possible the mass motions in the source as projected on the sky. This means that by measuring the polarisation of the waves from a binary system, one can infer its inclination angle: waves from a binary seen face-on will be circularly polarised, waves from one see edge-on will be linearly polarised.

1.2   The Dynamics of Gravitational Wave Sources

A key to the physics of gravitational wave sources is their frequency. For the strongest sources, the radiation will be produced by motions involving essentially all the mass. Then the motions' frequency will be the natural frequency of a self-gravitating system, $\omega = (\pi G \rho)^{1/2}$. The gravitational waves will typically come out with twice this frequency.

In Fig. 2 I plot, on a graph of the mass M of the system versus its size R, three lines of constant natural frequency f: $f = 10^{-4}$ Hz, $f = 1$ Hz, $f = 10^4$ Hz. They define two bands: the high-frequency band in which ground-based detectors will work, and the low-frequency band that is accessible only from space. Another important line is the black-hole line; there are no objects below this. It is clear from these simple arguments that ground-based detectors will see only stellar-mass objects, not the supermassive black holes.

The figure also illustrates the back-reaction effects of emitting gravitational radiation. By dividing the gravitational potential energy GM^2/R of a system by its gravitational luminosity L, one gets a time τ called the *chirp time*, because the emitted radiation will increase in frequency and amplitude on this timescale: it will produce a signal called a "chirp". This is the timescale on which the system will change significantly due to the loss of energy to gravitational waves. Binaries whose chirp time is one year are illustrated in the diagram; any system below this line will coalesce during the time we observe it. This includes all binaries observable from the ground. Of course, we can measure the changes in an orbit even if τ is longer than a year; all we require is that the orbital frequency change by more than the frequency resolution of a one-year observation, $\delta f = 3 \times 10^{-8}$ Hz. Systems whose chirp can be observed lie below the "binary chirp line" in the diagram. This includes all supermassive black hole binaries with masses above $10^5 M_\odot$.

1.3   Typical Gravitational Waves

One can put simple numbers into the above formulas to get order-of-magnitude estimates to guide detector development. If we assume our source is a typical neutron star in the Virgo cluster, then the mass of $1.4 M_\odot$ and size of 10 km

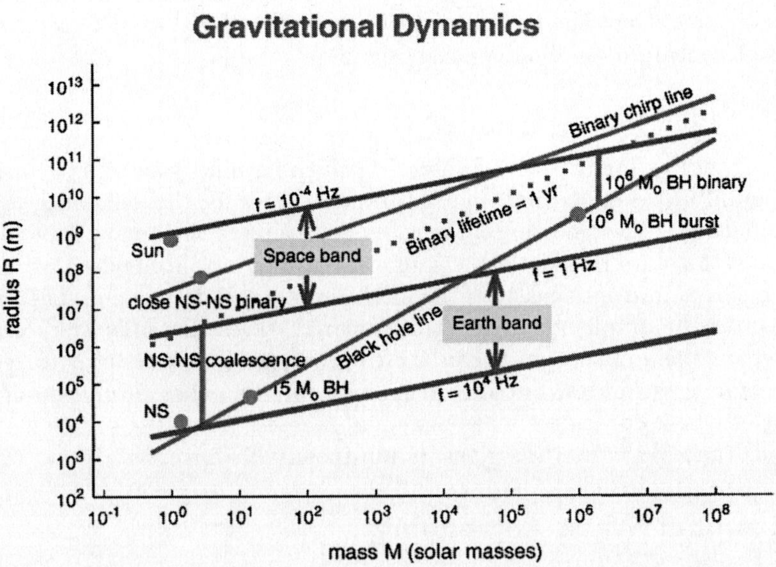

Gravitational Dynamics

Fig. 2. The mass-radius plane for sources of gravitational waves. The two bands of frequencies accessible from space and the ground are crossed by the black-hole line (no sources below this), the coalescence line (binaries that coalesce through emitting gravitational radiation in one year), and the chirp line (binaries that change their frequency measurably in one year)

imply a typical frequency of 2 kHz. This is the fundamental vibration frequency of such a star, and is also the highest frequency we can expect from orbital radiation in the last phase of inspiral. The luminosity in gravitational waves is so large that the time-scale for damping out the radiation is only 20 ms. At a distance of 20 Mpc, the energy flux is $0.6 \mathrm{Wm}^{-2}$, which is larger than the light flux from a full moon! The amplitude h is, however, only 10^{-21}, so all this energy passes through is with hardly a trace.

This amplitude is the goal of first-generation ground-based gravitational wave detectors. From the calculation, one can see that there is no guarantee that such detectors will have an abundance of events. In fact, there may well be no radiation at the 10^{-21} level. Only observations will tell.

As a low-frequency example, consider a pair of $10^6 M_\odot$ black holes at a distance of 4 Gpc (redshift $z = 1$), at their last stable circular orbit (separation of about 10^{10} m). The frequency is around 15 mHz, the amplitude is $h = 10^{17}$, and the timescale (to merger) is around 200 s. This is the last phase of merger: for months before this the frequency will be larger than 0.1 mHz.

These very weak amplitudes tell us that it will be hard to construct detectors sensitive enough to see the waves directly. Ground-based detectors all will rely on matched filtering to extract the signals. LISA, in space, will actually have sufficient sensitivity to see the black-hole mergers without filtering, but filtering will be necessary to extract the full information from the signal (and to remove it so it does not contaminate other, weaker, signals).

1.4 Chirping Binaries are Standard Candles

One of the most interesting aspects of observing chirping binary systems is that their gravitational waves contain enough information to infer their distance [6]. This happens because of a coincidence: the rate of chirping of the binary depends on the masses of the stars in exactly the same way as the amplitude does, so by measuring the chirp time one measures the intrinsic mass of the system. Then if one also measures the amplitude of the gravitational waves, the only unknown is the distance to the source, which can then be solved. It is not hard to see this from our above estimates, but it is true even for the exact formulas, and even for binaries whose components have unequal masses. When the source is at cosmological distances, the method gives its luminosity distance.

2 Gravitational Wave Astronomy

2.1 Overview of Detector Projects

Today four major interferometer projects – LIGO, VIRGO, GEO600, and TAMA – are building and commissioning first-generation detectors that will be able to observe at high frequencies (above 40 Hz). These detectors will begin observing during 2003, and they may well make the first detections of gravitational waves. But there is no guarantee that they will see anything at the their first sensitivity levels.

Plans for second-generation interferometers (to operate after 2006) are well advanced. These should improve the sensitivity by a factor of 10, and at this level it would be very surprising if they did *not* detect anything. The goal of the LIGO project, for example, has always explicitly been to reach this higher sensitivity; the first phase has always been seen as a step toward the interesting second level.

Plans for a third generation of ground-based detectors (2010+) are now being developed, but only in a preliminary way. New technologies will be needed, and the difficulties of this extra level of improvement (another factor of 10) will be great.

In space, LISA is being developed for a launch in 2011 as a joint ESA-NASA mission. It will open the low-frequency window (below 1 Hz), where it must make many detections. In contrast to the ground-based projects, LISA will be observing some specific systems – galactic binaries – that are already known from optical observations and whose radiation can be confidently predicted. Moreover, some potential LISA sources – coalescences of supermassive black holes in distant galaxies – will be at very high signal-to-noise ratios. The observations LISA can make have particular interest to this meeting, because they address fundamental physics issues in three areas: cosmological parameters, testing general relativity, and gravitational waves from the Big Bang.

2.2 LISA and Cosmological Parameters

Here I want to discuss the kind of scientific return LISA can provide from its observations of black hole coalescences. LISA will open up a huge unexplored area of astronomy, with payoffs in many fields. But for this meeting, the implications for cosmology are particularly interesting: LISA could, in principle, make measurements of the acceleration history of the Universe, and thereby illuminate the time-dependence of the dark energy.

LISA will survey whole universe for mergers of supermassive black holes: any equal-mass merger with $(1+z)M$ between about 10^3 and $10^7 M_\odot$ should be detected. This is a large range of masses, embracing a variety of source histories.

At the low end of this mass range is the population of $1000 M_\odot$ black holes that may have been formed in the first generation of star formation, at redshifts between 5 and 10. LISA may record their collapse events, if these events were sufficiently asymmetrical [7]. It is possible that these objects then merged to form the larger black holes that are now observed in the centers of most galaxies; if so, then LISA should record thousands of chirps from such events. Their signals chirp upwards from 1.5 mHz to 0.1 Hz in around 10 years. A one-year observation by LISA of a merger at $z = 1$ will have signal-to-noise ratio of around 60. At this low-mass end of the spectrum, LISA will illuminate the way that the supermassive black holes formed and grew.

At the higher-mass end of this range, galaxy-galaxy mergers could lead to a few $10^6 M_\odot$ black-hole chirp observations at z 1. The signal-to-noise ratios could be above 1000, and this means that the angular positions provided by LISA will be very accurate when compared with the "beam width" of LISA (the whole sky): of order 30 arc-minutes. But at a redshift of 1, this corresponds to an error box 40 Mpc on a side. In fact, since the error in the estimate of the amplitude of the gravitational waves from these sources is limited by this angular error, the inferred standard-candle distance to the sources will also have a similar error. Therefore LISA will locate such sources within a cube of side 40 Mpc. This is unfortunately not good enough to do cosmology: the distance error is too large to give interesting accuracy for the Hubble constant or other parameters.

To do better, we will need to identify the host galaxy for the merger event. This is likely to be difficult, but perhaps not impossible. The merger event itself is electromagnetically quiet for ideal black holes, but in realistic cases this may not be true. For example, if the holes retain accretion disks until their last plunge, then the hydrodynamics and magnetohydrodynamics of these remnant disks may lead to outflows, jets, or bursts of electromagnetic radiation that precede the merger event by up to a few months. Even if these discs are disrupted much earlier, the galaxies may show their original jets, and the inner regions of these jets may show disrupted features. If the galaxy is too far away for such studies, there may still be features in is morphology that give it away. For example, the central black holes are able to merge because they have lost energy to the stellar population; this energy may show up in the velocity dispersion of the central inner core, and especially in its anisotropy, with radial velocities dominating.

If host galaxy of the merger can be identified, then the uncertainty in the distance improves dramatically. The principal limit on accuracy then becomes gravitational lensing, which can magnify a source and therefore raise its amplitude [8]. The inference of the Hubble constant or acceleration measures will also be limited by velocity errors, such as the proper velocity of the host galaxy relative to the Hubble flow, and errors in measuring its redshift.

Within these error bounds, only two events at redshift 1 with identified galaxies would be sufficient to measure the Hubble constant and the present acceleration q_0 to accuracies of 1% or better. With a few more events at larger distances one can begin to watch the time-dependence of the acceleration of the Universe. One might be able to see the turn-over, where the acceleration of the dark energy began to dominate the deceleration of the dark matter. One might even be able to go back further, and see if the dark energy remains constant, or depends on time as predicted by quintessence models.

2.3   LISA and Testing General Relativity

Black holes are remarkable objects, not just because of their extreme gravitational fields and their ubiquity in the Universe, but even more because of their incredible simplicity. A time-independent vacuum black hole is described by just three parameters: its mass, spin, and charge. The entire metric outside the hole is a function of just these three numbers. A realistic black hole, surrounded by an accretion disk feeding it matter and supporting a magnetic field, is of course more complicated, but the complications make only small differences to the gravitational field. Only when black holes merge does the metric differ substantially from the Kerr-Newman form, and these deviations are radiated away as gravitational waves in a few light-crossing times of the horizon.

All this is theory, of course. LISA can test it. By observing "small" black holes (of say $10 M_\odot$) falling into supermassive black holes in galactic centers, LISA will be able to map in detail the metric of the supermassive black hole. The infalling hole is nearly a test particle in the background spacetime, a probe that measures the metric and sends us information in gravitational waves. The reason the probe is so effective is that such encounters are expected to have high eccentricity, with the result that the object may execute hundreds of thousands of orbits within the LISA frequency band before it is finally captured by the black hole. With such information, the map can be highly accurate. We can expect to test the black-hole uniqueness theorems of relativity stringently.

And the event rate is significant: a few events per year from distances of about 1 Gpc [9]. This, however, leads to a problem. LISA is able to see only the events in the "nearby" universe. There are many more at larger distances that are individually too weak to detect, but whose superposition adds up to a background noise. The problem is that LISA cannot point, it cannot exclude signals from unwanted directions. While it is observing (for a year) one nearby event at $z = 0.25$, the volume out to $z = 1$ contains 64 more events too weak to be seen individually, but whose total received power equals 4 times the power received from the nearby signal! In effect, LISA is running into an Olbers limit:

the Olbers argument of why the sky should not be black at night works very well for these gravitational wave sources.

The way to see the nearby events is to "point" LISA in data analysis, *i.e.* to do matched filtering on the source to remove all this excess uncorrelated noise. However, it is an open question, currently under study, whether improving the instrumental sensitivity of LISA will make this extraction easier. The distant astrophysical background will set the ultimate limit on sensitivity.

Nor will matched filtering be easy. The infalling objects finally get captured because they lose energy to gravitational radiation. But theorists have not yet worked out the radiation-reaction equation of motion of a small point mass in the Kerr metric. Until this problem is solved, it will be difficult to estimate the complexity of the filtering problem and the likelihood that good detections can in fact be made.

2.4   LISA and Primordial Gravitational Waves

The Big Bang should have generated gravitational waves that today form a random background, but there is great uncertainty about their strength. Standard inflationary models predict low energy densities and flat spectra, with $\Omega_{gw} <$ 10^{-14}. As an upper bound, the energy density in gravitational waves must be less than 10^{-5} of closure density to avoid disturbing nucleosynthesis. Between these extremes, models based on symmetry-breaking phase transitions (for LISA, at an energy of 1 TeV, where the electroweak transition occurred), branes, strings, or cosmic defects can be made to predict almost any level and spectrum.

LISA can only detect a background if it stands up above its own instrumental noise, which sets a lower limit of about Ω_{gw} 10^{-10}. Because LISA is triangular, it is possible to distinguish a cosmological background from instrumental noise: there is a consistency signal that can be extracted from the LISA data that almost cancels the gravitational-wave signal, while preserving instrumental noise [10]. LISA can also distinguish a cosmological background from a local galactic source population, because its limited angular sensitivity will be enough to determine whether the radiation is isotropic or from a disk population. But LISA will not, of course, be able to tell whether an isotropic background is primordial or simply a cosmological source population.

Although LISA will not reach the inflation bound, its observations will be extremely important. If it detects a background, that will probably be ranked as its greatest achievement. If it simply limits the background, it will exclude a wide range of (perhaps speculative) early-universe models.

2.5   LISA and the Unexpected

In addition, there is one class of potential sources that we can't say much about: the unexpected ones. Since LISA will open a new window on the universe with very good sensitivity, the chances are high that it will find things that astronomers were not able to predict. This is particularly likely because LISA will be sensitive in principle to the 90% of the material universe that emits no electro-magnetic radiation, the dark matter. Even if the dark matter is largely smoothly distributed, as required by CDM models, it is possible that the dark matter is a mixture of a variety of species of particles, and some minority is self-interacting, capable of forming structures. LISA may well see gravitational waves from some such structures. It is an exciting but at the same time frightening prospect! How would we have the confidence to know what we we seeing?

3   The Future of Space-Based Gravitational Wave Detection

LISA will be the first detector in space, but certainly not the last. Since we already know the limitations of LISA, it is not unreasonable to ask what we would like to do better: what should a LISA follow-on do? To some extent this will depend on what LISA finds or excludes. But these projects have long lead-times, so LISA scientists are already sketching out their proposals for the next generation.

I would identify two goals as important: improving angular resolution for locating coalescing supermassive black hole binary events, and detecting a prim-ordial background. The reason for the first is apparent in our earlier discussion: estimates of distance, and inferences of cosmological parameters, depend strongly on the accuracy with which sources can be located on the sky. With a smaller gravitational-wave error box, follow-up optical studies will have fewer candid-ates and correspondingly greater chance of finding the host galaxy for a merger. The reason for wanting better sensitivity to the gravitational wave background is also evident: it is our most fundamental probe of the Big Bang, and could in principle reveal the nature of the Universe only 10^{-30} s after the Big Bang and of the laws of physics at GUTs or even Planck energy scales.

It appears today that a mission to detect the gravitational wave background should aim at the frequency gap between ground- and space-based detectors, the range 0.1–10 Hz. High frequencies are in principle undesirable, because the energy density in gravitational waves is proportional to $h^2 f^2$, so to reach a given value of Ω_{gw} at a larger f then one must go to a smaller h. But below 0.1 Hz, astrophysical backgrounds are likely to be larger than the inflation bound on the primordial radiation [11].

At this frequency, the angular resolution of an instrument is also better, because it is basically limited by the number of gravitational wavelengths that span the orbit of the instrument. However, the strongest cosmological binary sources have the large masses, and this pushes their signals down in frequency. It is not clear, therefore, whether these two scientific goals can both be served by a single follow-on mission. This is something that will be discussed in detail over the next years, as we wait for LISA to fly and to bring in the first information about the low-frequency gravitational wave universe.

References

1. C. M. Will: Living Rev. Relativity **4**, 4 (2001). [Online article]: cited on 13 June 2002 http://www.livingreviews.org/Articles/Volume4/2001-4will/
2. S. Perlmutter *et al*: Astrophys. J. **517**, 565 (1999)
3. A.G. Riess *et al*: Astronom. J. **116**, 1009 (1998)
4. L. Blanchet: Living Rev. Relativity **5**, 3 (2002). [Online article]: cited on 13 June 2002 http://www.livingreviews.org/Articles/Volume5/2002-3blanchet/
5. B. F. Schutz: Am. J. Phys. **52**, 412 (1984)
6. B. F. Schutz: Nature **323**, 310 (1986)
7. K.C.B. New, S.L. Shapiro: Class. Quant. Grav. **18**, 3965 (2001)
8. S. Hughes: Mon. Not. Roy. Astr. Soc., **331**, 805 (2002)
9. S. Sigurdsson, M.J. Rees: Mon. Not. Roy. Astr. Soc., **284** (1997)
10. J.W. Armstrong, F.B. Estabrook, M. Tinto: Astrophys. J., **527**, 814 (1999)
11. C. Ungarelli, A. Vecchio: Phys. Rev. D, **63**, 064030 (2001)

Status and Prospect of Laser-Interferometric Gravitational Wave Astronomy

Karsten Danzmann[1,2] and Albrecht Rüdiger[3]

[1] Institut für Atom- und Molekülphysik, Universität Hannover, Callinstr. 38,
and Max-Planck-Institut für Gravitationsphysik, Albert-Einstein-Institut
[2] Callinstr. 38, D–30167 Hannover
[3] Hans-Kopfermann-Str. 1, D–85748 Garching, Germany

Abstract. The existence of gravitational waves is the most prominent of Einstein's predictions that has not yet been directly verified. The space project LISA shares its goal and principle of operation with the ground-based interferometers currently under construction: the detection and measurement of gravitational waves by laser interferometry. Ground and space detection differ in their frequency ranges, and thus in the detectable sources. Ground-based detection will allow detection only from a few Hz upwards, and up to a few kHz. On five sites worldwide, detectors of armlengths from 0.3 to 4 km are nearing completion. They will supply first scientific data in 2003. Only in space, detection of signals below, say, 1 Hz is possible. The project LISA consists of three spacecraft in heliocentric orbits, forming a triangle of 5 million km sides. Launch for LISA is scheduled for 2011, following a technology demonstrator LTP in 2006.

1 Introduction

The talk on which this paper is based dealt with a new method of astronomic observation presently being opened: the detection and measurement of gravitational waves. This is one of the great challenges to modern physics. Although predicted by Einstein in 1916, a direct observation of these waves has yet to be accomplished. Their detection would open a new window to the Universe [1].

Great hopes of such detection lie in the ground-based laser-interferometric detectors currently nearing completion. These ground-based detectors are sensitive in the 'audio' frequencies of a few Hz up to a few kHz. Perhaps even more promising are the space-borne interferometers, where we will mainly have to think of the joint ESA-NASA project LISA (Laser Interferometer Space Antenna), which would cover the frequency range from about 10^{-4} Hz to 1 Hz.

Gravitational waves (GW) share their elusiveness with neutrinos: they have very little interaction with the measuring device. But that feature also is a great advantage: due to their exceedingly low interaction with matter, gravitational waves can give us an unobstructed view into astrophysical events that will forever be obscured in the electromagnetic window. The price we have to pay is that, in order to detect and measure these gravitational waves, we will require the most advanced technologies in optics, lasers, and interferometry.

Efforts to observe these gravitational waves with ground-based interferometers have gone into their final phase of commissioning, and next-generation detectors are already in the discussion. Furthermore, an international collaboration on placing a huge interferometer, LISA, into an interplanetary orbit is close to reaching final approval.

2   Gravitational Waves

In two publications Albert Einstein has predicted the existence and estimated the strength of gravitational waves [2]. They are a direct outcome of his Theory of General Relativity, but a necessary consequence of *all* theories with finite velocity of interaction. A talk on the sources, the manifestation, and the significance of gravitational waves is given in this volume [1].

Gravitational waves of measurable strengths are emitted only when large cosmic masses undergo strong accelerations, for instance – as shown schematically in Fig. 1 – in the orbits of a (close) binary system. The effect of such a gravitational wave is an apparent strain in space, transverse to the direction of propagation, that makes distances ℓ between test bodies shrink and expand by small amounts $\delta\ell$, at twice the orbital frequency: $\omega = 2\,\Omega$. The strength of the gravitational wave, its "amplitude", is generally expressed by $h = 2\,\delta\ell/\ell$.

As is shown in Fig. 1, the strains $\delta\ell$ transverse to the direction of propagation are of opposite sign in the two orthogonal arms. An interferometer of the Michelson type, typically consisting of two orthogonal arms, is therefore an ideal instrument to register such differential strains in space.

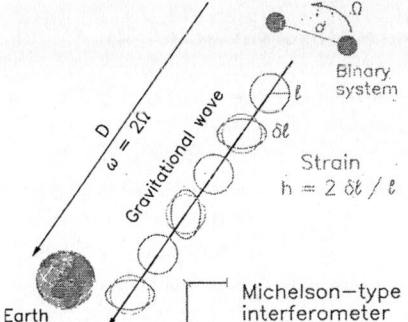

Fig. 1. Generation and propagation of a gravitational wave emitted by a binary system

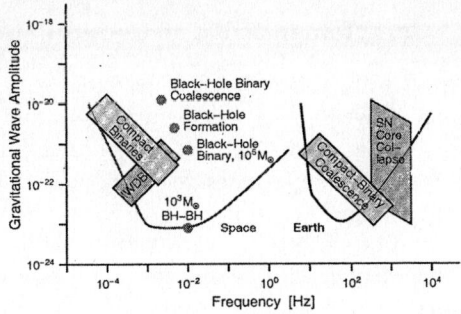

Fig. 2. Some sources of gravitational waves, with sensitivities of *Earth* and *Space* detectors

2.1   Strength of Gravitational Waves

But what appears so straightforward turns out to be an almost insurmountable task. That lies in the magnitude, or rather: the smallness, of the effect.

With a linearized approximation, the so-called 'quadrupole formula', the strength of the gravitational wave emitted by a mass quadrupole can be estimated, and for a binary with the distances d and D of Fig. 1, and with components of masses M_1 and M_2, or their respective Schwarzschild radii $R_i = 2GM_i/c^2$, the strain h to be expected is of the order

$$h \approx \frac{R_1 R_2}{d\,D}\,. \tag{1}$$

From such an in-spiral of a neutron star binary out at the Virgo cluster (a cluster of about 2000 galaxies, $D \sim 10\,\mathrm{Mpc}$ away), we could expect a strain of something like $h \approx 10^{-22}$. Similar (or even lower) strengths might be expected from supernovae out at Virgo cluster distances. That we insert such a large distance as the one to the Virgo cluster is to have a reasonable rate of a few events per year. Inside a single galaxy (as ours), we would not count more than a few supernovae per century. The seemingly hopeless task set out is to measure – in a Michelson interferometer of kilometer dimensions – path changes in the order of $10^{-19}\,\mathrm{m}$. However, the sensitivities already obtained with prototypes of ground-based interferometers bear evidence that this is within reach.

Electromagnetic radiation is easily obscured and absorbed by matter between source and observer; not so the gravitational waves: their interaction with matter is extremely weak, it reaches us effectively unblemished by any obstacles.

This is the great advantage: gravitational waves allow us to observe events that remain hidden "in the light" of electromagnetic radiation. Thus completely different types of information, and new insights in astronomy, astrophysics, fundamental physics can be obtained.

2.2 Complementarity of Ground and Space Observation

Figure 2 shows some typical sources of gravitational radiation. They range in frequency over a vast spectrum, from the kHz region of supernovae and final mergers of compact binary stars down to mHz events due to formation and coalescence of supermassive black holes. Indicated are sources in two clearly separated regimes: events in the range from, say, 5 Hz to several kHz (and only these will be detectable with terrestrial antennas), and a low-frequency regime, 10^{-5} to $1\,\mathrm{Hz}$, accessible only with a space project such as LISA. In the following sections we will see how the sensitivity profiles of the detectors come about. No detector covering the whole spectrum shown could be devised.

Clearly, one would not want to miss the information of either of these two (rather disjoint) frequency regions. The upper band ("Earth"), with supernovae and compact binary coalescence, can give us information about relativistic effects and equations of state of highly condensed matter, in highly relativistic environments. Binary inspiral is an event type than can be calculated to high post-newtonian order. This will allow tracing the signal, possibly even by a single detector, until the final merger. The ensuing phase of a ring-down of the combined core does again lend itself to an approximate calculation. Chances for detection are reasonably good, but not by wide margins.

The events to be detected by the space projects LISA, on the other hand, may have extremely high signal-to-noise ratios, and failure to find them would shatter the very foundations of our present understanding of the universe. The strongest signals will come from events involving (super-)massive black holes, during their formation as well when galaxies with their BH cores collide. But also the (quasi-continuous) signals from neutron-star and black-hole binaries are among the events to be detected ('Compact Binaries' in Fig. 2). Interacting white dwarf binaries inside our galaxy ('IWDB' in Fig. 2) may turn out to be so numerous that they cannot all be resolved as individual events. Catastrophic events such as the Gamma-ray bursts are not yet well enough understood to estimate their emission of gravitational waves, but there is a potential of great usefulness of GW detectors mainly at low frequencies.

The combined observation with electromagnetic and gravitational waves could lead to a deeper understanding of the violent cosmic events in the far reaches of the universe [3].

3   Ground-Based Interferometers

The underlying concept of all ground-based laser detectors is the Michelson interferometer (see schematic in Fig. 3), in which an incoming laser beam is divided into two beams travelling along different (perpendicular) arms. On their return, these two beams are recombined, and their interference (measured with a photodiode PD) will depend on the difference in the gravitational wave effects that the two beams have experienced. A gravitational wave propagating normal to the plane of the interferometer would give rise to a path difference δL between the two arms of

$$\delta L = h_+ \cdot L \cdot \frac{\sin(\pi f \tau)}{\pi f \tau} = h_+ \cdot L \cdot \frac{\sin(\pi L/\Lambda)}{\pi L/\Lambda} . \tag{2}$$

The changes δL in optical path become the larger the longer the optical paths L are made, optimally about half the wavelength Λ of the gravitational wave: e.g. to a seemingly unrealistic 150 km for a 1 kHz signal. Schemes were devised to make the optical path L significantly longer than the geometrical arm length ℓ, which is limited on Earth to only a few km. One way is to use 'optical delay lines' in the arms, the beam bouncing back and forth N times between two concave mirrors. The optical path in Eq. (2) is then $L = N\ell$. A version with a modest $N = 4$ is shown in Fig. 4.

The other scheme is to use Fabry-Perot cavities (Fig. 3), again with the aim of increasing the interaction time of the light beam with the gravitational wave. For GW frequencies beyond the inverse of the storage time τ, the response of the interferometer will roll off with frequency, as $1/f\tau$.

3.1   The Detector Prototypes

It is fortunate that on our way to the large-scale detectors we were able to go through generations of ever-improving prototypes. It was only with their positive results that the proposals for large-scale detectors attained sufficient credibility.

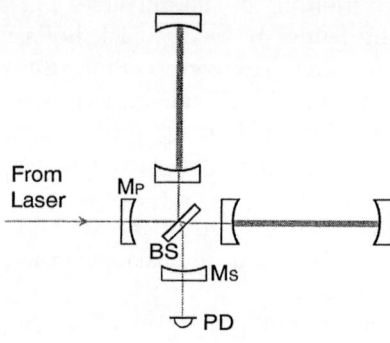

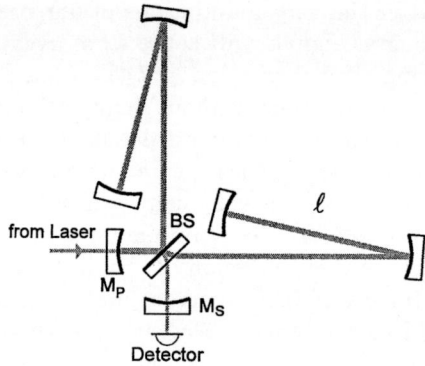

Fig. 3. Advanced Michelson interferometer with Fabry-Perots in the arms
and extra mirrors M_P, M_S for power
and signal recycling

Fig. 4. A delay-line configuration
(DL4) with dual recycling, as is to be
used in GEO 600

After pioneering work by Rai Weiss [4] at MIT (1972), groups at Munich/Gar-
ching, at Glasgow, then Caltech, Paris/Orsay, Pisa, and later in Japan and
Australia, also entered the scene. Their prototypes range from a few meters up to
30, 40, and even 80 m. They are all modelled after the Michelson interferometer.
As an alternative, a Sagnac configuration, is being investigated at Stanford [5].

Based on the idea of Weiss, Garching had also adopted the scheme of the
optical delay-line. The advantage was a very rapid attainment of sensitivities
that were limited only by shot noise [6], by gradually reducing newly discovered
noise mechanisms. It was only in the later 1990's that the (technologically more
challenging) prototypes with Fabry-Perot cavities reached similar, meanwhile
even better, sensitivities.

Even though some of these prototypes approached the sensitivities of cryogenic resonant-mass antennas, they were never meant to be used as detectors,
but rather as test-beds for verifying new schemes and configurations devised to
overcome otherwise limiting noise effects.

The "phase noise" reduction achieved in these prototypes already approaches
that required in full-fledged *terrestrial* interferometers, and it is by many orders
of magnitude better than that required (at low frequencies) for a *space mission*.

3.2 The Large-Scale Projects

Table 1 gives an impression of the wide international scope of the interferometer
efforts, listed according to size of detector. All of the large-scale projects will
use low-noise Nd:YAG lasers ($\lambda = 1.064\,\mu m$), pumped with laser diodes for high
overall efficiency. A wealth of experience has accumulated on highly stable and
efficient lasers, and also the space mission will profit from that.

Table 1. Current and future projects of ground-based GW detectors

Country:	USA		FRA ITA	GER GBR	JPN
Institute:	MIT, Caltech		CNRS INFN	AEI Glasgow	NAO, U-Tokyo, ICRR

Large Interferometric Detectors: the current generation

Project name:	LIGO		VIRGO	GEO 600	TAMA 300
Arm length ℓ:	4 km 2 km	4 km	3 km	600 m	300 m
Site (State)	Hanford (WA)	Livingston (LA)	Pisa ITA	Hannover GER	Mitaka JPN

Large Interferometric Detectors: the future generation

Planning (start):	1995		1999	1998
Arm length ℓ:	4 km	4 km	3 km	3 km
Site (State)	Hanford (WA)	Livingston (LA)	EUROPE	Kamioka JPN
Project name:	Advanced LIGO		EURO	LCGT
special features:	active isolation, suspension, RSE		high seismic rejection; cryogenic, diffractive optics, tunable	cryogenic, underground

LIGO   The largest is the US project named LIGO [7,8]. It comprises *two* facilities at two widely separated sites, in the states of Washington and Louisiana. Both will house a 4 km interferometer, Hanford an additional 2 km one. At both sites construction has long been completed, installation of the optics in the vacuum enclosures is done, and locking of the interferometers has now become routine. These three interferometers are designed for coincidence operation, allowing autonomous measurements inside the US project LIGO.

VIRGO   Next in size (3 km) is the French-Italian project VIRGO [9], being built near Pisa, Italy. An elaborate seismic isolation system, with six-stage pendulums (see Section 4.3), will allow measurement down to GW frequencies of 10 Hz or even below, but still no overlap with the space interferometer LISA. Currently, a short-arm interferometer, with all mirrors inside the central building, is being used for performance testing.

GEO 600   For the detector of the British-German collaboration, GEO 600 [10,11], with a de-scoped length of 600 m, construction and installation of the optics in the vacuum system are finished, and locking of the full power-recycled Michelson is routinely achieved. GEO 600 will employ the advanced optical technique of 'signal recycling', SR [12,13], to make up for the shorter arms. This interferometric scheme or its counterpart 'resonant sideband extraction' RSE [14] will later be transferred to the upgrades of LIGO and to the Australian detector.

TAMA 300   In Japan, on a site at the National Astronomical Observatory in Tokyo, construction, vacuum system, and optics installation of the detector called TAMA 300 [15] with 300 m armlength are completed, and several data runs of the Michelson have been successful. A recent run with a total of 1000 hours exhibited encouragingly long in-lock duty cycles [16]. The sensitivity-enhancing scheme of power recycling will have, however, yet to be added. Separate tests with power recycling appeared promising. TAMA is, just as LIGO and VIRGO, equipped with standard Fabry-Perot cavities in the arms.

ACIGA   Australia (not included in table) had to postpone earlier plans of a 3 km detector, due to lack of funding. Currently an 80 m prototype detector is being built near Perth, Western Australia, with the aim of investigating new interferometry configurations [17], follow-ons to the GEO schemes of signal recycling and/or RSE. The design and the site will allow later extension to 3 km.

3.3   International Collaboration

Most of these projects are rather well in synchronism. First scientific operation can be expected from the year 2003. For the received signal to be meaningful, coincident recordings from at least two detectors at well-separated sites are essential. A minimum of three detectors (at three different sites) is required to locate the position of the source, and it is only with at least four detectors that we can speak of a veritable gravitational wave *astronomy*, based on a close international collaboration in the exchange and analysis of the experimental data.

First common data run   At the turn of the year 2001/2002, a common data run between all three LIGO detectors and GEO 600 was undertaken, consisting of 17 days of mostly uninterrupted operation. It exhibited encouragingly high duty cycles of the interferometers being locked, in the GEO detector being improved considerably "on the fly" (up to better than 95 %), by upgrading the automatic mirror alignment.

With the detectors not yet being at the intended sensitivity level, the aim was not a search for gravitational waves, but rather to rehearse the activities of data acquisition and, then, also of data analysis. The analysis of the data has given further clues on the 'healthiness' of the interferometers.

(Note added in proof)   A second such data run, S 1, between LIGO and GEO 600, and partly also TAMA 300 joining in, was held in the period from 23 August to 9 September, 2002, with improved noise suppression. The data analysis is currently being carried out. A further science run, S 2, is tentatively scheduled for a duration of eights weeks from February to April, 2003.

4   Noise and Sensitivity

The measurement of gravitational wave signals is a constant struggle against the many types of noise entering the detectors. These noise sources have presented a great technological challenge, and interesting schemes of reducing their effects have been forwarded. The most prominent of such noise sources, with emphasis on the ground-based detectors, will be discussed in the subsections below.

4.1 Laser Noise

The requirements on the quality ('purity') of the laser light used for the GW interferometry are very stringent, but not insurmountable. As it happens, the light sources for the ground-based and the space-borne interferometers will both be Nd:YAG lasers, in the form of non-planar ring oscillators [18]. Pumped by laser diodes, they exhibit a high overall efficiency. Their good tunability allows efficient stabilization schemes.

Frequency stability A Michelson interferometer of exactly matching arms would be insensitive to frequency fluctuations of the light used. The detectors will, however, by necessity have unequal arms: the ones on the ground due to civil engineering tolerances and a particular modulation scheme chosen, the space detector due to unavoidable imperfections in the orbits of the individual spacecraft.

Thus, a very accurate control of the laser frequency is required, with (linear) spectral densities of the frequency fluctuations of the order $\widetilde{\delta\nu} = 10^{-4}\,\mathrm{Hz}/\sqrt{\mathrm{Hz}}$. Control schemes have been devised to reach such extreme stability, albeit only in the frequency band required, and not all the way down to DC (which would set an all-time record in frequency stability: $\widetilde{\delta\nu}/\nu = 3\times10^{-19}/\sqrt{\mathrm{Hz}}$).

Power stability Again due to asymmetries of the interferometer, the incoming laser beam needs to be closely controlled as to its power, in the frequency band of interest. Here, however, a power stability in the order of 10^{-7} is seen to be sufficient [18].

Beam purity Any geometrical asymmetry of the Michelson interferometer will make it prone to noise from geometrical fluctuations of the laser beam. Thus the illumination of the Michelson is required to be an almost pure TEM_{00} mode. For small light powers, below 1 W as in the space project, this purity can be gotten by passing the light through a single-mode fiber. For the laser powers needed in the ground-based interferometers, however, a 'mode-cleaner' is used: a non-degenerate cavity that is tuned for the TEM_{00} mode, but suppresses the (time dependent) lateral modes that represent fluctuations in position, orientation, and width of the beam [19].

4.2 Thermal Noise

All optical components – and in particular the mirrors – will cause fluctuations in the optical paths also due to their thermal vibrations, their *Brownian motion*. The noise coming from the pendulum modes of motion is most prominent at low frequencies, rolling off steeply towards higher frequencies. The noise due to the vibrational modes of the substrate rolls off less steeply and is thus a serious disturbance at intermediate frequencies. By choice of materials (high mechanical Q) and appropriate shaping of the substrates (to keep their resonant frequencies above our kHz range) the effect of these thermal motions can be reduced.

Intensive research is going into the development and choice of optimal materials for the mirror substrates (pure fused silica, sapphire), and the proper treatment for attaining the highest mechanical Q, e.g. several times 10^7. Such high values can be maintained only if the bonding to the suspension 'wires' does not introduce losses. Special bonding techniques are required using fibers of material identical to the substrate (monolithic suspension). Efficient collaboration between the European groups, under the leadership of the University of Glasgow, has given very promising results.

Thermoelastic noise   Only recently [20,21], the effect of thermoelastic losses in the substrate material was recognized as another serious noise contribution. It scales with the thermal expansion coefficient squared and linearly with the thermal conductivity (which so far we always wanted to be high). This noise has thus become an important issue in the choice of appropriate materials, and it may even rule out the otherwise ideal sapphire.

4.3   Seismic Noise

The mirrors between which the distances are to be monitored are suspended as pendulums in vacuum, to isolate them from extraneous vibrations: from seismic and acoustic noise. Combinations of various schemes (pendulum suspension, lead-and-rubber stacks, active position control) are used to reduce seismic noise by many powers of 10, which is relatively easy for frequencies above, say, 100 Hz. It is only with extreme effort that this lower frequency bound can be lowered to 10 Hz or less. Not only does the natural noise rise drastically towards low frequencies, but also the pendulum isolation becomes less effective. This causes the very steep rise to low frequencies in the righthand sensitivity curve in Fig. 2. VIRGO has developed an extremely powerful seismic isolation system, the 'super-attenuator', consisting of a series of 6 successive pendulum stages, in conjunction with an 'inverted pendulum', and an active isolation stage. This suspension will allow to extend GW search to lower frequencies than other terrestrial detectors.

For both of the *thermal* noise effects, the internal vibrations of the mirrors as well as the pendulation mode, and also for the *seismic* disturbances, the sensitivity goal in strain of $h = 2\delta\ell/\ell$ can only be reached if we choose the armlength ℓ long enough. This is where our need for kilometer dimensions comes from. The steep rise at the left-hand side of the sensitivity curve "Earth" in Fig. 2 is mainly due to the seismic and vibrational noise.

4.4   Shot Noise

Particularly at higher frequencies, the sensitivity is limited by another fundamental source of noise, the so-called shot noise, a fluctuation in the measured interference coming from the "graininess" of the light. These statistical fluctuations fake apparent fluctuations in the optical path difference ΔL that are inversely proportional to the square root of the light power P used in the interferometer. The spectral density (in the 'linear' form we prefer) of the fluctuations of the path difference, $\widetilde{\Delta L}$, are described by

$$\widetilde{\Delta L} = \left(\frac{\hbar c}{2\pi}\frac{\lambda}{\eta P}\right)^{1/2} \tag{3}$$

where η is the conversion efficiency of the photo diode, and λ the laser's wavelength. For measuring the minute changes of the order of $\Delta L \sim 10^{-18}$ m in our kilometric "advanced" detectors, as much as 1 MW of light power, in the visible or the near infrared, would be required. This is not as unrealistic as it may sound; it can be realized by the concept of "power recycling".

Power recycling   The laser interferometers are planned to monitor the (gravitational-wave induced) changes δL of the light path by observing the dark fringe of the interferometer in one output port. The (unused) light going out at the other port of the beam splitter can be fed back, via a mirror M_P, and in correct phase with the incoming light (Figs. 3, 4), so that the circulating light power is significantly enhanced. This scheme was proposed by Ron Drever in 1981, at the same time as Roland Schilling saw it come as a natural consequence in the Garching 30 m prototype, where the appropriate feedback had already been implemented for an efficient frequency stabilisation of the laser. The first implementations were done in that Garching prototype: 1987 with only short arms, in 1996 with the full 30 m arm length [22].

To achieve the sensitivity goals of the current generation of gravitational wave detectors, the light power circulating in the interferometer needs to be of the order 10 kW. With lasers of 10 to 100 W and power recycling gains of 100 to 1000, such values are within current technology. The 1 MW possibly needed for the "advanced" detectors will, however, require some technological break-through, particularly in the preparation of sufficiently loss-free substrate materials.

The shot noise limit Shot noise is a 'white' noise, but as the response in Eq. (2) rolls off as $1/f\tau$ at frequencies above the inverse storage time τ, the apparent strain noise rises proportional to frequency, as shown in the curves 'Space' and 'Earth' in Fig. 2. As we will see later, this frequency-proportional rise of the sensitivity curve will limit the space-borne interferometers in a similar way as here in the ground-based detectors.

4.5   Advanced Interferometry Configurations

An additional "recycling" scheme was later proposed by Brian Meers, and now forms the baseline for the GEO 600 interferometer: 'signal recycling' (SR) [12,13]. A further mirror, M_S, is added to the interferometer, this one in the output port (Figs. 3, 4). The microscopic position of this mirror can be adjusted such that the signal sideband is also resonant in the interferometer, providing an enhancement of the signal, with possibly reduced measuring bandwidth. Schemes like this "signal recycling (SR)", or the related "resonant sideband extraction (RSE)" [14], are expected to be employed in future upgrades also of the other detectors. They can be used to optimize and tune the detector bandwidth independently of the carrier storage time in the arm cavities. Experimental verification of 'dual recycling', the combination of power and signal recycling, was given at Garching in 1995, albeit only in a table-top setup [23]. The curve *"Earth"* in Fig. 2 indicates the sensitivities that will eventually be reached with the current large interferometers, at least in their advanced versions.

4.6   Next-Generation Ground-Based Detectors

Even though the current detectors are not yet in full operation, it is essential to develop a next generation of detectors early on. The study of new technologies to be employed, of new materials, of advanced interferometric configurations has to be pushed forward, so that the necessary new implementations can be undertaken in or around the year 2005.

Three plans for such next-generation detectors have been put forward, which are entered in the lower part of Table 1: Advanced LIGO, LCGT, and EURO. The status of these three future projects will be sketched below.

Advanced LIGO   The proposed US project, "Advanced LIGO", is furthest progressed [24]. It makes full use of the common efforts in the LIGO Scientific Collaboration, LSC. Advanced LIGO will rely on the existing facilities at the sites of Hanford and Livingston. The advantage is clear: no cost for new sites, for civil and vacuum engineering. A draw-back is that the incorporation of more "aggressive" approaches (cryogenics, all-refractive optics, Sagnac) is not so easy to realize, and it loses the option of lower seismic noise of underground sites.

Simulations by the Advanced-LIGO groups of LSC indicate that an operation limited only by the optics noise (shot noise, radiation pressure noise) appears possible. The suspension would have to be modeled after the GEO 600 triple pendulum concept, mirrors be made from large substrates of sapphire (or YAG), and the schemes of SR or RSE, developed at GEO, have to be used.

Buonanno and Chen [25] have shown that the 'detuned' implementation of SR/RSE can even lead to a (moderate) reduction of what is usually termed the 'Standard Quantum Limit', SQL, and Chen showed that a Sagnac configuration lends itself 'easily' to such a scheme of beating the SQL [26].

LCGT   The concept of the Japanese project 'Large Cryogenic Gravitational-Wave Telescope' (LCGT) is also rather well defined [27]; it will use super-cooled (cryogenic) mirrors, the armlength will be 3 km. The location of LCGT, deep inside the mountain that houses the neutrino detector Super-Kamiokande, has ground noise nearly two orders of magnitude lower than at ground level.

EURO   The four funding agencies (CNRS, MPG, INFN, PPARC) of France, Germany, Italy, and the UK, agreed to pursue the definition of a common European high-sensitivity detector, EURO [28]. However, with the completion and the commissioning of the current projects, GEO 600 and VIRGO, as the highest priority, the beginning of the project may be as late as 2008. A site deep underground (as for LCGT) would be preferred. Simulations showed that an operation limited only by the (quantum-)optical noise seems possible.

4.7   The Technological Challenge in Advanced Detectors

The technology required in the proposed future detectors is at the forefront of current state-of-the-art, and even beyond. Many new developments are being pushed just by this goal of improved gravitational wave detectors, but fortunately some developments required are also driven by commercial interest.

Optical properties   Of great importance is the effort to develop better optical materials. The high light powers (up to the order of Megawatt!) will require extremely low absorption losses, in the reflective and anti-reflective coatings, as well as in the bulk material of the substrate. Three materials have the greatest promise: (1) very pure fused silica, specially prepared to have low OH-content, for low absorption; (2) sapphire single crystals, having a naturally low absorption, but with a high constant of birefringence; (3) YAG (Yttrium aluminum garnet) as used for laser crystals.

For all three materials, the great technological challenge is to produce substrates of the order of half a meter in diameter, and of similar thickness, with a high level of homogeneity. For substrates of the end mirrors, which do not require light transmission, also silicon is a possible option. Proposals are also being made to use all-diffractive optics to avoid the problems of light transmission altogether.

At the same time, the optical components need to have extremely high mechanical quality Q. Investigation into the properties of the materials are carried out by several institutions, and values of the order $Q \sim 10^7$ at room temperature and 10^9 at cryogenic temperatures have been accomplished [29,30].

5   The Space Interferometer LISA

Only a space mission allows us to investigate the gravitational wave spectrum at very low frequencies. For all ground-based measurements, there is a natural, insurmountable boundary towards lower frequencies. This is given by the (unshieldable) effects due to varying gravity gradients of terrestrial origin: moving objects, meteorological phenomena, as well as motions inside the Earth. To overcome this, the only choice is to go far enough away, either into a wide orbit around the Earth, or better yet further out into interplanetary space. Once in outer space, we have some great benefits for free: to get rid of terrestrial seismic and gravity gradient noise, to have excellent vacuum along the arms, and in particular to be able to choose the arm length large enough to match the frequency of the astrophysical sources we want to observe.

5.1   The LISA Configuration

ESA (the European Space Agency), and NASA have agreed to collaborate on such a space mission called LISA, "Laser Interferometer Space Antenna" [31,32].

The LISA spacecraft   LISA consists of three identical spacecraft, placed at the corners of an equilateral triangle (Fig. 5). The sides are to be 5 million

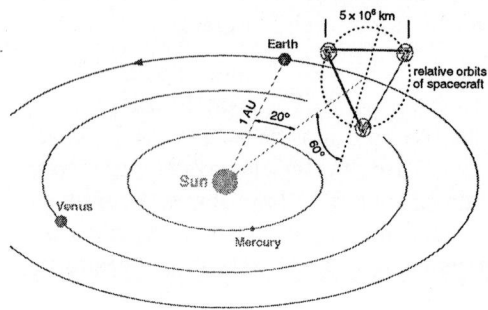

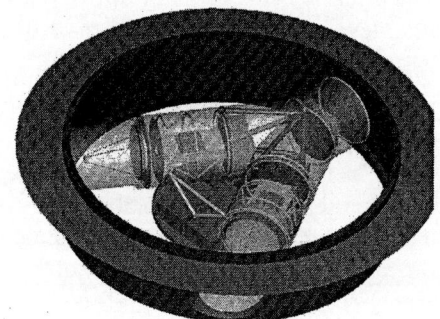

Fig. 6. View of one LISA spacecraft, housing two optical assemblies. The solar panel at top not shown, the thermal shield shown as semi-transparent

Fig. 5. Orbits of the three spacecraft of LISA, trailing the Earth by 20°. The triangle arms are scaled by factor 10

km long (5×10^9 m). This triangular constellation is to revolve around the Sun in an Earth-like orbit, about 20° (i.e. roughly 50 million km) behind the Earth. The plane of this equilateral triangle needs to have an inclination of 60° with respect to the ecliptic to make the common rotation of the triangle most uniform. The small orbit-correction manoeuvres required can be made with field-effect ion thrusters. The three spacecraft form a total of three, but not independent, Michelson-type interferometers, here of course with 60° between the arms.

The spacecraft at each corner will have two optical assemblies subtending an angle of 60°, which are pointed to the two other spacecraft (indicated in Fig. 6, with the Y-shaped thermal shields shown semi-transparent). An optical bench, with the test-mass cage at its center, can be seen in the middle of each of the two arms, and a telescope of 30 cm diameter at the outer ends. Each of the spacecraft has two separate lasers that are phase-locked so as to represent the "beam-splitter" of a Michelson interferometer.

The annual orbit of LISA   During its yearly motion around the sun, the three spacecraft of LISA will 'roll' on a cone of half-angle 60°, each spacecraft moving on a slightly elliptic and slightly inclined orbit around the sun.

This configuration has a number of advantages that make several of the design requirements less stringent. (1) The spacecraft face the sun at a constant angle of incidence of 30°, which provides a very stable thermal environment for the sensitive parts (optical assembly, the sensors) of the spacecraft. (2) The rather unperturbed orbits of the three spacecraft provide a very stable configuration, very close to an equilateral triangle. Thus it is relatively easy to devise articulation schemes for the two 'telescopes' in each spacecraft to follow these small-angle deviations. (3) The center of the LISA triangle trails the Earth in its orbit by 20°, or about 50 million km. This makes the distance to Earth, for radio communication, also quite stable, which reduces the problems in radio antenna design and radio transmission power.

5.2   The 'Drag-Free' Operation

The distances between the different spacecraft are measured from test masses housed *drag-free* in these three spacecraft. The three LISA spacecraft each contain two test masses, one for each arm forming the link to another LISA spacecraft. The test masses, 4 cm cubes made of an Au/Pt alloy of low magnetic susceptibility, reflect the light coming from the YAG laser and define the reference mirror of the interferometer arm. These test masses are to be freely floating in space.

Gravitational sensor   For this purpose these test masses are also used as inertial references for the drag-free control of the spacecraft that constitutes a shield to external forces. Development of these sensors is done at various institutions [33,34]. These sensors feature a three-axis electrostatic suspension of the test mass with capacitive position and attitude sensing. A resolution of $10^{-9}\,\mathrm{m}/\sqrt{\mathrm{Hz}}$ is needed to limit the disturbances induced by relative motions of the spacecraft with respect to the test mass: for instance the disturbances due to the spacecraft self-gravity or to the test-mass charge.

FEEP thrusters   The very weak forces required to keep up drag-free operation, less than $100\,\mu\mathrm{N}$, are to be supplied by field-effect electrical propulsion (FEEP) devices: a strong electrical field forms the surface of liquid metal (Cs or In) into a cusp from which ions are accelerated to propagate into space with a velocity (of the order $60\,\mathrm{km/s}$) depending on the applied voltage. Such FEEP thrusters have been developed at various European institutions, namely Centrospazio, Italy [35] and Seibersdorf, Austria [36]; their characteristics will be studied in a technology demonstration mission (Section 5.5)

5.3   Noise in LISA

This section will cover some of the most worrying noise sources in the LISA project, which then will be relevant also for LISA follow-on missions and for the Chinese project ASTROD.

Figure 2 showed sensitivity curves for the ground-based interferometers, as well as for LISA. In both cases the shape is that of a trough, with a steeper slope at the left than on the right. The curve for LISA is again shown in Fig. 7, enlarged and in greater detail. That LISA sensitivity curve consists of three main parts, as indicated by the three differently shaded frequency regions, in which different noise mechanisms take hold.

Shot noise   With the 30 cm optics planned, from 1 W of infrared laser power transmitted, only some $10^{-10}\,\mathrm{W}$ will be received after 5 million km, and it would be hopeless to have that light reflected back to the central spacecraft. Instead, also the distant spacecraft are equipped with lasers of their own, phase-locked to the incoming laser beam [37,38].

Due to the low level of light power received, shot noise plays a major role in the total noise budget of spurious displacements. It is responsible for the flat middle part of the sensitivity curve.

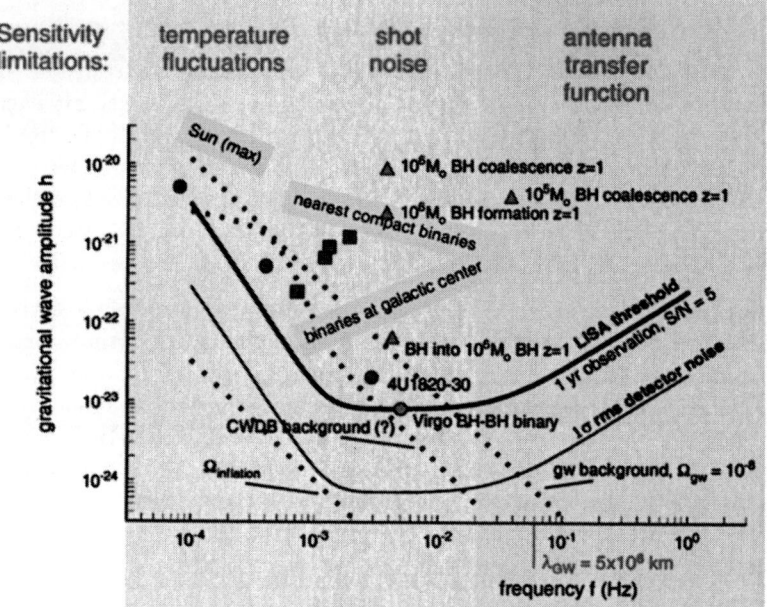

Fig. 7. Sensitivity of LISA: the heavy curve "LISA threshold" represents the signal strength that would provide a signal-to-noise ratio of 5 if averaged over one year, and over all possible directions and polarization angles. Major noise contributions are indicated by different shading

The effect of shot noise is a spurious 'path difference' $\widetilde{\delta L}$ inversely proportional to the square root of the power P available for interferometry, as given in Eq. 3. In the case of LISA, at arm lengths of 5 million km, this received power is of the order 10^{-10} W. With an increased armlength, perhaps to the order of 2 AU, i.e. 300 million km, the power would decrease by a factor of 60^2, and both the apparent spurious path differences $\widetilde{\delta L}$ and the optical path L would thus increase by an identical factor of 60. This means that the sensitivity of a space probe, other characteristics remaining the same, would have a shot noise limit for the strain $h \sim \delta L/L$ that is independent of the armlength, $L/2$. This fact will be of importance in the sensitivity estimates of LISA follow-on missions with different arm lengths and for ASTROD.

Antenna transfer function Again, as shown in Section 4.4, we have to consider that the antenna response rolls off as $1/f\tau$ at frequencies f above the inverse of the round-trip time τ. (See also Eq. (2).) Thus at these frequencies the shot noise leads to the frequency-proportional rise at the right-hand side of the sensitivity curve shown in Fig. 2 and, in more detail, in Fig. 7.

5.4   Acceleration Noise

At frequencies below 1 mHz, the noise is mainly due to accelerations of the test mass that cannot be shielded even by the drag-free scheme: forces due to gravitating masses on the spacecraft when temperature changes their distances, charging of the test masses due to cosmic radiation, residual gas in the test mass housing. Except for the cosmic ray charging, the acceleration noise contributions are dependent on temperature variations, and this is why in Fig. 7 they come under the heading 'temperature fluctuations'.

These accelerations have a rather 'white' spectral distribution, which thus results in position errors rolling off roughly as $1/f^2$.

Gravity gradient noise   The test mass, housed in the LISA spacecraft, is subject to the gravity field of the other masses that form part of the space-craft. These masses, though 'rigidly' connected to each other, will undergo small changes in their positions, due, e.g., to the changes in temperature distribution.

This thermal distortion of the spacecraft actually is one of the most prom-inent sources of 'acceleration noise'. Elaborate calculations on the temperature fluctuations to be expected (e.g. from variations in the solar radiation) and on the thermal behavior of the spacecraft's masses have resulted in a set of require-ments for the LISA design [32].

Noise due to charging of the test mass   Cosmic radiation will cause the test mass to acquire an electrical charge, which will result in a number of noise effects. A broad discussion is given in the LISA Pre-Phase A Study (PPA2) [31].

These charges will give rise to electrostatic forces of attraction to the cage walls. The charges will also, if not perfectly shielded by the cage and the space-craft shields, be subject to Lorentz forces due to LISA's motion in the inter-planetary magnetic field. And, similarly, changes in that magnetic field will also produce forces on the test mass.

As remedies, the test mass will be quite well shielded from outside fields, and in particular, the charge that has accumulated on the test mass will be monitored, and from time to time a discharge by shining ultraviolet light on the test mass, will be carried out [31].

Noise due to residual gas   A very wide field of acceleration noise contri-butions is due to the residual gas inside the sensor. Although the vacuum will have high quality, 10^{-8} mbar $= 10^{-6}$ Pa, the test mass will be subject to several non-negligible accelerations.

Foremost among these can be the stochastic noise due to the buffeting by the impinging residual gas molecules. This statistical noise is proportional to the square root of the residual gas pressure, p.

If the casing of the sensor has a temperature gradient, due, e.g., to changes in solar radiation or in the power dissipation in the spacecraft electronics, dif-ferences in gas pressure inside the sensor will build up. Here we must mention the so-called radiometer effect, but perhaps even more worrisome the effect of temperature-dependent outgassing of the cage walls.

Noise total   With a myriad of other, smaller, noise contributions the total apparent path noise amounts to something like $\widetilde{\delta L} \approx 40 \times 10^{-12}\,\mathrm{m}/\sqrt{\mathrm{Hz}}$ at the lowest part, the bottom of the trough. For signals monitored over a considerable fraction of a year, the best sensitivity is about $h \approx 3 \times 10^{-24}$, indicated in Fig. 2 by the curve marked "Space", and in more detail in Fig. 7.

The LISA prospects   Some of the gravitational wave signals are guaranteed to be much larger. Failure to observe them would cast severe doubts on our present understanding of the laws that govern the universe. Successful observation, on the other hand, would give new insight into the origin and development of galaxies, existence and nature of dark matter, and other issues of fundamental physics [1].

5.5   Status of LISA

LISA is approved by ESA as a cornerstone mission under Horizons 2000. A System and Technology Study [32] has substantiated that improved technology, lightweighting, and collaboration with NASA will lead to a considerable reduction of cost. Thus, a new, *faster, cheaper, and better* approach, together with NASA, is being pursued, under the auspices of an international LISA Science Team. Launch is foreseen for 2011, not very long after first operation of the next-generation ground-based detectors. LISA has a nominal lifetime of 2 years, but the equipment and thruster supply are chosen to allow even 10 years of operation.

A collection of papers given at the *Third International LISA Symposium, 2000*, is presented in a special issue of Classical and Quantum Gravity [39]. A similar special issue is forthcoming from the *Fourth LISA Symposium, 2002*.

Technology demonstrator   Some of LISA's essential technologies (gravitational sensor, interferometry, micro-newton thrusters) are to be tested in a mission LTP (LISA Technology Package) on board an ESA SMART-2 satellite.

LTP will contain, on a common optical bench, two gravitational sensors, similar to the one of Section 5.2. The relative motion between the two freely floating test masses will be monitored with high accuracy by interferometry. The sensitivity in this (scaled-down) experiment will come to within one power of ten to the proposed LISA sensitivity.

LTP is to be flown in a geocentric orbit relatively far away from Earth, so as to avoid the many disturbances near the Earth. The same mission SMART-2 might also host the NASA probe ST-7. Launch is definitely set for August 2006.

LISA follow-on   Even as early as now concepts are being discussed for a successor to LISA, on the possible enhancements in sensitivity and/or frequency band. One scheme would try to bridge the frequency gap between ground and space detectors, by reducing the arm lengths, leaving the general configuration unchanged. Another concept is to have a square constellation instead of the triangle, providing pairs of independent interferometers. These can be used to detect and measure a stochastic background of gravitational waves, similar to, but reaching much further back than the 3 K electromagnetic background radiation.

ASTROD, Astrodynamical Space Test of Relativity using Optical Devices, is the name of a Chinese space project that will, among other relativistic experiments, attempt to measure gravitational waves with arm-lengths of the order of 1 to 2 astronomical units. It will thus extend to a low-frequency range not fully covered by LISA, and thus it would – given similar sensitivity – be a further useful extension in the search for and measurement of gravitational waves [40].

5.6 LISA Data Analysis

Due to the low frequency band of the LISA detection, the data rate is rather low, and thus the total amount of data. Data will be collected on-board, and transmitted to Earth once per two to three days.

The low transfer rate from spacecraft to Earth would not allow much more than the envisaged 1.5 kbit/sec. But such a low data rate is possible only due to a great load of data analysis and data reduction being done on-board. That is particularly true if the schemes of 'time-delay interferometry' are to be utilized.

Directivity   LISA, as all interferometric GW detectors, has a preferred direction and a preferred polarization of the incoming gravitational wave. This would cause an antenna, fixed in space, to be particularly sensitive in some directions, and totally blind in others. The annual motion of LISA will, however, average out these types of directivity, as LISA is facing different locations at the sky, and with different preferred polarization directions at different times. This averaging is why the sensitivity curves in Fig. 7 for a signal-to-noise ratio of 5 is drawn by factor of $5\sqrt{5} = 11.2$ higher than the lower curve.

On the other hand, LISA's detection can make use of the 'signature' that continuous-wave signals will have, due to the changing response sensitivity, and due to the Doppler shifts that the signal will undergo as LISA approaches and recedes from the source during its annual orbit. A detailed analysis of the LISA sensitivity under these assumptions was made by Schilling [41,31]. One important result was that the drastic drops in sensitivity for gravitational waves with wavelengths fitting into the armlengths (see Eq. (2)) are benignly smoothed out in this averaging.

Noise due to fluctuating laser frequency   The strength of the Michelson-interferometer scheme is that the high symmetry between the two arms makes the interferometer insensitive to a number of fluctuations of the illuminating light source. The most serious of these is the fluctuation in laser phase, $\delta\phi$, or in frequency, $\delta\nu$. Any change in laser frequency will cause spurious signals proportional to the difference in arm lengths.

The celestial mechanics of the LISA orbits will cause relative armlength variations in the order of 10^{-2}, and these would produce spurious signals from the natural laser frequency fluctuations, well above the gravitational wave signals.

Unequal-armlength interferometry   Even if the laser frequency is well stabilized to the best of current technology, perhaps to $30\,\mathrm{Hz}/\sqrt{\mathrm{Hz}}$, a drastic further reduction of the effect is required. Here, schemes proposed by Giampieri et al. [42] (in the frequency domain), and then optimized with respect to the suppression of several LISA error sources [43–45] (in the time domain), promise significant improvement. The basic principle is to use a linear combination of the current read-out data s_i with data additionally delayed, in each arm by the travel time in the other:

$$X(t) = s_1(t) - s_2(t) - s_1(t - 2\tau_2) + s_2(t - 2\tau_1)\,, \tag{4}$$

where the delays τ_i are chosen to equal the true travel times T_i. It is easily verified that this algorithm can fully cancel the laser phase noise $\delta\phi(t)$.

The LISA analysis algorithms   How powerfully the more sophisticated methods of this *time delay interferometry* cancel out not only laser phase noise, but also other instrumental errors is shown in various papers [43–45]. These form the baseline for the LISA procedure [32]. It is assumed that phase measurements are made in all three spacecraft, each equipped with independent lasers, with independent highly stable clocks (USOs), and with an intraspacecraft link between the two lasers on board each spacecraft. An extension of the method to intentional gross armlength differences is in print [46].

6   Conclusion

The difficulties (and thus the great challenges) of gravitational wave detection stem from the fact that gravitational waves have so little interaction with matter (and space), and thus also with the measuring apparatus. Great scientific and technological efforts, large detectors, and a working international collaboration are required to detect and to measure this elusive type of radiation.

And yet – just on account of their weak interaction – gravitational waves (just as neutrinos) can give us knowledge about cosmic events to which the electromagnetic window will be closed forever.

This goes for the processes in the (millisecond) moments of a supernova collapse, as well as of the many mergers of binaries that might be hidden by galactic dust. Such high-frequency events (a few Hz up to a few kHz) will be accessible from the detectors on Earth. For the signals to be significant, a number of ground-based detectors should be operated in coincidence, and only such joint analyses will allow to locate the source in the sky.

The perspective of detecting events with non-electromagnetic radiation also holds for the distant, but violent, mergers of galaxies and their central (super)massive black holes. The low frequencies ($10^{-5}\,\mathrm{Hz}$ to $1\,\mathrm{Hz}$) characteristic of such sources are accessible only from space, e.g. with LISA. The expected high signal-to-noise ratios will allow unquestionable detection with only one detector. The very significant 'signature' of the signal, due to the LISA orbits around the sun, will even allow to locate the source in a narrow region in the sky.

A LISA follow-on mission, but also combinations of terrestrial detectors, might probe the GW background from the very beginning of our universe, back to 10^{-14}s or even only 10^{-22}s after the big bang [47].

In this way, gravitational wave detection can be regarded as a new window to the universe, but to open this window we must continue on our way in building and perfecting our antennas. It will only be after these large interferometers are completed (and perhaps even only after the next generation of detectors) that we can reap the fruits of this enormous effort: a sensitivity that will allow us to look far beyond our own galaxy, perhaps to the very limits of the universe.

References

1. B.F. Schutz: this issue
2. A. Einstein: Sitzungsber. Preuss. Akad. Wiss., 688–696 (1916);   154–167 (1918).
3. B.F. Schutz: *Lighthouses of gravitational wave astronomy*, in: *Lighthouses of the Universe*, ESO Astrophysics Symposia (2002) 207–224.
4. R. Weiss: *Electromagnetically coupled broadband gravitational antenna*, in: *Quart. Progr. Rep., Research Laboratory of Electronics*, MIT **105**, 54–76 (1972).
5. P. Beyersdorf: *The Polarization Sagnac interferometer for gravitational-wave detection*, Ph.D. Thesis, Stanford University, 2001.
6. D. Shoemaker et al.: Phys. Rev. D **38**, 423–432 (1988).
7. A. Abramovici et al.: Science **256**, 325–333 (1992).
8. D. Sigg et al.: Class. Quantum Grav. **19**, 1429–1435 (2002).
9. F. Acernese et al.: Class. Quantum Grav. **19**, 1421–1428 (2002).
10. A. Rüdiger and K. Danzmann: *The GEO 600 gravitational wave detector – status, research, development*, in: *Gyros, clocks, interferometers: Testing relativistic gravity in space*, Lecture Notes in Physics **562**, 131–140 (2001).
11. B. Willke et al.: Class. Quantum Grav. **19**, 1377–1387 (2002).
12. B.J. Meers: Phys. Rev. D **38**, 2317–2326 (1988).
13. G. Heinzel et al.: Class. Quantum Grav. **19**, 1547–1553 (2002).
14. J. Mizuno et al.: Phys. Lett. A **175**, 273–276 (1993).
15. K. Kuroda et al.: Class. Quantum Grav. **19**, 1237–1245 (2002).
16. about 1000 hour run: http://tamago.nao.ac.jp/tama/daq/recom/recom3/
17. D.E. McClelland et al.: Class. Quantum Grav. **18**, 4121–4126 (2001).
18. I. Zawischa et al.: Class. Quantum Grav. **19**, 1775–1781 (2002).
19. A. Rüdiger et al.: Opt. Acta **28**, 641–658 (1981).
20. V.B. Braginsky, M.L. Gorodetsky, S.P. Vyatchanin: Phys. Lett. A **264**, 1 (1999).
21. Y.T. Liu, K.S. Thorne: Phys. Rev. D **62**, 122002 1–10 (2000).
22. D. Schnier et al.: Phys. Lett. A **225**, 210–216 (1997).
23. G. Heinzel et al.: Phys. Lett. A **217**, 305–314 (1996).
24. LIGO Laboratory document M000352-00-M (Dec 2000) 117–162.
25. A. Buonanno and Y. Chen: Class. Quantum Grav. **18**, L95–L101 (2001).
26. Y. Chen: submitted to Phys. Rev D; arXiv:gr-qc/0208051
27. K. Kuroda: *Large-scale cryogenic gravitational wave telescope and R&D*, in: Gravitational Wave Detection II, Universal Academy Press (2000) 45–50.
28. A. Rüdiger: ftp://ftp.rzg.mpg.de/pub/grav/geo/georep/euro.ps (1999)
29. M.V. Plissi et al.: Rev. Sci. Instrum. **69**, 3055–3061 (1998).
30. V.B. Braginsky et al.: Rev. Sci. Instrum. **65**, 3771–3774 (1994).

31. LISA Pre-Phase A Report, 2nd edition, Max-Planck-Institut für Quantenoptik, Report 233 (July 1998); often referred to as PPA2.

32. LISA: System and Technology Study Report, ESA document ESA-SCI(2000)11, July 2000, revised as `ftp://ftp.rzg.mpg.de/pub/grav/lisa/sts/sts_1.05.pdf`

33. V. Josselin, M. Rodrigues, P. Touboul: Acta Astronautica **49/2**, 95–103 (2001).

34. A. Cavalleri et al.: Class. Quantum Grav. **18**, 4133–4144 (2001).

35. González, et al.: IECP-91–103.

36. M. Fehringer, F. Rüdenauer, and W. Steiger: 33rd AIAA Joint Propulsion Conf., Seattle (1997), AIAA 97-3057.

37. P.W. McNamara, H. Ward, J. Hough: *Laser phase-locking techniques for LISA: Experimental status*, in: *Laser Interferometer Space Antenna*, W.M. Folkner ed., American Institute of Physics, Woodbury, NY, (1998), pp. 143–147

38. A.-C Liao, W.-T. Ni, J.-T. Shy: Int. J. Mod. Phys. **D11**, 000 (2002)

39. Proceedings Third International LISA Symposium, Golm/Berlin, July 2000, Class. Quantum Grav. **18**, 3965–4164 (2001).

40. W.-T. Ni ed.: *Collection of papers on ASTROD Mission Concept Study*, Center for Gravitation and Cosmology, GP-118.

41. R. Schilling: Class. Quantum Grav. **14**, 1513–1519 (1997).

42. G. Giampieri, R. Hellings, M. Tinto, J. Faller: Opt. Comm. **123**, 669–678 (1996).

43. M. Tinto, J.W. Armstrong: Phys. Rev. D **59**, 102003 (1999).

44. F.B. Estabrook, M. Tinto, J.W. Armstrong: Phys. Rev. D **62**, 042002 (2002).

45. M. Tinto et al.: Phys. Rev. D (2002), in press.

46. S.L. Larson, R.W. Hellings, W.A. Hiscock: Phys. Rev. D (2002) in press.

47. B. Allen: *The stochastic gravity-wave background,* in: *Relativistic gravitation and gravitational radiation,* Cambridge University Press (1997) 373–417 (p. 381/382).

Fundamental Physics with Space Experiments

S. Vitale

Department of Physics, University of Trento, I-38050 Povo, Trento, Italy

Abstract. I review a category of experiments in fundamental physics that need space as a laboratory. All these experiments have in common the need of a very low gravity environment to achieve as an ideal free fall as possible: LISA, the gravitational wave observatory, and its technology demonstrator SMART-2. The satellite tests of the equivalence principle Microscope, and the ultimate sensitivity one STEP, with its close heritage from GP-B, the experiment to measure the gravito-magnetic field of the Earth. Finally the entirely new field of cold atoms in space with its promise to produce the next generation of inertial gravitational and inertial sensors for general relativity experiments.

1 Introduction

Over the last decades a new category of experiments has been developed, that require space as their natural laboratory. These are the experiments in fundamental physics that require one sort or the other of free-fall. By free fall I mean here a purely geodetic motion of some test-body in the sense of Einstein's General Relativity. Thus for instance LISA [1], the interplanetary gravitational wave observatory, requires that the end mirrors of its interferometer arms are in pure, relative geodetic motion to a high accuracy to be able to detect the tiny relative accelerations due to curvature. STEP [2], the satellite test of the equivalence principle, compares the relative accelerations of pairs of test-bodies in the gravitational field of the Earth, in search for an apparent violation of universality of free-fall. Gravity Probe-B [3] or HYPER [4], search for the coupling of Earth spin to the orbital motion of solid spheres and atoms respectively, a curvature effect predicted by General Relativity that can only be detected if one may define the asymptotically flat reference frame.

From an experimental point of view the concept of free fall needs to be qualified: it is obviously impossible to reach a pure geodetic motion on all time scales. A rather impressive low level of non-gravitational differential acceleration can be created in a ground based laboratory environment at frequencies higher than ≈ 10 Hz. The typical goal of ground based gravitational wave experiments is to suppress any non-gravitational relative acceleration of test-bodies between 10 Hz and a few kHz, down to the range of $10^{-14} - 10^{-15}$ ms$^{-2}/\sqrt{\text{Hz}}$. Lower frequencies, at this level of sensitivity, are in practice inaccessible to ground based experiments because of the very steep rise, below ≈ 1 Hz, of the seismic noise of the lab. In addition, for gravitational wave detectors, the gravitational field due to large masses moving nearby also increases very steeply at low frequency and obscures the signal due to outer sources.

Digging into this "1 Hz wall" is very difficult and the best performance to date has been achieved by carefully compensated torsion pendulums. Thanks to their high rejection of the translational acceleration of the inertial member, and to their very low sensitivity to all gravitational moments, they approach a sensitivity around $10^{-12} - 10^{-13}$ N/$\sqrt{\text{Hz}}$ in the mHz range [5].

All the space experiments mentioned above, on the contrary, aim at reaching very low levels of non-gravitational acceleration at frequency well below 1 Hz, down to 0.1 mHz. For instance LISA and STEP requires a differential acceleration noise of $\approx 3 \times 10^{-15}$ ms^{-2}/$\sqrt{\text{Hz}}$, for kilogram size masses, between 0.1 mHz and ≈ 10 mHz. The LISA Technology Package (LTP) on board SMART-2 [6] (see below) wants to test this performance at 1 mHz within one order of magnitude, i.e. at $\approx 3 \times 10^{-14}$ ms^{-2}/$\sqrt{\text{Hz}}$. Microscope [7] has a less ambitious goal of $\approx 3 \times 10^{-12}$ ms^{-2}/$\sqrt{\text{Hz}}$ in the 0.1-5 mHz range very close to the performance of the unsupported gyroscope of GP-B.

It is worth noticing that to achieve these levels of inertial motion, one cannot just simply suppress the Earth gravity by going to orbit. This is just a first pre-requisite needed to avoid the leakage of such a large force in the degrees of freedom that need to be quiet.

In addition to the 0-g environment, all the missions above share the "drag-free" navigation technique, where the spacecraft follows actively the test-bodies to minimize the disturbance on them. With this I mean that the position of the test-mass relative to a spacecraft fixed reference frame is measured via a displacement sensor, and the information is fed back, within an active control loop, to a thrusters system that moves the spacecraft such to minimize the displacement error signal.

A back-of-the-envelope estimate of the residual acceleration for a spacecraft following a test-mass along one axis, gives:

$$a_p = \frac{f_p}{m} + \omega_p^2 \Delta x = \frac{f_p}{m} + \omega_p^2 \left[\Delta x_n + \frac{F_{S/C}}{M\omega_{fb}^2} \right] \tag{1}$$

where a_p is the parasitic acceleration of the test-mass relative to local inertial frame and f_p stands for any parasitic force acting directly onto the test-mass.

As the parasitic forces on the test-mass may depend on the position, a spring-like coupling with constant k_p, or coupling frequency $\omega_p^2 = k_p/m$, will convert the spacecraft/test-mass relative motion Δx, back to a force on the test-mass. Eq. (1) also shows that Δx is mostly contributed by the displacement sensor noise Δx_n, and by the external forces acting on the spacecraft. The effect of those can only be zero if the gain $M\omega_{fb}^2$ of the drag-free control loop goes to infinity.

2   LISA and SMART-2

Details on LISA appear in the accompanying paper by K. Danzmann in these proceedings. Here we only quickly remind that LISA will be the first gravitational wave antenna in space. LISA consists of 3 satellites in orbit around the Sun. The orbits are selected so that the three satellites, though moving independently of each other, form a stable (to within $\approx$ 5%) equilateral triangle with a side of 5 million km. Each spacecraft contains a pair of test-masses, of approximately 2 kg, nominally in pure free-fall. Each test mass is the end-mirror of a single arm interferometer, the other end-mirror being in another satellite.

As for all gravitational wave detectors, each LISA arm performs an Einstein's geodesic deviation experiment: in the presence of curvature, two particles, like the interferometer end mirrors, placed at different places in space, feel a difference of acceleration

$$\frac{d^2\delta x^i}{dt^2} = c^2 R^i{}_{0j0}\,\delta x^j + \frac{\delta F^i}{m} \qquad (2)$$

where δx^i are the components of the 3-vector that joins the particles, and $R^\mu{}_{\nu\lambda\sigma}$ is the curvature tensor. I have also added in eq. (2) a contribution $\delta F^i/m$ due to all non-gravitational forces that try to drive the particles, of mass m, out of their geodesic trajectories. As usual with this notation, Latin indexes span from 1 to 3 and indicate space components.

For weak signals one can turn the curvature tensor into second derivatives of the metric tensor $g_{\mu\nu} = \eta_{\mu\nu} + h_{\mu\nu}$, with $\eta_{\mu\nu}$ the Minkowsky tensor of flat space-time and $h_{\mu\nu}$ the small deviation due to curvature:

$$\frac{d^2\delta x^i}{dt^2} = \frac{1}{2}\frac{\partial^2 h_{ij}}{\partial l^2}\delta x^j + \frac{\delta F^i}{m} \qquad (3)$$

For many good reasons, in the field of gravitational wave research, the signal amplitude associated with a gravitational wave is expressed in terms of h_{ij}. For instance the triangular interferometer that can be formed by taking the signals from the 3 independent arms of LISA has a sensitivity goal of $S_h^{1/2} \approx 4 \times 10^{-21}$ Hz$^{-1/2}$ at around 3 mHz.

By moving to the frequency domain, eq. (3) allows one then to define an effective noise metric disturbance:

$$h_{ij}^{\text{eff}} \approx \frac{2\delta F^i}{m\omega^2 \delta x^j} \qquad (4)$$

As for LISA $\delta x \approx 5 \times 10^9$ m, one gets that, to achieve the performance goal, $\delta F/m \approx 3 \times 10^{-15}$ ms$^{-2}/\sqrt{\text{Hz}}$. This result is obviously only approximated. A detailed calculation of the noise performance is contained in [1]. Also at higher frequency noise becomes dominated by the laser interferometer phase noise and stray forces become quickly irrelevant.

At lower frequencies the contribution in eq. (4) grows fast at least as $1/f^2$. Thus the level of stray forces one can reach sets the low frequency roll-off of the instrument. The position of this roll-off is a key feature: for LISA many of the most interesting sources, including the guaranteed calibration sources constituted by the galactic binaries [6], are in the mHz range. Thus demonstrating the feasibility of such a low level of parasitic forces becomes a top priority for the whole LISA programme.

Unfortunately this demonstration, as discussed above, cannot be performed on ground. This is why, very early during the various studies for LISA, the need for an in-flight demonstration was recognised. The first definite proposal for such a mission had been studied by the scientific community in 1998 and was known as ELITE [8]. That concept has now been recognized as a sufficient demonstration of the key technological aspects of LISA, and is now planned by ESA for flying, as the LISA Technology Package (LTP), on board SMART-2 in 2006. A similar package, known as the disturbance reduction system (DRS), is being developed by NASA and will fly on SMART-2 also.

The basic idea behind the LTP is to squeeze one LISA arm from 5×10^6 km to a few centimetres and place it aboard a single S/C. Thus two test-masses are in free-flotation within the spacecraft and a laser interferometer measures their relative position (Fig. 1).

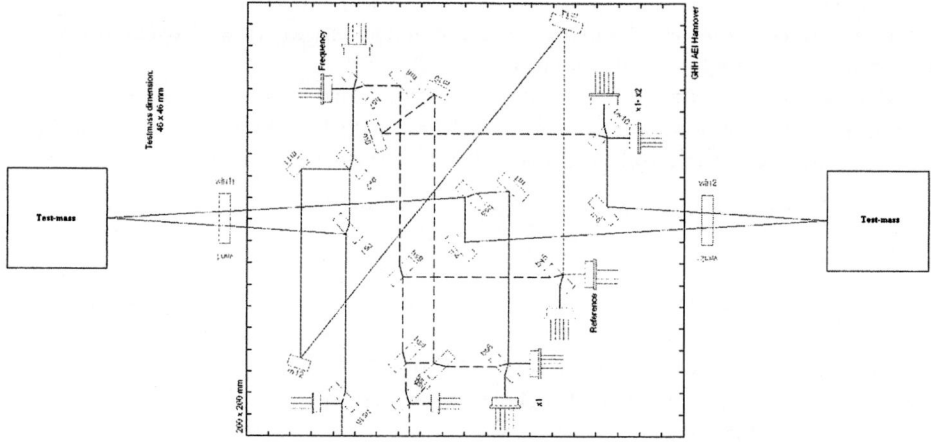

Fig. 1. The concept of the Lisa technology Package: two $46 \times 46 \times 46$ mm^3 gold-platinum test-masses are in free flotation within the spacecraft. An optical interferometer measures the differential displacement of the test-masses with 10pm/$\sqrt{\text{Hz}}$ resolution between 3 mHz and 30 mHz. In the selected baseline, the interferometer also measures the position of one of the test-masses relative to the spacecraft, and a few rotational degrees of freedom

Each test-mass is surrounded by a set of electrodes to measure the relative displacement or rotation of the test-mass relative to the spacecraft (Fig. 2).

This is made via a set of differential parametric bridges that measure the capacitances between the test-mass and the electrodes. Resolutions are different for different degrees of freedom, the best being that along the sensitive axis of

Fig. 2. The set of electrodes surrounding the test-mass. The test mass, not shown, fits into the cavity of the main body shown in the left picture. The lid shown on the left then closes the assembly. The test-mass leaves gap of 3 to 4 mm on all sides and has no mechanical contact to the spacecraft

the interferometer which is of ≈ 1 nm/$\sqrt{\text{Hz}}$. This level of resolution is achieved despite the necessary electric field bias are constrained to be low enough that the resulting spring-like coupling is kept below the requested $\approx 10^{-6}$ N/m level. The set of the test-mass and the sensing electrodes with their front-end electronics is called, within this field, the inertial, or gravitational sensor.

A view of the engineering model under development is given in Fig. 3 and a view of a partial integration of the whole package is shown in Fig. 4.

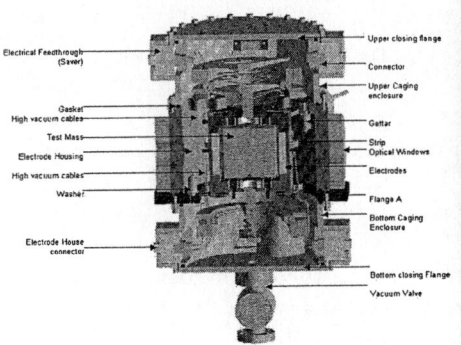
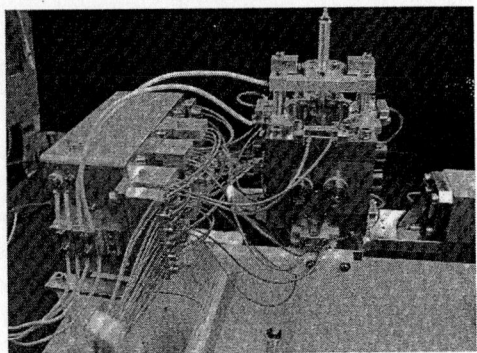

Fig. 3. The engineering model of the inertial sensor under development. Left: a general picture showing the electrode housing, the test-mass launch lock and the high vacuum enclosure. Right: a prototype of the core part under test

SMART-2 will fly in 2006 and will demonstrate that the forces on the test-mass can be reduced to within one order of magnitude of the LISA goal at mHz frequencies. This is a leap of approximately one order of magnitude from what

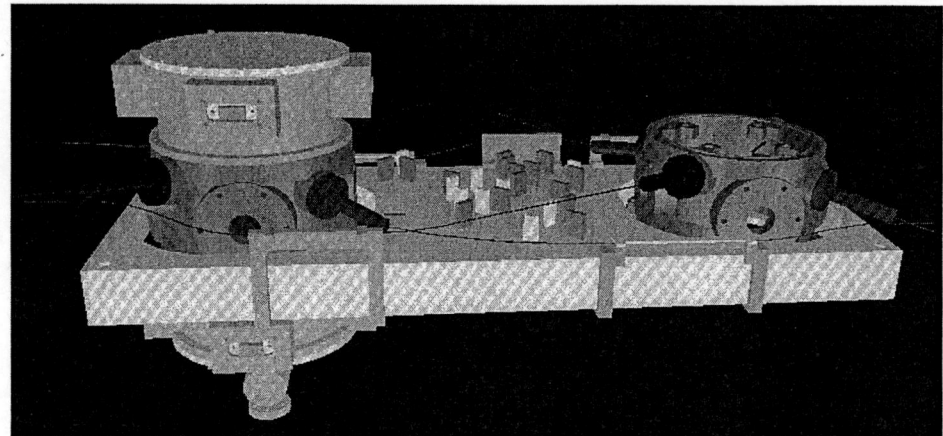

Fig. 4. A view of the LTP in an intermediate integration stage. The left inertial sensor is already integrated within its vacuum enclosure. The dark lines connected to the vacuum enclosure are the optical fibres that carry the UV light needed to discharge the test-masses by photoelectron emission, getting rid of the charge accumulated by cosmic rays. The optical layout of the interferometer is also visible

can be demonstrated on ground by means of torsion pendulums [5]. Thus at each stage, from ground studies, to SMART-2 and from this to LISA in 2011, the validation of the free-fall technology must improve by a reasonably achievable factor of 10.

3   Satellite Tests of Equivalence Principle

Almost all theories that try to reconcile gravitation and quantum physics are forced to call into play the existence of other quanta, besides the one that mediate the known interaction among particles. The action of these quanta may be described in many ways, but one of the most striking predictions is that they should produce new long range interactions between bodies.

For instance the Damour-Polyakov [9] string-based theory predicts that two neutral bodies in free fall in the field of the Earth, and virtually at the same place, would experience a difference of acceleration given by:

$$\eta = -\overline{\gamma}\left[C_B\left(\frac{B}{\mu}\right) + C_D\left(\frac{D}{\mu}\right) + 0.943 \times 10^{-5}\left(\frac{E}{\mu}\right)\right] \tag{5}$$

where $E = Z(Z-1)/(N+Z)^{1/3}$ is the nuclear electromagnetic energy, $B = N+Z$ the baryon number, and $D = N-Z$. $\overline{\gamma} = \gamma-1$, with γ the Eddington parameter. The prediction of eq. (5) contains unfortunately the unknown parameters C_B and C_D.

Ground based experiments have demonstrated that for all measured material pairs the acceleration due to the interaction either with the Sun or with Laboratory sources is the same to within some parts in 10^{13}. The accuracy of these measurements is again limited by the gravitational disturbance in the laboratory.

To go beyond this limit many different experiments in space have been proposed. Two have matured at a level sufficient to either be approved as a flight mission or to seriously compete for it: Microscope, that aims at improving the accuracy by a factor ≈ 100, and STEP with a $\approx 10^5$ times more ambitious goal. Any improvement in the measurements is obviously very interesting, however the chances to detect a violation at 10^{-15} are deemed low by the current wisdom of the theorists of the field.

Both experiments are based on the concept of measuring the difference of acceleration in the proximity of the Earth, between two free-falling test-masses, with their centres of gravity nominally located at the same place. This is meant to be an approximation of having two co-located gravitational monopoles, i.e. two point-like particles or two concentric spherical homogeneous masses.

As the violation of the Equivalence Principle is expected to be connected to the difference in the nuclear composition between the test-masses, ideally one would like to compare as many material pairs as one can carry on board. The compromise with reality limits the choice to 2 pairs for Microscope and to 4 pairs for STEP.

In reality each pair of test-masses constitute a differential accelerometer: the test-masses are elastically bound together and to the spacecraft. The spring may be an active one, as in Microscope where the position of the test-masses is measured by a capacitive sensor, as for SMART-2, and an active feedback loop applies a force to re-centre the test-mass relative to the spacecraft. As an alternative the spring may be passive, as the closed loop superconducting coil that in STEP is both used for measuring the test-mass position and to provide an elastic restoring force.

In both configurations, the science signal is the relative displacement of the test-masses that is simply proportional, at a given frequency, to their relative accelerations. As a dc-signal would be plagued by any sort of very low-frequency noise and inaccuracies, the spacecraft is not held earth-pointing. Instead it rotates around an axis normal to its orbit and to the sensing axes of the differential accelerometers. This way any signal connected to the Earth appears as modulated at the rotation frequency. Rotation frequencies of plus or minus a few times the orbit frequency of $\approx 2 \times 10^{-4}$ Hz are envisaged, including the case of inertial pointing of the spacecraft.

Both missions fly in low Earth orbit. The air drag is then a major source of random forces on the spacecraft that would propagate to the test-masses via their elastic coupling to the spacecraft. This problem is taken care of in a twofold way: first the displacement of the common centre of mass of both test-masses relative to the spacecraft is measured, and a drag-free control loop is closed by using an appropriate set of thrusters. This way the acceleration of this centre of mass is suppressed, for STEP, down to $\approx 10^{-12}$ ms$^{-2}/\sqrt{\text{Hz}}$. The second step is to isolate the differential mode of the test-masses form their common displacement with a rejection factor of 10^{-4}. This is achieved by matching the dynamical coupling

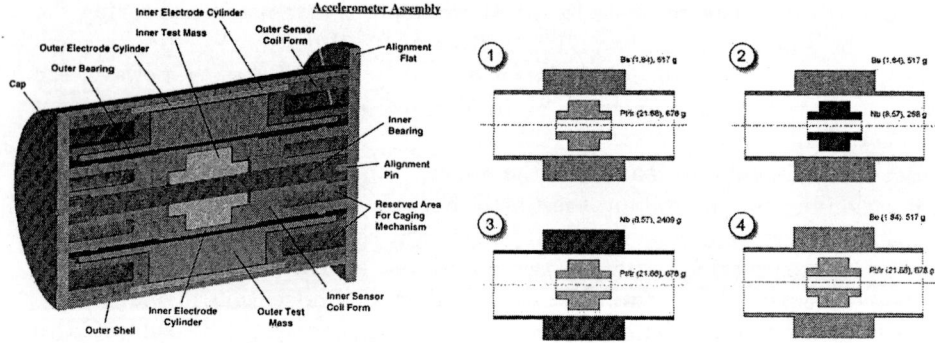

Fig. 5. The differential accelerometer of STEP. Left: schematic of the accelerometer. Test-mass motion along the measurement axis is read by a SQUID based superconducting displacement transducer that directly gives the differential displacement of the test-masses. Another SQUID based sensor reads the common displacement of the test-masses that constitute the drag-free control reference signal. Right: the selected 4 pairs of test-masses: Be-Pt/Ir, Be-Nb, Nb-Pt/Ir, Be-Pt/Ir. Two of the pairs are identical to test for systematics. The shape is a calculated close approximation to a monopole: the differential acceleration due to a point-like source within the spacecraft is $< 10^{-6}$

of the test-masses to the spacecraft. This way the requested residual differential acceleration level is achieved.

The more ambitious STEP is a cryogenic mission. Superfluid helium is used as the coolant, not only to make the SQUID-based superconducting test-mass motion readout to work, but also to achieve very high mechanical stability, high vacuum and in general the suppression of many spurious forces, like Brownian motion, that are connected to temperature.

Superfluid helium vents to the outer space via a set of proportional valves that are used as thrusters for the drag free-loop. Unfortunately helium venting leaves behind a helium gas bubble that, if unmanaged, would lock into the Earth gravitational gradient acting as a huge amplifier of gravitational gradient. To solve the problem, the cryostat is filled with silica Aerogel, a nanometer size porous material with the same density as helium. Capillary forces on helium dominate on gravity even in 1 g and prevent from moving the sub-micrometer size cavities that are formed upon helium boil-off. In addition test-mass shapes have been selected that represent a very close approximation to a monopole. This suppresses (and matches) by many order of magnitudes their coupling to all high order gravitational multipole, including the gradient.

Microscope is an approved mission with a launch in 2005. At the time this paper has been written, STEP, despite a very high rating of its science, failed to be selected for flight within NASA small explorers in 2005 and needs now to find another flight opportunity.

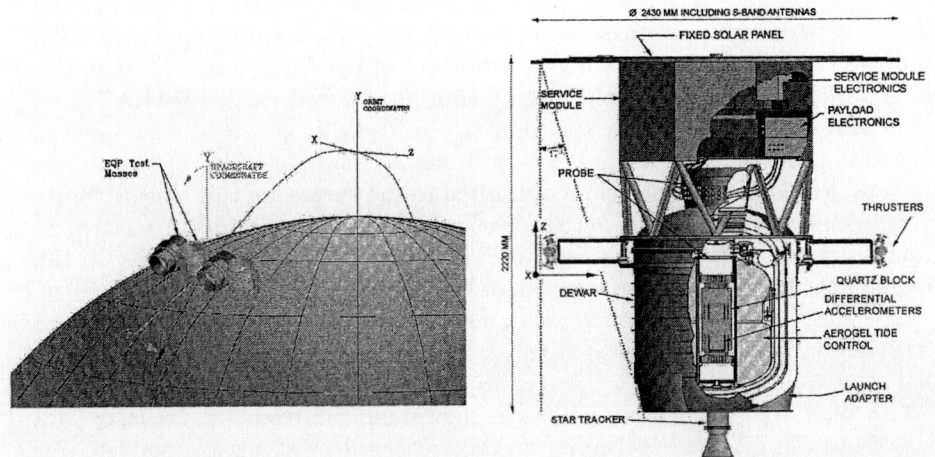

Fig. 6. Left: the configuration of STEP: the spacecraft (not shown), in a $\approx$ 500 km altitude, sun-synchronous orbit, rotates around the z-axis such that the sensitive axis of each accelerometer (the cylindrical symmetry axis of the Test-masses shown) rotates relative to the Earth to modulate the signal. Right: the overall configuration of the STEP spacecraft: the 4 differential accelerometers, hosted into a single high stability quarz-block, are kept at $\approx$ 2 K by the superfluid helium cryostat. The cryostat is filled with Silica Aerogel to prevent the helium tide from generating large gravitational gradient signals. The helium gas venting from the cryostat is used as the propellant of the drag-free loop

4 HYPER and Cold Atoms Physics in Space

Gravitational fields, in the linearised approximation, couple to the phase of quantum wave functions by the minimal coupling rule:

$$\hbar\partial_i\phi = p_i \rightarrow p_i + mch_{oi}$$
$$\hbar\partial_t\phi = E \rightarrow E - \frac{1}{2}mc^2 h_{oo} \tag{6}$$

where p_i is the momentum.

A variety of experiments have been proposed to exploit the consequences of eq. (6) to make quantum detectors of generalised gravitational fields using macroscopic quantum objects. For instance pretty impressive gyroscopes have been made by phase shifts induced in superfluid helium [10] in constrained geometries, in some analogy to what is made with light in gyro-lasers.

It is then of no surprise that the explosion of the field of cold atoms, with the variety of effects connected with phase interference in matter wave interferometer, has immediately produced ideas and practical experiments to detect the phase shifts induced by gravitational and inertial fields on cold atoms wave functions. Accelerometers and gyroscopes are the first natural application of this concept and test attempts have been immediately made, successfully and quickly reaching sensitivities comparable with state-of-art competing techniques.

4.1   Cold Atoms Clocks in Space

The first experiment proposed and approved to fly cold atoms in space, takes advantage in reality just of the 0-g environment of the International Space Station. ACES [11] will fly the cold atom Caesium fountain clock PHARAO, and the 0-g environment will allow the fountain to freely fly for ≈ 10 s as opposed to the ≈ 1 s time fountains achieve on ground. As the line-width of the atomic beat-note that is used for the clock is limited to the inverse of this time of flight, this represents an increase of on the clock accuracy of a factor 10 relative to the performance achievable on ground. The relative frequency stability of the PHARAO clock onboard the International Space Station is expected to be better than 10^{-13} for one second measurement time, 3×10^{-16} for one day and 1×10^{-16} for ten days. This is three orders of magnitude beyond the clocks which are currently flying in GPS satellites.

With such a highest stability clock in space one can push the accuracy of a series of tests in fundamental physics that are based on clock comparison. For instance ACES has a microwave link to ground. Actually providing a universal high stability time reference available everywhere on the Earth surface, through a high accuracy link, is the primary objective of the mission. The existence of a link allows to compare PHARAO to ground based atomic fountains to test for the gravitational red-shift due to the difference of gravitational potential. As both clocks have high absolute accuracy, the experiment will not be based on red-shift modulation due to orbit changes, as in the previous most accurate determination on Gravity Probe A [12]. On the contrary, the absolute measurement of the red-shift will be compared with the precise knowledge on the orbit of the International Space Station, with an expected accuracy improvement of ≈ 25.

Thanks to the ability to compare the ACES Caesium clocks to ground based ones that use atoms with a different Z value, ACES can also put an upper limit to the time derivative of the fine structure constant α, with a projected sensitivity for a 3 years mission of slightly better than $(1/\alpha)(d\alpha/dt) \leq 10^{-16}$/year.

4.2   HYPER

HYPER is a mission proposed to ESA in response to a call for medium size missions. Its main objective is to use a cold atom interferometer as a gyroscope to map the gravito-magnetic field of the Earth. Gravito-magnetism is a prediction of general relativity that associates a field to the intrinsic angular momentum of a body spinning relative to an asymptotically flat reference frame.

For instance in the vicinity of the Earth, in this asymptotically flat frame, the metric would take a term

$$h_o = \frac{2G}{c^3 r^3} \boldsymbol{J}_\oplus \times \boldsymbol{r} \tag{7}$$

where h_o is the 3-vector with components h_{oi} and $J_\oplus$ is the angular momentum of Earth. For reference the maximum effective angular velocity $2\Omega = c\nabla \times h_o$ corresponding to the field in eq. (7). Is of the order of $\Omega \approx 2 \times 10^{-14}$ rad/s near at 500 km altitude.

A closed loop interferometer is sensitive to any h_o, and in particular to rotation, as these cause a phase shift

$$\delta\phi = \frac{cm}{\hbar} \oint h_{oi} \cdot d\ell \qquad (8)$$

For comparison, the phase shift on a light interferometer, like a gyro-laser, would be:

$$\delta\phi = \frac{2\pi\nu}{c} \oint h_{oi} \cdot d\ell \qquad (9)$$

where ν is the frequency of light. The ratio $mc^2/\hbar\nu \approx 6 \times 10^{11}$ of these sensitivities shows how promising a matter wave interferometer is as gyroscope.

Unfortunately, or fortunately, an interferometer is also sensitive to acceleration. In particular, a uniform acceleration $(c^2/2)\nabla h_{oo}$ in-plane to the interferometer would cause a phase shift

$$\delta\phi \approx \frac{c^2 m}{2\hbar v_o} A\nabla h_{oo} \qquad (10)$$

where v_o is the velocity of the atom beam entering the interferometer of effective area A. To suppress this sensitivity, HYPER exploit the different signs that the effect of the acceleration and that of rotation have on two counter-propagating beams.

Thus the basic scheme of HYPER is a unit made of two counter-propagating interferometers as in Fig. 7.

The predicted sensitivity to rotation of such an interferometer is of order 3×10^{-12} rad/$\sqrt{s}$ and thus the Earth gravito-magnetic field needs approximately 3 hours integration to be detected. By the multiple passages given by the Sun-synchronous orbit, a full map of the field can be built up over the mission duration.

The counter-propagating interferometers can only cancel the effect of gravity field to the extent that they have the same area or that the areas are known well enough to allow for post-processing subtraction. This can only occur to some extent. Assuming that this common mode rejection can be done to 0.1%, one would then need to make the spacecraft inertial to $\approx 10^{-10}$ ms$^{-2}/\sqrt{\text{Hz}}$ at the signal frequency in the mHz range. In principle the necessary drag-free loop may be driven by the acceleration signal coming form the interferometer. However, for practical reasons, HYPER carries a set of independent inertial sensors to perform the drag-free control.

HYPER has further scientific objectives. For instance by operating the interferometer in another configuration, one can perform an high accuracy measurement of the fine structure constant. Even more important, a drag-free, inertial oriented platform is the only place where one can test the ultimate phase

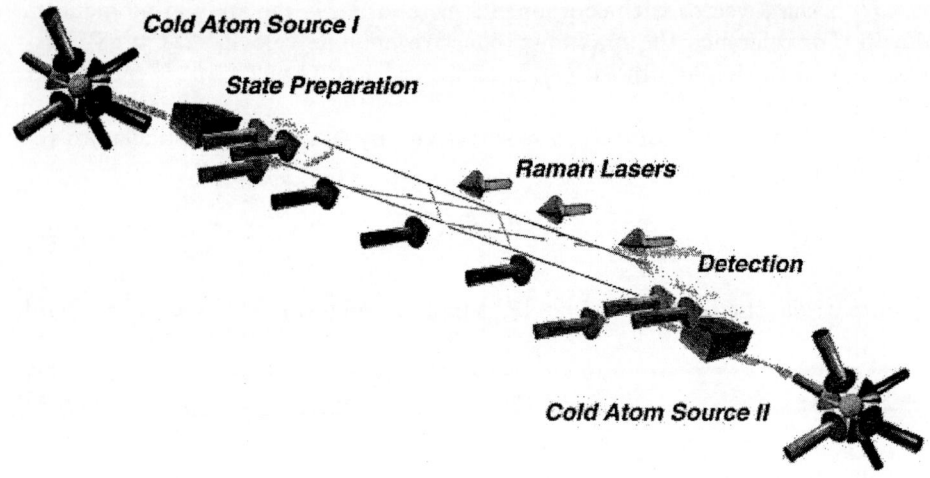

Fig. 7. Concept of the basic matter wave interferometer unit of HYPER, the Atomic Sagnac Unit (ASU). Two counterpropagating, pulsed atomic beams cross three pairs of counter-propagating light waves such that two Mach-Zehnder interferometers are formed. At both ends of the interferometer atoms are prepared in magneto-optical traps and the the fluorescence of one output port of each Mach-Zehnder interferometer is detected. HYPER payload includes two of these units lying within two normal planes

coherence of the matter wave, being inertial and gravitational fields the last, non-shieldable disturbance for a neutral atomic system.

HYPER is still a study. However the field of cold atoms is progressing so quickly that it is easy to predict that the technology of cold atoms interferometer may become ready for a space mission within a decade. Even more, it might well be that a miniaturised matter wave interferometer may become the next generation of inertial sensors to be used to achieve Einstein free-fall in Fundamental Physics experiments.

References

1. A. Hammesfahr, H. Faulks, K. Gebauer, K. Honnen, U. Johann, G. Kahl, M. Kersten, L. Morgenroth, M. Riede, H.-R. Schulte, M. Bisi and S. Cesare, O. Pierre, X. Sembely, L. Vaillon, D. Hayoun, S. Heys, B.J. Kent, F. Rüdenauer, S. Marcuccio, D. Nicolini, L. Maltecca, I. Butler, Jose Rodriguez-Canabal, R. Reinhard, T. Edwards, P. Bender, A. Brillet, A.M. Cruise, C. Cutler, K. Danzmann, F. Fidecaro, W.M. Folkner, J. Hough, P. McNamara, M. Peterseim, D. Robertson, M. Rodrigues, A. Rüdiger, M. Sandford, G. Schafer, R. Schilling, B. Schutz, C. Speake, R.T. Stebbins, T. Sumner, P. Touboul, J.-Y. Vinet, S. Vitale, H. Ward, W. Winkler. In "LISA Laser Interferometer Space Antenna: A Cornerstone Mission for the Observation of Gravitational Waves. System and Technology Study Report" ESA-SCI (2000) 11, (2000)

2. J. Mester, R. Torii, P. Worden, N. Lockerbie, S. Vitale and C.W.F. Everitt. Classical and Quantum Gravity, **18**, 2475 (2001)
3. S. Buchman, C.W.F. Everitt, B. Parkinson, J. P. Turneaure, D. DeBra, D. Bardas, W. Bencze, W. Bencze, R. Brumley, D. Gill, G. Gutt, D.H. Gwo, G.M. Keiser, J. Lipa, J. Lockhart, J. Mester, B. Muhlfeld, M. Taber, S. Wang, Y. Xiao, and P. Zhou, Advances in Space Research, **25**, 1177 (2000)
4. R. Bingham, C. Bordé, P. Bouyer, M. Caldwell, A. Clairon, K. Danzmann, N. Dimarcq, W. Ertmer, J. Helmcke, C. Jentsch, B. Kent,C. Lämmerzahl, A. Landragin, I. Percival, E.M. Rasel, C. Salomon, M. Sandford, W. Schleich, P. Tourrenc, S. Vitale, P. Wolf, R. Reinhard, M. Novara, L. Gerlach, R. Grünagel, S. Santandrea, G. Janin, R. Biesbroek, B. Schreiber, T. Bieler, P. Villar, F. Tonicello, M. Bacchetti, A. Santovincenzo, G. Barbagallo, E. Daganzo Eusebio, P. Plancke, S. Airey, G. Ulbrich, B. Harnisch, J.R. Alarcon Rodriguez, M. Lang, L. Serafini, P. Emanuelli, L. Appolloni, L. Fanchi, R. Henderson, A. Cotellessa, M. Braghin, B. Gardini, M. Bandecchi, in: "HYPER: Hyper-Precision Cold Atom Interferometry in Space, E.M. Rasel et al., ESA Assessment Study Report", ESA-SCI **(2000)** 10, (2000)
5. M. Hueller, A. Cavalleri, R. Dolesi, S. Vitale and W.J. Weber, Class. Quantum Grav. **19**, 1757 (2002)
6. S. Vitale, P. Bender, A. Brillet, S. Buchman, A. Cavalleri, M. Cerdonio, M. Cruise, C. Cutler, K. Danzmann, R. Dolesi, W. Folkner, A. Gianolio, Y. Jafry, G. Hasinger, G. Heinzel, C. Hogan, M. Hueller, J. Hough, S. Phinney, T. Prince, R. Reinhard, D. Richstone, D. Robertson, M. Rodrigues, A. Rüdiger, M. Sandford, R. Shilling, D. Shoemaker, B. Shutz, R. Stebbins, C. Stubbs, T. Sumner, K. Thorne, P. Touboul, H. Ward, W. Weber and W. Winkler, Nuclear Physics **B** (Proc. Suppl.) **110**, 209 (2002)
7. http://www.onera.fr/microscope/index.html
8. ELITE, European LISA Technology Demonstration Satellite, Proposal May 1998 (Unpublished)
9. T. Damour and A. Polyakov, Nucl. Phys. **B 423**, 532 (1994)
10. R. Packard and S. Vitale, Phys. Rev. **B46**, 3540 (1992)
11. C. Salomon, N. Dimarcq, M. Abgrall, A. Clairon, P. Laurent, P. Lemonde, G. Santarelli, P. Uhrich, L.G. Bernier, G. Busca, A. Jornod, P. Thomann, E. Samain, P. Wolf, E. Gonzalez, P. Guillemot, S. Leon, F. Nouel, C. Sirmain, S. Feltham. Comtes Rendus Académie des Sciences, série IV 1-17 (2001)
12. R.F.C. Vessot and M.W. Levine, Journal of General Relativity and Gravitation **10**, 181 (1979)

Astro Particle Physics from Space

Roberto Battiston

Dipartimento di Fisica and Sezione INFN, I-06121 Perugia, Italia

Abstract. We review how some open issues on Astro Particle physics can be studied by space borne experiments in a complementary way to what is being done at underground and accelerators facilities.

1 At the Beginning It Was Astro Particle

Figure 1 shows the conference photo taken at the 1939 Chicago Symposium on Cosmic Rays. Among the participants we find a quite exceptional group of physicists: Kohloster, Bethe, Shapiro, Compton, Teller, Eckart, Gouldsmit, Anderson, Oppenheimer, Hess, Wilson, Rossi, Auger, Heisenberg, Wheeler and many others who are the among the fathers of the modern physics, based on Quantum Mechanics, Elementary Particles and Fundamental Forces. Why Cosmic Rays,

Symposium on Cosmic Ray, 1939 (The University of Chicago, U.S.A.)

Fig. 1. Conference photo from the 1939 Chicago Cosmic Rays Symposium

discovered by Hess nearly 30 years before were, still in 1939, such an interesting topic for these distinguished scientists?

The answer lies in Table 1. In the years preceding 1937 both the first anti-particle (the positron) and the first unstable elementary particles ($\mu^{\pm}$) were discovered in Cosmic Rays. Many more particles were to be discovered during the following years analyzing the Cosmic Radiation, making Cosmic Rays symposia very exciting until at least 1953, when experiments at accelerators started to systematically discover new elementary particles while Cosmic Rays experiments suddenly stopped finding them.

Table 1. Discovery of elementary particles

Particle	Year	Discoverer (Nobel Prize)	Method
e^-	1897	Thomson (1906)	Discharges in gases
p	1919	Rutherford	Natural radioactivity
n	1932	Chadwick (1935)	Natural radioactivity
e^+	1933	Anderson (1936)	Cosmic Rays
$\mu^{\pm}$	1937	Neddermeyer, Anderson	Cosmic Rays
$\pi^{\pm}$	1947	Powell (1950), Occhialini	Cosmic Rays
$K^{\pm}$	1949	Powell (1950)	Cosmic Rays
π^0	1949	Bjorklund	Accelerator
K^0	1951	Armenteros	Cosmic Rays
Λ^0	1951	Armenteros	Cosmic Rays
Δ	1952	Anderson	Cosmic Rays
Ξ^-	1953	Armenteros	Cosmic Rays
$\Sigma^{\pm}$	1953	Bonetti	Cosmic Rays
p^-	1955	Chamberlain, Segre' (1959)	Accelerators
anything else	1955 $\Longrightarrow$ today	various groups	Accelerators
$m_\nu \neq 0$	2000	KAMIOKANDE	Cosmic rays

If Cosmic Rays have been instrumental to give birth to particle physics during the first half of the past century, starting from the fifties, however, accelerators have been the tools for the experimental triumph of the Standard Model of Particle Physics, including the discovery of the electro-weak bosons at CERN or of the heavy sixth quark at Fermilab.

During the last ten years, however, the rate of discoveries at accelerators seems significantly reduced, possibly because of the limited energy scale which can be tested at existing or future facilities. A growing number of physicists is then turning again to CR with new experimental techniques aiming to extend by orders of magnitude the sensitivities reached by past experiments.

In particular a number of space borne experiments have been proposed to measure, with unrivalled accuracy, the composition of primary high energy CR, searching for new phenomena not accessible to present accelerators.

In this paper we will review these experiments and their physics potential.

The paper is organized in two parts. In the first part I will discuss the characteristics of the CR flux, the beam nature gives us, reviewing the status of our knowledge of their energy spectrum and composition. In the second part I will discuss some of the space borne experiments planned in the next future which will contribute to the quest for answers to various unsolved questions in Astro Particle physics.

2   Understanding Nature's Beam

The Universe communicates with us by sending to Earth a continuous flow of radiation of different kinds. Here we are interested in the high energy part of the spectrum $(E > O(GeV))$.

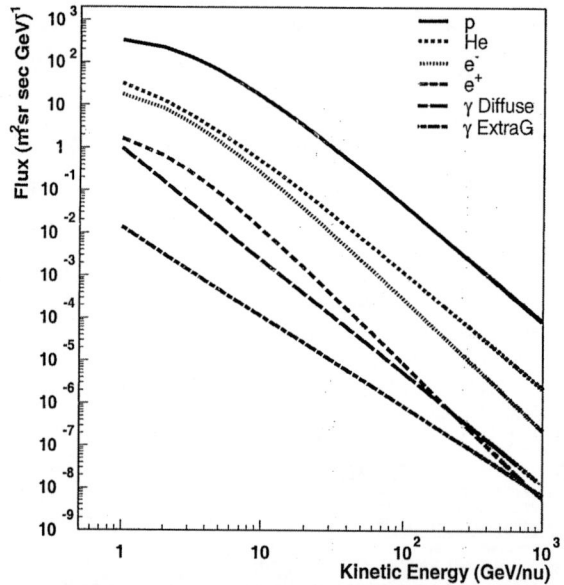

Fig. 2. High Energy Cosmic Rays composition [1]

Cosmic Rays traditionally refers to the charged component of the high energy particles travelling through the galaxy (Fig. 2): in order of abundance $p \sim 80\%$, $^4He \sim 15\%$, $e^- : O(1\%)$, $e^+ : O(0.1\%)$, $\bar{p} : O(0.01\%)$). In addition to p and

^4He, there is a composite hadronic component including all other nuclei and long lived isotopes, totalling about a few % of the total CR flux.

There are however two other forms of energetic radiation which are relevant for Astro Particle physics. The first is the high energy part of the electromagnetic spectrum, gamma rays above $\sim 100\ MeV$. Their flux is at the level of 10^{-5} of the CR flux. Gamma rays have an important property that charged CR do not have, namely they travel along straight lines, undisturbed by the magnetic field and reproducing the images of their sources.

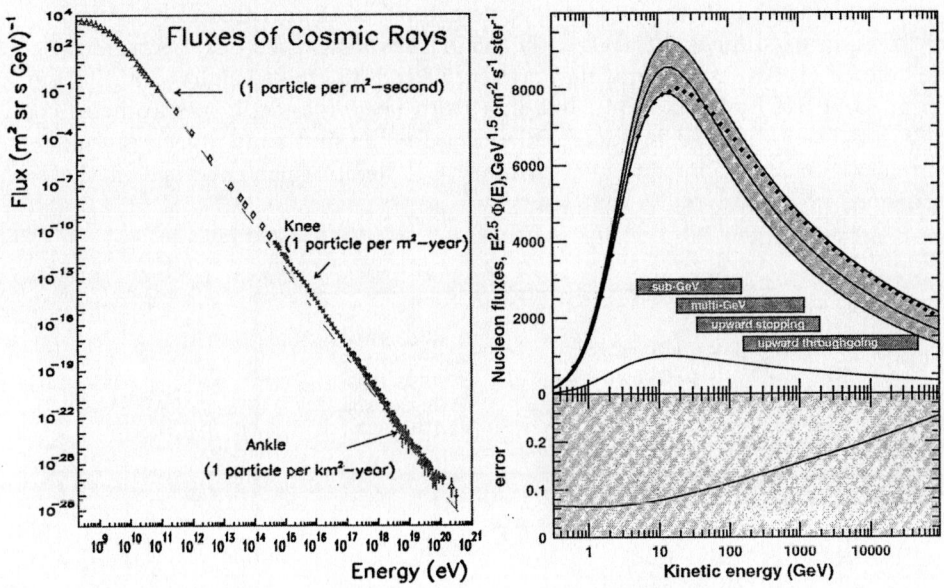

Fig. 3. Left: the flux of charged cosmic rays. Right: measurement and uncertainty on the primary CR spectrum (proton and He).The accurate knowledge of this part of the CR spectrum is important for the understanding of the atmospheric neutrino flux [2]

The second component are the neutrinos, which would also allow source imaging: at high energy the primary neutrino flux is much lower and steeper with increasing energy than for charged CR, with the exception of the secondary atmospheric component induced by high energy CR hitting the atmosphere. Around 1 GeV the neutrino flux is of the order of the flux of high energy gamma rays, but because of their low cross section, ν's are much harder to detect.

If we plan to use this energetic radiation to search for new particles or new effects, we must know the properties of Nature's beam with the best possible accuracy. Which is our current level on the knowledge of the Cosmic Radiation and what we can expect in the coming future?

3   The Charged CR Component

3.1   Charged Hadrons

Over the last 40 years the hadronic CR component has been measured systematically, using balloons (mostly), space borne (sometimes), and ground based detectors. The most complete information is obtained by particle spectrometers, experiments able to measure directly the CR composition through the determination of the charge, the sign of the charge, the momentum and the velocity of each particle. Often, however, simpler apparatus were used, based on calorimeters, emulsion chambers or Cerenkov detectors: in these cases only partial information were obtained. Until recently the most sophisticated CR spectrometers were flown on balloons operating between 30 and 35 km of height (BESS[3], MASS[4], CAPRICE[5], IMAX[6]) but in 1998 a large magnetic spectrometer, AMS[7], has been operated in space, providing for the first time a very precise measurement on the composition of primary CR before their entrance in the atmosphere.

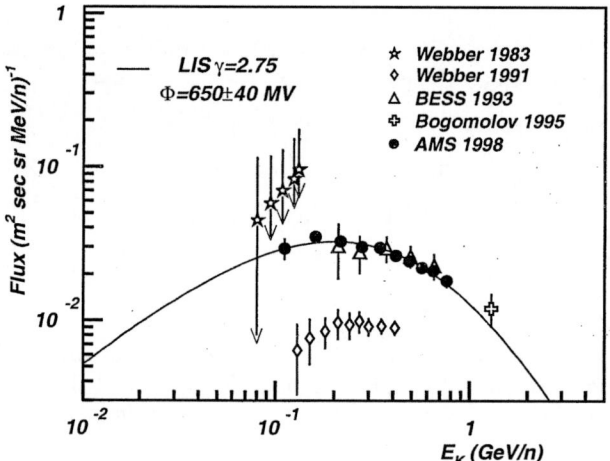

Fig. 4. Cosmic deuteron measurements[8–12]

The flux of the main CR component, the protons, is known with $5 - 10\%$ accuracy up to $\sim 200\ GeV$, and with $10 - 30\%$ accuracy up to $\sim 100\ TeV$ (Fig. 3) [2]. Helium flux is known with 10% accuracy up to $\sim 10\ GeV$ but above this energy the measurements are rather poor. Light $Z > 2$ nuclei have been measured with about 5% accuracy only up to $\sim 35\ GeV$. It would be important to extend the energy range for precise measurements of hadrons up to about $10\ TeV$ because they are the source of the atmospheric neutrinos used for the determination measurement of the neutrino mass in underground experiments.

For the study of CR composition, precise measurements of some stable isotopes like Deuterium, Boron and Carbon or long-lived isotopes like ^9Be are particularly important to understand the propagation and trapping mechanisms of CR in our Galaxy. Accurate measurements of D are available only up a few GV of rigidity (Fig. 4) while the knowledge of the cosmologically important ratio $^9Be/^{10}Be$ is very poorly known above about hundred MV of rigidity.

3.2   Antiprotons

Antiprotons are a rare but interesting hadronic component of high energy CR, because they could be produced by exotic sources like antimatter dominated regions or by the decay or annihilation of new particles. Their flux ratio to protons is at the level of $O(10^{-4})$ at kinetic energies around 1 GeV. This rate is in agreement with the expectation that $\bar{p}$ are produced in high energy CR interactions with the interstellar medium. However in more than 40 years of experiments with balloons, only a few thousands $\bar{p}$ have been measured, mostly with energies below 10 GeV (Fig. 5). Statistical errors are then quite large, in particular below 1 GeV and above 10 GeV: also systematic errors due to uncertainties in the modelling of the propagation in the Interstellar Medium or of the solar modulation are still at the level of $20 - 30\%$.

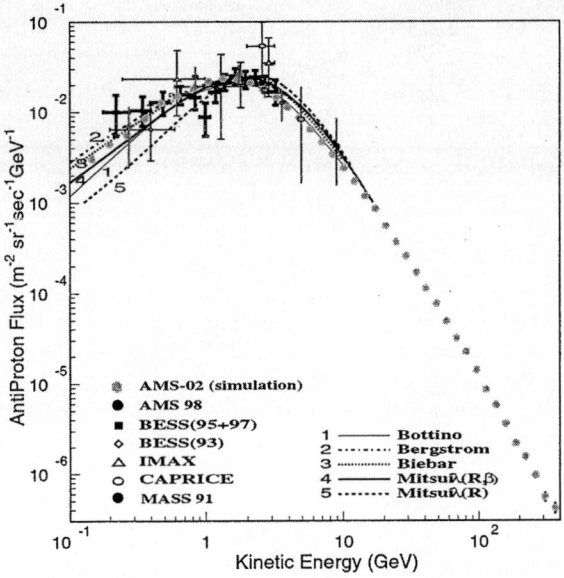

Fig. 5. Compilation of antiproton measurements. A simulated measurement of the AMS-02 experiment on the ISS is also reported. For all balloon data and models see[13] and reference therein. For AMS data see[7]

3.3   The Charged Leptons

Due to their lower flux ($\sim 0.5\%$ for e^- and $\sim 0.05\%$ for e^+ around 1 GeV) and their steeply falling spectrum, statistical uncertainties on the measurement of the spectra of the two stable leptons are larger; the experimental situation gets very confused above a few *GeV* (Fig. 6). Precision measurement of electrons and positrons are important since, being their fluxes quite low and with a strong charge asymmetry, contribution from exotic sources like supersymmetric particles annihilation could distort the e^+/e^- ratio at a level which could be detectable by a high precision experiment[15].

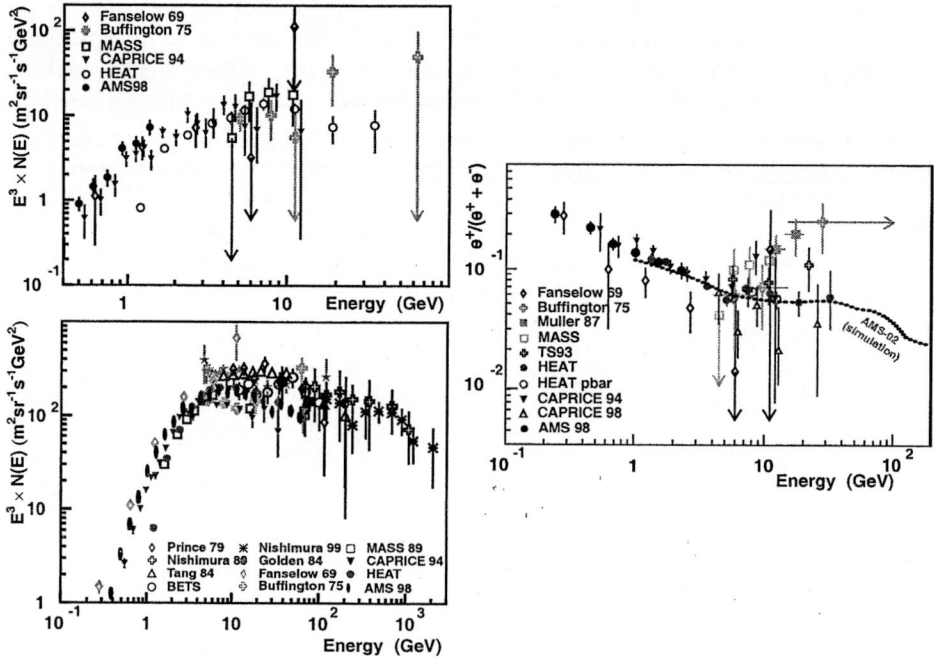

Fig. 6. Compilation of e^+ (top left), e^- (bottom left) and $e^+/(e^- + e^+)$ (right) data: labels correspond to the various references cited in [14]. A simulated long exposure result from AMS-02 is also included, for the case $m_\chi = 275\ GeV/c^2$[15]. All experiments are balloon borne spectrometers with the exception of AMS 98 and AMS-02 which are space borne spectrometers

3.4   Future Experiments

During the next 5 years our knowledge of the flux and composition of CR will be greatly improved thanks to two space borne spectrometers, PAMELA on a Russian rocket in 2003-2006[16] and AMS-02 on the International Space Station on 2006-2009[17]. These two experiments will increase the statistical samples of charged CR by a few to several orders of magnitude. AMS-02, in particular, thanks to its strong magnetic field and large geometrical factor, will be able to extend these measurements well in to the *TeV* region, covering a region which is interesting for various physics topics and which today is very poorly known. For a discussion on the improvement expected from AMS-02 on the hadronic CR component see [18]. Space borne experiments operating for three years or more are much more accurate than balloon flights lasting only about one day. Long duration balloon experiments lasting for 20 days or more, however, could be competitive, as is shown in Table 2.

Table 2. Future CR spectrometers

	Aperture $(cm^2 sr)$	Duration (days)	Altitude (km)	Latitude (degrees)	Launch (year)	Area*Time (AMS-01)
AMS-01	2300	10	320-390	< 51.7	1998	1.0
PAMELA	21	1000	690	70	2003	1.1
BESS Polar	3000	20	36	> 70	2004	2.6
AMS-02	5000	1000	320-390	< 51.7	2006	217.0

While at high energy the AMS-02 will be the most performing experiment, the Polar BESS[19] long duration flights, thanks to BESS large acceptance, will have a good sensitivity to the lower energy part of the CR spectrum and will perform a precise measurement of low energy $\bar{p}$. On the other side the space borne PAMELA spectrometer, in spite of its small aperture, is a timely experiment and will have a chance to perform high statistic CR measurements in the period between AMS-01 and AMS-02.

At energies around the knee ($\sim 10^{15} eV = 1\ PeV$) and above, all measurements come nowadays from large area ground based arrays. At these energies the CR elemental composition can be determined only by modelling the interaction of CR with the atmosphere and it is affected by large uncertainties. To solve this problem a long duration, large acceptance, space borne experiment, ACCESS, has been considered[20] either for deployment on the International Space Station or as a free flyer.

Ground based large area arrays have measured the CR spectrum (not the composition) up to Extreme Energies ($10^{21} eV = 1\ ZeV$) and they are at the moment the only experiments able to measure these very rare events. At these energies the flux is so low that eventually it will become unpractical to extend the surface or the time of exposure of ground based experiments. The largest experiment of this kind, Auger[21], is being built right now. The next step, however, will be to look to the fluorescence induced by the EECR showers from space, as proposed by EUSO[22], an experiment planned on the International Space Station and capable to collect 10-50 times more statistics than Auger or by KLYPVE[23] planned on a Russian free flyer.

4   The Neutral CR Component

4.1   High Energy Gamma Rays

High energy gamma rays are a rare $(O(10^{-5}))$ component of the cosmic radiation which is not traditionally included in a CR review paper. However high energy gamma rays are produced by the same sources producing high energy CR and carry complementary information. They should then be considered when discussing Astro Particle physics, in particular since their study could give important contributions to the understanding the problem of the origin of dark matter. Their energy spectrum could, in fact, be influenced by exotic sources like neutralino annihilations taking place at the center of the galaxy. Most of the high energy gamma rays data have been collected by the EGRET experiment on the CGRO satellite during the 90's. Since the end of the CGRO program (1999 − 2000) there are no experiments measuring high energy gamma rays in space. During the present decade there will be three space borne experiments which will be able to measure high energy gamma rays: AGILE[24] (2003) a small scientific mission of the Italian Space Agency, AMS-02 on the International Space Station (2006)[25] and GLAST[26] (2007). These experiments will be able to cover the region up to ∼ 300 GeV competing with ground base Cerenkov detectors which meanwhile will try to lower their threshold below ∼ 50 GeV.

4.2   High Energy Neutrinos

Neutrinos are a very important part of the Cosmic Radiation. Their spectrum (Fig. 7) extends over several orders of magnitude like for the charged CR component, but is rich in features coming from the various physical process at work.

Neutrinos have two great advantages and one disadvantage with respect to charged CR:

Advantages:

- 1) they travel on straight lines so neutrino astronomy is a possibility;
- 2) they have a small interaction cross section so they are basically unaffected by the GZK[27] cutoff and can reach us from the edges of the universe (see Fig. 8).

Disadvantage:

- 1) they have a small interaction cross section so their detection is problematic, requiring very large volumes of matter.

Many of the neutrinos spectral features have been measured (ν from the sun, from $SN1987$, from reactors, from the atmosphere), some may be on the verge of being seen as ν's of galactic origin[28] while other are expected to exist but are still undetected (relic neutrinos, neutrinos from AGN's). Some could be produced by exotic sources, like superheavy particles or topological defects. The interest of neutrinos is their capability to escape the GZK cutoff up to

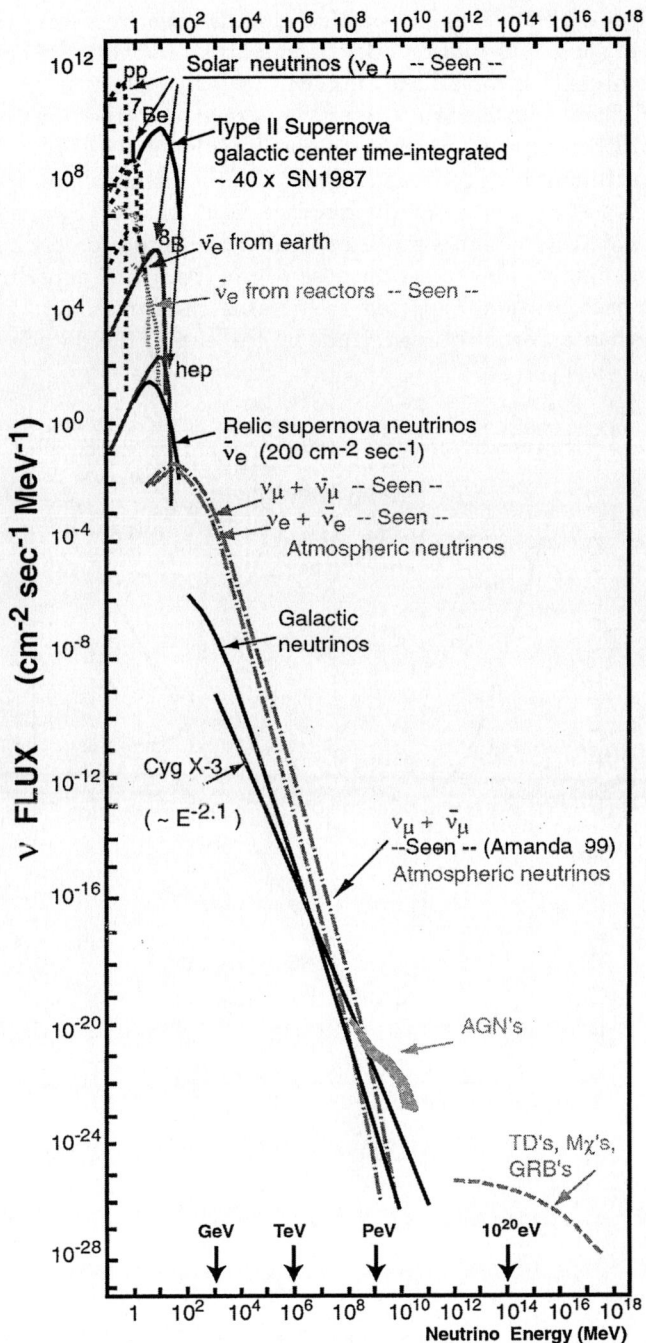

Fig. 7. The Cosmic neutrinos spectrum

extreme energies: they are the only particle that can reach us from the edges of the universe, carrying information about the status of space, time and matter, before the recombination took place (Fig. 8).

The main difficulty for neutrino detection is related to the mass required for their detection. The only way to access energies of the order of $10^{14}eV$ or above is to use the earth surface layers as targets, like the atmosphere, ice or water. In the case of the atmosphere the fluorescence light emitted by the developing shower could be detected from a space experiment able to monitor a sufficiently large area of our planet. This is the purpose or the EUSO[22] experiment on the International Space Station, expected to increase the sensitivity of the Auger array by more than an order of magnitude by the end of the decade.

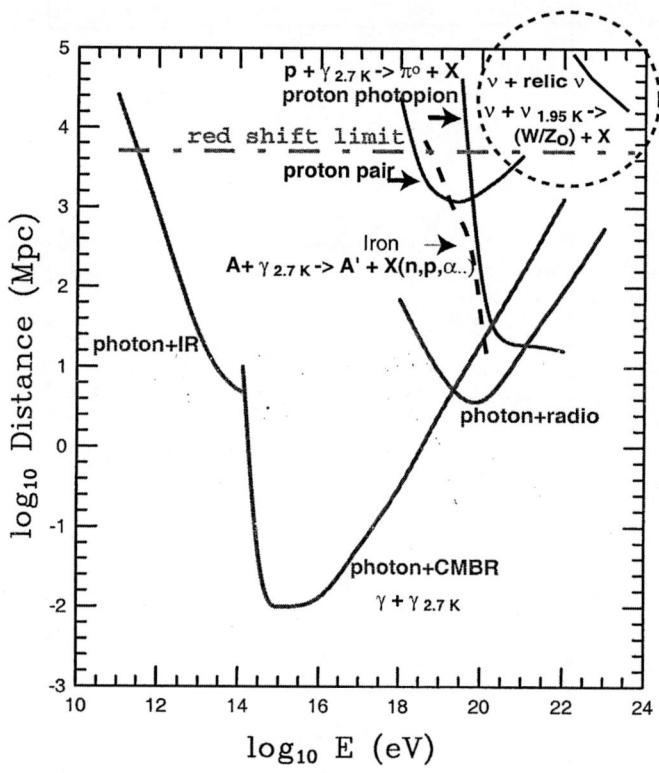

Fig. 8. Energy dependence of GZK cutoff for different CR species[29]

5   Searching for New Particles

In this section now briefly discuss the physics potential of the accurate study of the cosmic radiation. An area of clear interest for particle physics is obviously the search for new states of matter or for new particles.

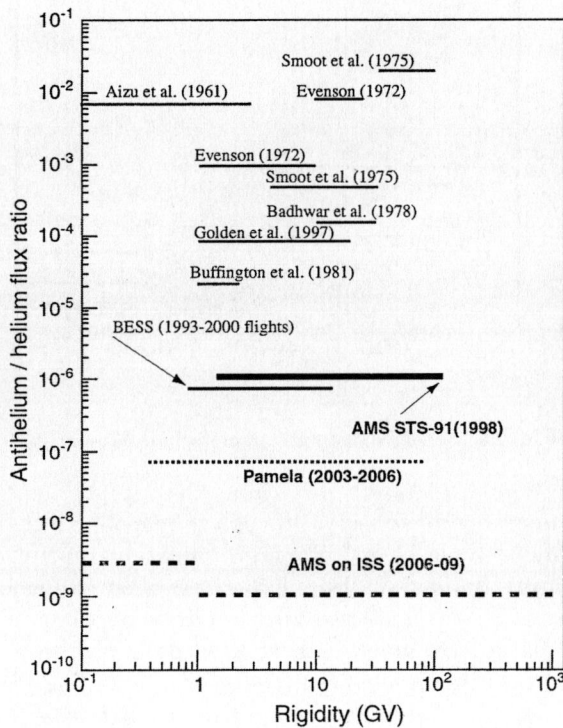

Fig. 9. Antimatter limits. For the references see [30]

5.1   Direct Search for Nuclear Anti-Matter

The disappearance of antimatter [31–33] is one of the most intriguing puzzles in our current understanding of the structure of the Universe. Absence of nuclear antimatter from the scale of our galaxy to the scale of the local supercluster is experimentally established at the level of 1 part in 10^6 by direct CR searches and indirect methods like the study of the energy spectrum of the diffuse gamma ray flux. For a genuine antimatter signal one should look to nuclei heavier than $\bar{p}$ since secondary $\bar{p}$ can be easily produced at the level 10^{-4} in high energy hadronic interaction of CR with the IM. This probability quickly vanishes with increasing atomic number. For $\bar{D}$ the secondary production is at the level of 10^{-8} or less and already for $^4\bar{H}e$ it is well below 10^{-12}.

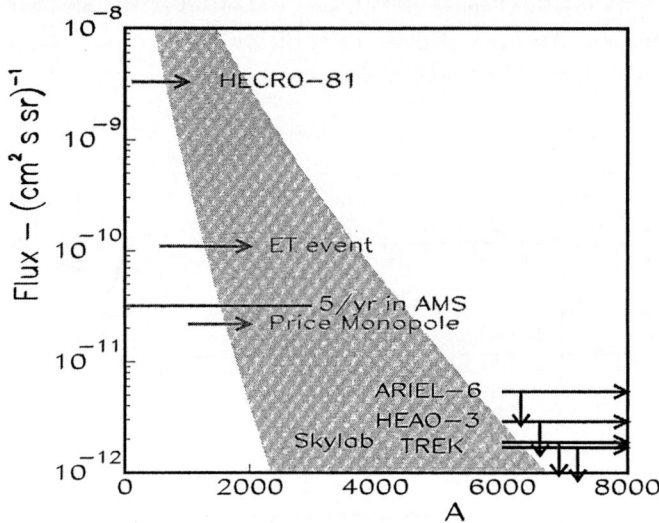

Fig. 10. Sensitivity to strangelets in AMS-02

This is why the unambiguous observation of a couple of $^4\bar{H}e$ at the level of one part in a billion or more would have profound implications on our understanding of baryogenesis. During the last 35 years experiments on balloons have pushed the limit on the $^4\bar{H}e$ at the level of less than about one part in a million (Fig. 9). Recently the AMS-01 spectrometer, during a 10 days precursor flight on the Shuttle, has reached the same level of sensitivity. In the coming years PAMELA, first, and AMS-02, later, will eventually reach a sensitivity a thousand times better, reaching rigidities of the order of a TV.

5.2   Direct Search for Strange States of Matter

Strangelets[34] are a stable state of nuclear matter containing a large fraction of strange quarks. Such states could have developed during the cosmological quark-hadron phase transition $10^{-5}s$ after the Big Bang or in the high density conditions of compact supernova remnants, which might then be strange stars composed of quark matter rather than neutron stars. These stars could send to space fragments of stable strange nuclear matter during catastrophic events like collisions with other strange stars. These fragments would have variable mass but a peculiar charge to mass ratio. For instance in the case of color-flavor locked strangelets we obtain a charge to mass relation of the type $Z = 0.3A^{2/3}$. These particles would look like high mass, low mass/charge ratio cosmic rays which could be easily identified in a space borne magnetic spectrometer like AMS-02. Fig. 10 shows the sensitivity to strangelets expected for AMS-02 after three years on the International Space Station[35].

5.3   Indirect Search for Dark Matter

The presence at all scales in our universe of a non luminous component of matter, Dark Matter (DM)[37,38], is possibly the most fascinating problem in Astro Particle physics. A viable solution, if not the most viable solution[36] to this problem, is given by the Lightest Supersymmetric Particle (LSP), the neutralino (χ). Supersymmetry links the existing Standard Model particles to a set of new, heavier, super particles through R-parity conservation, where R is a combination of the particle spin, lepton and baryon numbers, $R = (-1)^{3B-L+2S}$. The conservation of R-parity requires that the LSP is stable. LEP results suggest that the LSP is heavy ($m_\chi > 45\ GeV$)[39] and then these particles can be a good DM candidate in the Cold Dark Matter (CDM) scenario.

Unfortunately SUSY is a theory with many parameters still poorly constrained. The LSP (neutralino) can be expressed as superposition of the neutral gauge (g and W) and *Higgs* boson superpartners (H_{01}, H_{02}):

$$\chi = N_{11}B + N_{12}W_3 + N_{13}H_{01} + N_{14}H_{02} \tag{1}$$

The parameters of this superposition define the χ properties, like the mass, the annihilation cross section and branching ratios into detectable particles. These parameters can also be related to the χ cosmological density, which can be constrained in the interesting region $0.1 < \Omega_{DM} < 0.3$, in agreement with the recent astrophysical results[40].

χ annihilation would take place in the most dense regions of our galaxy, e.g. its center or in other existing DM clumps. These regions could be the source of prompt high energy CR without need of an acceleration mechanism. In order to detect these prompt CR we study rare CR components where the effects of the exotic χ contributions would be detectable against backgrounds due to the primary spectrum. Rare CR components like high energy $\bar{p}$[37], e^+[15], $\bar{D}$[42] and γ[43] have been suggested as potential indirect signatures for cold DM.

In Fig. 6 we give an example of spectral distortions induced by χ annihilations on the $e^+/(e^+ + e^-)$ ratio. Quite interesting is the case of $\bar{D}$ production, since it has been suggested[42] that a $\bar{D}$ signal at kinetic energies below $\sim 1\ GeV$ would be a strong indication for χ annihilation. In the case of high energy γ rays the spectral deformation due to χ annihilation is expected to have a strong spatial dependence, mimicking the DM halo structure which might have more than one clumps in our galaxy. For a subset of SUSY models this flux of HE γ rays could be detectable by space borne experiments (Fig. 11).

For many SUSY models the indirect search for χ in CR will be a difficult undertaking: SUSY predictions depend in fact on several parameters and the limited precision in the knowledge of CR composition and spectra will challenge the unambiguous detection of a χ signal. Before the advent of the LHC, however, the physics potential of SUSY searches in space is promising[41,39]. In fact most of the cosmologically relevant SUSY parameter space can be explored by CR experiments. Capability to measure at the same time all types of charged and neutral particles which can be signature of χ annihilation would clearly be an experimental advantage: starting from 2006, the AMS-02 experiment on the ISS

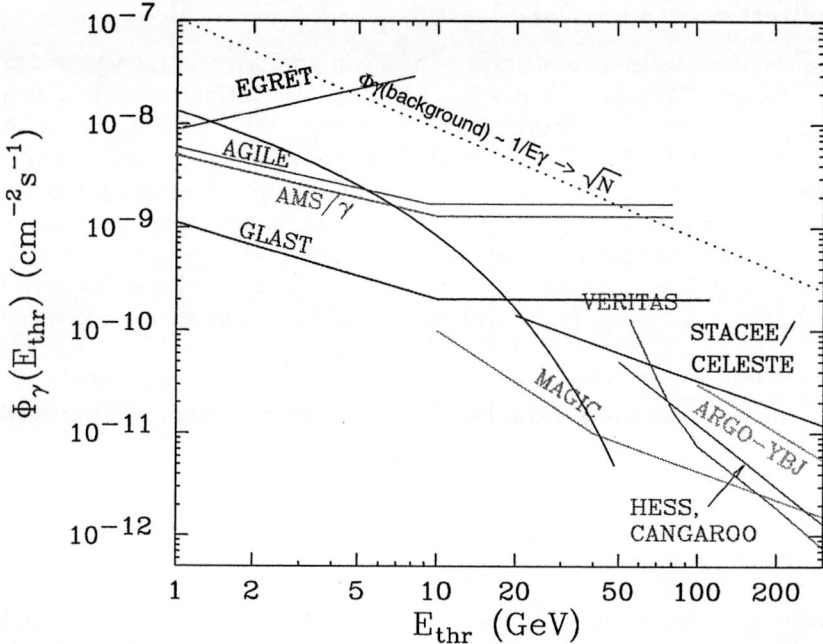

Fig. 11. Sensitivity to SUSY dark matter for various HE gamma ray experiments, either ground based (bottom right) or space based (top left). Vertical axis: gamma rays integral flux above the energy threshold E_{thr}. The dotted line represents the integral flux of HE gamma rays from known sources, which represent a source of background for this measurement. The continuous line represents the prediction for a SUSY model, with $m_\chi = 120~GeV$ and a boost factor $< J(0) >= 5000$[41]

will be the only experiment able to precisely measure at the same time all kinds of particles related to χ decay. Other experiments will also contribute to this search: PAMELA will contribute by precisely measuring the $\bar{p}$ component, in particular at low energy, starting in 2003, while GLAST will perform the most precise measurement of the high energy γ ray halo structure and spectra starting in 2007.

5.4   Indirect Search for Ultraheavy Particles

The existence of a measurable flux of Extreme Energies Cosmic Rays above $10^{19}eV$, represents a big puzzle in modern Astro Particle physics. The behavior of the GKS cutoff versus energy and particle type shown in Fig. 8 suggests that no particle except the ν's can travel for large distances at these energies. Above $10^{19}eV$ the volume of the region which can be traversed by hadronic EECR is then dramatically reduced.

One would then expect a steeper spectrum above the GZK cutoff. The experimental results suggest instead a smoother spectrum which would not be possible

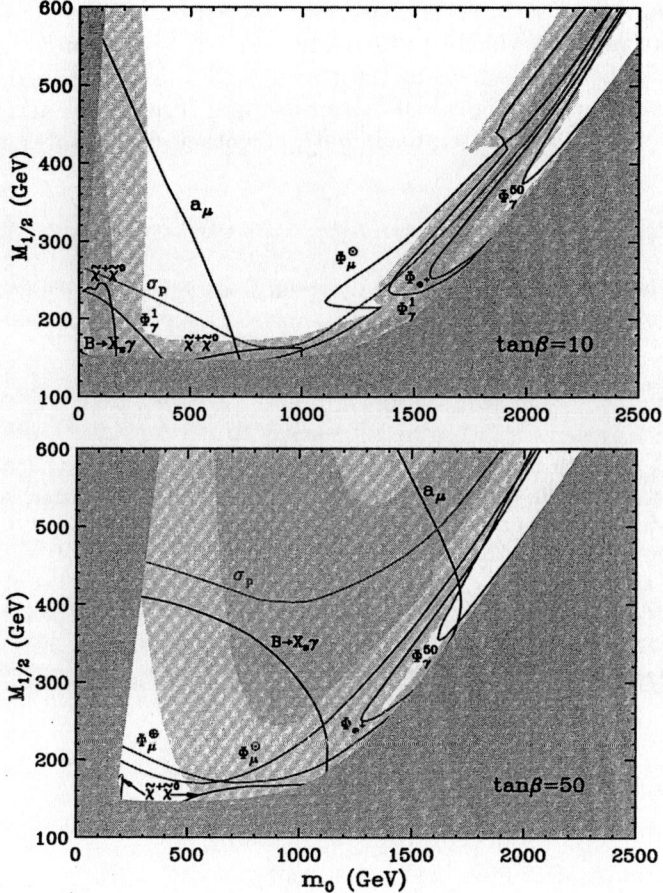

Fig. 12. SUSY parameter space explored by passive experiments. Before the advent of LHC most of the cosmologically interesting parameter space will be explored by ground based or space borne experiments. For more details see [41]

to explain using conventional physics. Although we are talking only about a few dozen events collected by various experiments over several years, their existence challenges standard explanations. One possible scenario is based on EE ν emitted in the decay of extremely heavy particles present in the very early phases of the universe. These particles travel until they reach regions close to us where they do interact with ordinary IS matter and produce EECR which in turn reach our planet where they can be detected. During this decade the advent of space experiments like EUSO[22] and KLYPVE[23] will improve by one or two orders of magnitude the sensitivity to EECR of the Auger experiment, hopefully clarifying the present situation.

6 Conclusion

One hundred years after their discovery Cosmic Rays have still an important potential for new physics. In order to exploit this potential new, more sensitive experiments are planned which can take advantage of the unique conditions of space for precision measurement of the primary CR flux. During the current decade these experiments might well deliver exciting surprises in Astro Particle physics, on issues like antimatter, dark matter or other exotic states of matter.

Acknowledgements

This work has been partially supported by the Italian Space Agency (ASI) under contract ARS-98/47.

References

1. V. Choutko, G. Lamanna, A. Malinin: Int.J. Mod. Phys. **A17**, 1817 (2002).
2. T. Stanev, Invited talk a the IInd International Workshop on Matter, Antimatter and Dark Matter,October 29-30 (2001), unpublished.
3. Y. Ajima, et al. Nucl. Instr. and Methods **A443**, 71 (2000).
4. R. Bellotti, et al., Phys. Rev. **D60**, p. 052002, (1999).
5. M. Boezio, et al., ApJ **518**, p. 457 (1999).
6. J. W. Mitchell, et al., Proc. 23rd ICRC, **1**, 519 (1993).
7. AMS Collaboration, Phys. Rep. **366** (2002).
8. W.R.Webber, et al., Ap.J. **275** 391 (1983).
9. W.R.Webber, et al., Ap.J. **380** 230, (1991).
10. J. Z. Wang, et al., Ap.J. **564** 244 (2002).
11. E.A. Bogomolov, et al., Proc. 23rd ICRC, **2**, 598 (1995).
12. G. Lamanna, et al., Proc. 27th ICRC, 1614 (2001).
13. T.Maeno, et al., Ap. Phys. **16**(2) 121 (2001).
14. B. Bertucci, Int.J. Mod. Phys. **A17**, 1613 (2002).
15. E. Diehl Phys. Rev. **D52**, 4223 (1995).
16. V. Bonvicini, et al., N.I.M. **A461**, 262 (2001).
17. R. Battiston, Frascati Physics Series, Vol XXIV, 261 (2002).
18. J. Casaus, Int.J. Mod. Phys. **A17**, 1603 (2002).
19. T. Sanuki, Int.J. Mod. Phys. **A17**, 1635 (2002).
20. O. Ganel, et al., AIP Conference Proceedings **458**, 272 (1999).
21. J. W. Cronin, AIP Conference Proceedings Volume **566**, 1 (2001).
22. L. Scarsi AIP Conference Proceedings Volume **566**, 113 (2001).
23. B. A. Khrenov, et al., AIP Conference Proceedings Volume **566**, 57 (2001).
24. M. Tavani, Int.J. Mod. Phys. **A17**, 1799 (2002).
25. R. Battiston, et al., Astropart.Phys. **13**, 51 (2000).
26. A. Morselli, Int.J. Mod. Phys. **A17**, 1829 (2002).
27. K. Greisen, Phys. Rev. Lett. **16**, 748 (1966). G.T.Zatsepin, V.A.Kuzmin, JETP Lett. **4**, 78 (1966).
28. F. Halzen, this conference.
29. A. Letessier-Selvon, astro-ph/0006111, (2000).

30. AMS Collaboration, Phys. Lett. **B461**, 387 (1999).
31. G. Steigmann, Ann. Rev. Astron. Astroph., **14**, 339 (1976).
32. E. W. Kolb, M. S. Turner, Ann. Rev. Nucl. Part. Sci. **33**, 645 (1983).
33. P. J. E. Peebles, Principles of Physical Cosmology, Princeton University Press, Princeton N.J. (1993).
34. E. Witten, Phys. Rev. D **30**, 272 (1984). J. Madsen, astro-ph/9809032, (1998). M. Alford, Ann. Rev. Nucl. Part. Sci. **30** 131 (2001). K. Rajagopal, F. Wilczek, hep-ph/0011333 (2000).
35. J. Madsen, astro-ph/0112153, (2001).
36. A. Masiero, S. Pascoli, Int.J. Mod. Phys. **A17**, 1723 (2002).
37. J. Ellis, et al., Phys. Lett. **B214**, 403, (1988).
38. M.S. Turner, F. Wilzek, Phys. Rev. D **42**, 1001 (1990).
39. J. Ellis, astro-ph/9911440, (1999).
40. P. de Bernardis, et al., Frascati Physics Series, Vol XXIV, 399 (2002).
41. J. L. Feng, K. T. Matchev, F. Wilczek, astro-ph/0008115 (2000).
42. F. Donato, P. Fornengo, P. Salati, Phys. Rev. D **62**, 043003 (2000).
43. L. Bergström, P. Ullio, J.H. Buckley, astro-ph/9712318, (1997).

Stars and Fundamental Physics

Georg G. Raffelt

Max-Planck-Institut für Physik (Werner-Heisenberg-Institut), Föhringer Ring 6, 80805 München, Germany

Abstract. Stars are powerful sources for weakly interacting particles that are produced by nuclear or plasma processes in their hot interior. These fluxes can be used for direct measurements (e.g. solar or supernova neutrinos) or the back-reaction on the star can be used to derive limits on new particles. We discuss two examples of current interest, the search for solar axions by the CAST experiment at CERN and stellar-evolution limits on the size of putative large extra dimensions.

1 Introduction

Astrophysics and cosmology provide a natural testing ground for virtually any new idea in the area of elementary particle physics. Usually one may first think of the early universe or perhaps high-energy cosmic rays when searching for astrophysical arguments in favor or against a new particle-physics model. However, there are a number of interesting cases where the low energies available in stars are quite sufficient for rather useful and restrictive tests of high-energy physics phenomena.

The basic idea is very simple. Stars are powerful sources for weakly interacting particles such as neutrinos, gravitons, hypothetical axions, and other new particles that can be produced by nuclear reactions or by thermal processes in the hot stellar interior. The solar neutrino flux is now routinely measured with such precision that compelling evidence for neutrino oscillations has accumulated. The measured neutrino burst from supernova (SN) 1987A has been used to derive many useful limits. Even when the particle flux can not be measured directly, the absence of visible decay products, notably x- or γ-rays, can provide important information. The properties of stars themselves would change if they lost too much energy into a new channel. This "energy-loss argument" has been widely used to constrain a long list of particles and particle properties. All of this has been extensively reviewed [1,2] and is now widely appreciated among particle physicists [3].

Therefore, instead of reviewing once more the general ideas I will rather focus on two topical examples of current interest that nicely illustrate the overall methods. One is the search for solar axions by the CAST experiment at CERN (Sec. 2). The other is the possibility that space-time has large extra dimensions. This hypothesis predicts a "tower" of graviton modes that can be produced in stars, notably in SN cores or neutron stars. The most restrictive limits on the size of the extra dimensions arises from the astrophysical arguments presented in Sec. 3. A brief summary and outlook is given in Sec. 4.

2   Axion-Like Particles

Axions are hypothetical particles that are predicted in the context of a theoretical scheme to solve the CP problem of strong interactions [4,5]. This is the problem that quantum chromodynamics (QCD) ought to violate the CP symmetry in that the neutron should have a large electric dipole moment, contrary to experimental evidence. This observation can be explained by a new symmetry, the Peccei-Quinn symmetry, that is spontaneously broken at some large energy scale f_a, the Peccei-Quinn scale or axion decay constant. Axions are the "almost" Nambu-Goldstone bosons of this new symmetry and as such nearly massless.

Phenomenologically one should think of axions as the neutral pion's little brother. Model-dependent details aside, the axion's mass and couplings are given by those of the π^0, scaled with f_π/f_a where $f_\pi = 93$ MeV is the pion decay constant. The axion decay constant f_a is a free parameter and thus can be very large. Therefore, axions can be very light and very weakly interacting even though they are fundamentally a QCD phenomenon.

There are other plausible solutions of the strong CP problem. However, the Peccei-Quinn approach is particularly elegant and predicts something new – in the guise of axions it provides a handle for a possible experimental verification. Moreover, axions can play the role of the cosmic cold dark matter [6]. Therefore, two fundamental problems would be solved by the existence of one new particle.

The experimental search for axions has focused on their predicted interaction with the electromagnetic field that would be of the form

$$\mathcal{L}_{a\gamma\gamma} = \tfrac{1}{4} g_{a\gamma\gamma} F_{\mu\nu} \widetilde{F}^{\mu\nu} a = -g_{a\gamma\gamma} \mathbf{E} \cdot \mathbf{B}\, a\,, \tag{1}$$

where F is the electromagnetic field-strength tensor, $\widetilde{F}$ its dual, and $\mathbf{E}$ and $\mathbf{B}$ the electric and magnetic fields, respectively. The coupling strength is

$$g_{a\gamma\gamma} = \frac{\alpha}{2\pi f_a} C_\gamma\,, \qquad C_\gamma = \frac{E}{N} - 1.92 \pm 0.08\,, \tag{2}$$

where E/N is the ratio of the electromagnetic and color anomalies, a model-dependent ratio of small integers. One popular case is the DFSZ model where $E/N = 8/3$, another the KSVS model where $E/N = 0$, but there are more general examples [7].

Assuming $m_a = 0.60$ eV $\times 10^7$ GeV$/f_a$ for the axion mass, Fig. 1 shows $g_{a\gamma\gamma}$ as a function of m_a. The diagonal band marked "Axion Models" is somewhat arbitrarily delimited by the DFSZ and KSVZ models. The role of axions or axion-like particles is frequently assessed in the full two-dimensional $g_{a\gamma\gamma}$-m_a-space rather than the narrow band defined by conventional axion models, although this band remains the best-motivated location in this parameter space.

The electromagnetic interaction allows for the two-photon decay $a \to \gamma\gamma$ with a rate $\Gamma_{\text{decay}} = g_{a\gamma\gamma}^2 m_a^3/64\pi$. This process is very slow if the axion mass is small and the coupling strength is weak. Therefore, it is more promising to consider the analogous process where one of the photons is virtual, i.e. an external electric or magnetic field. The $a \leftrightarrow \gamma$ conversion in the presence of an external E or B

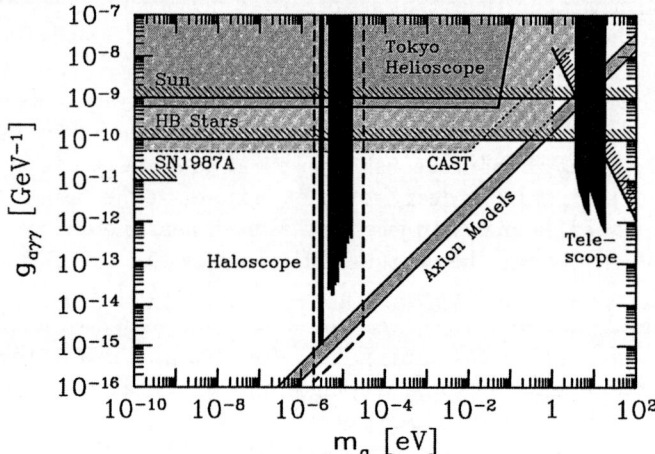

Fig. 1. Limits on the axion-photon coupling $g_{a\gamma\gamma}$ as a function of axion mass m_a. The limits apply to any axion-like particle except for the "haloscope" search which assumes that axions are the galactic dark matter; the dotted region marks the projected sensitivity range of the ongoing full-scale searches. Limits for higher masses than shown here are reviewed in Ref. [15]. The light-grey region marks the foreseen CAST sensitivity

field is known as the Primakoff process; it was first considered for neutral pions half a century ago [8].

If axions are the galactic dark matter, they can be detected in the laboratory by the "haloscope" technique [9]. One places a tunable high-Q microwave cavity in a strong magnetic field and measures the power output. If the resonance frequency matches m_a, the Primakoff-conversion can produce a measurable signal. Two pilot experiments [10,11] and a first full-scale search [12,13] exclude a range of coupling strength shown in Fig. 1 that is marked "Haloscope". The new generation of experiments in Livermore [13] and Kyoto [14] should cover the dashed area in Fig. 1, perhaps leading to the discovery of axion dark matter.

In a different region of masses and couplings axions are detectable with a related technique called the "helioscope" [9,16]. Thermal photons in the solar interior convert to axions by the Primakoff process in the microscopic electric fields of charged particles, producing a solar axion flux which peaks at energies of a few keV. If one views the Sun through a long dipole magnet, the axions partially back-convert into photons and become visible as x-rays at the far end of the magnet. A dedicated search for this effect by the Tokyo Axion Helioscope [17] excludes the dark-grey region in Fig. 1.

The conversion rate in the helioscope scales quadratically with the length L and field-strength B of the conversion region. Therefore, one can do much better in the new CAST project at CERN where a de-commissioned LHC test magnet is used as a "magnetic telescope" to search for solar axions [18,19]. Mounted on a movable platform (Fig. 2) allowing $\pm 40°$ horizontal and $\pm 5°$ vertical tracking, this instrument can achieve about 33 full days of alignment with the Sun per

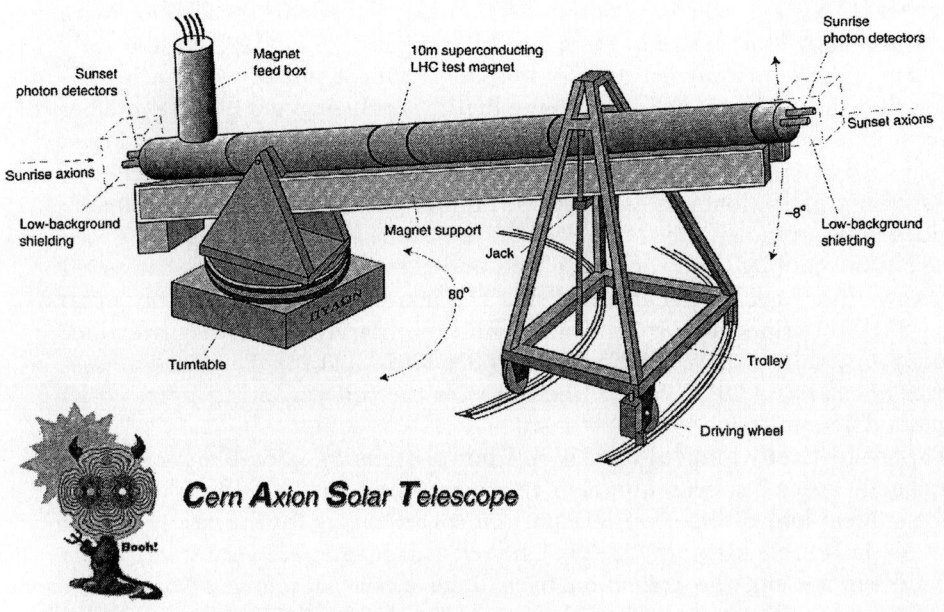

Fig. 2. Schematic view of the CAST experiment at CERN

year. If we express the coupling strength as $g_{a\gamma\gamma} = g_{10}\,10^{-10}$ GeV^{-1}, the solar axion flux at Earth is $g_{10}^2\,3.5 \times 10^{11}$ cm^{-2} s^{-1}. The conversion probability in the magnet is $g_{10}^2\,1.8 \times 10^{-17}(B/8.4\,\mathrm{T})^2(L/10\,\mathrm{m})^2$. For the two magnet bores with a cross section of 2×14 cm^2 we thus expect an x-ray event rate of $15\,g_{10}^4$ per day of exposure time.

In order to reach the sensitivity shown as a light-grey area in Fig. 1 one needs to make great efforts to suppress background counts. One way is to focus the x-rays to a small detector region. Specifically, an engineering model for the seven x-ray telescopes of the Abrixas satellite has become available for this purpose and has been tested to be in good working condition. CAST should be able to take first data shortly, i.e. in the summer or fall of 2002.

Figure 1 shows a loss of sensitivity for about $m_a > 10$ meV. The axion-photon conversion should be pictured as a phenomenon similar to neutrino oscillations [20]. For larger m_a the oscillation length becomes shorter than the magnet and the effective mixing angle is suppressed. This "momentum mismatch" between axions and photons can be overcome by giving the photons a refractive mass by virtue of a low-Z gas such as helium. This approach was successfully employed in the Tokyo Helioscope [17] and will be used in CAST as well. We may extend the sensitivity range to larger masses as shown in Fig. 1 and in the neighborhood of $m_a \sim 1$ eV actually bite into the parameter range of conventional axion models.

At somewhat larger masses axions are already ruled out by a telescope search for spectral lines from $a \to \gamma\gamma$ decay in galaxy clusters [21]. In the few-eV mass range axions would have been in thermal equilibrium in the early universe and contribute a small hot-dark matter component.

If we use the Sun as an axion source we must be sure that our sensitivity range is not excluded by an excessive modification of stellar properties by the axionic energy loss. An observable modification of the solar p-mode frequencies excludes $g_{a\gamma\gamma}$ values above the horizontal line in Fig. 1 marked "Sun" [22]. Significantly smaller couplings are excluded because the energy-loss of horizontal-branch (HB) stars would shorten their helium-burning lifetime, reducing the relative number of HB stars observed in globular clusters [1,2]; see the horizontal line in Fig. 1 marked "HB Stars". The CAST experiment advances into uncharted territory.

For very small axion masses, however, the CAST sensitivity range is already excluded by an argument involving SN 1987A. Axions would have been produced in the hot SN core by the Primakoff effect, and then would have back-converted into γ-rays in the galactic magnetic field. The non-observation of a γ-ray burst in the SMM instrument in coincidence with the observed SN 1987A neutrinos excludes $g_{a\gamma\gamma}$ values above the line marked SN 1987A [23,24]. This limit applies only for about $m_a < 10^{-9}$ eV; for larger masses the conversion is suppressed by the mass difference relative to photons.

The magnetically induced transition from photons to axion-like particles in intergalactic space has been proposed as a mechanism that would make distant photon sources look dimmer, with important consequences for the interpretation of the SN Ia Hubble diagram [25–29]. The relevant masses are very small, again to avoid suppressing the transition by a large axion-photon mass difference. Therefore, the relevant $g_{a\gamma\gamma}$ range is limited by the SN 1987A argument and thus falls outside the CAST sensitivity range.

3   Large Extra Dimensions

The Planck scale of about 10^{19} GeV, relevant for gravitation, is very much larger than the electroweak scale of about 1 TeV of the particle-physics standard model. A radical new approach to solving this notorious hierarchy problem holds that there could be large extra dimensions, the main idea being that the standard-model fields are confined to a 3+1 dimensional brane embedded in a higher dimensional bulk where only gravity is allowed to propagate [30–34]. This concept immediately puts stringent constraints on the size of the extra dimensions because Newton's law holds at any scale which has thus far been observed, i.e. down to about 1 mm. Extra dimensions can only appear at a smaller scale.

Following common practice the new dimensions are taken to form an n-torus of the same radius R in each direction. The Planck scale of the full higher dimensional space, $M_{P,n+4}$, can be related to the normal Planck scale,[1] $M_{P,4} = 1.22 \times 10^{19}$ GeV, by Gauss' law [30]

$$M_{P,4}^2 = R^n M_{P,n+4}^{n+2}. \qquad (3)$$

---

[1] Some authors define the Planck mass as $M_{P,4} = 1.22 \times 10^{19}$ GeV$/(8\pi)^{1/2} = 2.4 \times 10^{18}$ GeV. Limits on $M_{P,n+4}$ in this system of units have been reviewed in Ref. [35].

Therefore, if R is large then $M_{P,n+4}$ can be much smaller than $M_{P,4}$. If this scenario is to solve the hierarchy problem then $M_{P,n+4}$ must be close to the electroweak scale, i.e. $M_{P,n+4} < 10$–100 TeV. This requirement already excludes $n = 1$ because $M_{P,n+4} \simeq 100$ TeV corresponds to $R \simeq 10^8$ cm. However, $n \geq 2$ remains possible, and particularly for $n = 2$ there is the intriguing perspective that the extra dimensions could be accessible to experiments probing gravity at scales below 1 mm.

The most restrictive limits on $M \equiv M_{P,n+4}$ obtain from supernovae and neutron stars. The first example is the SN 1987A energy-loss argument. If large extra dimensions exist, the usual 4D graviton is complemented by a tower of Kaluza-Klein (KK) states, corresponding to new phase space in the bulk. These KK gravitons would be emitted from the SN core after collapse by nucleon bremsstrahlung $N + N \to N + N + KK$. The KK gravitons interact with the strength of ordinary gravitons and thus are not trapped in the SN core. However, this energy-loss channel can compete with neutrino cooling because of the large multiplicity of KK modes and shorten the observable signal [36–39]. This argument has led to the tight bound $R < 0.66$ μm ($M > 31$ TeV) for $n = 2$ and $R < 0.8$ nm ($M > 2.75$ TeV) for $n = 3$ [39].

The KK gravitons emitted by all core-collapse SNe over the age of the universe produce a cosmological background of these particles. Later they decay into all standard-model particles which are kinematically allowed; for the relatively low-mass modes produced by a SN the only channels are KK $\to 2\gamma$, e^+e^- and $\nu\bar{\nu}$. The relevant decay rates are $\tau_{2\gamma} = \frac{1}{2}\tau_{e^+e^-} = \tau_{\nu\bar{\nu}} \simeq 6 \times 10^9$ yr $(m/100$ MeV$)^{-3}$ [33]. Therefore, over the age of the universe a significant fraction of the produced KK modes has decayed into photons, contributing to the diffuse cosmic γ-ray background observed by EGRET. This argument implies that if the number of extra dimensions $n = 2$ or 3, their radius R must be about a factor of 10 smaller than implied by the SN 1987A cooling limit, i.e. for $n = 2$ one finds $R < 0.9 \times 10^{-4}$ mm or $M \geq 84$ TeV. For $n = 3$ the new limit is $R < 0.19 \times 10^{-6}$ mm or $M > 7$ TeV [40].

This, however, is not the end of the story. We later realized that the KK gravitons emitted by the SN core will stay gravitationally trapped because most of them are produced near their kinematical threshold, i.e. with barely relativistic velocities [41]. Therefore, every neutron star is surrounded by a halo of KK gravitons which is dark except for the decays into $\simeq 100$ MeV neutrinos, e^+e^- pairs and γ-rays. In principle, this radiation can be directly observed. Conversely, the non-observation allows one to set stringent limits. In addition, the radiation impinges on the neutron star, keeping it hot, above the observed temperature in some cases such as the pulsar PSR J0953+0755. One obtains the limit $M > 1680$ TeV for $n = 2$ and $M > 60$ TeV for $n = 3$. In view of these limits one expects that if large extra dimensions solve the hierarchy problem, their number n should probably exceed 4.

Similar arguments can be applied to other particles than gravitons that may exist and may be able to propagate in the bulk of the larger-dimensional space. The hypothetical majorons are one case in point [42].

Of course, there are loop holes to such limits. The size of the extra dimensions need not be equal, or there can be other than toroidal compactifications. The KK gravitons may be able to decay fast into invisible channels, and so forth. However, our main point is that straightforward astrophysical arguments lead to non-trivial and restrictive limits on the structure of this new theory.

4 Summary and Outlook

Stars continue to provide some of the most restrictive limits on new particle-physics ideas. The much-discussed hypothesis that our space-time has extra dimensions that are compactified on the sub-millimeter scale is a recent case in point. In addition to deriving limits, there are opportunities for new discoveries. The CAST experiment at CERN searching for solar axions will have a sensitivity range that for the first time pushes beyond stellar-evolution limits and thus has a realistic chance of actually finding axion-like particles emitted by the Sun.

In future the observation of solar neutrinos will continue to provide valuable information. The ongoing efforts in neutrino physics virtually guarantee that large detectors will operate for many years to come; even a megatonne detector may be built to search for proton decay and to perform precision measurements at laboratory neutrino beams. Therefore, chances are that one will measure the neutrino burst from a galactic supernova, providing high-statistics information both on the SN event and a host of information of particle physics interest.

The recent excitement about the possible discovery of strange-matter stars [43], even though not conclusive, illustrates that compact stars offer one of the few opportunities to discover the true ground state of nuclear matter.

Astroparticle physics is now an established research activity at the interface between inner space and outer space. The physics and observation of stellar objects continue to offer a number of intriguing opportunities in this multi-faceted and interdisciplinary field.

Acknowledgements

This work was partly supported by the Deutsche Forschungsgemeinschaft under grant No. SFB 375 and the ESF network Neutrino Astrophysics.

References

1. G.G. Raffelt: *Stars as Laboratories for Fundamental Physics* (Chicago University Press, Chicago, 1996)
2. G.G. Raffelt: "Particle physics from stars", Ann. Rev. Nucl. Part. Sci. **49**, 163 (1999).
3. D.E. Groom et al. (Particle Data Group): "Review of particle physics", Eur. Phys. J. C **15**, 1 (2000).
4. J.E. Kim: "Light pseudoscalars, particle physics and cosmology", Phys. Rept. **150**, 1 (1987).

5. H.Y. Cheng: "The strong CP problem revisited", Phys. Rept. **158**, 1 (1988).

6. E.W. Kolb and M.S. Turner: *The Early Universe* (Addison-Wesley, Redwood City, 1990).

7. J.E. Kim: "Constraints on very light axions from cavity experiments", Phys. Rev. D **58**, 055006 (1998).

8. H. Primakoff: "Photo-production of neutral mesons in nuclear electric fields and the mean life of the neutral meson", Phys. Rev. **81**, 899 (1951).

9. P. Sikivie: "Experimental tests of the 'invisible' axion", Phys. Rev. Lett. **51**, 1415 (1983), Erratum ibid. **52**, 695 (1984).

10. W.U. Wuensch et al.: "Results of a laboratory search for cosmic axions and other weakly coupled light particles", Phys. Rev. D **40**, 3153 (1989).

11. C. Hagmann, P. Sikivie, N.S. Sullivan and D.B. Tanner: "Results from a search for cosmic axions", Phys. Rev. D **42**, 1297 (1990).

12. S.J. Asztalos et al.: "Experimental constraints on the axion dark matter halo density", Astrophys. J. **571**, L27 (2002).

13. S. Asztalos et al.: "Large-scale microwave cavity search for dark-matter axions", Phys. Rev. D **64**, 092003 (2001).

14. K. Yamamoto et al.: "The Rydberg-atom-cavity axion search", hep-ph/0101200.

15. E. Masso and R. Toldra: "New constraints on a light spinless particle coupled to photons", Phys. Rev. D **55**, 7967 (1997).

16. K. van Bibber, P.M. McIntyre, D.E. Morris and G.G. Raffelt: "A practical laboratory detector for solar axions", Phys. Rev. D **39**, 2089 (1989).

17. Y. Inoue et al.: "Search for sub-electronvolt solar axions using coherent conversion of axions into photons in magnetic field and gas helium", Phys. Lett. B **536**, 18 (2002).

18. K. Zioutas et al.: "A decommissioned LHC model magnet as an axion telescope", Nucl. Instrum. Meth. A **425**, 482 (1999).

19. CERN Axion Solar Telescope homepage at http://axnd02.cern.ch/CAST/

20. G. Raffelt and L. Stodolsky: "Mixing of the photon with low mass particles", Phys. Rev. D **37**, 1237 (1988).

21. M.A. Bershady, M.T. Ressell and M.S. Turner: "Telescope search for multi-eV axions", Phys. Rev. Lett. **66**, 1398 (1991).

22. H. Schlattl, A. Weiss and G. Raffelt: "Helioseismological constraint on solar axion emission", Astropart. Phys. **10**, 353 (1999).

23. J.W. Brockway, E.D. Carlson and G.G. Raffelt: "SN 1987A gamma-ray limits on the conversion of pseudoscalars", Phys. Lett. B **383**, 439 (1996).

24. J.A. Grifols, E. Masso and R. Toldra: "Gamma rays from SN 1987A due to pseudo-scalar conversion", Phys. Rev. Lett. **77**, 2372 (1996).

25. C. Csaki, N. Kaloper and J. Terning: "Dimming supernovae without cosmic acceleration", Phys. Rev. Lett. **88**, 161302 (2002).

26. C. Csaki, N. Kaloper and J. Terning: "Effects of the intergalactic plasma on supernova dimming via photon axion oscillations", Phys. Lett. B **535**, 33 (2002).

27. J. Erlich and C. Grojean, "Supernovae as a probe of particle physics and cosmology", Phys. Rev. D **65** (2002) 123510.

28. C. Deffayet, D. Harari, J.P. Uzan and M. Zaldarriaga: "Dimming of supernovae by photon–pseudoscalar conversion and the intergalactic plasma", hep-ph/0112118.

29. E. Mörtsell, L. Bergström and A. Goobar: "Photon axion oscillations and type Ia supernovae", astro-ph/0202153.

30. N. Arkani-Hamed, S. Dimopoulos and G. Dvali: "The hierarchy problem and new dimensions at a millimeter", Phys. Lett. B **429**, 263 (1998).

31. I. Antoniadis, N. Arkani-Hamed, S. Dimopoulos and G. Dvali: "New dimensions at a millimeter to a Fermi and superstrings at a TeV", Phys. Lett. B **436**, 257 (1998).
32. N. Arkani-Hamed, S. Dimopoulos and G. Dvali: "Phenomenology, astrophysics and cosmology of theories with sub-millimeter dimensions and TeV scale quantum gravity", Phys. Rev. D **59**, 086004 (1999).
33. T. Han, J.D. Lykken and R. Zhang: Phys. Rev. D **59**, 105006 (1999).
34. G.F. Giudice, R. Rattazzi and J.D. Wells: "Quantum gravity and extra dimensions at high-energy colliders", Nucl. Phys. B **544**, 3 (1999).
35. Y. Uehara: "A mini-review of constraints on extra dimensions", hep-ph/0203244.
36. S. Cullen and M. Perelstein: "SN 1987A constraints on large compact dimensions", Phys. Rev. Lett. **83**, 268 (1999).
37. V. Barger, T. Han, C. Kao and R.J. Zhang: "Astrophysical constraints on large extra dimensions", Phys. Lett. B **461**, 34 (1999).
38. C. Hanhart, D.R. Phillips, S. Reddy and M.J. Savage: "Extra dimensions, SN 1987A, and nucleon nucleon scattering data", Nucl. Phys. B **595**, 335 (2001).
39. C. Hanhart, J.A. Pons, D.R. Phillips and S. Reddy: "The likelihood of GODs' existence: Improving the SN 1987A constraint on the size of large compact dimensions", Phys. Lett. B **509**, 1 (2001).
40. S. Hannestad and G.G. Raffelt: "New supernova limit on large extra dimensions: Bounds on Kaluza-Klein graviton production", Phys. Rev. Lett. **87**, 051301 (2001).
41. S. Hannestad and G.G. Raffelt: "Stringent neutron-star limits on large extra dimensions", Phys. Rev. Lett. **88**, 071301 (2002).
42. S. Hannestad, P. Keranen and F. Sannino: "A supernova constraint on bulk majorons", hep-ph/0204231.
43. C. Seife: "If it quarks like a star, it must be ... strange?", Science **296**, 238 (2002).

First Science Results from the VLT Interferometer

F. Paresce, A. Glindemann, P. Kervella, A. Richichi, M. Schoeller, M. Tarenghi, R. van Boekel, and M. Wittkowski

European Southern Observatory, Garching, Germany

Abstract. The VLT interferometer has been operating since the time of first fringes in March 2001 with a pair of 40 cm diameter siderostats at baselines of 16 and 66 m and, since October 2001, with a pair of 8 m diameter telescopes (UT1 and UT3) with a baseline of 103 m using the test camera VINCI operating in the K band. A fair fraction of its commissioning time has been devoted to observing a number of objects of scientific interest around the southern sky bright enough to allow high precision visibilities to be obtained on a routine basis. A large number of stellar sources with correlated magnitudes brighter than K $\sim$ 7 and K $\sim$ 3 with the 8 m and 40 cm telescopes respectively have been observed over this time period with limited (u,v) plane coverage. In this paper, we will briefly report on the present status of the VLTI and review the most interesting results on sources such as Eta Carinae, R Aquarii, and a number of Cepheid variables for which the VLTI data allow the establishment of tighter constraints on theoretical models.

1 Present Status

Since the corresponding times of first fringes in March and October of last year, the configuration we have been using is shown schematically in Figure 1. In particular, this means the use of the UT1 and UT3 8 m telescopes on a 103 m roughly NE-SW baseline both equipped with Coudé optics and tip/tilt sensors and 2 40 cm diameter siderostats on 16 and 66 m baselines oriented almost orthogonally as shown in Figure 1. In the 120 m long delay line tunnel, 3 60 m long stroke delay lines allow access to most of the available AT stations and the required tracking and OPD compensation. In the beam combination laboratory, pupil plane combination of the two beams is implemented by means of the test instrument VINCI operating in the K band at 2.2 μ. The whole system together with its scientific objectives is described in greater detail in the ESO web site: http://www.eso.org/projects/vlti/.

The currently achieved delay line precision is remarkable: flatness of rails better than 25 μm over 65m, an absolute position accuracy of the carriages of $\sim$ 30 μm and a relative position error of $\sim$ 20 nm RMS over 50 ms. It is this phenomenal precision and stability that makes the VLTI possible and unique.

2 First Science Results

First fringes with the siderostats were obtained March 17, 2001 and with UT1 and 3 on October 30, 2001. Technical commissioning is ongoing with highest pri-

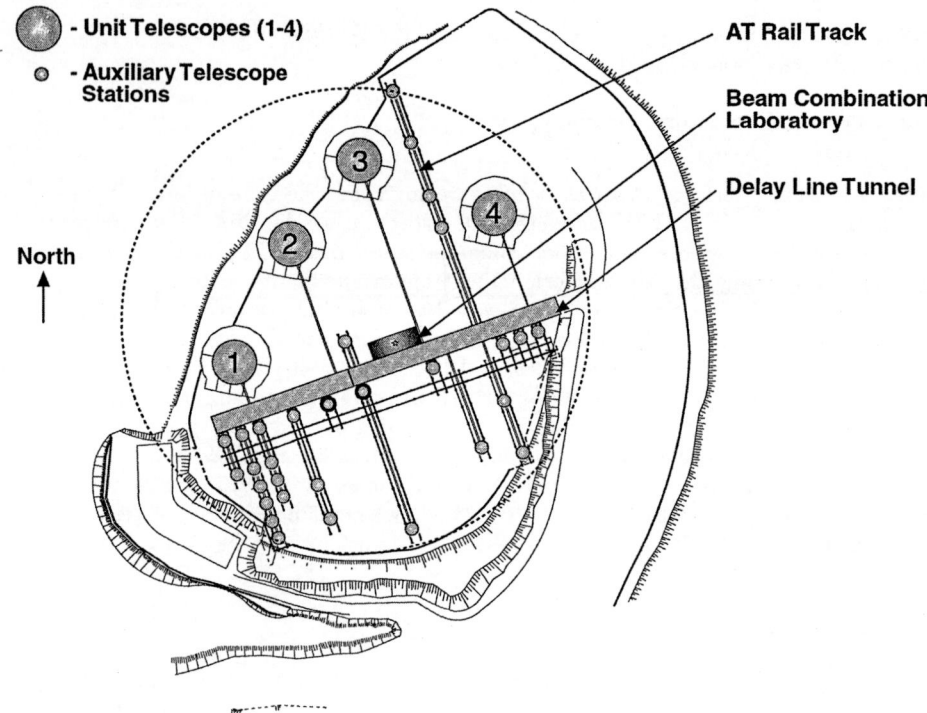

Fig. 1. Schematic layout of the VLT Interferometer facility on Cerro Paranal. The dotted circle has a diameter of 200 m for scale. The small circles indicate the auxiliary telescope (AT) and siderostat stations. The original siderostat positions with a 16 m almost E-W baseline are marked in bold relief. The current positions correspond to the W siderostat remaining where it is and the E siderostat moved to most Southerly station on the same track for a 66 m almost N-S baseline

ority. During natural pauses, observations of scientifically interesting sources take place. An internal science group decides on sources to be observed and the list is approved by the project manager responsible for commissioning. All scientifically interesting data taken in the period from March 17, 2001 to March 25, 2002 have been released to the community and are currently available from the ESO archive:
`http://www.eso.org/projects/vlti/instru/vinci/vinci_data_sets.html`.

About 25 ESO community scientists have availed themselves of the opportunity and are presently working on data taken so far. We encourage everyone to try! Data release is expected about every 3 months. VINCI and the siderostats have also been made available in service mode to the community on a shared risk basis starting on October 1, 2002 (Period 70). The deadline was April 3, 2002 and 39 proposals were received by that date.

The total number of objects measured so far is 140 (most of them repeatedly), 57 for the first time! The breakdown: 1 AGN, 1 W-R star, 1 LBV, 1 symbiotic nova, 1 S star, 3 Cepheids, 3 YSO, 3 emission line stars, 3 shell stars (IR excess), 6 C stars, 10 MS dwarfs, 12 Spectroscopic binaries, 39 Late-type giants, and 56 Miras.

The Cepheids are used to obtain a distance determination as precise as 1% or better by accurately measuring the variation of the star's diameter through the full pulsation cycle. The corresponding measurements of the radial velocity variations yield the distance to the object since both the physical and the angular sizes of the motion are determined simultaneously. Observations of the Cepheid Zeta Gem ($K = 2.1$, $D \sim 2.2\,\mathrm{mas}$, $d = 360\,\mathrm{pc}$, $P = 10\mathrm{d}$), for example, have shown that the required observing precision to track the pulsation has been achieved ($1.78 \pm 0.02\,\mathrm{mas}$) and that all that remains to be done is to follow the 10d period to extract the distance.

K velocimetry is limited mainly by the projection factor which is model and limb darkening dependent. Currently, the precision with which this parameter can be obtained is $\sim 1\%$ which corresponds approximately to the final accuracy on the distance. Sampling over many baseline orientations and beyond the first null for limb darkening effects should push this accuracy down to $\sim 0.1\%$. With these accuracies, the anchor of the distance ladder that is based mainly on the Cepheids will take a huge leap forward in usefulness and confidence.

The LBV Eta Carinae was also observed at the 3 different baseline lengths and orientations available with excellent results as the measured visibilities were all $> 20\%$ and relatively easy to measure. Less straightforward is the interpretation of the measurements. Some success has been obtained with wind models coupled to a disk to reconcile the observations with ISO spectra and high resolution AO images of the nuclear region.

A similar study is ongoing on the symbiotic nova R Aquarii whose diameter is being monitored carefully to detect variations due to the Mira's 387d pulsations. Measurements taken at different times cluster around the value of 16.13 mas as expected for a typical Mira of this type. The variations of the visibilities in time will be carefully monitored for signs of the possible hot white dwarf secondary.

3 Future Prospects

Future enhancements of the VLTI's capabilities will allow an enormous increase in the quantity and quality of the VLTI's scientific output. These will include, for next year (2003), the implementation of an axis fringe tracker (FINITO), AO with MACAO on the UTs and an extension to the N band ($10\,\mu$) with the MIDI instrument and 3-way simultaneous beam combination at J,H,K with AMBER. In addition, the first 3 ATs are expected to become operational in the next few years allowing 100% time coverage for interferometry. A little further down the line, phase referenced imaging and μas astrometry will be available thanks to the PRIMA facility now in an advanced planning stage.

Thus, the future looks bright for interferometry at Cerro Paranal. Thus, the entire European community should make full use of this ground-breaking facility.

Acknowledgements

The VLTI is where it is now thanks to the efforts of a large number of people. The work described very briefly here could not have been reported without the critical contributions of C. Cesarsky, S. Correia, F. Delplancke, F. Derie, E. DiFolco, Duc Than Phan, A. Gennai, R. Giacconi, P. Gitton, B. Koehler, S. Menardi, S. Morel, J. Spyromilio, A. Wallander, ... and many others at ESO, Fokker, Halfmann, Meudon etc.

Formation of Planetary Systems

Ewine F. van Dishoeck

Leiden Observatory, P.O. Box 9513, NL-2300 RA Leiden, The Netherlands

Abstract. A brief summary of our current understanding of the physical structure and evolution of circumstellar disks – the sites of planet formation – is given. The different scenarios for planet formation through agglomeration of particles or gravitational instabilities in disks are discussed. Recent examples of observations of disks in the embedded phase and in the transitional phase to debris disks are given. The chemical evolution of the gas and dust from protostars to new solar systems is illustrated. Prospects for future ESO instrumentation and ESA missions are summarized.

1 Introduction

The formation of stars and planets occurs deep inside clouds and disks of gas and dust with hundreds of magnitudes of extinction. In the standard scenario (see Fig. 1), gravitational collapse of part of an interstellar cloud leads to the formation of a protostar at the center which derives most of its luminosity from accretion. Because the cloud has some initial rotation, the material cannot continue to fall in radially but will end up in a rotating circumstellar disk through which accretion onto the growing star occurs. Shortly after its formation, the protostar develops a wind which will gradually blow away the surrounding envelope. Initially, the wind can escape only in a direction perpendicular to the disk, resulting in the bipolar CO outflows and jets observed to be widely associated with young stellar objects. When most of the envelope has been removed, a pre-main sequence star becomes visible at optical wavelengths but the star is still surrounded by the disk which is heated by the star light and emits at longer wavelengths (> 1 μm). It is this disk from which planets and other solar-system bodies may subsequently form over a period of ~ 100 Myr.

In this brief review, some aspects of the structure and evolution of disks are discussed, especially as they relate to scenarios for planet formation. Recent reviews of circumstellar disks include Beckwith (1999), Koerner (2001) and several chapters in the *Protostars & Planets IV* book, eds. Mannings et al. (2000).

2 Circumstellar Disks

2.1 Disks Around Pre-Main Sequence Stars

The first observational evidence for disks around low-mass pre-main sequence stars (ages $\sim$ few Myr, masses $<$ few $M_{\odot}$) came from observations at infrared

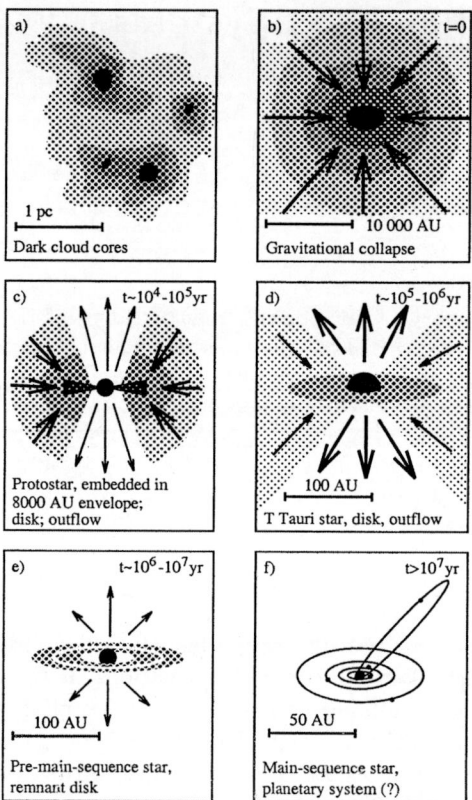

Fig. 1. Stages in the star- and planet-formation process for the case of a single, isolated low-mass star (Hogerheijde 1998, after Shu et al. 1987)

and submillimeter wavelengths (see Beckwith & Sargent 1996 for a review). Excess emission over the stellar photosphere was detected at near-infrared (e.g., Strom et al. 1989) and mid-infrared wavelengths (e.g., Lada & Wilking 1984) using ground-based telescopes and the InfraRed Astronomical Satellite (IRAS). Even longer wavelength radiation at ~1 millimeter due to optically thin thermal emission from dust was observed from a large sample of objects by Beckwith et al. (1990). Although these data could not resolve or image the disks, the lack of obscuration of the star and the presence of bipolar outflows in the younger stages strongly suggested that the material had to be distributed in a flattened, disk-like geometry. Pioneering millimeter interferometry observations of the CO molecule showed that the gas motions are indeed consistent with rotating disks (e.g., Koerner & Sargent 1995, Dutrey et al. 1996). In some special cases, the disks have been beautifully revealed by high-resolution optical and infrared images, either as silhouettes against a bright background nebula (e.g., McCaughrean & O'Dell 1996), or in nearly edge-on situations against scattered light (e.g., Padgett et al. 1999).

The main results of these observations are: (a) most (>50%) of pre-main sequence stars are surrounded by disks; (b) the disks have radii up to a few hundred AU, comparable to the size of our own solar system (40 AU out to Pluto); and (c) the masses of the disks are typically 0.01 $M_\odot$ (~10 M_{Jup}) of gas and dust, sufficient to form a planetary system like our own. Thus, the ingredients and conditions for planet formation appear to be widely available in our solar neighborhood, and presumably throughout much of the Universe.

2.2   Disks Around Main-Sequence Stars

Disks have also been detected around much older nearby main-sequence stars (ages >10 Myr). IRAS discovered strong excess far-infrared emission from the A0V star Vega (Aumann et al. 1984) and a handful of other objects (see Backman & Paresce 1993, Lagrange et al. 2000 for reviews). The masses of these disks are orders of magnitude lower than those of disks around the younger pre-main sequence stars, with only ~ 10^{-6} $M_\odot$ (~ 1 M_{Earth}) of dust in particles with sizes up to a few millimeter. Because of the tiny amount of dust, ultraviolet radiation can penetrate throughout the disks. Surveys with the Infrared Space Observatory (ISO) indicate that ~15% of the A-stars are surrounded by such tenuous disks (Habing et al. 2001).

Spatially resolved images in scattered light have been obtained with coronographs at optical and near-infrared wavelengths (e.g., Smith & Terrile 1984, Weinberger et al. 1999), and in thermal emission at mid-infrared (e.g., Koerner et al. 1998) and submillimeter wavelengths using the bolometer array camera SCUBA on the James Clerk Maxwell Telescope (JCMT) (Holland et al. 1998, Greaves et al. 1998). As Fig. 2 shows, the sizes of these disks are similar to those in the younger stages (few hundred AU radii), but they often show lumps, gaps and holes. These features have been interpreted as indirect evidence for planet formation, in which the protoplanet clears out a ring in the disk by accreting and sweeping up surrounding material (e.g., Bryden et al. 1999 and references cited). Contrary to the case for pre-main sequence stars, the gas and dust in these so-called 'debris disks' are not the remains of the interstellar material, but they are thought to originate from collisions and evaporation of larger planetary bodies. The original interstellar dust and gas are predicted to disappear on timescales which can be as short as 10^5 yr in the inner disk through radiation pressure, Poynting-Robertson drag, or collisions leading to destruction of the grains (Backman & Paresce 1993).

3   Theories of Planet Formation

There are two main theories for the formation of planets: the agglomeration scenario and the gravitational instability scenario.

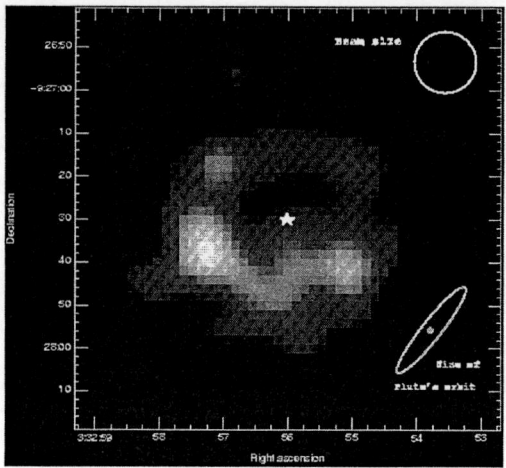

Fig. 2. Submillimeter continuum emission at 850 μm of the cold ($T_{\mathrm{dust}} \approx 30$ K) dust disk around the K2V star ϵ Eridani (d=3 pc). The emission has a radius of 115 AU, comparable to the size of the Kuiper Belt of icy bodies in the outer part of our solar system. Note the irregular structure with 'lumps' and 'holes' (Greaves et al. 1998)

3.1   Agglomeration/Accretion Scenario

The mechanism in which the rocky terrestrial planets and the gaseous Jovian planets form through successive agglomeration of dust particles and subsequent accretion of gas was first put forward by Safronov (1969) and subsequently elaborated by many others (see reviews and papers by Wetherill 1990, Weidenschilling 1997, Ruden 1999, Beckwith et al. 2000). In this scenario, grain growth to $\sim \mu$m- and cm-size particles first occurs through sticking and coagulation of ~ 0.1 μm-size interstellar dust particles. Once the turbulence in the disk has decreased to the point that it becomes comparable to the thermal and rotational motions, such growth can occur very rapidly on timescales of $< 10^5$ yr. Although the precise details of the growth to larger particles with sizes of ~1 m are not yet fully understood, it is thought that this process can eventually lead to ~km-size planetesimals. These planetesimals are large enough that they can interact with other particles through gravity, which enhances the attraction and leads to lunar-size bodies (~10^3 km, 10^{26} gr) on time-scales of ~ 10 Myr.

Numerical simulations of the interactions between these protoplanets show that the evolution can actually take two routes (e.g., Wetherill & Stewart 1989): (i) a steady-state orderly growth in which Earth-like planets with sizes of 10^4 km and masses of $\sim 6 \times 10^{27}$ gr are formed in 100 Myr; (ii) a run-away accretion in which a rocky core of $\sim$ M$_{\mathrm{Earth}}$ is formed in less than 10 Myr. If this core is sufficiently massive, > 10 M$_{\mathrm{Earth}}$, it can subsequently accrete gas by gravitational attraction (Mizuno 1980). Note that the formation of such gas-rich Jovian planets must occur early in the disk evolution, when the H$_2$ gas is still present.

The main problem with this popular scenario is that the time scale in the standard minimum-mass solar nebula of Hayashi (1981) to form the large rocky Earth-like cores is close to, or even larger than, the time scale on which the gas is thought to disappear from the disk (see §4.2). Thus, only disks more massive than about ~0.01 $M_\odot$ can produce gas-rich Jovian planets.

3.2 Gravitational Instabilities

In this scenario, massive gas-rich disks are subject to instabilities caused by gravitational torques, which can cause part of a disk to collapse and form a giant gas-rich planet. This scenario, first put forward by Kuiper (1951), Cameron (1978) and Goldreich & Ward (1973) and most recently elaborated by Boss (1997, 2001), can occur already in the embedded phase of star formation, where the interaction with the envelope or with companion stars can enhance the instabilities. The required disk surface density and mass is again fairly high, >0.01 $M_\odot$. Fragmentation can lead to multiple systems.

The two theories clearly differ in their predictions for the formation of giant gaseous planets. The time scales range from more than 10 Myr to less than 1 Myr in the agglomeration and disk instability theories, respectively. The location varies from the inner to the outer disk. Future observations will be able to distinguish between these scenarios through statistical studies, both through high resolution imaging of gaps in disks in different evolutionary stages and through the distribution of exo-planets with radius.

4   Examples of Recent Issues

4.1   Disks in the Deeply-Embedded Protostellar Stage

The advent of large bolometer arrays on single-dish submillimeter telescopes has allowed unbiased searches for the earliest, deeply embedded protostars (see André et al. 2000 for a review). In these sources, which have ages of only a few$\times 10^4$ yr, the star is still being assembled through accretion of material from the circumstellar disk. High resolution imaging (e.g., Looney et al. 2000) has shown that many of these sources are binary or multiple systems. Compact continuum emission is seen around most sources, but the lack of spatial resolution and spectral line sensitivity prevents the disk and envelope contributions to be disentangled.

A rare example of a clear detection of a large (2000 AU radius) rotating disk in the embedded phase is shown in Fig. 3 (Hogerheijde 2001). This object, L1489 IRS, is in the transitional phase to the pre-main sequence stage at an age of a few $\times 10^5$ yr. According to the theory of the formation of rotating disks by Terebey et al. (1984), the disk radius grows with time as t^3 as long as matter continues to fall in from the envelope. Infall motions are detected in the disk, indicating that it may contract to the typical size of a few hundred AU on a time scale of a few times 10^4 yr, once accretion from the envelope ceases.

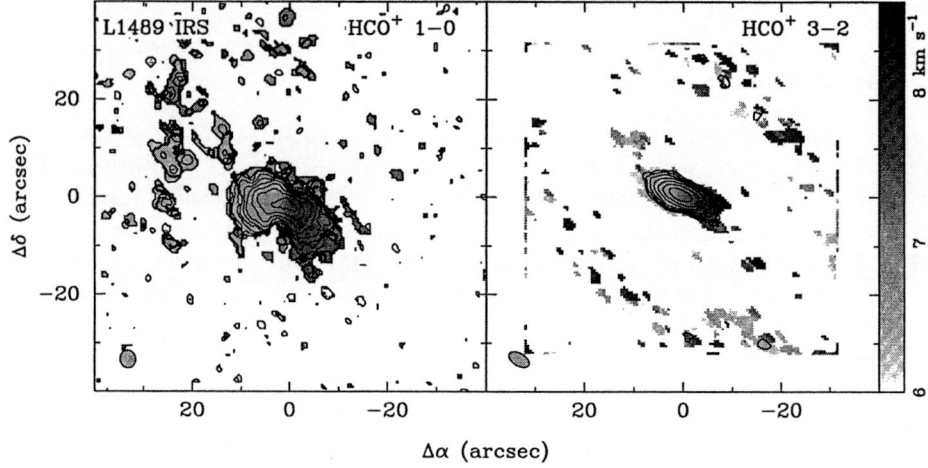

Fig. 3. Detection of a 2000 AU radius rotating disk around the embedded young stellar object L1489 IRS, which is in transition to the pre-main sequence phase. The image shows the HCO^+ 1–0 and 3–2 emission and velocity pattern due to Keplerian rotation obtained with the Berkeley-Illinois-Maryland Array (BIMA) (Based on: Hogerheijde 2001)

4.2   Time Scales for Disk Dissipation

The time scale for the evolution of the massive gas-rich disks seen around pre-main sequence stars to the tenuous dust debris disks around mature stars, and the related time scales of the gas and dust dissipation, are key ingredients for testing models of planet formation (see §3). Several statistical studies have recently been carried out to address these issues.

Dust dissipation: Imaging and photometry of a large set of pre-main sequence stars at near-infrared wavelengths has been performed most recently by Haisch et al. (2001). A set of objects in nearby young clusters with well-determined ages has been targeted, thereby reducing the (large) uncertainty in the age of individual objects. The fraction of sources with excess near-infrared emission drops rapidly to nearly zero after ∼ 5 Myr (see Fig. 4, left). These studies are only sensitive to the warm, few hundred K dust in the inner disk (< few AU), however.

The bulk of the cold dust in disks can only be probed at submillimeter wavelengths. Surveys at 1 millimeter by Osterloh & Beckwith (1995) of a set of isolated classical T Tauri stars in Taurus show no significant decrease in disk mass up to ∼10 Myr, although the older so-called weak-line T Tauri stars have weaker millimeter fluxes. Statistics on sources with higher ages and in other star-forming regions are still poor, but the general lack of detection of submillimeter emission from older objects suggests a decrease in the disk mass beyond 10 Myr by at least one order of magnitude (Meyer et al., in preparation).

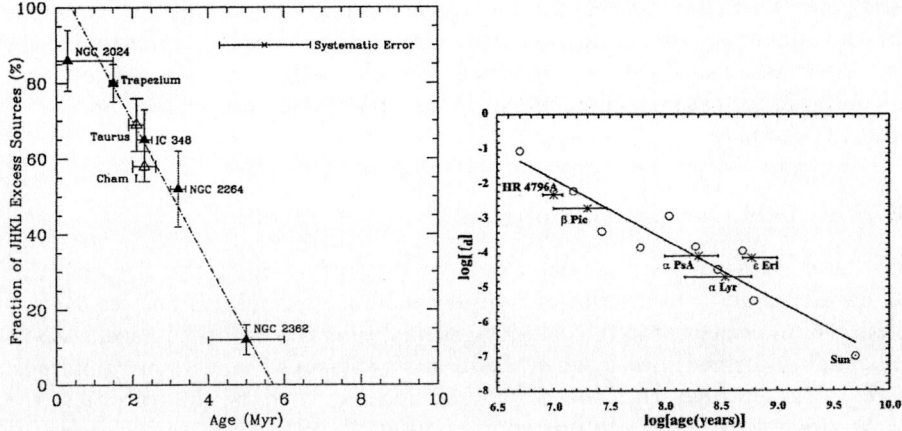

Fig. 4. Left: Fraction of disks which show near-infrared 3 μm emission in excess of the stellar photospheric emission due to thermal radiation from warm dust in the inner disk as a function of age of the stellar cluster (Haisch et al. 2001); Right: Fractional excess luminosity f_d of dust with respect to the stellar bolometric luminosity based on far-infrared 60–100 μm emission from cool dust (Spangler et al. 2001). Note the different age scales in the two plots

ISO far-infrared statistics indicate that the debris disks around nearby main-sequence stars disappear after ~400 Myr, comparable to the timescale of heavy bombardment in our own solar system (Habing et al. 2001), and that the fraction of disks with excess emission decreases proportional to t^{-2} (Spangler et al. 2001, Fig. 4, right).

Gas dissipation: The principle molecule used to trace gas in disks is CO, which has been imaged with millimeter interferometry toward a few dozen low- and intermediate mass pre-main sequence stars (Dutrey et al. 1996, Mannings & Sargent 1997, Natta et al. 2000). Searches for CO in a few disks with higher ages (>10 Myr) have yielded largely negative results (e.g., Duvert et al. 2000, Greaves et al. 2000), suggesting that the gas disappears on time scales of < 10 Myr, similar to the conclusion from the warm dust. However, chemical modeling of disks has shown that CO can be destroyed by photodissociation in the outer regions of disks and by freeze-out in the cold midplane (e.g., Aikawa et al. 2002), reducing the value of this molecule as a probe of gas mass.

The main gaseous constituent of disks is H_2, which does not suffer from freeze-out and can protect itself against photodissociation (e.g., Kamp & Bertoldi 2000). H_2 does not have any transitions at millimeter wavelengths, because it is a light symmetric molecule with no dipole moment. The lowest-lying quadrupole transitions occur at mid-infrared wavelengths, at 28 μm ($J=2\rightarrow0$) and 17 μm ($J=3\rightarrow1$). Deep searches for these lines have been made with ISO by Thi et al. (2001) and some tentative detections have been found, perhaps even in objects with ages >10 Myr. Because the energy levels lie at more than 500 K above ground, only warm gas (>50 K) can be probed by this method, but the advantage is that no abundance estimates or conversion factors are needed to determine the warm gas mass. The H_2 data, combined with theoretical models, suggest that the time scale for gaseous giant planet formation may be longer than indicated by the CO data.

4.3   Chemical Composition of Disks

From the collapse of a dense cloud core to the formation of the circumstellar disk, molecules undergo a series of complex chemical changes. Processes such as depletion of molecules onto the cold icy grains during the collapse phase, evaporation of newly-formed species when the protostar starts to heat its surroundings, and high temperature reactions in shocked zones created by the impact of the outflow, cycle molecules from one compound into another (see van Dishoeck & Blake 1998, Langer et al. 2000, Ehrenfreund & Charnley 2000 for overviews).

Observations of gas-phase molecules other than CO in disks have become possible only recently with single-dish submillimeter telescopes (e.g., Dutrey et al. 1997, Kastner et al. 1997, van Zadelhoff et al. 2001, Thi et al. 2002) or interferometers (Qi 2000). These observations are only sensitive to the outer disk at radii greater than 50 AU. Several simple molecules such as HCO^+, HCN, CN and H_2CO have been detected, indicating the importance of ion-molecule reactions (HCO^+) and photodissociation (CN) in the disks. Chemical modeling has shown that most of the active chemistry occurs in an intermediate layer of the disk just below the surface, which is warm enough to prevent freeze-out of the molecules and sufficiently shielded from ultraviolet radiation to prevent rapid photodissociation (e.g., Aikawa et al. 2002).

Solid-state species including silicates, ices and large molecules such as polycyclic aromatic hydrocarbons (PAHs) can be probed in disks through emission and absorption features at mid-infrared wavelengths. A gas-phase species can be readily distinguished from a molecule in the solid phase by its characteristic ro-vibrational spectrum. Emission features due to various species have been detected in disks around intermediate mass Herbig Ae stars by ISO (e.g., Waters & Waelkens 1998, Meeus et al. 2001) and show a bewildering variation from source to source. Not only amorphous silicates, but also crystalline material has been identified. The emission arises predominantly from the inner warm disk where the features can be excited.

5   Prospects for Future ESO-ESA Facilities

Current observational studies of circumstellar disks and planet formation suffer greatly from lack of spatial resolution and sensitivity, and have therefore only revealed the 'tip of the iceberg'. A wealth of new observational facilities will remedy this situation in the next decade. A brief summary of instruments relevant for studies of planet formation is given below; details can be found at the respective WWW sites.

Table 1. Characteristic scales of protoplanetary disks around young stars

Stage	Component	Typical distance of objects (pc)	Radius (AU)	Radius ('')
Deeply embedded Class 0	Entire disk	300	100	0.3
Pre-main sequence Class I & II	Entire disk	140	100	0.7
	Solar system/Kuiper Belt		30	0.2
	Planet migration		0.1–15	0.007–0.1
	Gaps by planets		2	0.015
Old pre-main sequence & ZAMS	Entire disk	10–60	100–1000	10–1.7
				100–17

ALMA: The Atacama Large Millimeter Array (ALMA) is a joint project between North America and Europe to build an array of 64×12m submillimeter telescopes on the Chajnantor 'altiplano' at 5000 m in Chile to be completed by $\sim$2011 (Kurz et al. 2002). Japan may also join the project. It will be equipped with heterodyne receivers covering the 30–900 GHz atmospheric windows at a resolving power $R > 10^6$. Its observing speed will surpass that of existing interferometers by up to a factor of 10^4.

Protoplanetary disks are prime targets for ALMA. Because of their small angular size (see Table 1), current millimeter arrays have only probed the outer disks. ALMA will image disks down to the 1 AU scale, with masses as low as a fraction of an Earth mass in the nearest star-forming regions. By using the highest angular resolution and multifrequency observations, ALMA will be able to search for gaps in disks and measure the changes in dust properties, which may be interpreted as signatures of planet formation and dust settling in the midplane. By observing a large sample, ALMA will be able to constrain the frequency and time scales for these processes as functions of stellar mass, luminosity and environment. ALMA will have the sensitivity to map lines down to a few AU resolution, providing information about the gas content, chemistry and kinematics. ALMA will also be able to image the dust and CO gas in debris disks with an order of magnitude higher spatial resolution than current facilities, in systems which are more than an order of magnitude weaker.

VLT, VLTI: Various new instruments on the ESO-VLT will be well suited for studying (bright) circumstellar disks. At near-infrared wavelengths (1–5 μm), ISAAC is currently providing data on thermal emission and scattered light from warm dust (e.g., Brandner et al. 2000) and CO emission and absorption (e.g., Pontoppidan et al. 2002). The arrival of CRIRES in 2004 with much higher spectral resolving power up to 10^5 is eagerly awaited for gas-phase chemistry and kinematic studies. In the mid-infrared (8–22 μm), VISIR will be able to image the thermal emission from dust and perform spectroscopic observations of selected solid-state and gaseous species. The VLTI will allow the highest angular resolution imaging of warm dust and features down to 1 AU, which will provide stringent tests of different models for the location and origin of the dust emission.

NGST, SIRTF, SOFIA: The Next Generation Space Telescope (NGST), the 6-m class successor of the Hubble Space Telescope, is a joint NASA-ESA-CSA project optimized for near-infrared wavelengths with a planned launch in 2010. The recommended suite of instruments includes a mid-infrared camera/spectrometer, with a sensitivity surpassing that of previous facilities by orders of magnitude and with a spatial resolution that is a factor of 10 higher than that of previous space missions. The mid-infrared instrument on NGST will have the sensitivity to detect and image debris disks with masses down to a fraction of a lunar mass around main-sequence stars and will thus provide, together with ALMA, a census of disks over the entire planet building phase. NGST will also be able to probe the wealth of solid-state features in disks, and, with sufficient spectral resolution, search for the H_2 mid-infrared lines with a sensitivity down to a fraction of a Jupiter mass for gas at $T \approx 50$ K, thus constraining the gas-dissipation time scales.

Prior to NGST, NASA will launch in 2003 the Space InfraRed Telescope Facility (SIRTF), a 85 cm cooled telescope. It has two cameras covering the 3–180 μm range, and one spectrometer covering 5–40 μm at $R=\lambda/\Delta\lambda=60$–120 and 10–38 μm at $R=600$. NASA, in collaboration with Germany, will also operate the Stratospheric Observatory For Infrared Astronomy (SOFIA), a 2.5 m telescope in a B747 airplane starting routine flights in 2005. Although its sensitivity will be lower than that of SIRTF because of the higher temperature, it will have a phenomenal set of state-of-the-art instrumentation (cameras, low- and high-resolution spectrometers) covering the full 2–500 μm range at $R = 3 - 10^6$.

Herschel: The Herschel Space Observatory, an ESA cornerstone mission in collaboration with NASA, is a 3.5 m telescope to be launched in 2007. It covers the ~80–500 μm wavelength range with three instruments, including an imaging spectrometer with $R \approx 3000$ from 80–200 μm and heterodyne receivers with $R > 10^6$ from 200–500 μm. It will be well suited to determine the total luminosity of protostars and disks, which, combined with spatially resolved imaging using ALMA, can be used to tightly constrain disk models. Herschel will also have unique spectroscopic capabilities at THz frequencies, which will allow searches for H_2O and O_2, two of the dominant oxygen-bearing molecules, in protostellar objects and disks.

GAIA: The GAIA mission of ESA will make an essential contribution to studies of planet formation by providing accurate distances, and thereby ages and masses, of the pre-main sequence and main-sequence stars. Without GAIA, distance uncertainties may well be the limiting factor in the interpretation of the ALMA and NGST data and statistics.

References

1. Aikawa, Y., van Zadelhoff, G.J., van Dishoeck, E.F., Herbst, E.: A&A **386**, 622 (2002)
2. André, P., Ward-Thompson, D., Barsony, M.: In *Protostars & Planets IV*, ed. by V. Mannings et al. (Univ. of Arizona, Tucson 2000), p.59

3. Aumann, H.H., Beichman, C.A., Gillett, F.C. et al.: ApJ **278**, L23 (1984)
4. Backman, D.E., Paresce, F.: In *Protostars & Planets III*, ed. by E.H. Levy, J.I. Lunine (Univ. of Arizona, Tucson 1993), p. 1253
5. Beckwith, S.V.W.: In *The Origin of Stars and Planetary Systems*, ed. by C.J. Lada, N.D. Kylafis (Kluwer, Dordrecht 1999), p. 579
6. Beckwith, S.V.W., Sargent, A.I.: Nature **383**, 139 (1996)
7. Beckwith, S.V.W., Henning, T., Nakagawa, Y.: In *Protostars & Planets IV*, ed. by V. Mannings et al. (Univ. of Arizona, Tucson 2000), p. 533
8. Beckwith, S.V.W., Sargent, A.I., Chini, R.S., Guesten, R.: AJ **99**, 924 (1990)
9. Boss, A.P.: Science **276**, 1836 (1997)
10. Boss, A.P.: ApJ **563**, 367 (2001)
11. Bryden, G., Chen, X., Lin, D.N.C., Nelson, R.P., Papaloizou, J.C.B.: ApJ **514**, 344 (1999)
12. Brandner, W., Sheppard, S., Zinnecker, H., et al.: A&A **364**, L13 (2000)
13. Cameron, A.W.G. Moon Planets, **18**, 5 (1978)
14. Dutrey, A., Guilloteau, S., Duvert, G., et al.: A&A **309**, 493 (1996)
15. Dutrey, A., Guilloteau, S., Guélin, M.: A&A **317**, L55 (1997)
16. Duvert, G., Guilloteau, S., Ménard, F., Simon, M., Dutrey, A.: A&A **355**, 165 (2000)
17. Ehrenfreund, P., Charnley, S.B.: ARA&A **38**, 427 (2000)
18. Goldreich, P., Ward, W.R.: ApJ **183**, 1051 (1973)
19. Greaves, J.S., Coulson, I.M., Holland, W.S.: MNRAS **312**, L1 (2000)
20. Greaves, J.S., Holland, W.S., Moriarty-Schieven, G., et al.: ApJ **506**, L133 (1998)
21. Habing, H.J., Dominik, C., Jourdain de Muizon, M., et al.: A&A **365**, 545 (2001)
22. Haisch, K.E., Lada, E.A., Lada, C.J.: ApJ **553**, L153 (2001)
23. Hayashi, C.: Prog. Theor. Phys. Suppl. **70**, 35 (1981)
24. Hogerheijde, M.R.: PhD Thesis, University of Leiden (1998)
25. Hogerheijde, M.R.: ApJ **553**, 618 (2001)
26. Holland, W., Greaves, J.S., Zuckerman, B. et al.: Nature **392**, 788 (1998)
27. Kastner, J.H., Zuckerman, B., Weintraub, D.A., Forvcille, T.: Science **277**, 67
28. Kamp, I., Bertoldi, F.: A&A **353**, 276 (2000)
29. Koerner, D.W.: In *Tetons 4: Galactic Structure, Stars and the Interstellar Medium*, ASP Conf. series 231, ed. by C.E. Woodward et al. (ASP, San Francisco 2001), p. 563
30. Koerner, D.W., Sargent, A.I.: AJ **109**, 2138 (1995)
31. Koerner, D.W., Ressler, M.E., Werner, M.W., Backman, D.E.: ApJ **503**, L83 (1998)
32. Kuiper, G., Proc. Natl. Acad. Sci. USA, **37**, 1 (1951)
33. Kurz, R., Guilloteau, S., Shaver, P.: Messenger **107**, 7 (2002)
34. Lada, C.J., Wilking, B.: ApJ **287**, 610 (1984)
35. Lagrange, A.-M., Backman, D.E., Artymowicz, P.: In *Protostars & Planets IV*, eds. V. Mannings et al. (Univ. Arizona, Tucson 2000), p. 639
36. Langer W.D., van Dishoeck E.F., Blake G.A. et al.: In *Protostars & Planets IV*, ed. by V. Mannings et al. (Univ. Arizona, Tucson 2000), p. 29
37. Looney, L.W., Mundy, L.G., Welch, W.J.: ApJ **529**, 477 (2000)
38. Mannings, V.G., Sargent, A.I.: ApJ **490**, 792 (1997)
39. Mannings, V.G., Boss, A., Russell, S., eds.: *Protostars & Planets IV*, (Univ. Arizona, Tucson 2000)
40. McCaughrean, M.J., O'Dell, R.C.: AJ **111**, 1977 (1996)
41. Meeus, G., Waters, L.B.F.M., Bouwman, J., van den Ancker, M.E., Waelkens, C., Malfait, K.: A&A **365**, 476 (2001)

42. Mizuno, H.: Prog. Theor. Phys **64**, 544 (1980)
43. Natta, A., Grinin, V., Mannings, V.: In *Protostars & Planets IV*, ed.by V. Mannings et al. (Univ. Arizona, Tucson 2000), p. 559
44. Osterloh, M., Beckwith, S.V.W.: ApJ **439**, 288 (1995)
45. Padgett, D.L., Brandner, W., Stapelfeldt, K.R., Strom, S.E., Terebey, S., Koerner, D.: AJ **117**, 1490 (1999)
46. Pontoppidan, K.M., Schöier, F.L., van Dishoeck, E.F., Dartois, E.: A&A, in press (2002)
47. Qi, C.: PhD Thesis, California Institute of Technology (2000)
48. Ruden, S.: In *The Origin of Stars and Planetary Systems*, ed. by C.J. Lada, N.D. Kylafis (Kluwer, Dordrecht 1999), p. 643
49. Safronov, V.S. *Evolution of the Protoplanetary Cloud and Formation of the Earth and Planets* (Nauka, Moscow 1969)
50. Shu, F.H., Adams, F.C., Lizano, S.: ARA&A **25**, 23 (1987)
51. Smith, B.A., Terrile, R.J.: Science **226**, 1421 (1984)
52. Spangler, C., Sargent, A.I., Silverstone, M.D., Becklin, E.E., Zuckerman, B.: ApJ **555**, 932 (2001)
53. Strom, K.M., Strom, S.E., Edwards, S., Cabrit, S., Skrutskie, M.F.: AJ **97**, 1451 (1989)
54. Terebey, S., Shu, F.H., Cassen, P.: ApJ **286**, 529 (1984)
55. Thi, W.F., van Dishoeck, E.F., Blake, G.A., et al.: ApJ **561**, 1074 (2001)
56. Thi, W.F., van Dishoeck, E.F., van Zadelhoff, G.J.: A&A, submitted (2002)
57. van Dishoeck, E.F.: In *Origins of Stars and Planets: The VLT View*, ed. by J. Alves, M. McCaughrean (Springer, Heidelberg 2002), in press
58. van Dishoeck E.F., Blake G.A.: ARA&A **36**, 317 (1998)
59. van Zadelhoff, G.J., van Dishoeck, E.F., Thi, W.F., Blake, G.A.: A&A **377**, 566 (2001)
60. Waters, L.B.F.M., Waelkens, C.: ARA&A **36**, 233 (1998)
61. Weidenschilling, S.J.: Icarus **127**, 290 (1997)
62. Weinberger, A.J., Becklin, E.E., Schneider, G. et al.: ApJ **525**, L53 (1999)
63. Wetherill, G.W.: Ann. Rev. Earth Planet Sci. **18**, 205 (1990)
64. Wetherill, G.W., Stewart, G.R.: Icarus **77**, 330 (1989)

Extrasolar Planetary Systems

Michel Mayor and Nuno C. Santos

Geneva Observatory, 51 ch. des Maillettes, CH-1290 Sauverny, Switzerland

Abstract. Radial Velocity surveys have revealed up to now about 100 extra-solar planets ($M \sin i < 10\, M_{Jup}$) and 7 multi-planetary systems. The discovered planets present a wide variety of orbital elements and masses, which are raising many problems and questions regarding the processes involved in their formation. But the analysis of the distributions of orbital elements, like the period and eccentricity distributions is already giving some constraints on the formation of the planetary systems. Furthermore, the study of the planet host stars has revealed the impressive role of the stellar metallicity on the giant planet formation. The chemical composition of the molecular cloud is probably the key parameter to form giant planets. In this article we will review the current status of the research on this subject.

1 Introduction

Although only recently the presence of planets around stars other than the Sun was verified, the existence of such bodies was not unexpected. Indeed, the planetary formation is believed to be a simple by-product of the stellar formation process. As a cloud of gas and dust contracts to give origin to a star, conservation of angular momentum leads to the formation of a flat disk of gas and dust around the central newborn "sun". In this flattened disk, dust particles and ice grains are then gathered to form the first planetary seeds. This picture, today widely accepted, is confirmed by the discovery of numerous disks around young stars in the last few years (c.f. [22]).

But it was only in 1995, following the discovery of the planet orbiting the solar-type star 51 Peg [21], that the search for extra-solar planets finally had its first success. Today, about 100 extra-solar planetary systems have been unveiled around stars other than our Sun[1]. These discoveries, that include 7 multi-planetary systems, have brought to light the existence of planets with a huge variety of characteristics, opening unexpected questions about the processes of giant planetary formation.

Since planets are cold bodies, their visible spectrum results basically from reflected light of the parent star. As a result, the planet/stellar luminosity ratio is of the order of 10^{-9} (at visible wavelengths), and seen from a distance of a few parsec, a planet is no more than a small "undetectable" dot embedded in the diffraction and/or aberration of the stellar image. This facts make it very difficult to look for planets by direct imaging.

---

[1] See, e.g., table at
http://obswww.unige.ch/~naef/who_discovered_that_planet.html

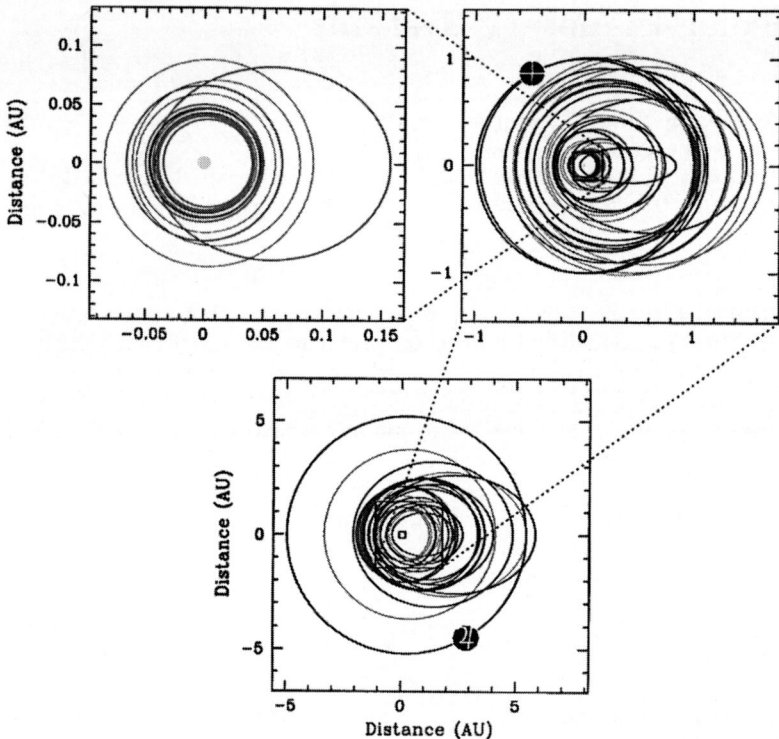

Fig. 1. Schematic orbital configurations for some of the newly found extra-solar planets. In the upper-left panel, the Sun is drawn to scale. In this same panel we can perfectly see the completely lack of planetary companions orbiting closer to a certain distance from their host stars (see text for more details). The orbits of the Earth and Jupiter are also drawn for comparison in the upper-right and lower panels, respectively

The discovery of planets around other stars was thus (up to now) only possible due to the development of high-precision radial-velocity techniques. These techniques have permitted astronomers to look for the tiny periodic motion of a star as it moves around the center-of-mass of the star-planet system. The biggest challenge, however, is that one needs to measure the stellar velocity with a very high-precision. For example, Jupiter induces a periodic perturbation with an amplitude of only $13 \, \mathrm{m s^{-1}}$ on the Sun. For comparison, current techniques have already achieved precisions of about $3 \, \mathrm{m s^{-1}}$ (corresponding to $\Delta\lambda/\lambda = 10^{-8}$). This precision permitted the discovery of extra-solar giant planets with masses lower than the one of Saturn ($\sim$95 times the mass of the Earth).

The diversity of the discovered extra-solar planets is well illustrated in Fig. 1. Unexpectedly, they don't have much in common with the giant planets in our own Solar System. Contrarily to these latter, the "new" worlds present an enormous and unexpected variety of masses and orbital parameters (astronomers were basically expecting to find "jupiters" orbiting at $\sim$5 A.U. or more from their

host stars in *quasi*-circular trajectories). In fact they were not even supposed
to exist according to the traditional paradigm of giant planetary formation [28].
Their masses vary from sub-saturn to various times the mass of Jupiter. Some
have orbits with semi-major axis smaller than the distance from Mercury to
the Sun, and except for the closest companions, they generally follow eccentric
trajectories, contrary to the case of the giant-planets in the Solar System.

But the relatively large number of discovered planets is already permitting
us to undertake the first statistical studies of the properties of the exo-planets,
as well of their host stars. This is bringing new constraints to the models of
planet formation and evolution. In the rest of this article we will review the
current results on the planetary searches, and in particular we will focus on the
observational constraints the new discoveries are bringing.

2   Statistical Properties of the Extra-Solar Planets

2.1   The Period Distribution

One of the most interesting problems that appeared after the first planets were
discovered has to do with the proximity to their host stars. Giant planets were
previously thought to form (and be present) at distances of a few A.U. from their
host stars [28]. In order to explain the newly found systems, several mechanisms
have thus been proposed. Current results show that *in situ* formation is very
unlikely [1], and we need to invoke inward migration, either due to gravitational
interaction with the disk [9,17,39,24] or with other companions [31,16] to explain
the observed orbital periods.

Although still strongly biased for the long period systems, the period distribution of the extra-solar planetary companions can already tell us something
about the planetary formation and evolution processes. This is particularly true
for the short period systems, for which the biases are not so important. In particular, one of the most impressive features present in the current data is the clear
pile-up of planetary companions with periods around 3 days, and the absence of
any system with a period shorter than this. This means that somehow the process involved in the planetary migration makes the planet "stop" at a distance
corresponding to this orbital period. To explain this fact, several ideas have been
presented, invoking e.g. a magnetospheric central cavity of the accretion disk,
tidal interaction with the host star, Roche-lobe overflow by the young inflated
giant planet, or evaporation.

2.2   The Eccentricity

One of the most enigmatic results to date is well illustrated in Fig. 2. A look
at the figure shows that there are no clear differences between the eccentricity
distributions of planetary and stellar binary systems. How then can this be fit
into the "traditional" picture of a planet forming in a disk? For masses lower
than $\sim 20\,M_{\rm Jup}$, it has been shown that the interaction (and migration) of a

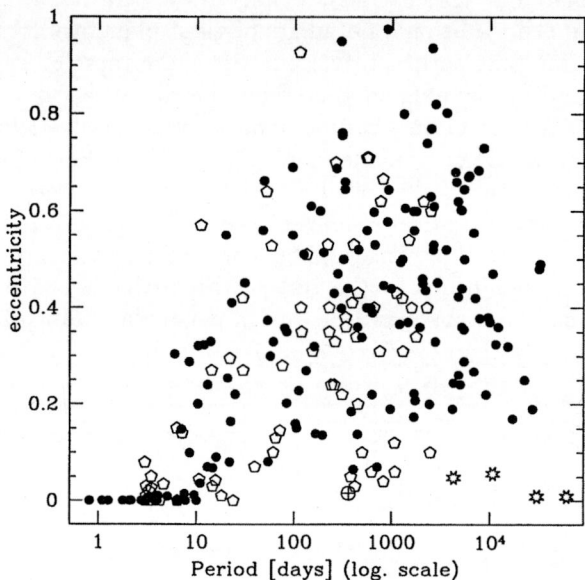

Fig. 2. The $e - \log P$ diagram for planetary (open pentagons) and stellar companions (filled circles) to solar type field dwarfs. Starred symbols represent the giant planets of our Solar System, while the "earth" symbol represents our planet

companion within a gas disk has the effect of damping the eccentricity [9,39,26]. This suggests that other processes, like the interaction between planets in a multiple system [31] or between the planet and a disk of planetesimals [24], the simultaneous migration of various planets in a disk [23], or the influence of a distant stellar companion, may play an important role in defining the "final" orbital configuration. In this respect, one particularly interesting case of very high eccentricity (above 0.9) amongst the planetary companions is the planet around HD 80606 [25].

Although still not clear, however, a close inspection of the Fig. 2, permits to find a few differences between the eccentricities of the stellar and planetary companions. For example, for periods in the range of 10 to 30 days (clearly outside the circularization period by tidal interaction with the star), there are a few stars with planets having very low eccentricity, while no stellar binaries are present in this region. The same and even more strong trend is seen for longer periods, suggesting the presence of a group of planetary companions with orbital characteristics more similar to those of the planets in the Solar System. On the other hand, for the very short period systems, we can see some planetary companions with eccentricities higher than those found for stellar companions of similar period. These facts may be telling us that different formation and evolution processes took place: for example, the former group may be seen as a sign of formation (and evolution) in a disk, and the latter one as an evidence of

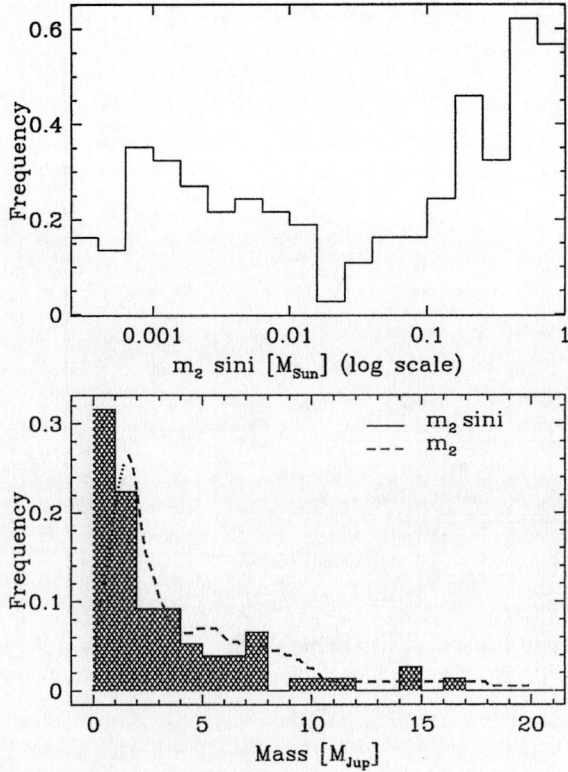

Fig. 3. Mass function of companions to solar-type stars in *log* (top) and linear (bottom) scales. In the lower panel, the dashed line represents the result of a deconvolution of the observed distribution in order to take into account the effect of the orbital inclination. As in [13]

the gravitational influence of a longer period companion on the eccentricity – cf. case of HD 217107 [8].

2.3 The Mass Distribution

Another important clue concerning the nature of the now discovered planetary systems comes from their mass distribution. Although the radial-velocity technique is more sensitive to massive companions than to their lower mass counterparts, a look at the mass distribution (Fig. 3) shows that this distribution strongly rises towards the low mass regime – see Fig. 3, lower panel.

Several conclusions may be taken from the plots. The gap in the distribution, separating low mass stellar companions from the lower mass planets (often called the "brown dwarf desert") represents a strong evidence that these two populations are the result of different formation processes. Furthermore, we can see that the planetary mass distribution has a sharp cutoff for masses around

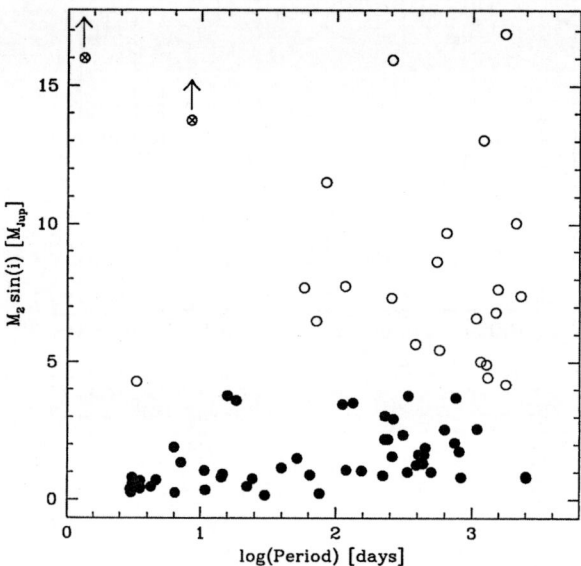

Fig. 4. Planetary minimum mass vs. logarithm of the orbital period for the very low mass companions to solar-type stars. This plot shows the absence of high-mass low-period planetary companions

$\sim 10\,M_{\rm Jup}$ [13]. This limit is clearly not related to the Deuterium-burning mass limit of $\sim 13\,M_{\rm Jup}$, sometimes considered as the limiting mass for a planet. As it was recently shown by [13], this result is not an artifact of the fact that for most of the targets we only have minimum masses[2], but a real upper limit for the mass of the planetary companions discovered so far, since it is clearly visible in a deconvolved distribution (where the effect of the unknown orbital inclination was taken into account).

It is also very interesting to note that recent results strongly suggest that there is some relation between the mass of the companion and its orbital period: there seems to be a paucity of high-mass planetary companions ($M > 2\,M_{\rm Jup}$) orbiting in short period (lower than ~ 40-days) trajectories [40,37] – Fig. 4. This trend, clearly significant, is nonetheless not found for those planets orbiting stars that have other stellar companions. These results are indeed compatible with the current ideas about planetary orbital migration (either due to an interaction with the disk or with other companions) – [40], and although still not very constraining, the observed correlations will probably permit to help decide between the different models of planetary formation.

---

[2] The unknown orbital inclination implies that we can only derive a minimum mass for the companion from the radial-velocity measurements.

3   The Metallicity of Planet-Host Stars

Up to now we have been reviewing the results and conclusions we have obtained directly from the study of the orbital properties and masses of the discovered planets. But another particular fact that is helping astronomers understand the mechanisms of planetary formation as to do with the planet host stars themselves. In fact, they were found to be particularly metal-rich, i.e. they have, in average, a metal content higher than the one found in stars without detected planetary companions [10,34]. The most recent results seem to favour that this metallicity "excess" is original from the cloud that gave origin to the star/planetary system [27,34]. A possible and likely interpretation of this may pass by saying that the higher the metallicity of the cloud that gives origin to the star/planetary system (and thus the dust content of the disk), the faster a planetesimal can grow, and the higher the probability that a giant planet is formed before the proto-planetary disk dissipates. In other words, the metallicity seems to be playing a key role in the formation of the currently discovered extra-solar planetary systems. However, it is not known precisely how the influence of the metallicity is influencing the planetary formation and/or evolution; for example, the mass of the disks themselves, that can be crucial to determine the efficiency of planetary formation, is not known observationally with enough precision.

But recent observations suggest also that planets might in fact be engulfed by their parent stars, whether as the result of orbital migration, or e.g. of gravitational interactions with other planet or stellar companions [10,12,15]. Probably the most clear evidence of such an event came recently from the detection of the lithium isotope ^{6}Li in the atmosphere of the planet-host star HD 82843 [12]. This fragile isotope is easily destroyed (at only 1.6 million degrees, through (p,α) reactions) during the early evolutionary stages of star formation, when the proto-star is completely convective, and the relatively cool material at the surface is still deeply mixed with the hot stellar interior (this is not the case when the star reaches its "adulthood"). ^{6}Li is thus not supposed to exist in stars like HD 82843, and the simplest and most convincing way to explain its presence is to consider that planet(s), or at least planetary material, have fallen into HD82843 sometime during its lifetime.

The question of knowing whether this case is isolated or else if the fall of planetary material is a frequent outcome of the planetary formation process is still under debate, but the current results seem to suggest that at least the degree of stellar "pollution" is not incredibly high [34,15,33].

4   Transiting Planets

Current observations are also starting to teach us something about the planets themselves. Recently, astronomers have detected the dip in the luminosity of the star HD 209458 as a previously detected planet crossed its disk [6,11,18,2]. This detection not only represents an independent confirmation of the nature of the discovered body, but also permits to precisely constrain some planetary physical properties, like its mass, radius, or mean density.

The results show that the planet around HD 209458 has a radius of about 1.35 times the radius of Jupiter, for a mass of only $0.7 \, M_{Jup}$. This corresponds to a mean density lower than the one of Jupiter, a result that is expected given the proximity of the planet to its host star (0.045 AU, i.e., more than 20 times as close as the Earth is to the Sun), and thus its higher temperature [5].

Subsequent studies have also revealed that the planet's orbit is perpendicular to the stellar rotation axis [30]. This result perfectly fits into the picture of planetary formation: the planet around HD 209458 was most probably formed in a proto-planetary disk.

In this context, many hopes are coming now from large surveys that are looking for planetary transits in the galactic bulge. The first results have now been announced, and a few planetary-transit candidates are known [36]. The growing number of detections will permit to study the relation between various physical properties of the extra-solar giant planets, like the dependence of their density on the mass, orbital separation, or even on the metallicity of their host stars.

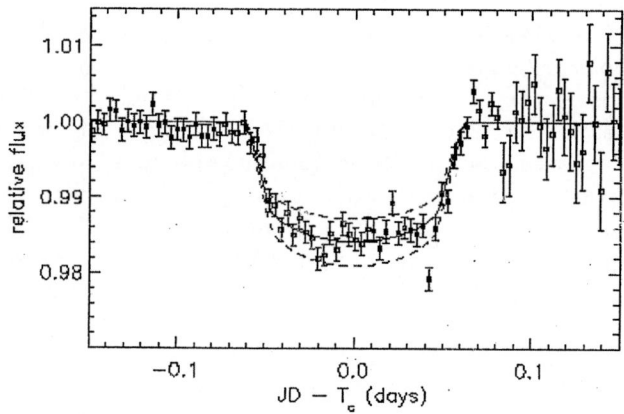

Fig. 5. Light curve of HD 209458 as the planetary companion crosses its disk. From [6]

5   Multi-Planetary Systems

Another possible source of information about the formation of giant planets may come from the multi-planetary systems. To date, 7 such systems are known. A few cases, like the resonant planets around HD 82943 [20] – see Fig. 6 – and Gl 876 [19], or the planet-brown dwarf pair around HD 168443 [38], are of particular interest; their orbital configurations may provide new constraints on the planetary migration and eccentricity pumping mechanisms (c.f. [23]).

Further information can come from those systems for which the giant planets are orbiting quite close to each other (this is particularly true for the resonant

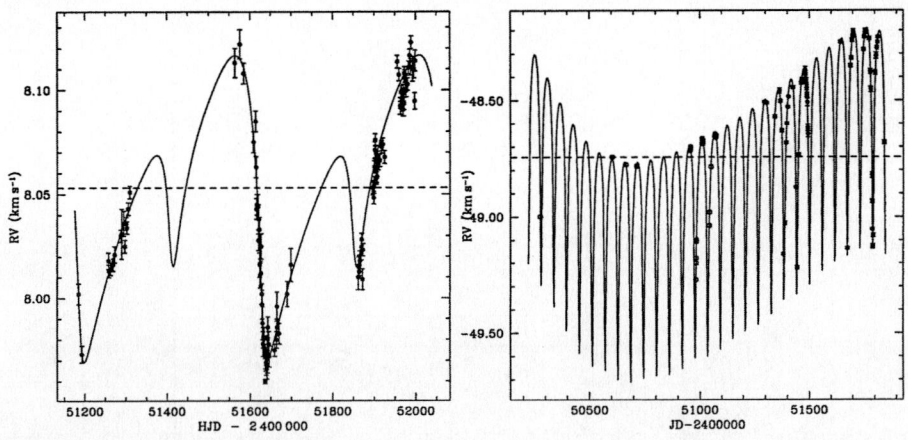

Fig. 6. *Left*: Radial-velocity measurements and best Keplerian solution for HD 82943, a system of two resonant planets in eccentric orbits [20]; *right*: The same for the system around 168443. This latter "odd" system is formed of a 7.4 and 16.9 M_{Jup} companions orbiting their host star in ∼60 and 1750 days, respectively

systems). In these cases we can expect some orbital evolution to occur. Since this evolution will strongly depend on variables like the mass of the two planets and the relative inclination of the orbital planes, the followup with radial-velocity measurements of the orbital changes will permit, when compared with n-body simulations, to determine precisely the real mass (and not only the minimum mass) of the systems, as well as their orbital inclinations.

6 Prospects for the Future

The study of extra-solar planetary systems is just giving its first steps. After only 7 years, we can say that at least 5% of the solar type dwarfs have giant planetary companions with masses as low as the mass of Saturn and orbital separations of a few Astronomical Units (the limits imposed by the current planetary search techniques). But the understanding of how giant planets are formed is still shaded in many points.

To help solve some of the problems, several projects are currently in the pipeline. From one side, future space missions, like the photometric missions like COROT and Eddington, or the astrometric satellite GAIA, will definitely permit to unveil hundreds (or even thousands) of "new" planets (see contribution by M. Perryman in this volume). But ground based astronomy will also give enormous steps in the next few years.

Huge developments are expected in the radial-velocity surveys by the development of instruments like HARPS [29], capable of measuring velocities down to a $1\,\mathrm{m\,s^{-1}}$ precision. It is easy to conclude that these instruments will definitely

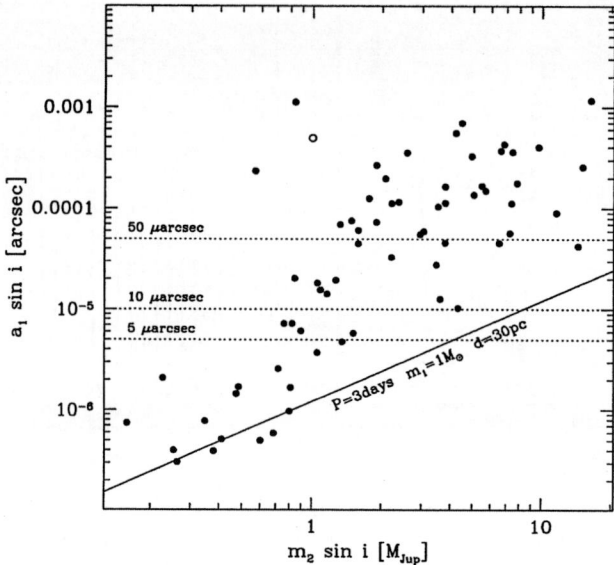

Fig. 7. Mass vs. astrometric motion diagram for stars with very low mass companions discovered by radial-velocity surveys. The dotted lines represent the limits for astrometric precisions of 50, 10 and 5 μarcsec. The open circle illustrates the position of the Sun as seen from a distance of 10pc. The solid line indicates an approximate limit imposed by the fact that no planets were found with periods shorter than ~3 days

unveil the presence of many more giant planets around stars in the solar neighborhood. In particular, they will permit to discover lower mass companions in long period orbits, maybe more similar to the ones we find in our Solar System. However, it is still not clear what is the limiting precision that we can obtain with the radial-velocity technique. When measuring a stellar spectrum we are obtaining information coming from the integrated light of the stellar disk, i.e. the sum of the photons coming from every single surface element of the star at the same time. It is well known that a stellar atmosphere presents convective motions with velocities of the order of $1\,\mathrm{km\,s^{-1}}$. When active regions are present, dark-spotted regions appear, and convective motions are changed locally, changing thus the observed spectrum. The way these features are able to induce radial-velocity "jitter" is still not well studied [32,35], and thus the limitations caused by these kind of phenomena are still not known.

On the astrometric point of view, instruments like the VLTI interferometer (ESO) will permit to combine the light of the four VLT 8-m telescopes (plus 4 Auxiliary Telescopes), enabling astronomers to obtain astrometric observations with a resolution down to 10-50 micro-arcsec. Similar capabilities, also achieved by the Keck interferometer and even exceeded by the GAIA and SIM space missions (see contribution by M. Perryman), will hopefully enable the detection of the astrometric motion of many of the now known planet host stars around the center-of-mass of the star-planet system (see Fig. 7). The measurement of this displacement, together with the radial-velocity measurements, will permit to obtain real masses for the planetary companions.

The direct detection of an extra-solar planetary companion has been up to now just a remote possibility. As mentioned above, the planet/stellar luminosity ratio is of the order of 10^{-9} at visible wavelengths, making the detection of planetary companions impossible with current technology. However, the contrast ratio is much lower in the Infra-red. Particularly for planets around young stars, much hotter and brighter than when they are older [5] and thus having a stellar-to-planet light contrast that is much more favorable, state-of-the-art infra-red imaging facilities like NAOS/CONICA at the VLT will eventually permit to make the first direct images of such bodies.

On the other hand, it will also be possible to look for planetary companions around white-dwarf stars. These stellar remnants, that result from the "death" of a moderate mass star, are 10^3–10^4 times fainter than their progenitors, making it easier thus to directly detect and image a planet that might have survived the last stages of the stellar evolution [4].

All these steps will give the opportunity to add hundreds or even thousands of new giant exo-planets to the lists, permitting to improve the statistical analysis of their characteristics. This will lead to a better understanding of the processes involved in the various stages of planetary formation, and consequently to revise the current theories dealing with the formation of planetary systems. It is important, for example, to understand exactly what is the main process involved in the formation of the giant planets: do they form by the "traditional" grain accretion process, where a (previously) formed solid core accretes a gas envelope (c.f. [28]), or is it also possible to form giant planets as a result of disk instabilities [3]? The key to answer this question might pass by the study (and search) of planets around binary stars [14,7], a programme that is currently being done at the Geneva planet search group [7].

Although today we know that giant planets are common throughout universe, the search for terrestrial-like planets is still not possible with current techniques. But the development of new instruments, like the space interferometer Darwin, will soon open the possibility of finding and even studying the composition of exo-earths. This will represent a first step towards the discovery of life signatures in planets around other stars.

We would like to thank the members of the Geneva extra-solar planet search group, D. Naef, F. Pepe, D. Queloz, S. Udry, our French colleagues J.-P. Sivan, C. Perrier and J.-L. Beuzit, as well as G. Israelian and R. Rebolo (from the IAC), who have largely contributed to the results presented here. We wish to thank the Swiss National Science Foundation (Swiss NSF) for the continuous support to this project. Support from Fundação para a Ciência e Tecnologia, Portugal, to N.C.S. in the form of a scholarship is gratefully acknowledged.

References

1. Bodenheimer P., Hubickyj O., Lissauer J.J., 2000, Icarus 143, 2
2. Brown T., Charbonneau D., Gilliland R., Noyes R., Burrows A., 2000, ApJ 552, 699
3. Boss A.P., 2000, ApJ 536, L101
4. Burleigh M.R., Clarke F.J., Hodgkin S.T., 2002, MNRAS 331, L41
5. Burrows A., Hubbard W.B., Lunine J.I., Liebert J., 2001, Review of Modern Physics 73, 719
6. Charbonneau D., Brown T.M., Latham D.W., Mayor M., 2000, ApJ 529, L45
7. Eggenberger A., Udry S., Mayor M., 2002. In: "Scientific Frontiers in Research on Extrasolar Planets", ASP Conf. Series, in press
8. Fischer D.A., Marcy G.W., Butler R.P., et al., 2001, ApJ 551, 1107
9. Goldreich P., Tremaine S., 1980, ApJ 241, 425
10. Gonzalez G., 1998, A&A 334, 221
11. Henry G.W., Marcy G.W., Butler R.P. Vogt S.S., 2000, ApJ 529, L41
12. Israelian G., Santos N.C., Mayor M., Rebolo R., 2001, Nature 411, 163
13. Jorissen A., Mayor M., Udry S., 2001, A&A 379, 992
14. Kley W., 2001. In: Zinnecker H., Mathieu R.D. (Eds.), "The Formation of Binary Stars", in press
15. Laws C., & Gonzalez G., 2001, ApJ 553, 405
16. Lin D.N.C., Ida S., 1997, ApJ 477, 781
17. Lin D.N.C., Bodenheimer P., Richardson D.C., 1996, Nat. 380, 606
18. Mazeh T., Naef D., Torres G., et al., 2000, ApJ 532, L55
19. Marcy G.W., Butler R.P., Fischer D., et al., 2001 ApJ 556, 296
20. Mayor M., et al., 2001a, ESO press-release 07/01
21. Mayor M., Queloz D., 1995, Nature 378, 355
22. McCaughrean M.J., O'dell C.R., 1996, AJ 111, 1977
23. Murray N., Paskowitz M., Holman M., 2002, ApJ 565, 608
24. Murray N., Hansen B., Holman M., Tremaine S., 1998, Science 279, 69
25. Naef D., Latham D.W., Mayor M., et al., 2001, A&A 375, L27
26. Papaloizou J.C., Nelson R.P., Masset F., 2001, A&A 366, 263
27. Pinsonneault M.H., DePoy D.L., Coffee M., 2001, ApJ 556, L59
28. Pollack J.B., Hubickyj O., Bodenheimer P., et al., 1996, Icarus 124, 62
29. Queloz D., Mayor M., et al., 2001, The Messenger 115, 1
30. Queloz D., Eggenberger A., Mayor M., et al., 2000, A&A 359, L13
31. Rasio F.A., Ford E.B., 1996, Science 274, 954
32. Saar S.H., Butler R.P., Marcy G.W., 1998, ApJ 498, L153
33. Santos N.C., García López R.J., Israelian G., et al., 2002, A&A 386, 1028
34. Santos N.C., Israelian G., Mayor M., 2001, A&A 373, 1019
35. Santos N.C., Mayor M., Naef D., et al., 2000, A&A 361, 265
36. Udalski A., Paczynski B., Zebrun K., et al., 2002, Acta Astronomica 52, 1
37. Udry S., Mayor M., Naef D., et al., 2002, A&A, in press (also at astro-ph/0202458)
38. Udry S., Mayor M., Queloz D., 2001. In: Penny A., Artimowicz P., Lagrange A.-M., Russel S., "Planetary Systems in the Universe: Observations, Formation and Evolution", ASP Conf. Ser., in press
39. Ward Wm. R., 1997, ApJ, 482, L211
40. Zucker S., Mazeh T., 2002, ApJL, in press (also at astro-ph/0202415)

Towards the Detection
of Earth-Like Extra-Solar Planets

M.A.C. Perryman

Astrophysics Division, European Space Agency, ESTEC, Noordwijk, The Netherlands

Abstract. The discovery of the first extra-solar planet surrounding a main-sequence star was announced in 1995, based on very precise radial velocity (Doppler) measurements. A total of about 80 such planets were known by the end of March 2002, and their numbers are growing steadily. Space missions under development or consideration by the European Space Agency and NASA are expected to contribute to the detection and characterization of Earth-like systems over the next 10–20 years: via astrometry through the detection of systems like our own Solar System (Jupiter mass objects at a few AU from their parent star) in very large numbers (GAIA); via Earth-mass object detection through photometric transit measurements (Eddington and Kepler); and via interferometric detection of Earth-like systems (Darwin and TPF). Efforts to characterise planets occupying the 'habitable zone', in which liquid water may be present, and indicators of the presence of life, are meanwhile advancing quantitatively.

1   Introduction

There are hundreds of billions of galaxies in the observable Universe, with each galaxy such as our own containing some 10^{11} stars. Surrounded by this seemingly limitless ocean of stars, mankind has long speculated about the existence of planetary systems other than our own, and the possibility of the development of life elsewhere in the Universe. Only recently has evidence become available to begin to distinguish the extremes of thinking that has pervaded for more than 2000 years, typified by opinions ranging from *'There are infinite worlds both like and unlike this world of ours'* (Epicurus, 341–270 BC) to *'There cannot be more worlds than one'* (Aristotle, 384–322 BC). The last 10–20 years has seen rapid advances in theoretical understanding of planetary formation, the development of a variety of conceptual methods for extra-solar planet detection, the implementation of observational programmes to carry out targeted searches and, within the last few years, the detection of a number of planets beyond our own Solar System.

A number of theories of the origin of our Solar System have been advanced. In the most widely considered 'solar nebula theory' planet formation in our Solar System (and, by inference, planetary formation in general) follows on from the process of star formation, through the agglomeration of residual protoplanetary disk material. In this paradigm, planet formation proceeds through several sub-stages characterised by differences in the respective particle interactions. First, the dust grains settle into a dense layer in the mid-plane of the disk, and begin to stick together as they collide. In a second stage, further collisions lead to the formation of 'planetesimals', objects up to a km or so in size, driven by gravitational interactions, and leading to the concentration of objects into particular orbits, with nearly empty gaps between them. In the third phase, the mutual gravitational interaction between planetesimals leads to small changes in their Keplerian orbits, resulting in subsequent collisions, some of which would shatter the planetesimals, but most of which occur at velocities producing a single larger object, or embryo.

Based on present knowledge from radial velocity surveys, about 5% of solar-type stars may harbour massive planets, perhaps formed in this way, and an even higher percentage may have planets of lower mass or with larger orbital radii (currently, seven main-sequence extra-solar planetary system are known which contains more than one planet). If these numbers can be extrapolated, the number of planets in our Galaxy alone would be of order 1 billion.

For the future, experiments capable of detecting tens of thousands of extra-solar planetary systems, lower mass planets down to around 1 $M_\oplus$, and spectral signatures which may indicate the presence of life, are now underway or are planned. The most substantial advances may come from space observatories over the next 10–20 years.[1]

2   Detection Methods

Shining only by reflected starlight, extra-solar planets comparable to bodies in our own Solar System should be typically billions of times fainter than their host stars and, depending on their distances from us, at angular separations from their accompanying star of, at most, a few seconds of arc. This combination makes direct detection extraordinarily demanding, particularly at optical wavelengths where the star/planet intensity ratio is large, and especially from the ground given the perturbing effect of the Earth's atmosphere. Alternative detection methods, based on the dynamical perturbation of the star by the orbiting planet, on planetary transits, and on gravitational lensing, have therefore been developed, although these effects are also extremely subtle. A review of these methods (a number of which are not considered further here), the observational status as of mid-2000, and an overview of the formation and evolution of extra-solar planetary systems, can be found in [1].

Figure 1 summarises detection possibilities so far discussed for the detection of extra-solar planets. In the following discussion, parameters used are mass M, radius R, and luminosity L, with subscripts $*$ and p referring to star and planet respectively. Systems are characterised by their orbital period P, semi-major

---

[1] The text employs a number of standard astronomical terms and units. Planetary systems are conveniently characterized in terms of corresponding Solar System quantities ($\odot$ = Sun; $\oplus$ = Earth; J = Jupiter): $M_\odot \simeq 2.0 \times 10^{30}$ kg; $M_J \simeq 9.5 \times 10^{-4}\, M_\odot$; $M_\oplus \simeq 3.0 \times 10^{-6}\, M_\odot \simeq 3 \times 10^{-3}\, M_J$. 1 AU = 1 astronomical unit (mean Sun-Earth distance) $\simeq 1.5 \times 10^{11}$ m. For Jupiter, $a = 5.2$ AU and $P = 11.9$ yr. Stellar distances are conveniently given in parsec (pc), defined as the distance at which 1 AU subtends an angle of 1 second of arc (or arcsec); 1 pc $\simeq 3.1 \times 10^{16}$ m $\simeq 3.26$ light-years.

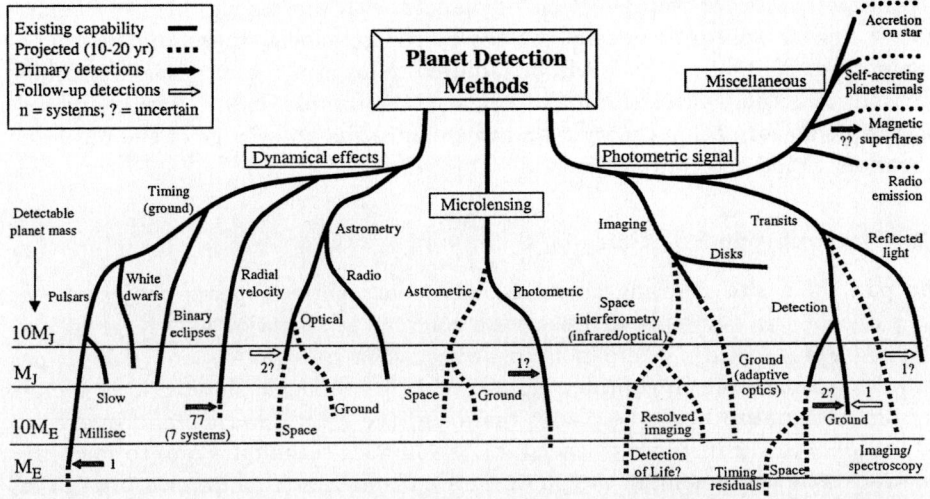

Fig. 1. Detection methods for extra-solar planets. The lower extent of the lines indicates, roughly, the detectable masses that are in principle within reach of present measurements (solid lines), and those that might be expected within the next 10–20 years (dashed). The (logarithmic) mass scale is shown at left. The miscellaneous signatures to the upper right are less well quantified in mass terms. Solid arrows indicate (original) detections according to approximate mass, while open arrows indicate further measurements of previously-detected systems. '?' indicates uncertain or unconfirmed detections. The figure takes no account of the numbers of planets that may be detectable by each method. Numbers are as of March 2002

axis a, eccentricity e, orbital inclination with respect to the plane of the sky i ($i = 0°$ face-on, $i = 90°$ edge-on), and distance from the Solar System d. Figure 2(a) illustrates the parameter regions probed by radial velocity, astrometric, and transit measurements at current and projected accuracy levels.

2.1   Dynamical Perturbation of the Star

The motion of a single planet in a circular orbit around a star causes the star to undergo a reflex circular motion about the star-planet barycentre, with orbital radius $a_* = a \cdot (M_p/M_*)$ and period P. This results in the periodic perturbation of three observables, all of which have been detected (albeit in different systems): in radial velocity, in angular (or astrometric) position, and in time of arrival of some periodic reference signal.

All of the known extra-solar planets around normal main-sequence stars have been discovered, starting with the first in 1995, using radial velocity techniques. Current state-of-the-art measurements reach around 3 m s^{-1}, corresponding to an accuracy of about 1 part in 10^8 in wavelength. Intrinsic accuracy limits of around ± 1 m s^{-1} may arise from the effects of star spots and convective inhomogeneities on the stellar surface, even in older less-active.

The prospects for the detection of planets with masses significantly below that of Jupiter appear somewhat limited by radial velocity measurements. The development of other methods will be required to lower planetary mass detection limits towards the 'habitable zone', to enlarge the sample sizes to provide better constraints on formation theories, and to enhance the knowledge of the physical properties of the detected systems.

2.2   Astrometric Position

The path of a star orbiting the star-planet barycentre appears projected on the plane of the sky as an ellipse with angular semi-major axis α given by $\alpha = (M_p/M_*) \cdot (a/d)$, where α is in arcsec when a is in AU and d is in pc (and M_p and M_* are in common units). This 'astrometric signature' is therefore proportional to both the planet mass and the orbital radius, and inversely proportional to the distance to the star. Astrometric techniques aim to measure this transverse component of the photocentric displacement. Jupiter orbiting the Sun viewed from a distance of 10 pc would result in an astrometric amplitude of 500 microarcsec (μas), while the effect of the Earth at 10 pc is a one-year period with 0.3 μas amplitude (the motion of the Sun over, say 50 years, is complex due to the combined gravitational effect of all the planets). The astrometric technique is particularly sensitive to relatively long orbital periods ($P > 1$ yr), and hence complements radial velocity measurements.

Astrometric measurements can be made more accurately from above the Earth's atmosphere. Only a single astrometric space mission has been carried out to date, Hipparcos [2,3] which provided $\sim$ 1 milliarcsec accuracy for about 120 000 stars. For known planetary systems, the Hipparcos data have provided some weak constraints on planetary masses. It is evident from Fig. 2(b), however, that milliarcsec astrometry can contribute only marginally to extra-solar planet detection. In contrast, large-scale acquisition of microarcsec astrometric measurements in the future promises at least three important developments. First, measurements significantly below 0.1 milliarcsec offer planet detection possibilities well below the Jupiter mass limit, out to 50–200 pc. Second, in combination with spectroscopic measurements they provide direct determination of the planet mass (in terms of the star mass), independent of the orbital inclination. Third, the relative orbital inclination of multi-planet systems can be determined.

A future space astrometry experiment which will have a major impact on planet detection is GAIA, an ESA project accepted in 2000 for launch around 2010–12 [4]. GAIA will survey approximately a billion stars to $V \sim 20$ mag as part of a census of the Galactic stellar population. On the assumption that 4–5 per cent of solar-type stars have Jupiter-mass companions it will detect (and provide orbital parameters in many cases) for upwards of 10 000 planetary systems of mass $\sim 1\,M_J$ and periods around 1–10 years [5], for stars as faint as about 15 mag.

In the context of longer-term efforts to detect and characterise Earth-mass extra-solar planets, GAIA's principle contribution is expected to be twofold: (i) by detecting many thousands of planetary systems, it will provide statistically significant data for modeling formation and evolution of such systems; (ii) by detecting large numbers of systems containing Jupiter-mass planets and Jupiter-like orbital radii, it will pinpoint systems that could be candidates for harbouring Earth-mass inner planets in multiple planetary configurations like that of our own Solar System.

3   Photometric Transits

Detection of extra-solar planets by measuring the photometric signature of the eclipse of the star by a planet was first considered by [6]. Given a suitable alignment geometry, star light is attenuated by the transit of the orbiting planet across its disk, with the effect repeating at the orbital period of the planet. For a Sun/Jupiter system at 10 pc, the resulting luminosity change is of order 2%, or 0.02 mag. Detection probabilities depend on the transit geometry and on the luminosity drop produced by an object on the line of sight to the star, which approximates to $(\Delta L/L_*) \simeq (R_p/R_*)^2$ under the assumption of a uniform surface brightness of the star.

Values of $\Delta L/L_*$ for the Earth, Mars, and Jupiter transiting the Sun are 8.4×10^{-5}, 3×10^{-5}, and 1.1×10^{-2} respectively. If the radius of the star can be estimated from, say, spectral classification, then R_p can be determined. With knowledge of P and an estimate of M_* (also from spectral classification or via evolutionary models), a can be derived from Kepler's law. Other observational parameters are given, to first order, by simple geometry. Even satellites of extra-solar planets, and planetary rings appear detectable by this method [7].

The principal disadvantage of the method is that it requires configurations in which the viewing direction (to the Earth) lies in the orbital plane of the planet. This time-independent geometrical alignment therefore occurs with only very low probability, such that surveys without a priori knowledge of system geometry or orbital period and orbital phase are characterised by very low detection probabilities.

Ground-based photometry to better than about 0.1% accuracy is complicated by variable atmospheric extinction, while scintillation, the rapidly varying turbulent refocusing of rays passing through the atmosphere, imposes limits at about 0.01%. Extension of the transit method to space experiments, where very long uninterrupted observations can be made above the Earth's atmosphere, therefore holds particular promise. Eddington was selected for study by ESA in early 2000 [8]. It has an telescope of 1 m^2 collecting area and 6 deg^2 field of view, and a CCD focal plane detector array. The first 2–3 years is to be devoted to stellar seismology, aiming to detect solar-type acoustic oscillations in stars in the nearest open clusters. A further 2–3 years would be dedicated mainly to planetary transit detection in up to 700 000 stars in about 20 fields observed for one month each. Reaching a photometric precision of about 10^{-6}, the expected statistics of planetary detections are as follows: out of 500 000 observed stars to $V = 18$ mag, some 20 000 planets with $R < 15 R_\oplus$ are predicted, of which 2000 might be terrestrial-mass planets, of which some dozen may be Earth-mass

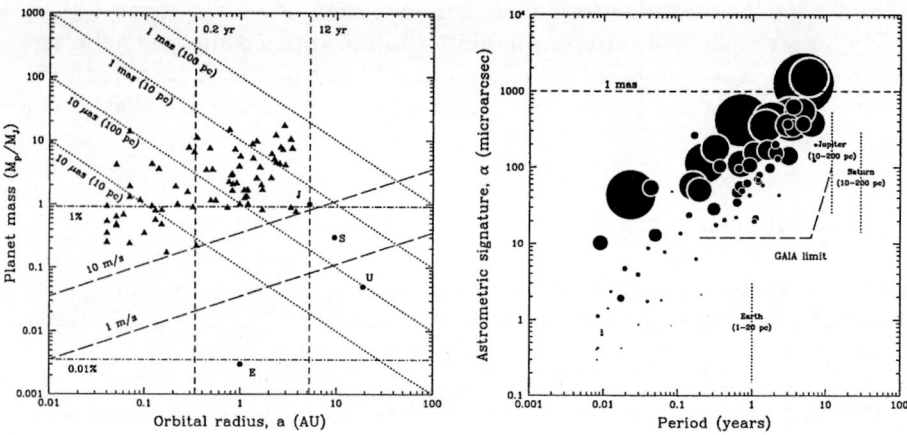

Fig. 2. (a) Detection domains for methods exploiting planet orbital motion, as a function of planet mass and orbital radius, assuming $M_* = M_\odot$. Lines from top left to bottom right show the locus of astrometric signatures of 1 mas and 10 μarcsec at distances of 10 and 100 pc. Very short and very long period planets cannot be detected by planned astrometric space missions: vertical lines show limits corresponding to orbital periods of 0.2 and 12 years. Lines from top right to bottom left show radial velocities corresponding to 10 and 1 m s^{-1}. Horizontal lines indicate photometric detection thresholds for planetary transits, of 1% and 0.01%, corresponding roughly to Jupiter and Earth radius planets respectively. The positions of Earth (E), Jupiter (J), Saturn (S) and Uranus (U) are shown, as are the lower limits on the masses of known planetary systems (triangles). **(b)** astrometric signature, α, induced on the parent star for the known planetary systems as a function of orbital period. Circles are shown with a radius proportional to $M_p \sin i$. Astrometry at the milliarcsec level has negligible power in detecting these systems, while the situation changes dramatically for microarcsec measurements. Short-period systems to which radial velocity measurements are sensitive are difficult to detect astrometrically, while the longest period systems will be straightforward for microarcsec positional measurements. Effects of Earth, Jupiter, and Saturn are shown at the distances indicated. Known planets are as of March 2002

planets with the 'habitable zone' (see Section 6). A similar US space mission, Kepler [9] has recently been selected as part of NASA's Discovery Program, and is specifically designed to detect and characterise Earth-class planets in and near the 'habitable zone'. COROT [7] is a small pre-cursor mission led by the French space agency CNES, scheduled for launch in 2004.

An important result from ground observations has been the detection of the first extra-solar planetary transit event, observed by [10] and independently by [11] for HD 209458, a planet with small a. The former observed two well-defined transits of duration about 2.5 hours, providing an orbital period consistent with the radial velocity determination, and confirming beyond any doubt that its radial velocity variations arise from an orbiting planet. The precise shape of the transit curve is determined by five parameters: the planetary and stellar radii, the stellar mass, the orbital inclination i, and the limb-darkening parameter. Best-fit parameters yield $R_p = 1.27 \pm 0.02\,R_J$ and $i = 87.1 \pm 0.2°$ which, in combination with $M_p \sin i = 0.63\,M_J$ from the radial velocity solution, yields $M_p = 0.63\,M_J$ ($\sin i \sim 1$). Being the first extra-solar planet of known radius and mass, several important physical quantities can be derived for the first time [10]: they estimate $\rho \sim 0.38\,\mathrm{gm\,cm}^{-3}$, significantly less dense than Saturn, the least dense of the Solar System gas giants. The surface gravity is $g \sim 9.7\,\mathrm{m\,s}^{-2}$. Assuming an effective temperature for the star of 6000 K, the effective temperature of the planet is $T_p \sim 1400(1-p)^{1/4}\,\mathrm{K}$ where p is the albedo. This implies a thermal velocity for hydrogen of $v_t < 6.0\,\mathrm{km\,s}^{-1}$, a factor 7 less than the calculated escape velocity of $v_e \sim 42\,\mathrm{km\,s}^{-1}$, confirming that these planets should not be losing significant amounts of mass due to the effects of stellar insolation.

4   Imaging

Imaging of an extra-solar planet generally refers to the detection of a point source image of the object seen in the reflected light from the parent star. Large ground-based telescopes with adaptive optics, and interferometric arrays in space, particularly in the infrared, have been proposed for this.

In addition to the important issue of improving the planet/star contrast, an infrared space interferometer would provide access to the spectral region in which molecular species considered as indicators of life, in particular O_3 and H_2O, are present. The European Space Agency is considering the Darwin Infrared Space Interferometer as a high-priority but longer-term programme [12]. It would employ nulling interferometry in the infrared to detect and obtain spectra of Earth-like planets around 100–200 stars out to distances of 15–20 pc. It would comprise 4–6 free-flying 1-m class telescopes, passively-cooled to 40 K, separated by up to 50 m, and operating between 6–17 μm to cover spectral lines including H_2O, CH_4, O_3 and CO_2. NASA is considering a 75–1000 m baseline infrared interferometer TPF (Terrestrial Planet Finder, [13] as part of its Origins Program. Both Darwin and TPF (or some combination) are targeted for launch some time after 2010.

In the more distant future, ground- or space-based (or lunar) interferometric arrays of 10–100 km baseline could start to tackle resolved planetary imaging. A partial design of a separated spacecraft interferometer which could achieve visible light images with 10×10 resolution elements across an Earth-like planet at 10 pc was made by [14]. This called for 15–25 telescopes of 10-m aperture, spread over 200 km baselines, with the dominant problem being that of suppressing starlight to the necessary levels. Reaching 100×100 resolution elements would require 150–200 spacecraft distributed over 2000 km baselines, and an observation time of 10 years per planet. In the approach of [15] a 30-min exposure using a hyper-telescope comprising 150 3-m diameter mirrors in space with separations up to 150 km, would be sufficient to detect green spots similar to the Earth's Amazon basin on a planet at a distance of 10 light-years.

5   Atmospheres

Predicted spectral properties of extra-solar planets were made in advance of their discovery, giving brightness versus mass and age [16] and including sensitivity to parameters such as deuterium and helium abundance, rotation rate, and presence of a rock-ice core. Models have been updated subsequently, including effects of the outer radiative zones caused by the strong external heating which inhibits atmospheric convection. For giants at the very small orbital distances of 51 Peg, one hundred times closer to its primary than Jupiter, [17] calculated radii and luminosities for a range of compositions (H/He, He, H_2O, and Mg_2SiO_4). They showed that such a planet is stable to classical Jeans evaporation, and to photodissociation and mass loss due to extreme ultraviolet radiation, even for tidally-locked planets, finding a mass loss rate for a gas giant at 0.05 AU of about $10^{-16} M_\odot$ yr^{-1}.

Hydrostatic evolution calculations and atmospheric models, including radiative effects, are being used to determine structures, radii, equilibrium temperatures, luminosities, colours, and spectra of objects with temperatures from 100–1300 K. These will help to classify and characterise the planets as more information about them becomes available. With increasing temperature, chemical species likely to condense near the photosphere range from NH_3, H_2O, and NH_4Cl at the lowest temperatures, up to MnS, $MgSiO_3$, and Fe at the highest.

6   Habitability and the Search for Life

The search for other planets is motivated by efforts to understand their formation mechanism and, by analogy, to gain an improved understanding of the formation of our own Solar System. Search accuracies will progressively improve to the point that the detection of telluric planets in the 'habitable zone' will become feasible, and there is presently no reason to assume that such planets will not exist in very large numbers. Improvements in spectroscopic measurements, whether from Earth or space, and developments of atmospheric modelling, will lead to searches for planets which are progressively habitable, inhabited by micro-organisms, and ultimately by intelligent life (these searches may or may not prove fruitless). Search strategies will be assisted by improved understanding of the conditions required for development of life on Earth, combined with observational feasibility. This will be a cross-disciplinary effort, with the participation of astronomers, chemists and biologists. The last few years has seen the establishment of a number of exo-biology initiatives and numerous conferences on the search for life, beginning to quantify philosophical debate that has been ongoing for centuries [18,19].

Assessment of the suitability of a planet for supporting life, or habitability, is based on our knowledge of life on Earth. With the general consensus among biologists that carbon-based life requires water for its self-sustaining chemical reactions [20], the search for habitable planets has therefore focused on identifying environments in which liquid water is stable over billions of years. The habitable zone is defined by the range of distances from a star where liquid water can exist on the planet's surface. This is primarily controlled by the star-planet separation, but is affected by factors such as planet rotation combined with atmospheric convection. For Earth-like planets orbiting main-sequence stars, the inner edge is bounded by water loss and the runaway greenhouse effect, as exemplified by the CO_2-rich atmosphere and resulting temperature of Venus. The outer boundary is determined by CO_2 condensation and runaway glaciation, but it may be extended outwards by factors such as internal heat sources including long-lived radionuclides as on Earth, and tidal heating due to gravitational interactions (as in the case of Jupiter's moon Io). These considerations result, for a $1\,M_\odot$ star, in an inner habitability boundary at about 0.7 AU and an outer boundary at around 1.5 AU or beyond. The habitable zone evolves outwards with time because of the increasing Sun's luminosity with age, resulting in a narrower width of the continuously habitable zone over ~ 4 Gyr of around 0.95–1.15 AU.

Within the ~ 1 AU habitability zone, Earth 'class' planets can be considered as those with masses between about 0.5–$10\,M_\oplus$ or, equivalently, radii between 0.8–$2.2\,R_\oplus$. Planets below this mass in the habitable zone are likely to lose their life-supporting atmospheres because of their low gravity and lack of plate tectonics, while more massive systems are unlikely to be habitable because they can attract a hydrogen-helium atmosphere and become gas giants. Habitability is also likely to be governed by the range of stellar types for which life has enough time to evolve, i.e. stars not more massive than spectral type A. Habitability may be further confined within a narrow range of [Fe/H] of the parent star, by the the specific motion of the Sun with respect to its Galactic orbit, and by many other factors.

Large-scale biological activity on a telluric planet necessarily produces a large quantity of O_2 [20]. Photosynthesis builds organic molecules from CO_2, with the help of H^+ ions which can be provided from different sources. In the case of oxygenic bacteria on Earth, H^+ ions are provided by the photodissociation of H_2O, in which case oxygen is produced as a by-product. However, this is not the case for anoxygenic bacteria, and thus O_2 is to be considered as a possible but not a necessary by-product of life. Indeed, Earth's atmosphere was O_2-free until about 2 billion years ago, suppressed for more than 1.5 billion years after life originated. Plate tectonics and volcanic activity provide a sink for free O_2, and are the result of internal planet heating by radioactive uranium and of silicate fluidity, both of which are expected to be generic whenever the mass of the planet is sufficient and when liquid water is present. For small enough planet masses, volcanic activity disappears some time after planet formation, as do the associated oxygen sinks.

O_3 is itself a tracer of O_2 and, with a prominent spectral signature at 9.6 μm in the infrared where the planet/star contrast is significantly stronger than in the optical, should be easier to detect than the visible wavelength lines. These considerations are motivating the development of infrared space interferometers for the study of lines such as H_2O at 6–8 μm, CH_4 at 7.7 μm, O_3 at 9.6 μm and CO_2 at 15 μm. Higher resolution studies might reveal the presence of CH_4, its presence on Earth resulting from a balance between anaerobic decomposition of organic matter and its interaction with atmospheric oxygen; its highly disequilibrium co-existence with O_2 could be strong evidence for the existence of life.

We now know that other worlds – large ones at least – are common. Developments have been so rapid over the last few years that many significant developments, and many new surprises, can be predicted with confidence. Concerns such as 'who will speak for Earth' when or if contact with other civilisations is made [21], are far from today's scientific mainstream, but perhaps not as far as they were 10 years ago.

References

1. Perryman, M. A. C. 2000, Rep. Prog. Phys., 63, 1209
2. ESA. 1997, The Hipparcos and Tycho Catalogues, ESA SP–1200
3. Perryman, M. A. C. et al. 1997, A&A, 323, L49
4. Perryman, M. A. C. et al. 2001, A&A, 369, 339
5. Lattanzi, M. G. et al., MNRAS, 317, 211
6. Struve, O. 1952, Observatory, 72, 199
7. Schneider, J. et al. 1998, ASP Conf. Ser. 148, 298–303
8. Roxburgh, I. et al. 2000, Eddington, Proposal to ESA
9. Borucki, W. J. et al., 1997, ASP Conf. Ser. 119, 153–173
10. Charbonneau, D. et al. 2000, ApJ, 529, L45
11. Henry, G. W. et al. 2000, ApJ, 529, L41
12. Fridlund, C. V. M. 2000, in Darwin and Astronomy, ESA SP–451, 11–18
13. Beichman, C. A. 1996, Tech. Rep. 96-22, JPL
14. Bender, P. L. & Stebbins, R. T. 1996, J. Geophys. Res., 101(E4), 9309
15. Labeyrie, A. 1999, Science, 285, 1864
16. Burrows, A. et al., 1995, Nature, 375, 299
17. Guillot, T. et al. 1996, ApJ, 459, L35
18. Crowe, M. J. 1986, The Extraterrestrial Life Debate, CUP
19. Dick, S. J. 1996, The Biological Universe, CUP
20. Owen, T. 1980, in Strategies for the Search for Life, Reidel, 177
21. Goldsmith, D. 1988, in Bioastronomy, Kluwer, 425–428

Future Perspectives at CERN

John Ellis

Theoretical Physics Division, CERN, Geneva, Switzerland

Abstract. Current and future experiments at CERN are reviewed, with emphasis on those relevant to astrophysics and cosmology. These include experiments related to nuclear astrophysics, matter-antimatter asymmetry, dark matter, axions, gravitational waves, cosmic rays, neutrino oscillations, inflation, neutron stars and the quark-gluon plasma. The centrepiece of CERN's future programme is the LHC, but some ideas for perspectives after the LHC are also presented.

1 Outline

The scientific mission of CERN is to provide Europe with unique accelerators for the study of the fundamental particles of matter and the interactions between them. The scientific programme of CERN for the next decade is centred on the LHC accelerator, which is scheduled for completion in 2006, so that its experiments can start taking in 2007. A description of the LHC scientific programme is the centrepiece of this talk. The motivations for this and other new accelerators provided by ideas about possible physics beyond the Standard Model were discussed earlier at this meeting [1].

Between now and the startup of the LHC, CERN has a very limited programme of running experiments. However, scientific diversity at CERN is enhanced by a number of recognized experiments, that do not use the CERN accelerators and are not supported by CERN, but whose scientists are allowed to use other CERN facilities. In parallel with the construction of the LHC, CERN is also preparing to send a long-baseline neutrino beam to the Gran Sasso underground laboratory in Italy, in a special programme largely supported by extra contributions from interested countries. These are described before the LHC programme.

In the longer term, CERN has started thinking about its future prospects beyond the LHC. Various options have been proposed, including upgrades of the LHC, a concept for a multi-TeV electron-positron collider called CLIC, a neutrino factory, and a possible role in space experiments. These are mentioned at the end of this talk, with particular emphasis on CLIC.

At each stage in this talk, the relevances of CERN experiments to astrophysics, cosmology and space science are emphasized. The symbiotic relationships between particle physics and these subjects, that motivated this meeting, are amply reflected in the many connections between microphysics and macrophysics revealed in this brief survey.

2   Present CERN Experiments

In addition to the LHC accelerator that is currently under construction, as seen in Fig. 1, CERN has a number of lower-energy accelerators operating at energies between a few hundred MeV and several hundred GeV. Protons at the lowest energies feed the ISOLDE facility, CERN's source of radio-active ions. Several experiments at ISOLDE address astrophysical issues, including a search for axions and massive neutrinos, studies of neutron-rich isotopes relevant to the supernova r-process, etc. [2].

Protons from CERN's next-lowest-energy accelerator, the PS, are used partly to make antiprotons. These are in turn slowed down in CERN's antiproton decelerator (AD) and used to manufacture antihydrogen atoms [3] in the world's first antimatter factory [4], illustrated in Fig. 2. Their numbers are insufficient by many orders of magnitude to drive spaceships à la Star Trek, but they can be used to test matter-antimatter asymmetry in the form of CPT violation, with unprecedented accuracy. Scenarios for generating the baryon asymmetry of the Universe usually rely on the breaking of matter-antimatter symmetry via CP violation, but it has sometimes been suggested that the violation of CPT might also play a role. The AD and its associated experiments will provide some pointers on such ideas, by using laser spectroscopy to probe for differences in the energy levels of hydrogen and antihydrogen atoms.

CERN's current highest-energy accelerator, the SPS, is mainly used to test components for the LHC experiments, but also has a limited research programme. For example, the COMPASS experiment [5] is contributing to the understanding of the proton spin, which is relevant to calculations of the scattering of cold dark matter particles. The NA48 experiment [6] has been studying CP violation in kaon decays, establishing its presence directly in decay amplitudes [7], as postulated in many scenarios for baryogenesis. This an important proof of principle, but baryogenesis would require analogous direct CP violation the decays of different particles.

CERN is also conducting one non-accelerator experiment of interest to astrophysics, namely the CAST experiment [8] that is searching for axions from the Sun. Shown in Fig. 3, it uses a surplus superconducting LHC magnet to look for axion-to-photon conversion in a strong magnetic field, and hopes to achieve a sensitivity beyond indirect astrophysical limits on axions [9].

3   Experiments Recognized at CERN

As has already been mentioned, recognized experiments do not use CERN accelerators, and receive no financial support from CERN, but are allowed officially to use other CERN facilities, such as office space and the computer network. These concessions were originally requested by physicists who were splitting their time between some experiments that use CERN accelerators and others that do not. CERN accepts the principle of such time-sharing, and is grateful for the scientific diversity that it provides, particularly during the pre-LHC period when there are relatively few new CERN data.

CERN Accelerators
(not to scale)

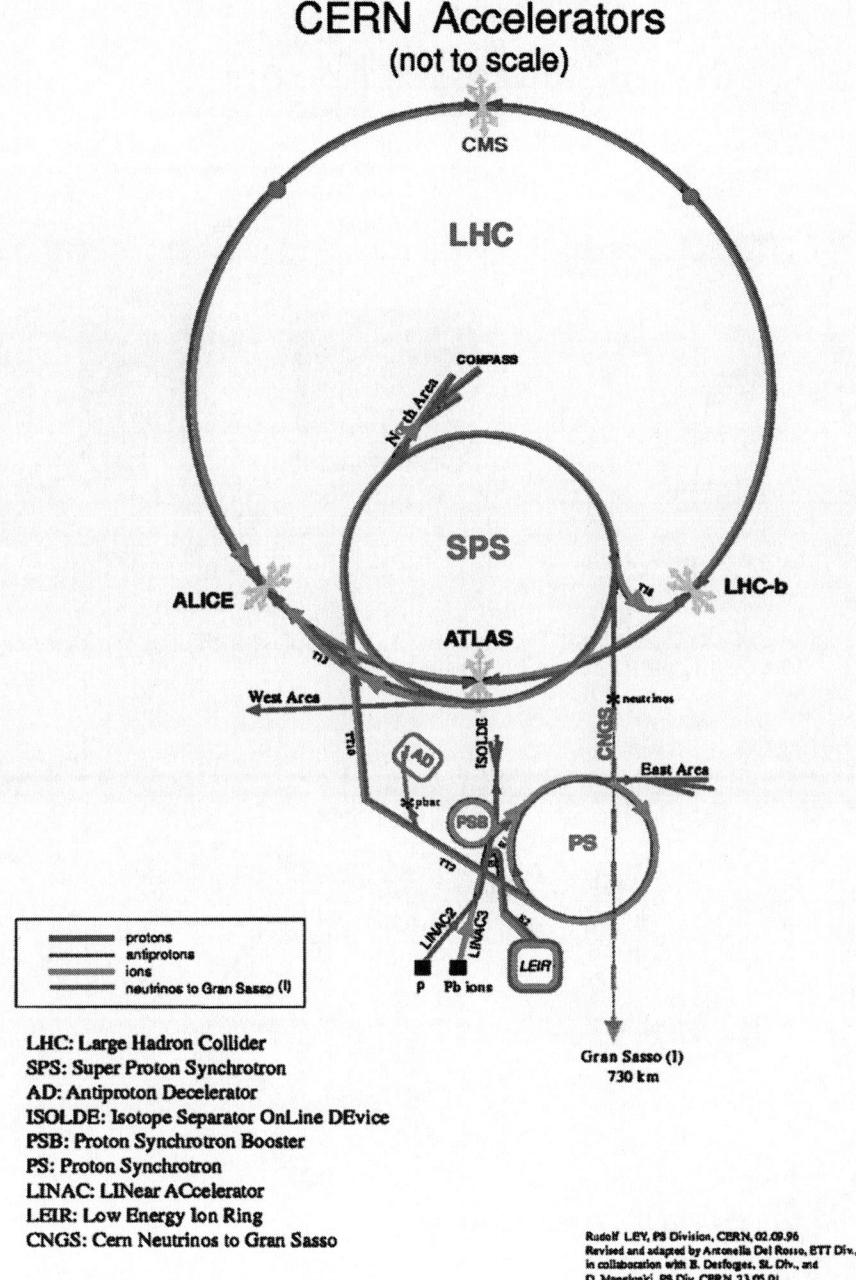

protons
antiprotons
ions
neutrinos to Gran Sasso (I)

LHC: Large Hadron Collider
SPS: Super Proton Synchrotron
AD: Antiproton Decelerator
ISOLDE: Isotope Separator OnLine DEvice
PSB: Proton Synchrotron Booster
PS: Proton Synchrotron
LINAC: LINear ACcelerator
LEIR: Low Energy Ion Ring
CNGS: Cern Neutrinos to Gran Sasso

Gran Sasso (I)
730 km

Rudolf LEY, PS Division, CERN, 02.09.96
Revised and adapted by Antonella Del Rosso, ETT Div.,
in collaboration with B. Desforges, SL Div., and
D. Manglunki, PS Div. CERN, 23.05.01

Fig. 1. Map of the CERN accelerators, showing how the LHC will be fed from the
smaller SPS and PS rings via the transfer lines TI 2, TI 8. Also shown are the locations
of the ALICE, ATLAS, CMS and LHCb experiments, as well as the starting-point of
the CNGS beam and the antoproton decelerator (AD)

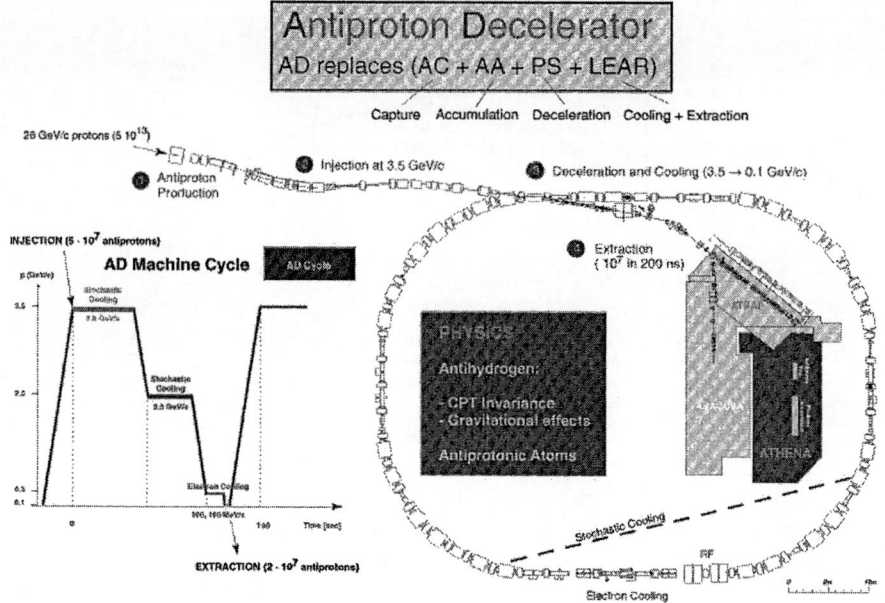

Fig. 2. Layout of the Antiproton Decelerator (AD) complex, including the three experiments ASACUSA, ATHENA and ATRAP

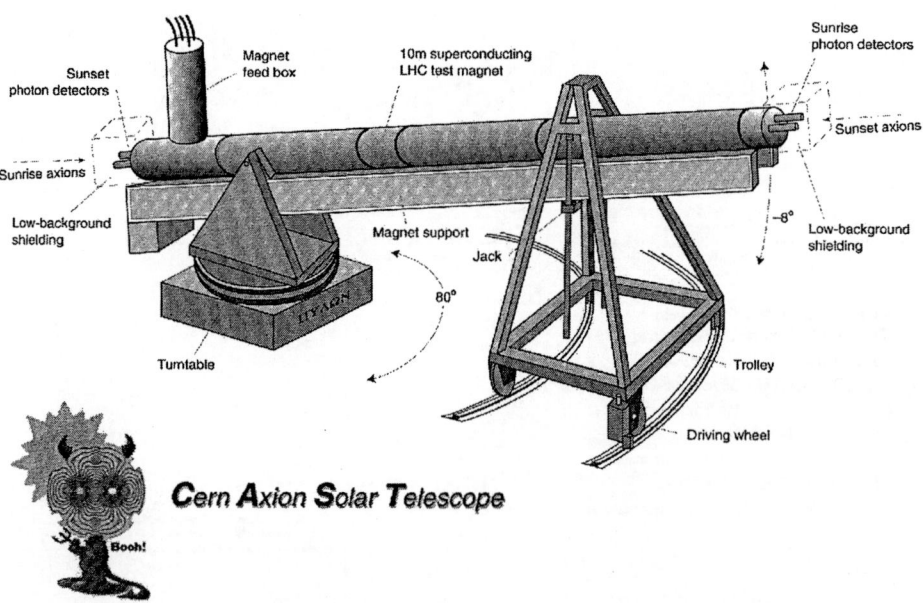

Fig. 3. Design of the CAST experiment, which will use a surplus LHC magnet to search for solar axions

3.1 Gravitational Waves

EXPLORER [10] was the first experiment to be recognized at CERN. It uses CERN's cryogenic facilities to keep cold its large bar detector for gravitational waves. More recently, CERN has recognized the LISA experiment [11], a trio of spacecraft that is now being designed to look for lower-frequency, longer-wavelength gravitational waves in space.

3.2 Astrophysical Antimatter

AMS [12] was the first particle spectrometer to have been sent into space, on the space shuttle in 1998, and an improved configuration is currently being prepared to fly again for several years on the International Space Station. AMS looks, in particular, for antimatter particles such as positrons, antiprotons and antinuclei, that might be signatures of antimatter in the Universe, or of dark matter annihilations. Its first flight made interesting measurements of the Earth's cosmic-ray albedo [13], and set a new upper limit on anti-helium in the primary cosmic-ray flux [14] (see Fig. 4) that constitutes further evidence against a matter-antimatter symmetric cosmology. Most of the financial support for the AMS detector comes from European funding agencies, particularly in France, Germany, Italy and Switzerland, so it is natural to have a centrally-located ground base in Europe. Two other recognized space experiments that will also look for astrophysical antimatter are CAPRICE [10] and PAMELA [15]. GLAST [16] is a satellite gamma-ray spectrometer able to look, for example, at gamma-ray bursters and/or energetic photons from the core of our galaxy that might be generated by the annihilations of dark matter particles.

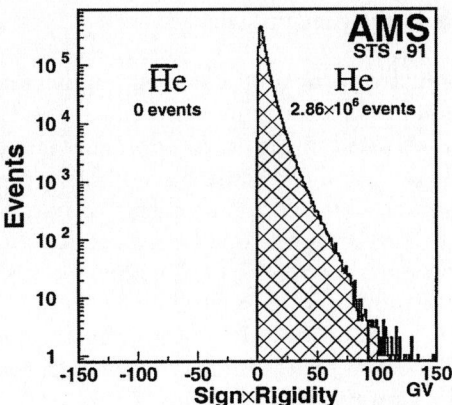

Fig. 4. The AMS experiment looked for antihelium nuclei during its first flight on the space shuttle. It found many conventional helium nuclei, but no antihelium

technical, financial, organizational, computational and physical challenges. This project provides us with us with exciting scientific perspectives for the rest of this decade and (most of) the next, providing us with plenty of time to develop one of the above ideas, or perhaps another, into a longer-term perspective worthy of CERN's mission.

References

1. J. Ellis, talk at this meeting, see also arXiv:hep-ph/0110192.
2. For a listing of ISOLDE experiments, see: http://greybook.cern.ch/, under the ISOLDE rubric.
3. The production of antihydrogen atoms was first reported in G. Baur *et al.* [PS210 Collaboration], Phys. Lett. B **368**, 251 (1996); Nucl. Instrum. Meth. A **391**, 201 (1997); and subsequently in G. Blanford, D. C. Christian, K. Gollwitzer, M. Mandelkern, C. T. Munger, J. Schultz and G. Zioulas [E862 Collaboration], Phys. Rev. Lett. **80**, 3037 (1998).
4. For a listing of AD experiments, see: http://greybook.cern.ch/, under the PS rubric.
5. The home page of this experiment is: http://wwwcompass.cern.ch/.
6. The home page of this experiment is: http://na48.web.cern.ch/NA48/.
7. V. Fanti *et al.* [NA48 Collaboration], Phys. Lett. B **465**, 335 (1999) [arXiv:hep-ex/9909022]; A. Lai *et al.* [NA48 Collaboration], Eur. Phys. J. C **22**, 231 (2001) [arXiv:hep-ex/0110019]; see also A. Alavi-Harati *et al.* [KTeV Collaboration], Phys. Rev. Lett. **83**, 22 (1999) [arXiv:hep-ex/9905060].
8. The home page of this experiment is: *http://axnd02.cern.ch/CAST/*.
9. For a review, see: G. Raffelt, Nucl. Phys. Proc. Suppl. **77**, 456 (1999) [arXiv:hep-ph/9806506].
10. For information about this and other experiments recognized at CERN, see: http://greybook.cern.ch/, under the 'Recognized Experiments' rubric.
11. The home page of this experiment is: http://lisa.jpl.nasa.gov/.
12. The home page of this experiment is: http://hpl3tri1.cern.ch/.
13. J. Alcaraz *et al.* [AMS Collaboration], Phys. Lett. B **472**, 215 (2000); Phys. Lett. B **484**, 10 (2000).
14. J. Alcaraz *et al.* [AMS Collaboration], Phys. Lett. B **461**, 387 (1999).
15. The home page of this experiment is: http://wizard.roma2.infn.it/pamela/fram_des.htm.
16. The home page of this experiment is: http://www-glast.stanford.edu/.
17. The home page of this experiment is: http://antares.in2p3.fr/.
18. P. K. Grieder [NESTOR Collaboration], Nuovo Cim. **24C**, 771 (2001).
19. The home page of this experiment is: http://l3cosmics.cern.ch:8000/l3cosmics/index.htm.
20. The home page of this experiment is: http://www.auger.org/auger.html.
21. A. Watson, talk at this meeting.
22. N. Hayashida *et al.*, Astrophys. J. **522**, 225 (1999) [arXiv:astro-ph/0008102].
23. K. Greisen, Phys. Rev. Lett. **16**, 748 (1966); G. T. Zatsepin and V. A. Kuzmin, JETP Lett. **4**, 78 (1966) [Pisma Zh. Eksp. Teor. Fiz. **4**, 114 (1966)].
24. K. Benakli, J. R. Ellis and D. V. Nanopoulos, Phys. Rev. D **59**, 047301 (1999) [arXiv:hep-ph/9803333].

3.3 Neutrino Telescopes

ANTARES [17] and NESTOR [18] are two underwater neutrino telescopes that will be looking for high-energy cosmic neutrinos, as might come from astrophysical sources or the annihilations of dark matter particles in the Sun and the core of the Earth.

3.4 Cosmic Rays

L3+C [19] is an extension of the L3 experiment, that made measurements with CERN's LEP accelerator, using its muon detectors and additional counters to study cosmic rays. L3+C has provided some interesting measurements of the cosmic-ray muon flux, that help constrain calculations of the atmospheric neutrino flux, and hence refine the interpretation of neutrino oscillation experiments. AUGER [20] is an ultra-high-energy cosmic-ray experiment being constructed in Argentina by a team from several continents, with a strong European participation. As discussed here by Watson [21], if the AGASA experiment [22] is correct, AUGER should be able to gather large numbers of events from beyond the Greisen-Zatsepin-Kuzmin cutoff [23], and tell whether they are due to compact astrophysical sources, or to the decays of very heavy metastable dark matter particles (cryptons [24]). There is currently some controversy in the energy calibrations of ultra-high-energy cosmic rays, which accelerator experiments at CERN could in principle help resolve, e.g., by a PS or SPS experiment to remeasure the spectrum of fluorescence light from nitrogen at different frequencies [21], and/or by constraining models of high-energy particle showers using data on hadron production at the LHC.

4 Neutrino Beam from CERN to the Gran Sasso Laboratory

CERN is currently constructing a beamline for sending neutrinos the 730 km to the Gran Sasso underground laboratory in Italy, called the CNGS project [25], whose starting-point is illustrated in Fig. 1. Experiments on atmospheric neutrinos suggest strongly that μ neutrinos oscillate mainly into τ neutrinos [26], but direct experimental proof is still lacking. The energy $E_\nu \sim 20$ GeV of the beam to be sent from CERN to the Gran Sasso is optimized for τ production via the charged-current reaction $\nu_\tau + N \to \tau + X$. Civil engineering was started in 2000, and the beam may be commissioned in 2006 [27].

The first experiment to be approved for the long-baseline beam was OPERA [28], which will use emulsion techniques with high spatial resolution to identify events in which τ leptons are produced. The ICARUS collaboration [29] proposes to build a 3000-tonne liquid argon calorimeter in the Gran Sasso underground laboratory, having already demonstrated the feasibility of the technique using a 600-tonne pilot module, and got approval to install it at Gran Sasso.

In the longer run, other experiments might want to use the CERN-Gran Sasso beam. For example, it has recently been proposed [30] to place a detector in the

Gulf of Taranto, just off the beam axis [31], where it would be very sensitive to $\nu_\mu \to \nu_e$ transitions.

5 LHC

The Large Hadron Collider (LHC), under construction for installation in the 27 km tunnel previously used for LEP, is primarily designed to deliver proton-proton collisions with a centre-of-mass energy of 14 TeV with a luminosity of 10^{34} cm^{-2}s^{-1} [32]. It will also be able to collide lead ions at a centre-of-mass energy of 1.2 PeV with a luminosity of 10^{27} cm^{-2}s^{-1}.

The main objective of the proton-proton programme is to explore physics in a new energy range from 100 GeV to a TeV and beyond. This is the energy range where the origin of particle masses is expected to be revealed, and one of the principal preys of the LHC will be the Higgs boson, or whatever accomplishes its task of giving masses to the elementary particles. The prospects of finding the Higgs boson are good, as seen in Fig. 5 [33]. Discovery of such an elementary scalar boson would also be interesting for cosmologists, since it is the prototype for the inflaton [34].

Most particle theorists believe that the Higgs boson will have to be accompanied by other new particles, such as those predicted by supersymmetry, which should also be accessible to the LHC, as seen in Fig. 6 [33,35]. The lightest supersymmetric particle is a leading candidate [36] for the cold dark matter thought by astrophysicists to infest the Universe.

LHC's lead-lead collisions will probe nuclear matter at temperatures and pressures typical of those in the early Big Bang when the Universe was less

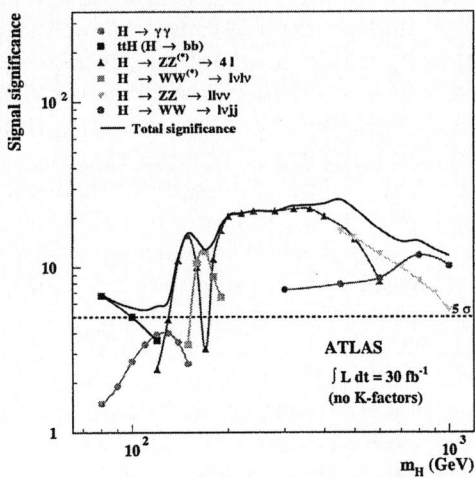

Fig. 5. Either of the major LHC experiments will be able to discover the Higgs boson of the Standard Model in a variety of different decay modes, with a total significance exceeding 10 standard deviations

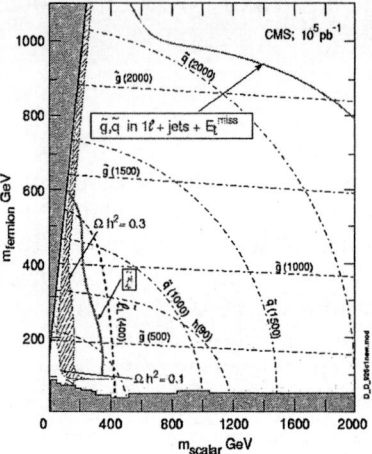

Fig. 6. Either of the major LHC experiments will be able to discover supersymmetry with high significance throughout most of the parameter space where the lightest supersymmetric particle could constitute the cold dark matter, indicated by the diagonal hatching

than about 10^{-6} s old, when it is thought to have taken the form of a quark-gluon plasma. Experiments at the LHC experiment will also continue studies of matter-antimatter asymmetry in decays of particles containing the bottom quark, probing whether they can be described by the Standard Model, and seeking to cast light on cosmological baryogenesis. If history is a reliable guide, the best-remembered discovery of the LHC will probably be none of these!

The LHC experimental programme will consist of two major detectors, ATLAS [37] and CMS [38], that are designed to look for new particles and other novel phenomena at high energies, another experiment ALICE [39] directed towards studies of heavy-ion collisions, and the smaller LHCb experiment [40] looking at matter-antimatter asymmetry. New underground caverns have been dug to accommodate the ATLAS and CMS experiments, whereas ALICE and LHC-b will be housed in caverns used previously by LEP experiments, as seen in Fig. 1. Another small experiment called TOTEM [41] will be attached to CMS to measure the total and elastic proton-proton cross sections, and a dedicated search for magnetic monopoles is also being proposed.

For the first time, the LHC accelerator is being built as a true global collaboration, with important contributions from many other laboratories besides CERN. Outside CERN's member states, important components are being provided by the United States, Russia, Japan, Canada and India. The contracts for the 1200 main-ring dipole magnets have now been placed. Pre-production models have already been delivered to CERN, and have met the specifications for accelerating protons to 7 TeV. These magnets are very challenging, since they must operate at 1.9 K and achieve fields of 9 Tesla. A test string of LHC magnets has been ramped successfully up to the field required to reach this design energy.

Some delays were incurred in the civil engineering for the experimental caverns and the tunnels used to transfer particles into the LHC ring, which turned out to be quite complicated. For example, the access pit to the CMS cavern had to be excavated through an underground stream, that had to be frozen with liquid nitrogen before digging was possible. The cavern itself consists of two large caves side-by-side. To support the rock above, first the wall between these caves was excavated and filled with concrete, and only subsequently could the caves themselves be dug out. The ATLAS cavern is so large that its concrete roof has to be held up by steel stays bolted into the rock above.

The civil engineering is now nearly complete, and no further delays are anticipated from this source. The principal items still on the critical path are procuring the superconducting cable for the main dipole magnets, which has also incurred significant delays, and the cryolines in the LHC tunnel. Because of the delays suffered so far, the scheduled completion of the LHC ring has been pushed back to the end of 2006, with first collisions in 2007.

The major detectors resemble onions, as illustrated in Fig. 7, whose concentric layers measure different types of particles produced in the collisions, such as charged particles, photons, electrons, strongly-interacting particles and muons. They include enough silicon detectors to cover a football field, (in the case of CMS) tens of thousands of lead tungstate crystals, and large superconducting magnets. The ATLAS detector is so large that its cavern could accommodate CERN's central administration building - and some wags have suggested that might be a good place to put it. Large parts of these detectors have already been constructed and delivered to CERN, ready for assembly and installation in their respective caverns. The collaborations are on course for having operational detectors ready to take useful data as soon as the LHC starts colliding protons.

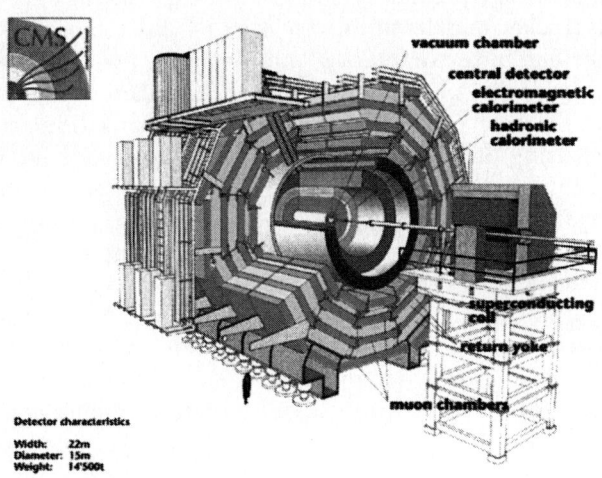

Fig. 7. Conceptual drawing of the CMS experiment, illustrating its 'onion-skin' structure, with different layers designed to detect and measure different types of particles

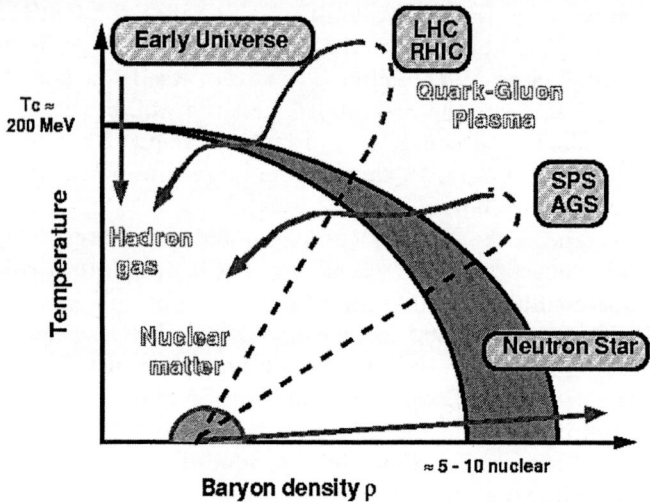

Fig. 8. Relativistic heavy-ion experiments such as ALICE at the LHC attempt to produce a hot and dense state of matter, which is thought to have been a quark-gluon plasma. The interiors of neutron stars are thought to be relatively cool but with a large baryon density, SPS and Brookhaven AGS experiments have probed higher temperatures, and the ALICE and Brookhaven RHIC experiments will probe conditions closest to those in the first microsecond of the Big Bang

The ALICE detector not only reuses an old LEP experimental cavern, it also reuses the magnet used previously for the L3 experiment. Inside it also has a large particle tracker and features specialized subdetectors for photons, electrons and muons. The putative quark-gluon plasma is not expected to have a single distinctive signature, or 'smoking gun', but rather is expected to be identified using a number of convergent indicators, such as Hanbury-Brown-Twiss interferometry using particles emitted from the last-scattering surface of the expanding fireball produced by each 'Little Bang', characteristic abundances of heavier particles, and energetic photons emerging directly from inside the Little Bang, reproducing conditions early in the Big Bang, as seen in Fig. 8.

Previous experiments at CERN and BNL colliding heavy-ion beams with fixed targets provided matter under high pressures, but with relatively low temperatures and high baryon densities, reminiscent of neutron stars. More recent experiments with the RHIC heavy-ion collider at BNL have pushed to higher temperatures [42], but the LHC will come closest to reproducing conditions in the Big Bang, where the baryonic chemical potential was negligible compared to the temperature.

6   CERN's Global Network

The important contributions of non-European countries to the LHC accelerator have already been underlined. The significance of their contributions to the

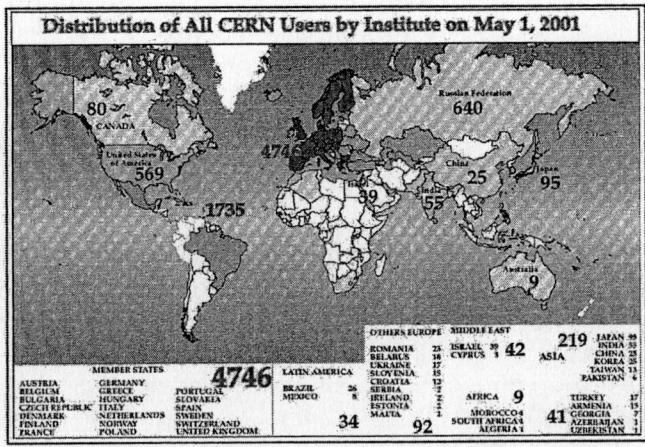

Fig. 9. High-energy physicists from around the world use CERN for their experiments. Most come from CERN's European member states, but around 2000 scientists come from elsewhere

LHC experiments is even greater: about 50% of the physicists and engineers in the experimental collaborations come from institutions outside CERN's member states, and about 30% of the total value of the components of the detectors. Each of the major LHC detectors has about 1800 participating physicists and engineers, and about fifty countries are represented officially. As seen in Fig. 9, the total number of scientists registered as making scientific use of CERN's facilities is between 6000 and 7000, with 4000 to 5000 coming from its member states. The largest external contingents Russia and the United States, around 600 each, followed by Japan, Canada, Israel, Brazil, China and Korea. We collaborate with physicists from every continent except Antarctica.

The age distribution of these scientific users is particularly interesting: the mode is below 30, corresponding to the large numbers of students and postdoctoral research associates that pass through CERN every year. However, more than half of these students subsequently move out of academic research, into industry, finance, etc..

7   From the Web to the Grid

The World-Wide Web was invented at CERN to enable all the far-flung collaborators in its LEP experimental programme to share information. Nobody at CERN anticipated the social phenomenon it would become. Now the LHC project is confronting CERN with a staggering new computing challenge.

Each of the major LHC experiments will take data at a rate of about a Petabyte (10^{15} bytes) per second. This is equivalent to about a billion people surfing the Web simultaneously, or everybody on the planet making dozens of simultaneous mobile phone calls. The great majority of these data are not interesting: perhaps only one collision in 10^{12} will contain a Higgs boson, for example. Therefore, the detectors' data acquisition systems are designed to discard all except one interesting candidate in every 10^7 collisions, approximately. This still leaves several Petabytes of data to be recorded each year and analyzed, a task that would require 100,000 or more of today's PCs, which CERN cannot afford.

CERN plans to tackle this problem by deploying [43] the Grid computing technology, with which the user of a desk-top or portable computer should be able to analyze data stored anywhere in the world, using CPU power wherever it is available, just as we can switch a light on without wondering where the electricity comes from. The Grid objective is transparent user access to data, programs and computing power. Applications of Grid technology to many other sciences, such as biology (e.g., the human genome project), space science and environmental science (e.g., earth observation) are also being developed, but the advent of the LHC provides a definite time-frame over which particle physicists must get a solution in place.

The amount of computing power required to simulate and analyze LHC data is set to grow much faster than Moore's Law, the rate at which the computing power of a single chip has grown historically. In the past, CERN has solved this problem by establishing farms of commodity PCs, rather than using monolithic supercomputers. The Grid project carries this decentralization to its logical conclusion, linking together farms of farms around the planet.

CERN has many collaborating partners in developing the Grid. One of the largest European projects is DataGrid, funded by the European Union and including ESA, PPARC in the United Kingdom, CNRS in France, INFN in Italy and NIKHEF in the Netherlands as well as CERN. Other European projects include CrossGrid, which involves new partners from Ireland and Central Europe, there is the DataTag project to link up with the United States, and parallel projects exist there and elsewhere. In the unique OpenLab venture, we have also attracted industrial partners to provide hardware for Grid development at CERN.

During Phase I of the LHC computing Grid project extending to 2005, the plan is to write the basic software and middleware (system management software), and to demonstrate its functioning with simulated data at a level that is a significant fraction of the eventual LHC requirement. Installation of the full LHC Grid is scheduled for the years 2006 to 2008.

8   Options for CERN after the LHC

Even though the LHC will only start taking data in 2007, and will continue to provide exciting data for a decade or more, the preparation times needed for new accelerator projects are so long that we at CERN have already started thinking about possible new projects after the LHC [44,45]. Several laboratories around the world have for some years already been developing plans for linear electron-positron colliders capable of centre-of-mass energies up to about 1 TeV, hoping

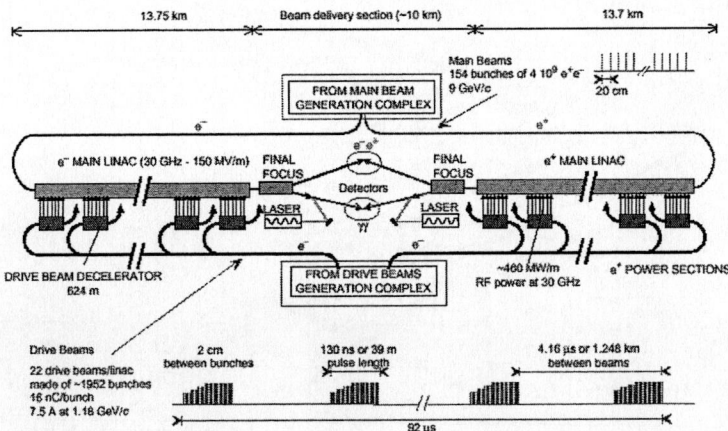

Fig. 10. Conceptual design for a linear e^+e^- collider capable of reaching a centre-of-mass energy of 3 TeV or more, called CLIC, which uses high-power but low-energy drive beams to accelerate less-intense colliding beams to high energies

to start one of these projects in parallel with the operation of the LHC. One of the options being considered at CERN is a possible next-generation electron-positron collider called CLIC [46], capable of higher centre-of-mass energies up to about 5 TeV: see Fig. 10. If supersymmetry exists, this would almost guarantee accurate measurements of the properties of all the supersymmetric particles, enabling us, for example, to calculate better the density of supersymmetric relics from the Big Bang, and the rates at which they would scatter on conventional matter.

In order to avoid a very long device, this energy objective would require a very high accelerating gradient. CERN proposes to achieve this by an innovative double-beam technique, in which an intense low-energy drive beam is used to generate RF power that is then transferred to a lower-intensity, higher-energy beam. This is the origin for the name CLIC, for 'Compact Linear Collider'. The double-beam principle has been demonstrated in a couple of test facilities, and another now under construction [47] is intended to provide an engineering demonstration of the CLIC concept. Another step towards demonstrating its feasibility might be a Higgs factory using $\gamma\gamma$ collisions generated by shining laser beams on e^-e^- beams colliding at around centre-of-mass energies around 150 GeV [48].

Any new accelerator project is sure to require a global collaboration even more widely spread than the LHC. However, other projects considered at CERN have even more planetary dimensions. One possibility is a neutrino factory [49], designed to produce a controlled beam two or three orders of magnitude more intense than current long-baseline projects, using the decays of muons captured in a storage ring, as seen in Fig. 11. One of the primary objectives of such a project would be to measure matter-antimatter asymmetry in neutrino oscillations.

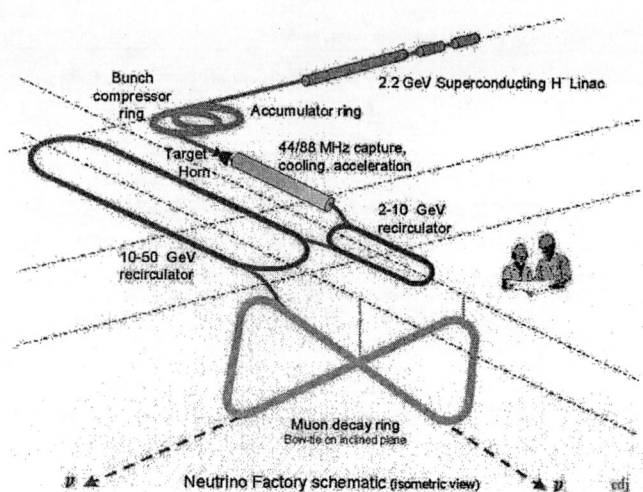

Neutrino Factory schematic (isometric view)

Fig. 11. Conceptual layout for a neutrino factory, based on an intense superconducting proton linac that produces many pions, whose decay muons are captured, cooled in phase space and stored in a 'bow-tie' ring. Their subsequent decays send neutrinos with known energy spectra and flavours to a combination of short- and long-baseline experiments

This might require sending a neutrino beam several thousand kilometres from one continent to another, making the experiments truly global. Such experiments might cast important light on one of the favoured scenarios for baryogenesis [50], in which the decays of massive neutrinos first provide a lepton asymmetry, that Standard Model interactions subsequently convert into a baryon asymmetry. Following steps in this programme could include $\mu^+\mu^-$ colliders with various centre-of-mass energies, possibly including one of more Higgs factories and/or a high-energy frontier machine [49].

Another future possibility for CERN that has attracted some interest is a more active role in space experiments. As already mentioned, CERN currently recognizes several space experiments, including AMS, CAPRICE, GLAST and PAMELA. It has a similar interest in AUGER, which is looking for ultra-high-energy cosmic rays, and EXPLORER, looking for gravitational waves. The next steps in these directions may be EUSO [51] or OWL, monitoring from space cosmic-ray impacts in even larger volumes of the atmosphere, and LISA. We at CERN certainly have legitimate scientific interests in the physics of these experiments: once the problem of mass has been sorted out, grand unification and gravity may be next on the particle physicists' agenda, and space experiments such as these might offer good ways to test them. But would CERN add significant value to such projects?

CERN will not take any irrevocable decision concerning its future before we have at least some results from the LHC, providing clearer hints where to head next. For now, CERN has its hands full constructing the LHC, with all its

25. For more information about this project, see:
    http://proj-cngs.web.cern.ch/proj-cngs/.
26. Y. Fukuda *et al.* [Super-Kamiokande Collaboration], Phys. Rev. Lett. **81**, 1562 (1998) [arXiv:hep-ex/9807003].
27. The home pages of other long-baseline experiments include:
    K2K, http://neutrino.kek.jp/;
    NUMI-MINOS, http://www-numi.fnal.gov/.
28. The home page of this experiment is:
    http://operaweb.web.cern.ch/operaweb/index.shtml.
29. The home page of this experiment is: http://www.aquila.infn.it/icarus/.
30. F. Dydak, private communication.
31. G. Barenboim, A. De Gouvea, M. Szleper and M. Velasco, arXiv:hep-ph/0204208.
32. For more information about this project, see:
    http://lhc-new-homepage.web.cern.ch/lhc-new-homepage/.
33. ATLAS Collaboration,
    http://atlasinfo.cern.ch/Atlas/GROUPS/PHYSICS/TDR/access.html;   CMS
    Collaboration,
    http://cmsinfo.cern.ch/Welcome.html/CMSdocuments/CMSdocuments.html; F.
    Gianotti, talk at this meeting.
34. A. H. Guth, Phys. Rev. D **23**, 347 (1981).
35. M. Battaglia *et al.*, Eur. Phys. J. C **22**, 535 (2001) [arXiv:hep-ph/0106204].
36. J. Ellis, J.S. Hagelin, D.V. Nanopoulos, K.A. Olive and M. Srednicki, Nucl. Phys. B **238** (1984) 453; see also H. Goldberg, Phys. Rev. Lett. **50** (1983) 1419.
37. The home page of this experiment is:
    http://atlasinfo.cern.ch/ATLAS/internal/Welcome.html.
38. The home page of this experiment is: http://cmsinfo.cern.ch/Welcome.html/.
39. The home page of this experiment is: http://alice.web.cern.ch/Alice/.
40. The home page of this experiment is: http://lhcb.web.cern.ch/lhcb/.
41. The home page of this experiment is: http://totem.web.cern.ch/Totem/.
42. For more information about this project, see: http://www.bnl.gov/RHIC/.
43. For more information about this project, see:
    http://lhcgrid.web.cern.ch/LHCgrid/.
44. J. R. Ellis, E. Keil and G. Rolandi, *Options for future colliders at CERN*, CERN-EP-98-03 (1998).
45. A. De Roeck, J. R. Ellis and F. Gianotti, *Physics motivations for future CERN accelerators*, arXiv:hep-ex/0112004.
46. For more information about this project, see:
    http://ps-div.web.cern.ch/ps-div/CLIC/Welcome.html.
47. For more information about this project, see:
    http://ctf3.home.cern.ch/ctf3/CTFindex.htm.
48. D. Asner *et al.*, arXiv:hep-ex/0111056.
49. For more information about this project, see:
    http://muonstoragerings.web.cern.ch/muonstoragerings/Welcome.html.
50. J. R. Ellis, J. Hisano, S. Lola and M. Raidal, Nucl. Phys. B **621**, 208 (2002) [arXiv:hep-ph/0109125]; J. R. Ellis, J. Hisano, M. Raidal and Y. Shimizu, Phys. Lett. B **528**, 86 (2002) [arXiv:hep-ph/0111324].
51. The home page of this experiment is: http://www.ifcai.pa.cnr.it/~EUSO/.

Future Perspectives at ESO

Catherine Cesarsky

European Southern Observatory, Karl-Schwarzschild-Str. 2,
D-85748 Garching, Germany

Abstract. An overview is given of ESO's current facilities and its plans for the future.

1 Introduction

In the year of its fortieth birthday, ESO, the prime European organisation for ground astronomy, finds itself at a crossroads. With the Very Large Telescope (VLT) in its early stage of active life, with the VLT Interferometer soon to engage in regular observations of unequalled sharpness and depth, new goals are being set, and further exciting projects are being launched or studied.

2 The VLT

The Very Large Telescope project provides the European community with access to over 1400 nights per year of 8-metre telescope time with the most advanced instrumentation. The project, approved by the ESO council in 1986, reached its first critical milestone in May of 1998 with the first light of the first of the four 8-metre telescopes, with the other three following in rapid succession, in March of 1999, January of 2000 and September of 2000. All four telescopes are in routine scientific operations.

The VLT is located on top of Cerro Paranal, a 2600-metre high mountain in the northern Atacama desert of Chile. This site provides excellent seeing (0.65 arcseconds median), a very large number of clear nights (total weather down time is less than 15%) and no discernible light pollution. These advantages come at a cost, as there is no supply of power, water and the site can only be accessed via an unpaved road. All utilities for the telescopes and for the personnel supporting the operations are either brought into the observatory (e.g. water) or generated on site (e.g. power).

The 8-m telescopes use active optics, as first demonstrated at the ESO NTT telescope in the late 1980s at La Silla. The wavefront sensing and closed loop corrections of the force settings and positions of the telescope mirrors provide the VLT with exceptional image quality. Each unit telescope of the VLT provides 4 foci at which scientific observations can be made (Cassegrain, 2 Nasmyth and a Coude focus which is used for interferometry). The telescopes provide the guiding and wavefront sensing facilities to control the active optics. The four 8-m telescopes are all equipped with state of the art instruments. ISAAC, FORS1 and FORS2, UVES, NACO, VIMOS and FLAMES are currently installed at the telescopes. These instruments provide a wide variety of resolutions, both spatial (0.013 arcseconds to 0.2 arcseconds) and spectral ($\lambda/\Delta\lambda$ from 300 to 120,000), at wavelengths ranging from 3000 Å to 5-μm. In addition, NACO is equipped with adaptive optics to correct the atmospheric aberrations and provides imaging at the diffraction limit of the telescope.

The operations at Paranal observatory are supplemented by user support and quality control groups at the ESO headquarters in Garching. All data obtained at the telescopes are archived and are associated with the necessary calibrations such that they can be repeatedly used for a variety of scientific programmes.

The observatory runs two modes of operation. Visitor mode observations involve the astronomers travelling from Europe to the site and participating directly to the observations. In service mode observations are obtained by the ESO staff on behalf of the astronomers in the ESO community, who supply the necessary information and constraints (e.g. seeing and weather) for their observations. In both visitor and service observing, the complexity of the VLT and its instrumentation require that expert operations staff command all critical actions. The data obtained at the observatory are pipeline processed to remove the instrumental signatures and are made available to the users in both a raw and calibrated form.

The observatory runs at extremely high efficiency, both in terms of telescope availability (technical down time is approximately 2 per cent) and in terms of time spent on target (shutter open times of order 80 per cent).

3   VLT Instrumentation

Each of the four 8-m diameter Unit Telescopes is delivering in-focus stable sub arc-second images at three different possible locations. The role of the up to 12 focal instruments at the VLT is to analyze the light in a small patch of the sky and in particular its spectral composition. The large instrument complement of the VLT is being built by ESO, or by extensive Consortia, involving in total some fifty European Institutes, in close collaboration with ESO. The performance and versatility of the instrumentation define largely the Observatory competitiveness.

Our strategy is based on 10 to 11 different instruments installed permanently, each at its own focus, with one focus left for innovative visiting instruments. This gives highly stable systems that can be fully calibrated. This high instrumental variety reflects the many different astrophysical needs of our community, in particular to:

1. Cover the wide accessible wavelength range from 0.3 μm to 28 μm, albeit with a number of missing regions inside which the Earth atmosphere is totally opaque. This corresponds to about six octaves and cannot be filled efficiently by single instruments.

2. Ensure different levels of sky field coverage and angular resolution (Fig. 1a). On one hand, survey type instruments cover relatively large fields, up to

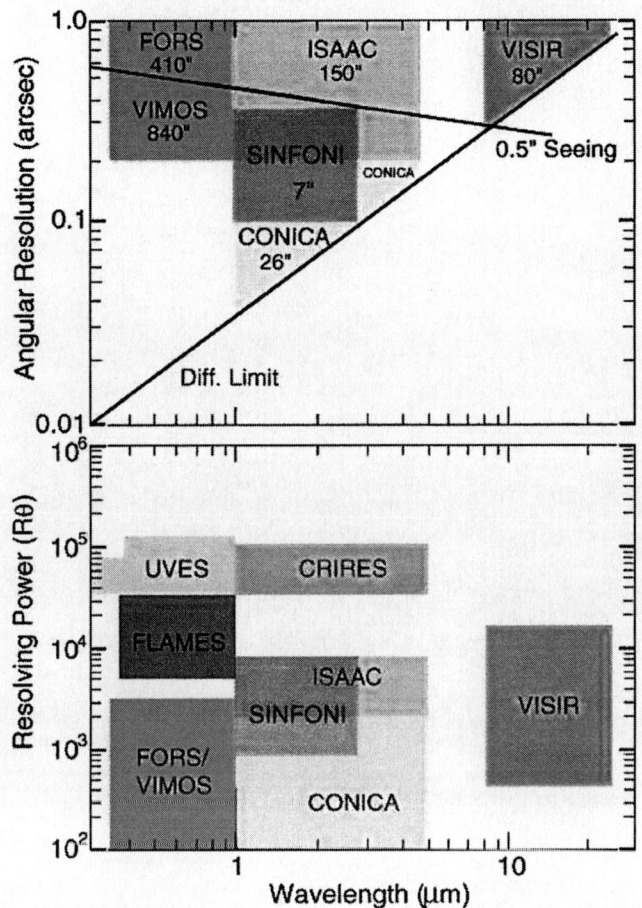

Fig. 1. VLT Instrumentation.   **a.** Spatial resolution versus wavelength coverage.
b. Spectral resolution versus wavelength coverage

14 × 14 arc-minute at the VLT (roughly one third of the full moon area)
with ∼ 0.25 arc-second angular resolution, just adequate for images taken
under good seeing conditions. At the other extreme, one gets much smaller
fields, of the order of half an arc-minute, but with adaptive optics corrected
images sampled at the diffraction limit of the Telescope (0.025 arc-second at
1 μm).

3. Provide a large palette of spectral resolutions (Fig. 1b), each optimised for
different astrophysical targets. Imaging with broad ($\Re = \lambda/\delta\lambda \sim 15$) or
narrow ($\Re \sim 100$) filters remains a fundamental tool for practically all type
of sources. Spectroscopy with $\Re \sim 200$-2,000 is essential for surveys of faint
distant galaxies. Higher resolutions ∼ 7,000-20,000 are used to study galaxy
internal structure, stellar aggregates in nearby galaxies and regions of stellar

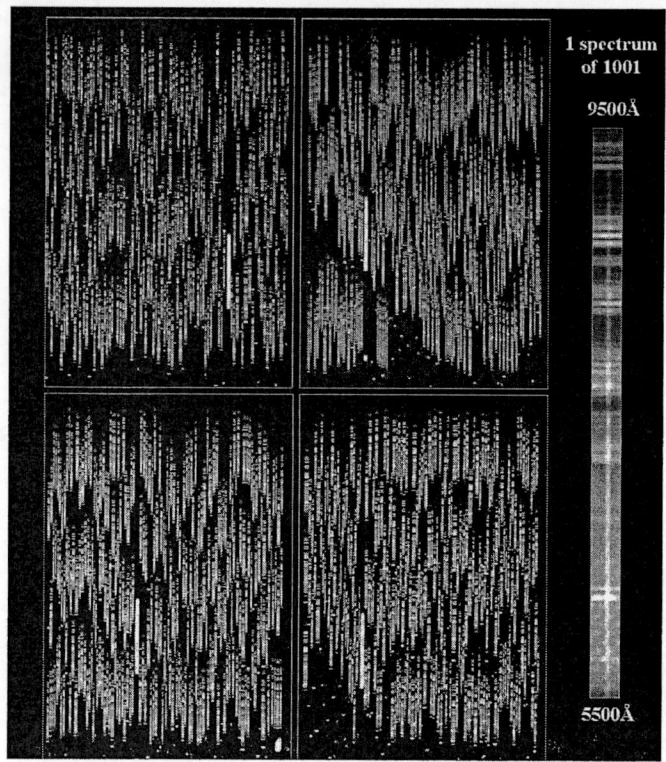

Fig. 2. 1001 galaxy spectra in a $14' \times 14'$ field obtained in one exposure with VIMOS at the VLT

formation in our own. Finally, $\Re$ up to 100,000 is needed for detailed studies of stars in our Galaxy, as well as of the tenuous Intergalactic Medium from line of sight absorption features seen in the spectrum of distant quasars.

4. Even if direct imaging and long-slit spectroscopy remain basic capabilities available on most astronomical instruments, large gains in observing efficiency are obtained by developing special modes, directly tailored for given classes of astrophysical objects. In the multi-object mode, sparsely distributed sources, e.g. the tapestry of distant galaxies, are collected by multi-slits or multi-fibers positioned precisely on the most interesting ones. With hundreds of galaxy spectra made simultaneously (Fig. 2), this speeds us considerably the very large surveys needed to understand the early evolution of the Universe. In the Integral Field mode (Fig. 3), a small portion of the sky is sliced in thousands of sub arc-second spatial pixels which, fed to a spectrograph, give as many spectra. A small compact object from a satellite in our solar system to a distant galaxy can then be entirely studied in one shot.

VLT INTEGRAL FIELD SPECTROGRAPHY

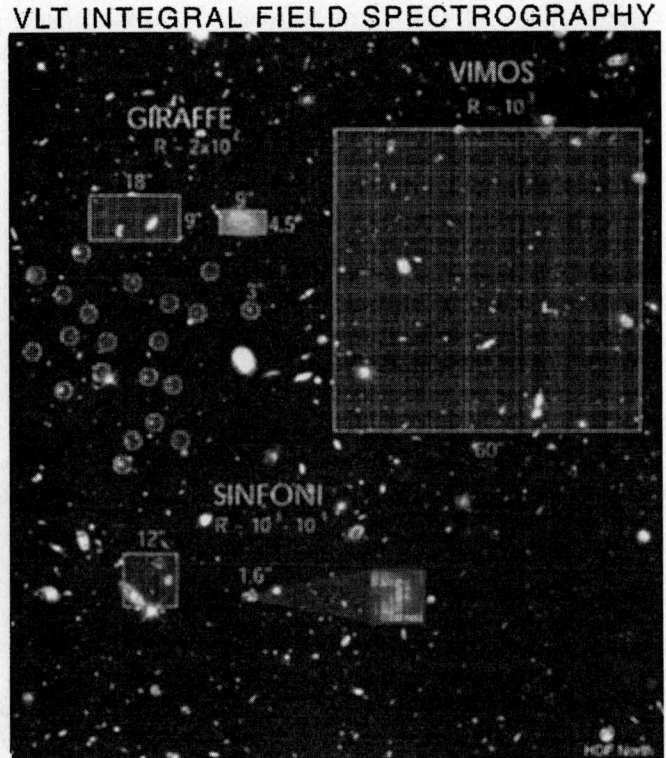

Fig. 3. Integral Field capabilities at the VLT

Four instruments are presently in routine operation at the VLT. FORS1 is an optical (0.35-0.70 μm) spectro-imager with up to 19 slitlets positioned in the $7' \times 7'$ field, and features extensive spectro-polarimetric capabilities. Spectroscopic spectral resolution ranges from a few 100s to 3,000. The similar FORS2 instrument is optimised for slightly longer wavelengths (0.5-1.0 μm) and features laser-cut multi-slit masks with up to 100 slits per mask. ISAAC is a near-infrared (1.0 to 5 μm) imager, with a $2\farcs5 \times 2\farcs5$ field, and a long slit spectrometer. Its spectral resolution, up to 3,000, is large enough to get faint object spectra between the extremely bright upper atmosphere lines, especially between 1.0 and 1.8 μm. UVES is a high spectral resolution (up to 100,000) spectrometer covering the whole 0.32 μm to 1.0 μm "optical" range in just two exposures.

Three more instruments are at an advanced stage of realization and commissioning in Paranal. NAOS-CONICA (NACO) is a very high spatial resolution (spectro)-imager in the near infrared, reaching down to the telescope diffraction limit, thanks to a state of the art adaptive optics system that corrects for atmospheric turbulence. It will be put in regular operation in October 2002. VIMOS is a large optical spectrometer, basically the equivalent of four FORS; regular operation is foreseen to start in April 2003. FLAMES is a complex multi-object system, featuring a large field (25′ diameter) robot positioner that put fibers bundles on astrophysical objects. Up to 120 single fibers can be connected to the Giraffe spectrograph which gives a spectral resolution of up to 25,000; 8 fibers can also be directed to UVES. Alternatively 24 small bundles (mini-integral field units) can be put on more extended galaxies. FLAMES is also scheduled to be offered in April 2003.

The VISIR mid-infrared (8 to 28 μm) spectro-imager had first light in the Lab in the last few months; it will be installed in Paranal in mid-2003. SINFONI is an AO-assisted integral field spectrometer in the 1.0 to 2.4 μm range. It is in full construction and scheduled to arrive in Paranal end of 2003. CRIRES is a high spectral resolution (100,000) 1 to 5 μm spectrometer, also in full construction and scheduled for mid-2004. In the meantime, extensive preparatory studies have started to build the so-called 2nd generation VLT instruments that will gradually replace the present ones as they wear off and/or become increasingly non-competitive, with e.g. progress in adaptive optics systems and especially in detectors size and performance.

4   The VLTI

The Very Large Telescope Interferometer (VLTI) on Cerro Paranal is approaching completion. After the four 8-m Unit Telescopes (UT) individually saw first light in the last years, two of them were combined for the first time on October 30, 2001 to form a stellar interferometer. The remaining two UTs were integrated into the interferometric array in September 2002. In the first half of 2003, the instruments MIDI and AMBER and the fringe sensor unit FINITO will arrive. In the course of 2003, the UTs will be equipped with the adaptive optics system MACAO, and three 1.8-m Auxiliary Telescopes (AT) will be integrated, with a fourth in construction. Three more Delay Lines for a total of six will complete the first phase of the VLTI. In 2005, the dual feed facility PRIMA will extend the capabilities of the VLTI to faint objects (K = 20) and will allow for high precision astrometry. Last but not least, ESO and ESA are jointly studying GENIE, a DARWIN ground demonstrator, which may become a second generation instrument for VLTI. The VLTI is described in detail in Glindemann et al. [4].

The interferometric array of the VLT observatory is unique in offering the possibility to combine four 8-m UTs with a maximum baseline of 130 m, and to combine a maximum of eight 1.8-m ATs if the Delay Line tunnel is equipped with eight Delay Lines. The ATs can be moved to 30 different stations with a maximum baseline of 200 m providing an excellent uv-coverage. The Delay Line System is one of the most spectacular subsystems of the VLTI, moving the 2.25 m long carriages with the Cat's Eye reflector at speeds up to 0.5 m/sec in the 130m long tunnel. While moving the carriage, the reflected beam is tilted less than 1.5 arcsec at all times, the absolute position accuracy is 30 μm over the full range of travel of 65 m and the position error is of the order of 20 nm.

All data of scientific interest taken between First Fringes on March 17, 2001 and September 22, 2002 have been released to the community via the ESO archive. A summary of the scientific results of the first year can be found in Paresce et al. in these proceedings and in Paresce et al. [7]. The VLTI was included for the first time in the ESO Call for Proposals for Period 70, starting in October 2002. Part of the VLTI commissioning time was opened for shared risk observing programmes in service mode with VINCI and two 40 cm siderostats. 150 hours were offered and 40 proposals were received. The result of the observations, i.e. the output of the data pipeline (visibility and accuracy), as well as the raw data and the data reduction software will be released to the community.

At the end of 2002, the science instrument MIDI, and, in the first half of 2003, AMBER and the fringe sensor unit FINITO will arrive. The integration of the first two MACAOs (the UT adaptive optics systems) and of the Auxiliary Telescopes will start early in 2003. MIDI will operate in the N-band (8-12 μm). The details of the instrument are described in Leinert et al. [5]. MIDI will be delivered to Paranal in October 2002; first light with the siderostats is planned for December 2002. Regular science operations are planned to start in October 2003.

The near-infrared science instrument, AMBER, will operate between 1 and 2.5 μm, at first with two telescopes, with a spectral resolution up to 10,000 (Petrov et al. [9] and Malbet et al. [6]). AMBER has been designed for three beams to enable imaging through phase closure techniques. It is planned to start commissioning AMBER with the siderostats in the second quarter of 2003.

The VLTI fringe sensor unit is called FINITO for 'Fringe sensing Instrument NIce TOrino', since the concept was developed and tested in a prototype at Nice Observatory (OCA). At the Observatory of Torino (OATo) the concept of the prototype is converted into a VLTI style instrument according to the VLT standards (Gai et al. [3]). FINITO operates in the H-band using fibers as spatial filters. The delivery to Paranal is planned for the first quarter of 2003.

The adaptive optics system MACAO will have a 60-actuator bimorph mirror and a curvature wavefront sensor in the visible (Arsenault et al. [1]). The curvature wavefront sensor is placed in the Coude focus of the UTs picking the reference star in a field of 2 arcmin. MACAO is essential for all near-infrared instrumentation including FINITO when observing with the Unit Telescopes. This means that also a mid-infrared instrument like MIDI needs adaptive optics in order to improve the limiting magnitude by using FINITO. It is planned to have MACAO ready for interferometric observations with two UTs in July 2003.

The first two 1.8-m Auxiliary Telescopes (AT) will be ready for the VLTI in November 2003, the third in April 2004 (Flebus et al. [2]). The telescopes are relocatable on 30 stations providing baselines between 8 and 200 m. Using three telescopes with AMBER and, thus, three baselines at the same time will allow the application of closure phase techniques eliminating the influence of atmospheric turbulence on fringe position. Each AT will be equipped with a tip-tilt system correcting for the fast image motion induced by atmospheric turbulence. Under the seeing conditions at Paranal tip-tilt correction on a 1.8-m telescope in the near infrared means almost diffraction limited image quality. One should note that the ATs are available exclusively for the VLTI, forming an observatory that can be operated independently of the UTs.

The Phase Referenced Imaging and Micro-arcsec Astrometry (PRIMA) facility is the third VLTI instrument. It is a dual feed system adding a faint object imaging and an astrometry mode to the VLTI. PRIMA is the key to access higher sensitivity, imaging of faint objects with high angular resolution, and high precision astrometry (~ 10 μarcsec over a 10 arcsec field). The scientific objectives of the PRIMA facility are presented in Paresce et al. [8]. As a detector for PRIMA, either one of the two scientific instruments MIDI or AMBER can take advantage of the fringe stabilisation provided by PRIMA, or a dedicated PRIMA detector will be used for high precision astrometry. PRIMA enables simultaneous interferometric observations of two objects – each with a maximum size of 2 arcsec – that are separated by up to 1 arcmin, without requiring a large continuous field of view. Then, the sensitivity of the VLTI is improved by using a bright guide star for fringe tracking – similar to the guide star in adaptive optics for wavefront sensing – in one of the two feeds, allowing to increase the exposure time on the science object in the other feed up to 10-30 minutes depending on the position in the sky.

5   ESO End-to-End Operations

The VLT project has seen a realization of ESO's vision for a new epoch of observatory operations. The financial investment of the ESO community, the technical achievements of the VLT project and the expectations of astronomers demanded that the VLT maximise its scientific return. ESO chose three pillars on which to deliver a maximal return – flexible scheduling, calibrated and tracked instruments and an archive to enable data reuse. These three elements were wrapped into an end-to-end science operations system consisting of software tools, hardware and software infrastructures and human operational processes. Many of the concepts of this system were borrowed from space missions and adapted to the demands of a ground-based array of 8 m telescopes with an arsenal of diverse and demanding instrumentation. The first three years of life of this new operations scheme have seen an outstanding and unequaled level of operational efficiency, fully vindicating this approach.

6   La Silla

La Silla was the *raison d'être* of ESO from its foundation until the Organization was unified in Garching in 1980. By then the initially small scientific group that Lo Woltjer started in Geneva had grown to be more than a small group of scientists supporting the construction of the 3.6 m telescope, to a group that organised workshops and symposia, supported data reduction activities, and turned ESO into one of the important research centres in Europe. And as the VLT project started to take shape in Europe, La Silla also underwent important changes. The completion of the 3.6 m was followed in 1982 by the 2.2 m Max Planck telescope which had been waiting in boxes for a site in the Southern hemisphere to be assembled. Italy and Switzerland joined ESO in 1984 and their entrance was essential to finance the construction of the NTT, which pioneered many of the technologies that were later to mature at the VLT. Much later, Riccardo Giacconi realised that the NTT could be used not only to test some of the hardware innovations for the VLT, but also to test the full end-to-end operations model for the very large telescope. Thus the NTT was the first telescope where the VLT software was (success)-fully tried. So, by 1990 La Silla had grown from having a number of small telescopes, the largest 1.5 m in diameter, to having two 4 m-class telescopes, a wide-field 2 m telescope, and also the only sub-mm antenna in the southern hemisphere – SEST.

Along with these activities, ESO developed a strong instrumentation programme building state-of-the-art optical and IR instruments for all telescopes on La Silla. Thus, the instrumentation effort at La Silla formed the base of the instrumentation programme for the VLT.

La Silla reached the peak of its glory in the early 90's when it was one of the largest optical observatories in the world. Besides the two 4 m class telescopes and the SEST, La Silla operated four 2 m class telescopes (including the CAT), three 1 m telescopes (including the Schmidt camera), and a number of smaller instruments including several from institutes in member countries. On a clear night one could count up to 18 telescope gazing the sky.

With the advent of Paranal observatory La Silla began to shrink in the mid 90's. One by one the Schmidt, the ESO 50, the CAT, the 1 m, the 1.5 m, and some of the national telescopes were taken out of operation. ESO's mission had changed. With the VLT ESO could only afford to provide the European astronomers with facilities no single country or institute could build and, this, of course, means large telescopes. But La Silla continues to provide infrastructure and technical support to individual institutes in Europe and continues to offer opportunities for experiments on the mountain.

Curiously, as it shrank La Silla also became stronger. After the NTT, the 3.6 m, and the 2.2 m telescopes were fully renovated. A new suite of powerful instruments was developed (SOFI, WFI, TIMMI2, FEROS, SIMBA), and the VLT end-to-end model of operations was implemented. Thus, the demand for La Silla telescopes remains strong in spite of the smaller number of telescopes, and there is no sign of decline.

7   The ESO/ST-ECF Science Archive Facility

The ESO/ST-ECF Science Archive Facility (SAF) is a joint operational and developmental activity of ESO and the ESA Space Telescope European Coordinating Facility. The SAF currently holds data from both ESO observatories as well as the Hubble Space Telescope. Figure 4 shows that currently the SAF holds more than 10 Terabytes of data representing more than one million individual observation frames.

One of the fundamental mechanisms ESO chose to maximise the return from the VLT was the tracking of instrument performance and the maintenance of data quality. The result of this approach to maximizing return, is the creation of a data archive that remains usable in the future. In this way the scientific return on data can be multiplied by allowing its reuse for new projects through archival research. The increase in data requests (output) versus data input to the ESO archive as shown in Fig. 4 is perhaps an initial indication of the success and worth of this approach.

8   Virtual Observatories and the Astronomical GRID

Astronomy, like many other physical sciences, has reached a crisis point for the execution of large national and international research programmes. In the last decade of the 20th century, an array of new ground and space observatories were inaugurated which now collect data across large sections of the electromagnetic spectrum. The data explosion from these new observatories can no longer be readily processed, explored and exploited on the desktops of individual astronomers. Researchers must now turn to the GRID paradigm of distributed computing and resources to conduct new and innovative programmes. A necessary step to utilise this new IT paradigm is to join the existing astronomical data centres and archives into an interoperating and federated unit. This new astronomical data resource will effectively form a Virtual Observatory (VO) in which the digital Universe resident in the new archives can be seamlessly explored across the entire spectrum.

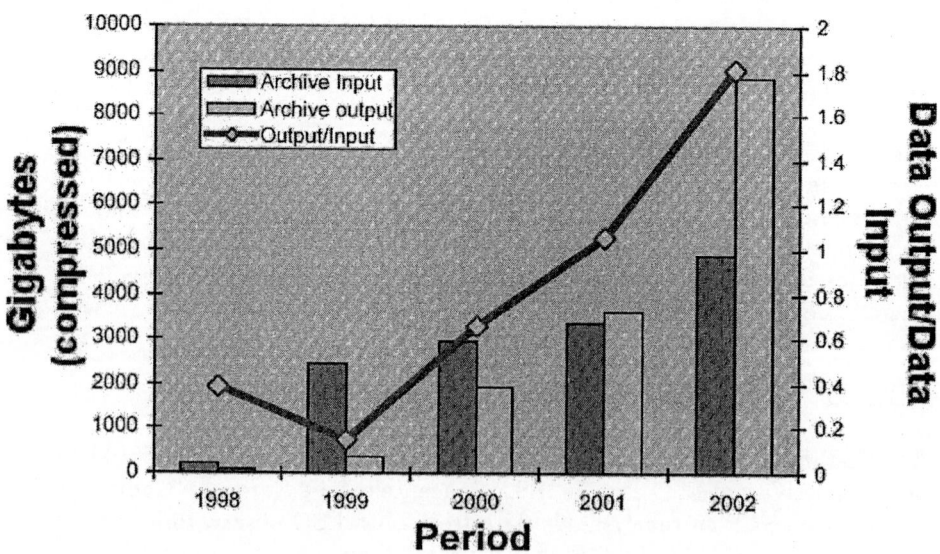

Fig. 4. Increase of data requests versus data input in the ESO archive

The Astrophysical Virtual Observatory Project (AVO) has been jointly funded by the European Commission and six European organizations (led by ESO and including ESA) for a three year Phase-A work programme valued at 5 million Euro. The Phase A programme will focus its work in three areas. Firstly, a detailed description of the science requirements for the AVO will be constructed following the experience gained in a smaller scale science demonstration programme called ASTROVIRTEL (Accessing Astronomical Archives as Virtual Telescopes) led by ST-ECF. Secondly, the difficult issue of data and archive interoperability will be addressed by new standards definitions for astronomical data and trial programmes of "joins" between specific target archives within the project team. Finally, the necessary GRID and database technologies will be assessed and trailed for use within a full AVO implementation.

9 The Atacama Large Millimeter Array

The Atacama Large Millimeter Array (ALMA) is ESO's next major project. It will be the largest ground-based astronomy project following the VLT/VLTI. It is a 50-50 Europe-North American project, with ESO leading Europe's participation. The project has now officially entered the construction phase, following formal approvals by the ESO Council and the U.S. National Science Board in July and August 2002 respectively. Japan hopes to enter the project as a third equal partner within the next year or two. As ALMA will be based in Chile, that country will also participate as host country.

Fig. 5. Artist's view of the Atacama Large Millimeter Array (ALMA), with 64 12-m antennae

For astronomy, ALMA is one of the highest priority projects in the world. A large array of antennas, it will image the Universe at millimeter and sub-millimeter wavelengths with unprecedented sensitivity and angular resolution. It represents a jump of two orders of magnitude in sensitivity and angular resolution, and will certainly be a giant step in astrophysics. ALMA will be a millimeter/submillimeter counterpart of the VLT and HST, with similar angular resolution and sensitivity but unhindered by dust opacity. Together with the Next Generation Space Telescope (NGST), it will be one of the two major new facilities for world astronomy coming into operation at the end of this decade.

The main scientific objectives are the origins of galaxies, stars and planets. ALMA will be able to detect the most distant forming galaxies, and it will be able to explore in detail the physical and chemical processes of star and planet formation in molecular clouds and protoplanetary disks that are hidden from view by dust obscuration in the normal optical/UV wavebands. ALMA will, however, go far beyond these major science objectives, and will make major contributions in virtually every area of astronomy.

ALMA will be comprised of 64 12-meter diameter antennas of very high precision, with baselines extending up to 14 km. The array of antennas will be reconfigurable, giving ALMA a zoom-lens capability. The highest resolution images will come from the most extended configuration, and lower resolution images of high surface brightness sensitivity will be provided by a compact configuration in which all antennas are placed close to each other. The instrument thus combines the imaging clarity of detail provided by a large interferometric array together with the brightness sensitivity of a large single dish. The large number of antennas provides over 2000 independent interferometer baselines, making possible excellent imaging quality with "snapshot" observations of very high fidelity. The receivers will cover the atmospheric windows at wavelengths from 0.3 to 10 millimeters. ALMA will be located on the high-altitude (5000 meter) Llano de Chajnantor, east of the village of San Pedro de Atacama in northern Chile. This is an exceptional site for millimeter astronomy, possibly unique in the world.

ALMA will be operated in pure service mode, and its data products will be images that are readily accessible to astronomers who are not specialists in the techniques of millimeter astronomy. Thus, ALMA will serve the entire astronomical community. In this respect, ALMA will benefit enormously from ESO's pioneering experiences in operating the VLT.

The construction phase of ALMA has now begun, and completion is foreseen in 2011. The first scientific operations with a partial array will begin as early as 2007. The total estimated construction cost is about 600 million Euro, to be shared equally between Europe and North America. Japanese participation would provide significant scientific enhancements to the bilateral project, and this possibility has been extensively studied. If it is realised, ALMA will become a truly global project, the first ever in ground-based astronomy.

10   OWL

Preliminary requirements and possible technological solutions for the next generation of ground-based optical telescopes were laid down at ESO in 1998. The project is currently in a conceptual study phase, the aims of which are to verify the feasibility of a 100-m class optical telescope (dubbed OWL for the eponymous bird's keen eye vision), explore potential science cases, define a baseline design and operation's scheme, and establish reliable cost, schedule, and performance estimates. Design, analysis, and industrial studies initiated since 1998 confirm the timely feasibility of the telescope within the estimated cost, and the readiness of key suppliers to respond to the underlying demand.

With milliarcsecond resolution and limiting magnitudes thousands of times fainter than what can be achieved today, OWL will advance our knowledge in most areas of astronomy. Within the solar system it will image objects at resolutions comparable to that offered by space probes (but over much longer time scales). In the nearby Universe it will unveil the intricate processes underlying the formation of stellar and planetary systems. It will resolve solar type stars at the distance of Virgo. It will be able to image exoplanets and exo-earths and determine their atmospheres' composition, possibly revealing the existence of biospheres. Peering into the deepest reaches of the far Universe, OWL will

Fig. 6. Layout of the OWL telescope and facilities. During daytime or adverse conditions, the telescope would be protected by a sliding enclosure (not shown)

witness the birth of the very first stars and galaxies, obtain spectroscopy of supernovae out to a redshift of 5, and study the evolution of the cosmological parameters (and ultimately of the Universe itself). Indeed, OWL may revolutionise our perception of the Universe as much as Galileo's telescope did.

The design of the OWL observatory (Fig. 6) capitalises on the VLT successful design, construction and operation, and on available industrial expertise in Europe. In contrast with virtually all previous attempts at making larger telescopes, with few exceptions it relies extensively on proven and available technologies, both in the factory and in the telescope. Another particular feature of the design is its strong reliance on cost-conscious standardisation. The telescope structure and the mirror cells, for example, are assembled from nearly all-identical structural modules; the kinematics is provided by a large number of identical, low-cost, distributed friction drives. Thanks to a somewhat unusual optical solution with spherical primary, flat secondary mirrors and a 4-elements active wavefront corrector, the primary and secondary mirrors are assembled from all-identical segments. This optical solution also ensures an unusually low sensitivity to external excitations such as wind and thermal changes, thereby simplifying enclosure requirements.

A notable exception to conservative, proven technologies is OWL adaptive optics, which requires substantial extrapolation from available techniques. A gradual approach towards adaptive optics is therefore foreseen, starting with a moderate extrapolation of current techniques (narrow-field, infrared adaptive optics with a partially filled, 50-m aperture) and culminating with relatively large field adaptive optics in the infrared and in the visible over the full 100-m aperture, three to five years after start of operations.

Acknowledgements

I thank P. Dierickx, R. Gilmozzi, J. Melnick, G. Monnet, F. Paresce, P. Quinn and P. Shaver for their help in preparing this manuscript.

References

1. Arsenault, R., et al. 2002, Proc. SPIE 4839: Adaptive Optical System Technologies II, in press
2. Flebus, C., et al. 2002, Proc. SPIE 4838: Interferometry for Optical Astronomy II, in press
3. Gai, M., et al. 2002, Scientific Drivers for ESO Future VLT/VLTI Instrumentation, eds. J. Bergeron & G. Monnet, Springer-Verlag Berlin-Heidelberg, p. 328
4. Glindemann, A., et al. 2002, Proc. SPIE 4838-11: Interferometry for Optical Astronomy II, in press
5. Leinert, Ch., et al. 2002, Proc. SPIE 4838: Interferometry for Optical Astronomy II, in press
6. Malbet, F., et al. 2002, Proc. SPIE 4838: Interferometry for Optical Astronomy II, in press

7. Paresce, F., et al. 2002a, Proc. SPIE 4838-68: Interferometry for Optical Astronomy II, in press
8. Paresce, F., et al. 2002b, Proc. SPIE 4838-113: Interferometry for Optical Astronomy II, in press
9. Petrov, R., et al. 2002, Proc. SPIE 4838: Interferometry for Optical Astronomy II, in press

Note Added in Proof

Since this paper was written (March to August 2002), ESO has witnessed many new events and successes, including the celebration of its fortieth anniversary in October 2002. The commissioning of the first VLT instrument with adaptive optics, NACO, was concluded and the instrument has been available to the community since October 2002. As NACO is the most advanced instrument in its category at the world level, it has already obtained several front line discoveries in various fields. VIMOS and FLAMES have been commissioned as well; FLAMES in fact entered into regular operations in February 2003, beating the anticipated schedule by two months. The first interferometric instrument, MIDI, had a very successful first light, or rather first fringes, in December 2002, and is currently under commissioning. The fringe sensor unit FINITO has been delivered to Paranal after tests at Garching had demonstrated its excellent performance. VLTI results obtained by the community with siderostats and/or the VLT telescopes and with the test instrument VINCI have started to appear in the literature.

The La Silla Observatory also received a new and very exciting instrument, the high resolution spectrometer HARPS, capable of measuring the radial velocities of stars with an accuracy of 1m/s. At the focus of the 3.6m telescope, HARPS promises to be one of the most efficient instruments in the world in the search for extrasolar planets. Early this year the first demonstration of the Astrophysical Virtual Observatory took place, where the prototype software was applied to an extended deep survey data base, GOODS.

In parallel, the development of ALMA proceeded at a fast pace. A centralized organization of the project is now in place, with a Director heading a Joint Alma Office, and all the subsystem teams in both continents are completing and testing prototypes. The bilateral agreement for the construction of ALMA was signed by ESO and the NSF in February 2003; Spain has now officially joined the project on the ESO side, and negotiations for the participation of Japan continue.

In a collaboration led by the Max Planck Institute and including the Onsala Space Observatory, ESO is presently installing at the ALMA site, Chajnantor, at an altitude of 5000m, the APEX antenna, similar to the ALMA antennae, as a pathfinder to ALMA.

Poster Papers

Inflationary Perturbations from Non-Scale Invariant Potentials

Jennifer Adams and Bevan Cresswell

University of Canterbury, Private Bag 4800, Christchurch, New Zealand

Abstract. We use a numerical code for computation of the linear density perturbations generated during an inflationary epoch in which the inflaton potential has features induced by a spontaneous symmetry breaking phase transition. A sharp step in the amplitude or the slope of the inflaton potential generate k dependent oscillations in the primordial spectrum of density perturbations. We constrain such features in the inflaton potential using observations of the cosmic microwave background anisotropy.

It is usually assumed that the primordial spectrum of density perturbations is scale invariant or nearly scale invariant, with the spectrum given by $\mathcal{P}_\mathcal{R} \propto k^{n-1}$ and n, the spectral index, close to 1. However there has been some observational support for a feature in the matter power spectrum, for example in the spatial distribution of Abell-ACO clusters [1] and in the power spectrum derived from the deep pencil beam survey of Broadhurst *et al* [2]. Moreover when one considers inflaton potentials with a particle physics basis then phase transitions in other scalar fields may introduce a break into the inflaton potential [3]. Such a break will necessarily induce characteristic features in the primordial density perturbation spectrum.

We consider two inflationary models which have inflaton potentials which cause a violation of slow roll. The first of these is the step potential

$$V(\phi) = \frac{1}{2}m^2\phi^2 \left[1 + c\tanh\left(\frac{\phi - \phi_{\text{step}}}{d}\right)\right] \tag{1}$$

which has a sharp decrease in the potential at $\phi = \phi_{\text{step}}$ with size dependent on c and gradient dependent on d. The second potential we consider is the potential first explored by Starobinsky [4] which has the form

$$V(\phi) = \begin{cases} V_0 + A_+(\phi - \phi_{\text{step}}) & \phi > \phi_{\text{step}} \\ V_0 + A_-(\phi - \phi_{\text{step}}) & \phi \leq \phi_{\text{step}} \end{cases} \tag{2}$$

This potential is often referred to as broken scale invariance (BSI).

The effect on the primordial density spectrum of a change in the inflaton potential amplitude was discussed in [5] and that of a BSI potential in [4,6] In the latter case an exact analytical expression for the initial power spectrum was obtained by Starobinsky [4].

In both cases k dependent oscillations are generated in the spectrum of primordial density perturbations.

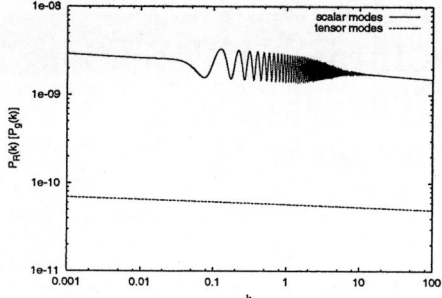

Fig. 1. The scalar and tensor power spectrum for the inflaton potential (1) with $c = 0.001$ and $d = 0.001$

In a forthcoming paper we use measurements of the cmb anisotropy to constrain the position and severity of discontinuities in the amplitude or slope of the inflaton potential. In Fig. (2) we show one example of the χ^2 obtained for varying positions of a step in the amplitude.

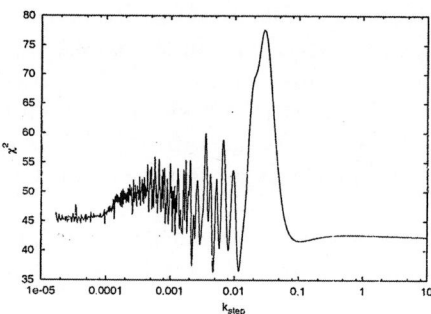

Fig. 2. χ^2 for fit with cmb anisotropy measurements for the potential (1) with $c = 0.001$, $d = 0.001$ and varying positions of the step position. The background cosmology and data used is described in our forthcoming publication

References

1. J. Einasto: Nature **385**, 139 (1997), J. Retzlaff, S. Borgani, S. Gottlober, A. Klypin, V. Muller: New Astron.**3**, 631 (1998)
2. T. Broadhurst, R. Ellis, D. Koo, A. Szalay: Nature **343**, 726 (1990)
3. J. Adams, G. Ross, S. Sarkar: Nucl. Phys.**B503** (1997)
4. A. Starobinsky: JETP **55**, (1992)
5. J. Adams, B. Cresswell, R. Easther: Phys.Rev. **D64**, (2001)
6. J Lesgourgues, D Polarski, A. Starobinsky: Mon.Not.Roy.Astron.Soc. **297**, 769 (1998), M. Gramann, G. Hűtsi MNRAS **316**, 631 (2000), S. Leach, M. Sasaki, D. Wands, A. Liddle: Phys.Rev.**D64**, (2001)

QSO Absorption Lines as a Cosmological Probe: Exploring the Lyman Forest with VLT/UVES

Simone Bianchi[1], Stefano Cristiani[2], Tae-Sun Kim[1], and Sandro D'Odorico[1]

[1] ESO, Karl-Schwarzschild-Strasse 2, D-85748 Garching, Germany
[2] ST-ECF, ESO, Karl-Schwarzschild-Strasse 2, D-85748 Garching, Germany

1 Observations and Data Reduction

The remarkable efficiency of the Ultra-Violet Echelle Spectrograph (UVES) at the VLT has made it possible to push high-resolution, high-S/N ground observations of the Lyα forest down to $z \sim 1.5$, gaining new insight into the physical state of the intergalactic medium (IGM) and its evolution over more than 90% of the cosmic time. The results presented here are based on recent UVES observations of the spectra of 8 QSOs, covering the Lyα forest at $1.5 < z_{\mathrm{Ly}\alpha} < 3.6$ with S/N $\sim$ 40–50 and resolution $R \sim 45\,000$ [1,2]. The absorption lines were fit to the Voigt profiles to derive the three line parameters: absorption redshift, z, the HI column density, N_{HI} in cm^{-2}, and the Doppler parameters, b in km s^{-1}.

2 The Number Density Evolution of the Lyα Forest

Figure 1 shows the evolution of the line number density per unit redshift, dn/dz, in the interval $N_{HI} = 10^{13.6-17}$ cm^{-2}. The maximum-likelihood fit to the $z > 1.5$ data is $dn/dz = 6.1\,(1+z)^{2.5\pm0.2}$ (dashed line). The evolution of the Lyα forest with z is mainly governed by two physical processes: the Hubble expansion and the ionizing ultraviolet background flux (UVB,[3]). At higher z, the Hubble expansion and the non-decreasing UVB cause a rapid evolution of dn/dz. At lower z, HST observations have shown a slow-down in dn/dz (solid line in Fig. 1,[4]), which can be explained with a decrease of the UVB flux in the local universe. The UVES observations imply that the turn-off in the evolution occurs at $z \sim 1$. The evolution and the redshift of the turn-off are consistent with an UVB to which galaxies contribute as much as QSOs or more, as long as a fraction $f_{\mathrm{esc}} \geq 0.05$ of the UV flux can escape the internal absorption in a galaxy [5].

3 The Cosmic Baryon Density

A lower bound to the cosmic baryon density can be derived from the distribution of Lyα optical depths [6]. For a UVB with a contribution from galaxies ($f_{\mathrm{esc}} = 0.1$), the effective optical depths measured in the UVES spectra at $1.5 < z < 4$ implies $\Omega_b h^{1.5} > 0.028$ (assuming IGM temperature $T = 2 \cdot 10^4$K, $\Omega_m = 0.3$, $\Omega_\Lambda = 0.7$). This value is consistent with the BBN value for a low D/H primordial abundance. Most of the baryons reside in the Lyman forest at $1.5 < z < 4$ with

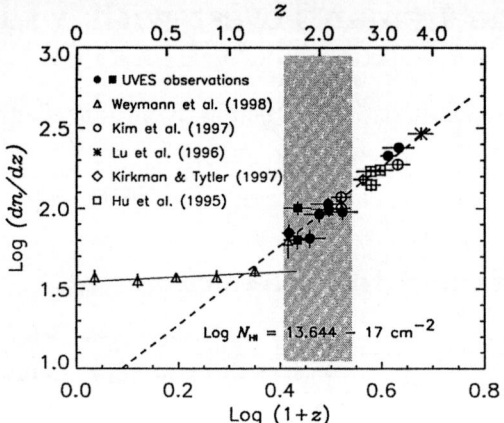

Fig. 1. The number density evolution of the Lyα forest

little change in the contribution to Ω as a function of z. Conversely, given the observed opacity, a higher UVB requires a higher Ω_b. As pointed out by Haehnelt et al. [7], values of f_{esc} as large as 0.4 [8], would result in too large Ω_b values.

4   The Temperature of the IGM

Absorption lines in the Lyα forest are broadened by gas thermal motion and other processes (cloud turbulence, residual Hubble expansion, Jeans smoothing [9]). Therefore, the minimum b value can provide an upper limit on the temperature. The minimum b value is found to increase as z decreases [10]. When the column densities are converted into over-densities, this implies that the temperature at the mean IGM density decreases with z. The large fluctuations in the minimum b value and in its dependence on N_{HI}, even at a similar redshift, suggest that the temperature of the intergalactic medium might fluctuate. A large fluctuation seen at $z \sim 3.1$ is probably due to the HeII reionization [10,11].

References

1. T.-S. Kim, S. Cristiani, S. D'Odorico: A&A, **373**, 757 (2001)
2. T.-S. Kim, et al.: A&A, in press (2002)
3. T. Theuns, A. Leonard, G. Efstathiou: MNRAS, **297**, 49 (1998)
4. R. J. Weymann et al.: ApJ, **506**, 1 (1998)
5. S. Bianchi, S. Cristiani, T.-S. Kim: A&A, **376**, 1 (2001)
6. D. H. Weinberg et al.: ApJ, **490**, 564 (1997)
7. M. Haehnelt et al.: ApJ, **549**, L151 (2001)
8. C. Steidel, M. Pettini, K. Adelberger: ApJ, **546**, 66 (2001)
9. T. Theuns, J. Schaye, M. Haehnelt: MNRAS, **315**, 600 (2000)
10. T.-S. Kim, S. Cristiani, S. D'Odorico: A&A, **383**, 747 (2001)
11. J. Schaye et al.: MNRAS, **318**, 817 (2000)

Higher Order Corrections to Lensing Parameters for Extended Gravitational Lenses

Salvatore Capozziello and Virginia Re

Universitá di Salerno, 84081 Baronissi (Sa), Italy;
Istituto Nazionale di Fisica Nucleare, Sez. di Napoli, Italy

Abstract. We discuss the contribution to the characteristic lensing quantities, i.e. the deflection angle and Einstein radius, due to the higher order terms (e.g. the gravito-magnetic terms) considered in the lens potential. The cases we analyze are the singular isothermal sphere and the disk of spiral galaxies. It is possible to see that the perturbative effects could be of the order 10^{-3} with respect to the ordinary terms of weak field and thin lens approximations. In both cases, the proper motions (i.e. the velocity dispersion and the circular velocity) have to be taken into account and, in some peculiar cases (e.g. AGN, or other kinds of extreme active galaxies) their contributions could be far to be trivially ignored.

1 The Singular Isothermal Sphere

Isothermal sphere is the simplest model used to describe the mass function of the haloes of galaxies and to derive the potential of elliptical galaxies [1]. As a lens model, it is the further step after the point-like Schwarzschild lens. The internal motions (e.g. the velocity dispersions, proper motions of the stars, etc.) give rise to nontrivial effects capable of supporting the dynamics of the real systems [1]. If the system described by an isothermal sphere acts as a lens, these effects can lead to gravitomagnetic corrections which could be quantitatively significant. Gravitational potential is given by:

$$\phi(x) = -2\sigma_v^2 \ln \left(\frac{x}{R}\right) = -2\sigma_v^2 \ln \left(\frac{|\,\boldsymbol{\xi} + l e_{in}\,|}{R}\right) \tag{1}$$

where σ_v is the velocity dispersion of the lens, $\boldsymbol{\xi}$ is the distance from the centre of the sphere, e_{in} is a unitary vector in the initial direction of light. R is a cut-off distance introduced to eliminate the singularity in the origin. A vector potential can be defined as

$$V = v\phi \tag{2}$$

where v is a velocity. The deflection angle of light, taking into account also the potential vector term [2],[3] is given by

$$\boldsymbol{\alpha} = \frac{2}{c^2} \int \boldsymbol{\nabla}_\perp \phi \, dl - \frac{4}{c^3} \int (e \wedge (\boldsymbol{\nabla} \wedge V)) dl \tag{3}$$

where dl is the Euclidean line element. The last term in the right-hand side is the gravitomagnetic correction. Solving for the k-component and evaluating the integrals between 0 and ∞, we get:

$$\alpha_k = -\frac{2\sigma_v^2 \pi}{c^2} \frac{\xi_k}{\xi} - \frac{4\sigma_v^2 v_k}{c^3} \ln\left[1 + \left(\frac{R}{\xi}\right)^2\right]. \tag{4}$$

The last term in (4) is due to the gravitomagnetic correction and it clearly depends on the ratios v_k/c (i.e. the kinematics) and R/ξ (i.e. the geometry). It is straightforward to see that the correction is significant only for $v_k/c \simeq 10^{-(2 \div 3)}$ and $R \sim \sqrt{2}\xi$. This means that high proper motions and the impact parameters of the light beams comparable to the physical sizes of the lenses can give rise to appreciable gravitomagnetic corrections.

2   The Disk of Spiral Galaxies

In this case the k-component of the gravitomagnetic correction to the deflection angle is:

$$\alpha_k^{grav}(\xi) = \frac{4}{c^3} v_k \psi_0 - \left(\frac{D_{ds}}{D_d D_s}\right) \frac{32 v_k \pi G \xi_c^2 \Sigma_0}{c^5} \left[e^{-\frac{\xi}{\xi_c}} - \sum_{n=1}^{\infty} (-1)^n \frac{1}{nn!} \left(\frac{\xi}{\xi_c}\right)^n\right]. \tag{5}$$

Also in this case, the role of the ratios $\dfrac{v_k}{c}$ and $\dfrac{\xi}{\xi_c}$ is leading to appreciate the correction. The above results tell that the effects of proper motions can be very relevant and then the vector potential terms in the perturbative expansion of gravitational field are not negligible.

In both the analyzed cases, we have found that the gravitomagnetic correction depends on the kinematics and the geometry of the system. Then it is possible to link this term to the redshift as in [3] in order to give them a quantitative evaluation. To give relevant effects, the second term in Eq.(4) should be of the order $10^{-(2 \div 3)}$ with respect to the first. These are the limits set by the forthcoming space and ground-based experiments. These constraints can be achieved considering exotic objects (e.g. AGN) as lenses. Furthermore such constraints could be used also to study exotic non-compact invisible bodies [4],[5] by taking into account their kinematics. The main result is that all lensing quantities are corrected and several observations, which cannot be fitted in General Relativity context, could be reinterpreted in this scheme.

References

1. Binney J., Tremaine S. "Galactic Dynamics", Princeton Univ. Press, Princeton 1987.
2. Schneider P., Ehlers J., Falco E.E.,1992, Gravitational lenses. Springer-Verlag, Berlin.
3. Capozziello S., Lambiase G., Papini G., Scarpetta G., 1999, Phys. Lett. A, 254, 11.
4. Gurevich A.V. and Zybin K.P., 1995 Phys. Lett. A **208**, 276.
5. Sazhin M.V., Yagola A.G., and Yakubov A.V., 1996 Phys. Lett. A **219**, 199.

Testing Cosmological Models with Negative Pressure

Alberto Cappi

Osservatorio Astronomico di Bologna, via Ranzani 1, I-40127, Bologna, Italy

Abstract. There is presently compelling evidence that the cosmic energy density is dominated by a dark component with negative pressure. We discuss how the Alcock–Paczyński test can discriminate between different cosmological models with such a component, including VCDM models (with an equation of state $w \equiv P/\rho c^2 < -1$).

1 The Alcock–Paczyński Test

What is the nature of dark energy? The most natural answer is the vacuum, acting as a positive cosmological constant Λ (ΛCDM model). As the theoretically expected value of Λ should be much larger than observed, alternative explanations have been explored. In particular, it has been suggested that the dark component could be "quintessence" (Caldwell et al. 1998), i.e. a scalar field evolving in a potential, coupled to matter through gravitation, spatially inhomogeneous and slowly evolving with time, with $-1 < w < 0$ (QCDM model). Caldwell (1999) has also suggested a component with $w < -1$, naming it "phantom energy", as in his model there is a negative kinetic term in the Lagrangian of the scalar field; however, $w < -1$ can also be obtained if the vacuum energy is due to a quantized free scalar field of low mass (VCDM model, Parker & Raval 2001). Alcock & Paczyński (1979) showed that the value of Ω_Λ can be inferred from a galaxy redshift survey through a geometric test, as the spatial distribution of galaxies derived from redshifts assuming an Einstein–de Sitter model is distorted if the universe has a Λ component, depending also on the equation of state. Other distortions are also induced by peculiar velocities $-\beta$ distortion ($\beta \equiv \Omega_M^{0.6}/b$) on linear scales and fingers of God on small scales–, but some variant of the AP test might be successfully applied to future redshift surveys, with the extraction of Λ and β from anisotropic power–spectrum data. The distortion can be quantified by a flattening factor $F(z)$ relative to the EdS model. At first order, for Q models we have: $F(z) = 1 + (1/4)[1 - \Omega_M - (1 + 3w_Q)\Omega_Q]z + O(z^2)$. In Fig. 1a–c $F(z)$ is shown for different flat models. There are 3 main systematic effects due to w, shown in Fig. 1d for flat models with $\Omega_M = 0.3$. With a more negative w we have that: **a)** the amplitude of the maximum anisotropy becomes (slightly) smaller; **b)** the maximum of $F(z)$ shifts towards smaller redshifts; **c)** after the maximum, $F(z)$ decays faster. The amplitude is not very large ($\sim 20\%$), but for the VCDM models the maximum is reached at low z: $z_m \sim 0.32$ if $w_V = -3$. The effect could be detected in on–going galaxy redshift surveys, depending on our ability to disentangle the geometric distortion from the redshift distortion due to peculiar velocities (for more details, see Cappi 2001).

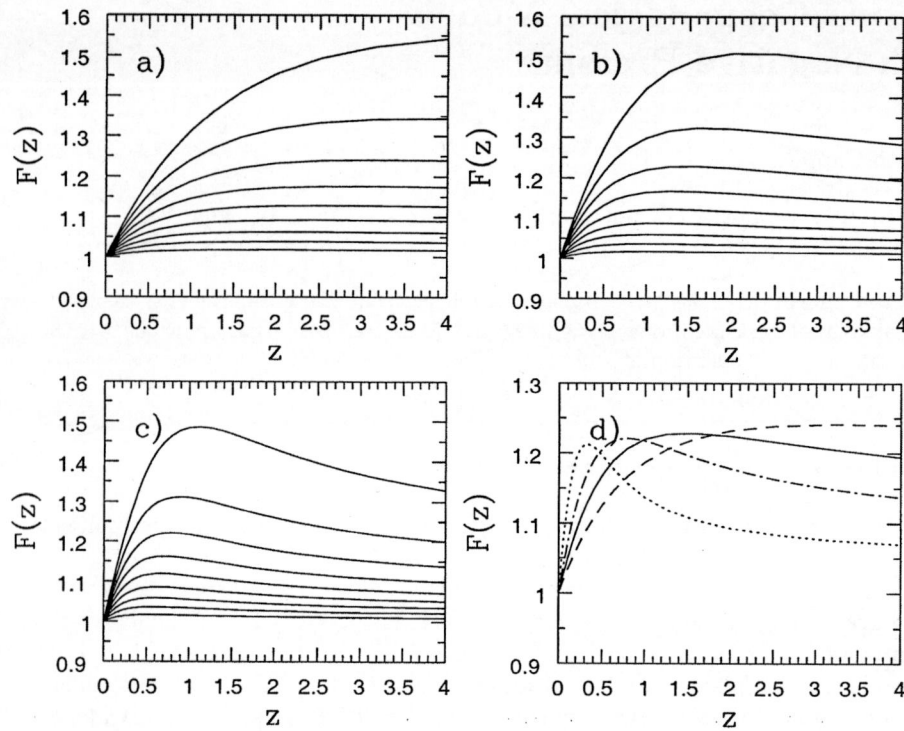

Fig. 1. Flattening factor $F(z)$ for flat models; in (a–c), Ω_M varies from 0.1 (top) to $\Omega_M = 0.9$ (bottom), step=0.1. a) QCDM ($\Omega_M + \Omega_Q = 1$, $w_Q = -2/3$); b) ΛCDM ($\Omega_M + \Omega_\Lambda = 1$); c) VCDM ($\Omega_M + \Omega_V = 1$, $w_V = -3/2$); d) 4 flat models with $\Omega_M = 0.3$. Dashed curve: QCDM, $w_Q = -2/3$; solid curve: ΛCDM; dashed–dotted curve: VCDM, $w_V = -3/2$; dotted curve: VCDM, $w_V = -3$

References

1. Alcock C., Paczyński B., 1979, Nature 281, 358
2. Ballinger W.E., Peacock J.A., Heavens A.F., 1996, MNRAS 282, 877
3. Caldwell R.R., 1999 (astro–ph/9908168)
4. Caldwell R.R., Dave R., Steinhardt P.J., 1998, Phys.Rev.Lett. 80, 1582
5. Cappi A., 2001, Astrop. Lett. & Comm. 40, 161 (astro–ph/0105382)
6. Parker L., Raval A., 2001, Phys. Rev. Lett. 86, 749

Determination of the Hubble Constant from Quadruply Imaged Gravitational Lens Systems

V.F. Cardone[1,2], S. Capozziello[1,2], V. Re[1,2], and E. Piedipalumbo[2,3]

[1] Dipartimento di Fisica "E.R. Caianiello", Via S. Allende,
    84081 - Baronissi (Salerno), Italy
[2] INFN, Sez. di Napoli, Compl. Univ. di Monte S. Angelo – Ed. N, Via Cinthia,
    80126 - Napoli, Italy
[3] INAF - OAC, Via Moiariello 16, 80131 - Napoli, Italy

Abstract. Determining H_0 from the time delays in quadruply imaged lens systems needs the knowledge of the deflecting potential $\psi(r, \theta)$. We present two different methods to solve this problem. On the one hand, we assume that the potential belongs to a broad class of non-elliptical models and we write down a system of equations which can be numerically solved to get all the parameters of the model using only the image positions and some physical constraints. On the other hand, we show how it is possible to estimate H_0 from the time delays without completely reconstructing the potential, provided that $\psi(r, \theta) = r^\alpha F(\theta)$, whatever the shape function $F(\theta)$ is. After some tests on simulated cases, we apply both methods to the real quadruple lenses PG1115+080 and B1422+231 finally obtaining $H_0 = 58^{+17}_{-15}$ km s^{-1} Mpc^{-1}.

It is well known that time delays in gravitational lens systems may be used as a tool to estimate the Hubble constant H_0 [3] avoiding the calibration problems which affect the classical local estimators (for instance Cepheids or low-redshift SNe Ia). The time delay among two images in a multiply imaged lens systems may be written as [4]:

$$\Delta t_{ij} = h^{-1} \left(\frac{D_{OL} D_{OS}}{D_{LS}} \right) \frac{1 + z_L}{c} \left[\frac{1}{2} r^2 - r r_s \cos(\theta - \theta_s) + \frac{1}{2} r_s^2 - \psi(r, \theta) \right]_{(r_i, \theta_i)}^{(r_j, \theta_j)}$$

with h the Hubble constant in units of 100 km s^{-1} Mpc^{-1} and the usual meaning for the distances. The only observable quantities in a lens systems are the time delays and the images positions (r_i, θ_i), while the source coordinates (r_s, θ_s) and the lensing potential $\psi(r, \theta)$ have to be recovered in order to get an estimate of H_0. In this contribution we present two different methods to solve these problems.

The usual technique to reconstruct the lensing potential needs the use of fitting algorithms to recover the parameters of a lens model fixed from the beginning. We prefer to follow a different approach assuming that:

$$\psi(r, \theta) = r^\alpha [1 - \delta \cos 2(\theta - \theta_p)]^\beta .$$

This kind of potential is quite general since one can recover many usual lens models by correctly choosing the slope α of the radial profile and the boxiness parameter β, while δ is a flattening indicator and θ_p is the position angle of the main axis of the potential. Including also the source coordinates (r_s, θ_s), to get the estimate of H_0 we have to recover now the values of six parameters. On the other hand, a quadruply imaged lens system allows us to write eight equations i.e. two for each image positions. In [1] we show how it is possible to rewrite this system of equations so that it can be numerically solved to get both the lensing potential parameters $(\alpha, \beta, \delta, \theta_p)$ and the source coordinates (r_s, θ_s). The system may have different solutions because of both the non linearity of the equations and the lensing potential degeneracy. To select among these we have used a set of criteria based on physical considerations. After having tested the method on simulated cases, we have applied it to the real lens PG1115+080 finally obtaining $H_0 = 56^{+12}_{-11}$ km s^{-1} Mpc^{-1}.

A different approach to the estimate of H_0 from time delays is to bypass the problem of reconstructing the lensing potential using non-parametric methods [5]. However these methods overestimate the systematics connected with the lens modelling since they are not able to discriminate among physically reasonable or unreasonable models. We have followed a different approach assuming that the lensing potential may be written as:

$$\psi(r, \theta) = r^\alpha F(\theta) - \frac{1}{2}\gamma r^2 \cos 2(\theta - \theta_\gamma)$$

being $F(\theta)$ a whatever shape function which we do not assign *a priori* and (γ, θ_γ) the external shear parameters. Now we have to determine the slope α of the radial profile, the values of $F(\theta)$ in the images positions, the shear parameters and the source coordinates summing up to nine unknown quantities for a system with four images. But a system with four images gives us up to ten equations: eight for the images positions and two for the time delay ratios. In [2] we present a method to numerically solve this set of equations. Since the system is highly nonlinear and, beside, we do not give an explicit expression for $F(\theta)$, we get a large sample of solutions; we use a set of physically motivated selection criteria to exclude all the unphysical models. The final sample of solutions is used to get the estimate of H_0. To further reduce the uncertainties, we have implemented a simple procedure to combine the results from different quadruple lenses. Tests of the method on simulated cases have confirmed that it indeed works. We then apply it to the real lenses PG115+080 and B1422+231 thus finally obtaining: $H_0 = 58^{+17}_{-15}$ km s^{-1} Mpc^{-1} (68% CL).

References

1. Cardone, V.F., Capozziello, S., Re, V., Piedipalumbo, E. 2001, A&A, 379, 72
2. Cardone, V.F., Capozziello, S., Re, V., Piedipalumbo, E. 2001, A&A, 382, 792
3. Refsdal, S. 1964, MNRAS, 128, 307
4. Schneider, P., Ehlers, J., Falco, E.E. 1992, *Gravitational lenses*, Springer-Verlag
5. Williams, L.L.R., Saha, P. 2000, AJ, 119, 439

The Sky Polarization Observatory (SPOrt) Programme

E. Carretti[1], S. Cortiglioni[1], G. Bernardi[1], S. Cecchini[1], C. Macculi[1],
C. Sbarra[1], J. Monari[2], A. Orfei[2], S. Poppi[2], G. Boella[3], S. Bonometto[3],
M. Gervasi[3], G. Sironi[3], M. Zannoni[3], M. Tucci[4], M. Baralis[5], O.A. Peverini[5],
R. Tascone[5], R. Fabbri[6], L. Nicastro[7], K.W. Ng[8], V.A. Razin[9],
E.N. Vinyajkin[9], M.V. Sazhin[10], and I.A. Strukov[11]

[1] I.A.S.F./C.N.R. Sezdi Bologna, Via Gobetti 101, I-40129, Bologna, Italy
[2] I.R.A./C.N.R., via P. Gobetti 101, I-40129 Bologna, Italy
[3] Dip. di Fisica, Univ. di Milano - Bicocca, P.za della Scienza 3, I-20126 Milano, Italy
[4] Instituto de Fisica de Cantabria, Fac. de Ciencias, Avda Los Castros s/n, 39005 Santander, Spain
[5] I.R.I.T.I./C.N.R., c.so Duca degli Abruzzi 24, I-10129 Torino, Italy
[6] Dip. di Fisica, Univ. di Firenze, Via Sansone 1, I-50019 Sesto Fiorentino (FI), Italy
[7] I.A.S.F./C.N.R. Sez. di Palermo, via U. La Malfa 153, I-90146 Palermo, Italy
[8] Academia Sinica, 11529 Taipei, Taiwan
[9] NIRFI, 25 B.Pecherskaya st., Nizhnij Novgorod 603600/GSP-51, Russia
[10] Schternberg Astronomical Institute, Moscow State University, Moscow 119899, Russia
[11] I.K.I., Profsojuznaja ul. 84/32, Moscow 117810, Russia

Abstract. SPOrt is an experiment aimed at making a polarization survey of the sky in the microwave range (22-90 GHz) on large angular scales (FWHM = 7°).

SPOrt (Sky Polarization Observatory) [3,2] is an experiment aimed at making a polarization survey of the sky in the microwave range on large angular scales (FWHM = 7°). It has been selected by ESA to be flown on board the International Space Station (ISS) in 2005 for a minimum lifetime of 18 months and it is funded by the Italian Space Agency (ASI).

The Cosmic Microwave Background (CMB) is a powerful tool to understand origin and evolution of the Universe, but only upper limits on the CMB Polarization (CMBP) have been set so far. In particular, the information contained in the CMBP can solve the degeneracies among cosmological parameters that CMB anisotropy alone is not able to remove [5,6]. For instance, on large angular scales, the polarized angular power spectra are much more sensitive to the optical depth τ of the re-ionized medium in the dark ages than the temperature spectrum does.

Surveys of the polarized diffuse emission in the microwave range are also crucial for the study of the Galactic contribution, which, besides its intrinsic interest, acts as a foreground for CMB experiments: only its accurate knowledge will allow clean measurements of CMB features. In spite of its importance, the Galactic polarized background has been mapped up to 2.7 GHz only, in a narrow belt around the Galactic plane. At high Galactic latitudes the available maps are widely undersampled.

The SPOrt experiment is aimed at filling the current gap in measurements of the diffuse polarized emission in the 22-90 GHz range. Its main goals are:

- the building of liner polarization maps of the Galactic emission at the lowest frequencies (22 and 32 GHz) over nearly 80% of the sky
- the tentative detection of the linearly polarized component of the CMB on large angular scales (expected to be $< 1\,\mu K$)

SPOrt is the first space mission devoted to Q & U Stokes parameter measurements in the microwave domain, and it has been designed to be as insensitive as possible to instrumental polarization.

Since the main information carried by CMBP is to be found on large angular scales (multipoles $\ell < 10$, corresponding to $\theta > 20°$), a very simple optics design has been adopted (corrugated feed horns), providing a resolution of 7° (FWHM). The need to detect the CMBP on large scale requires all–sky surveys, calling for a space mission.

Following this baseline, great care has been taken in optimising the instrument design with respect to systematics generation, long term stability and observing time efficiency. The main features of the SPOrt design are: four channels (22, 32, 60, 90 GHz); very simple on-axis optics; direct and simultaneous detection of both Q & U by correlation of the circularly polarized components; high performance horn, OMT and polarizer for low radiometric offset generation [1]; high rejection (> 60 dB) of the unpolarized component by the correlation unit [4]; active temperature control for both the cryogenic front–end ($\sim 80 \pm 0.1$ K) and the warm section (~ 300 K).

References

1. E. Carretti, R. Tascone, S. Cortiglioni, J. Monari, M. Orsini: NewA **6**, 173 (2001)
2. E. Carretti, et al.: In: *Astrophysical Polarized Backgrounds, Bologna October 9–12, 2001*, ed. by S. Cecchini, S. Cortiglioni, C. Sbarra, R. Sault, AIP Conf. Proc. **609**, 109 (2002)
3. S. Cortiglioni, et al.: In: *3 K Cosmology, EC–TMR Conference, Roma October, 1998*, ed. by L. Maiani, F. Melchiorri, N. Vittorio, AIP Conf. Proc. **476**, 186 (1999)
4. O.A. Peverini, M. Baralis, R. Tascone, D. Trinchero, A. Olivieri, E. Carretti, S. Cortiglioni: In: *Astrophysical Polarized Backgrounds, Bologna October 9–12, 2001*, ed. by S. Cecchini, S. Cortiglioni, C. Sbarra, R. Sault, AIP Conf. Proc. **609**, 177 (2002)
5. M.V. Sazhin, N. Benitez: Astrophys. Lett. Commun. **32**, 105 (1995)
6. M. Zaldarriaga, D.N. Spergel, U. Seljak: ApJ **488**, 1 (1997)

EUSO – Extreme Universe Space Observatory

Osvaldo Catalano, Maria Concetta Maccarone, Andrea Santangelo, and
Livio Scarsi
(on behalf of the EUSO Collaboration)

Ist. Astrofisica Spaziale e Fisica Cosmica, IASF-CNR, Via Ugo La Malfa 153,
90146 Palermo, Italy

Extended Abstract

The Extreme Universe Space Observatory *EUSO* is the first Space mission devoted to the exploration of the outermost bounds of the Universe through the detection of the Extreme Energy Cosmic Rays and Cosmic Neutrinos ($> 5 \times 10^{19}$ eV). *EUSO* will do astronomy by looking downward, from the International Space Station ISS, at the Earth Atmosphere which acts as a detector for the Extensive Air Showers induced by the Primary particles, exploiting the UV fluorescence signal excited in the air. The "*EUSO* detector" has a geometrical factor of several hundreds thousands km^2 sr and a target mass of the order of 10^{12} tons [1–3].

Astrophysics, Cosmology, and Fundamental Physics will get benefit from the EECR/ν investigation performed with *EUSO*:

- **Extreme Energy Cosmic Rays** - extend the measurement of the energy spectrum beyond the GZK limit with a significant statistical evidence. About 500 EECR events per year above 10^{20} eV will be available with *EUSO* allowing a quantitative energy spectral definition, together with an all sky map of arrival distribution; this will allow the detection of possible anisotropy effects and clustering (if any) for the directions of arrival
- **High Energy Cosmic Neutrinos** - detection of a possible flux and opening the way to the High Energy Neutrino Astronomy.

EUSO will measure the EECR/ν flux looking at the streak of fluorescence light produced by the shower particles; the Čerenkov signal, diffused when the shower hits ground or the top of a cloud, will be also imaged. The EAS will be imaged in *EUSO* by a couple of large Fresnel lenses (2.5 m diameter) onto a finely segmented focal plane detector; the very high time resolution and segmentation will enable the reconstruction of the energy ($\sim 25\%$ resolution) and of the arrival direction ($\sim 1°$) of the shower.

EUSO, located as external payload on the Columbus module of the ISS, will operate from about 400 km altitude, looking to the Nadir at the dark Earth atmosphere, with a 60° full field-of-view exploiting the UV wavelength band (330 − 400 nm). The properties of the atmosphere at the shower occurrence

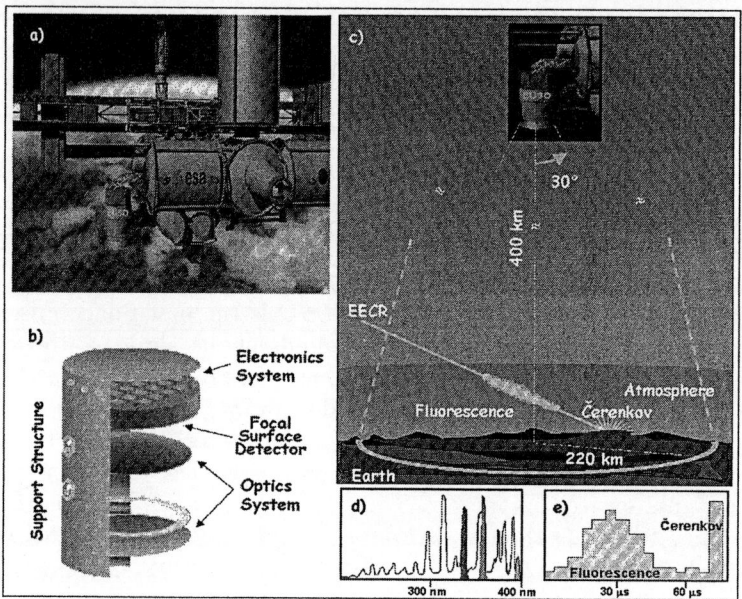

Fig. 1. Artistic view of *EUSO*.(**a**) Possible accommodation on board the ISS. (**b**) Overview of the *EUSO* main telescope. (**c**) The observational approach. (**d**) UV fluorescence spectrogram. (**e**) Image and Time structure

(transparency, cloud distribution) will be sounded by a dedicated Lidar. The detection of other atmospheric phenomena as meteors, lightning, elves, blue jets, will consent systematic investigation in the various fields.

EUSO is presently under Phase A study by ESA, with a goal for a three year mission starting in 2008. The scientific payload is a collaborative effort of research groups in Europe, Japan, and USA.

References

1. O. Catalano: Il Nuovo Cimento, **24-c,3**, 445 (2001)
2. L. Scarsi, O. Catalano, M.C. Maccarone, B. Sacco: 'EUSO - Doing Astronomy Looking downward the Earth Atmosphere'. In: *Proc. 27th ICRC, Hamburg, Germany, August 2001*, ed. by K.-H. Kampert, G. Hainzelmann, C. Spiering (Copernicus Gesellschaft, 2001), Vol. HE, pp.839-842
3. *"The Official EUSO Web Site"*, http://www.euso-mission.org

Hubble Constant or Hubble Variable H?: Why Is the "Exact" H Not Discovered?

Albert Chechelnitsky

Laboratory of Theoretical Physics, Joint Institute for Nuclear Research,
141980 Dubna, Moscow Region, Russia.   E-mail: ach@arcor.de ; ach@thsun1.jinr.ru

Abstract. For many decades outstanding researchers have been continuing the wasting search for the "Unique Right" value of the H Hubble Constant. We have not seen the end of such activity for a long time... According to representations of the Wave Universe Concept (WU Concept), such confrontation can be very durable (similar to Sizif's Work) so long as researchers follow the representations of the "Unique Right Doctrine" of modern cosmology – the Standard Cosmology – and, at last, until they understand the true physical meaning of the Hubble Variable H in the light of the Cosmological Distance Law

$$d = D \cdot 10^{m/5} Rz^2$$

Here d is the cosmological distance, m is the apparent magnitude of an astronomical object (radiation source), R is the radius of a source, z is the redshift, and

$$D = 0.58608 \cdot 10^{19}$$

is a constant. In the linear representation of Hubble Postulate the H Hubble "Constant" can be found as the Hubble Variable, generally speaking, depending on the variables m, R, z.

1   Unceasing Controversy

The search for the "True" value of Hubble Constant H has a long and dramatic History. The search for its "Exact" value has been continuing already for nearly 3/4 century, but outstanding researchers – theorists and observers – up to now cannot come to a unique opinion. By the present time Constant has decreased by some times, but hitherto has not calmed down yet... [1–5].

2   Other Horizons: Wave Universe Concept

The investigations realized in the framework of the *Wave Universe Concept (WU Concept)* have shed new light on the problem. New results may be found extremely unexpected and contradictory to conventional views on the problem becoming habitual. Instead, gradually become clear the perspectives and fate of the Unceasing Controversy. Many extremely important problems of Physics of the Universe and Cosmology, questions about the relation between objects of micro and megaworld have the answers in the framework of the *Wave Universe Concept* [6,7]. The WU Concept suggests that not only arbitrary systems of microworld, but also giant astronomical systems of megaworld possess the general fundamental aspect – all these dynamic systems, in principle, are the *WAVE dynamic systems* (WDS). If the Quantum Wave Mechanics describes the wave structure of microscopic scales objects, then the *Megaquantum Wave Astrodynamics* (Chechelnitsky, 1980–2002) analyzes and emphasizes the *MEGA-WAVE* structure of giant astronomical systems – presence of waves and rhythms of large length and periods in these planetary and galaxy systems. The fundamental, of principle, new approach argues its effectiveness in many Hot Points of Physics and Cosmology [*physics/0102036*] – from the analysis and discovery of the mass spectrum of neutrino [*physics/0103066*] to the redshift spectrum of extremely far quasars [*physics/0102089*]. The true physical structure of the Universe at large scales can be understood only in the context of the *Wave Hierarchy*, existence of very large astronomical objects, Unity and Universality of the wave structure of arbitrary objects of the Universe [*physics/0102008*].

3   Cosmological Distance Law As a Consequence of Astrophysical Photometrics

There are some ways of derivation of the Cosmological Distance Law (CD Law). Here we follow a simpler and chronologically the first way (Chechelnitsky 1986, 2000). In the frame of the WU Concept it is possible to consider some aspects of the Astrophysical (Megaquantum) Photometrics. A very important Temperature-Redshift $T - z$ Law, (Chechelnitsky 1986, 2000)

$$T = T_{z=1}z , \ T_{z=1} = (1/3k)m_e c^2 = 1.97662 \cdot 10^9 \text{K} = 0.170332 \text{ Mev} ,$$

where m_e is the electron mass, c is the velocity of light, k is the Boltzmann constant, naturally leads to the representation of $d = F(z)$ Cosmological Distance Law (CD Law), in particular, in the form $d = d_{z=1}z^q$. The representation of the cosmological distance d depending on (electronic) temperature follows from the relation

$$4\pi d^2 f - 4\pi R^2 \sigma T^4 = 0 .$$

Here $L = 4\pi R^2 \sigma T^4$ is the luminosity of the Stefan-Boltzmann law (σ – Stefan-Boltzmann Constant) equal to total luminosity $L = 4\pi d^2 f$ (f – radiation flux). In the explicit form

$$d = \left[R/(f/\sigma)^{1/2} \right] T^2$$

and further, with using the $T - z$ Law ($T = T_{z=1}z$), we arrive at the CD Law

$$d = d_{z=1}z^2 , \ (q = 2) , \ d_{z=1} = T_{z=1}^2 (\sigma/f)^{1/2} R .$$

4   General Form of CD Law

The dependence of the Cosmological Distance Law on the f-flux of energy can effectively be replaced by the dependence on *(bolometrical) apparent magnitude* $m = m_b$ by virtue of the relation (Lang, 1974 give in CGS)

$$f = \beta \cdot 10^{-(2m/5)} = (2.52 \cdot 10^{-5}) \cdot 10^{-0.4m} ,$$

where $\beta = 2.52 \cdot 10^{-5}$ erg/(cm$^2 \cdot$ s).

Taking into account $f^{1/2} = \beta^{1/2}/10^{m/5}$, we get the explicit representation – General Form for the CD Law (with the dependence on m, R, z) as

$$d = D \cdot 10^{m/5} R z^2 ,$$

where the coefficient D has a dimensionless value $D = T_{z=1}^2 (\sigma/\beta)^{1/2} = 0.58608 \cdot 10^{19}$.

5   Hubble Variable

In the linear representation of the Hubble Postulate

$$d = d_{z=1}^{(H)} z = (c/H)z, \quad d_{z=1}^{(H)} = c/H,$$

the H Hubble "Constant" can be found as the Hubble Variable

$$\left[D \cdot 10^{m/5} Rz \right] = c/H = \text{varia}, \quad H = c \, / \left[D \cdot 10^{m/5} Rz \right]$$

generally speaking, depending on the variables m, R, z.

6   Search for Hubble "Constant": The General Tendency

In connection with the general analysis of the WU Concept with the use of the CD Law it is possible to make a rather general statement connected with intensive search for a numerical value of Hubble "Constant" which has already been continued for almost 3/4 of the century: one and only one *Unique Right, True value of Hubble "Constant" does not exist! Because H is not Constant, but Variable!*

7   Beyond Geometrodynamics and Expanding Universe

Perspectives of Modern Cosmology depend on new ideas involved in the wide consideration by scientific community. In close connection with this Wave Universe Concept we insist on the assertion that genesis of the redshifts has the real physical (not geometrical) nature, has nonkinematic, nondoppler, nontransitional (no "galaxies scattering"), but endogenic, physical (temperature) character. The WU Concept fundamentally argues that the Hubble Law is consequent on the Astrophysical (Megaquantum) Photometrics and the Cosmological Distance Law (its linear approximation), but is not a result of the Geometrodynamics – (doppler) kinematics and scattering of the Universe.

References

1. G. De Vaucouleurs, G. Bollinger: ApJ 233 433 (1979)
2. G.A.Tammann: Publ. Astron. Soc. of Pacific, 108: 1083-1090 (1996)
3. W.L. Freedman et al., astro-ph/0012376; astro-ph/0202006
4. H. Arp, astro-ph/0106466
5. B. Leibundgut: Ann. Rev. Astron. Astroph. 89: 67-98 (2001)
6. A.M. Chechelnitsky: Extremum, Stability, Resonance in Astrodynamics and Cosmonautics, (Mashinostroyenie, Moscow, 312 pp., Monograph in Russian, 1980) (Library of Congress Control Number: 97121007; Name: Chechelnitskii A.M.).
7. A.M. Chechelnitsky: 'Hot Points of the Wave Universe Concept: New World of Megaquantization', In: Hot Points in Astrophysics, International Conference, Dubna, Russia, August 22-26, 2000, ed. by V. Belyaev (JINR, Dubna, 2000) pp. 391-403, [physics/0102036]

The NASA – MIDEX Swift Mission

G. Chincarini[1], for the Swift team (listed below)

[1] Universitá di Milano Bicocca, Via dell'Ateneo Nuovo 1, 20136 Milano - Italy

1 The Swift Mission

The Gamma Ray Bursts as powerful extragalactic sources are one of the major discovery which occurred in the last decade thanks to the BATSE satellite and, above all, to the Beppo-SAX Italian Dutch Mission. Indeed the GRBs represent one of the major challenges to astronomy and cosmology because of the very high energies involved, the super relativistic motion of the jets and the so far the completely unknown mechanism of energy production and nature of the progenitors. To understand a) the nature of the Bursts and the environment in which they occur, b) the physics of the Burst and the mechanism producing them and generating the super relativistic motion (Lorentz γ factor of about 100), c) the energy spectrum and the energy distribution function together with d) the possibility of using them to investigate the early Universe and the intergalactic medium, we designed a NASA MIDEX mission (USA lead with the collaboration of Italy and UK) which is now in development for launch in 2003.

A rapidly pointing and automatically controlled spacecraft has been equipped with three state of the art instruments: I) a Burst Alert Telescope (BAT) capable of detecting Bursts in the high energy band of the spectrum. Following detection and position estimate of the event (a few seconds time) the on board computer will give instruction to reorient the spacecraft to point on the target on a minute timescale. II) An X Ray Telescope (XRT) which will measure the X ray flux and the position of the source in matter of seconds and III) an Ultra-Violet Optical Telescope (UVOT) sensitive from the UV to the Red wavelengths and providing also spectroscopy and sub-arcsecond position.

The Mission is a big step forward due to the large improvement in wavelength coverage, in the fast detection and spacecraft response, sensitivity and rate of detection, telemetry and data collection. Indeed the on board instrumentation cover the electromagnetic spectrum from high energy, about 150 keV and beyond, to the visible at about 6000 Å. The BAT instrument, sensitivity $\sim 10^{-8}$ $erg\,cm^{-2}s^{-1}$ for an on axis burst of 5 sec duration and pass band in the range 10 - 150 keV with a field of view of 1.4 sr, will detect at least 150 GRBs per year with a position accuracy of 4 arcmin. The alert will be immediately transmitted to Earth and all the other instruments will be on the source in matter of second or minutes. The XRT, sensitive in the band 0.2 - 10 keV has an effective area of 110 cm^2 and can determine the position with an error of a few arcsec. UVOT, sensitive in the range 170 - 600 nm has provision for spectroscopy and filter photometry and reaches an angular resolution of 0.9 arcsec with a sensitivity of

24 mag. The instruments will be capable of time resolved flux measurements to study rapid variability.

The mission will use Malindi (ASI Kenia) as the primary Ground Station. This is the facility that has been used by the Beppo - SAX mission. The facility is at the moment being upgraded by the Italian Space Agency (ASI), in collaboration with Telespazio, for the Swift Mission. The communication network, from the ground Station to the Mission Operation Center, and from this back and for with the Italy and UK where we will also have the data archives and Data analysis, will also use of a recent set up by ASI, ASINet, which should facilitate the transfer of data and information.

The GRBs data will be public immediately for the use of all the scientific community and the collaboration and organization for the follow up is being carefully organized and supported with small to very large telescopes. As we stated at the very beginning it is the occasion to unripe the secrets, and to understand the physics, of the GRBs.

2   The Swift Team

Name	Affiliation	Role	Email
Lorella Angelini	GSFC USRA	Ground System Archive	angelini@Milkyway.gsfc.nasa.gov
Louis Barbier	GSFC	BAT Science Electronics	lmb@cosmicra.gsfc.nasa.gov
Scott Barthelmy	GSFC	BAT Lead	scott@lheamail.gsfc.nasa.gov
David Burrows	PSU	XRT Lead	burrows@astro.psu.edu
Patrizia Caraveo	IFCTR/CNR Milan	Ground System Scientist	pat@ifctr.mi.cnr.it
Margaret Chester	PSU	Ground System PSU Ops	chester@astro.psu.edu
Guido Chincarini	OAB & USMB	Italy Team Lead	guido@merate.mi.astro.it
Oberto Citterio	OAB	XRT mirror module Lead	citterio@merate.mi.astro.it
Tom Cline	GSFC	Science Theory Team	cline@apache.gsfc.nasa.gov
Lynn Cominsky	Sonoma State	Public Relations Lead	lyncc@charmian.sonoma.edu
Robin Corbet	GSFC USRA	Science Theory Team	corbet@milkyway.gsfc.nasa.gov
France Cordova	UCSB	Science Theory Team	cordova@omni.ucsb.edu
Mark Cropper	MSSL	UVOT Telescope	msc@mssl.ucl.ac.uk
Eric Feigelson	PSU	Education / Public Outreach Team	edf@astro.psu.edu
Ed Fenimore	LANL	BAT Science Flight Software	efenimore@lanl.gov
Dale Frail	NRAO/VLA	Science Theory Team	dfrail@nrao.edu
Neil Gehrels	GSFC	Principal Investigator	gehrels@gsfc.nasa.gov
Gordon Garmire	PSU	Science Theory Team	garmire@astro.psu.edu
Paolo Giommi	BeppoSAX/SDC	Italian Data Center	giommi@asi.it
Scott Horner	Lockheed-M.	UVOT DPU Heritage	scott.horner@lmco.com
Kevin Hurley	UC Berkeley	Follow-up Team Lead	khurley@sunspot.ssl.berkeley.edu
Keith Jahoda	GSFC	Ground System HX Survey	keith@rosserv.gsfc.nasa.gov
Francois Lebrun	CEN Saclay	BAT Detectors	flebrun@cea.fr
Frank Marshall	GSFC	Ground System Lead	frank.marshall@gsfc.nasa.gov
Keith Mason	MSSL	UVOT UK Lead	kom@mssl.ucl.ac.uk
Peter Meszaros	PSU	Science Theory Team Lead	pmeszaros@astro.psu.edu
Richard Mushotzky	GSFC	Science Theory Team	mushotzky@lheavx.gsfc.nasa.gov
Jay Norris	GSFC	Science Theory Team	norris@grossc.gsfc.nasa.gov
John Nousek	PSU	Narrow-Field Instrument Lead	nousek@astro.psu.edu
Bohdan Paczynski	Princeton	Science Theory Team	bp@astro.princeton.edu
David Palmer	GSFC USRA	BAT SciFl.Cen.	palmer@lheamail.gsfc.nasa.gov
Ann Parsons	GSFC	BAT Detector Scientist	parsons@lheavx.gsfc.nasa.gov
Jacques Paul	CEN Saclay	Science Theory Team	jpaul@cea.fr
Peter Roming	PSU	UVOT Lead	proming@astro.psu.edu
Tim Sasseen	UCSB	UVOT OM Liaison	tims@leo.physics.ucsb.edu
Alan Smale	GSFC USRA	Ground System Sci.Cen.	alan@rosserv.gsfc.nasa.gov
Luigi Stella	OAR	Science Theory Team	stella@coma.mporzio.astro.it
Gianpiero Tagliaferri	INAF-OAB	Italian Data Cen.ter &XRT	gtagliaf@merate.mi.astro.it
Leisa Townsley	PSU	UVOT Team	townsley@astro.psu.edu
Jack Tueller	GSFC	Ground System HX Survey	tueller@gsfc.nasa.gov
Martin Turner	U. Leicester	Science Theory Team	mjt@leicester.ac.uk
Mario Vietri	URoma3	Science Theory Team	vietri@coma.mporzio.astro.it
Martin Ward	U. Leicester	Ground System UK Data Center	mjw@star.le.ac.uk
Alan Wells	U. Leicester	UK Team Lead	aw@star.le.ac.uk
Nick White	GSFC	Science Working Group Chair	white@adhoc.gsfc.nasa.gov
Laura Whitlock	Sonoma State	Education / Public Outreach Lead	laura.whitlock@sonoma.edu
Richard Willingale	U. Leicester	XRT Calibration	rw@star.le.ac.uk
Will Zhang	GSFC	Science Theory Team	zhang@xancus10.gsfc.nasa.gov
Filippo Maria Zerbi	INAF-OABrera	Follow up - REM P.I.	zerbi@merate.mi.astro.it

The Latest on the Database of Galaxies at High Redshifts

Duilia de Mello[1,2], E.P.G. Johansson[2], and P. Markström[2]

[1] Onsala Space Observatory, Dept. of Astronomy & Astrophysics,
43992, Onsala, Sweden
[2] Chalmers University of Technology, Göteborg, Sweden

1 Introduction

We are now going through a golden era in the extragalactic field. It is now possible to see galaxies at redshifts beyond 5. Although more larger telescopes are becoming available, it is still very expensive to obtain good quality data of the high redshift universe. High resolution and high signal-to-noise data of only a few objects are available right now. It is the right time to start collecting what we have learned from these data in order to draw a picture of galaxy evolution.

1.1 The Database

Currently there are no databases dedicated to high-z galaxies. The extragalactic databases available such as the NASA/IPAC Extragalactic Database (NED) requires the name or coordinates of the object in order to search for information. Since most of the extragalactic astronomers are not familiar with the objects at high-z it is important to have the main objects studied so far in one specialized database. Our database is dedicated to galaxies at z>1 which have been analysed in detail in the literature.

We used keywords in the e-print archive for astrophysics preprints (astro-ph http:// arXiv.org/archive/astro-ph) and in the NASA Astrophysics Data System Abstract Service (ADS http://adswww.harvard.edu/) to search for the objects.

The database gives direct links to the original papers and summarizes some information such as:

- redshift,
- magnitudes,
- luminosities,
- masses,
- identified lines,
- star formation rate,
- metallicity and
- some special remarks

Currently there are 19 entries (i.e. galaxies) in the database from z=2 to z=5.74.

1.2 How to Contribute to the Database

We will update the database as new papers appear in the literature. However, we encourage the reader to send us via email any information that we have missed or that are to be published soon. The webpage manager email is duilia@oso.chalmers.se and the URL is http://www.oso.chalmers.se/~highz

1.3 The Database Webpage

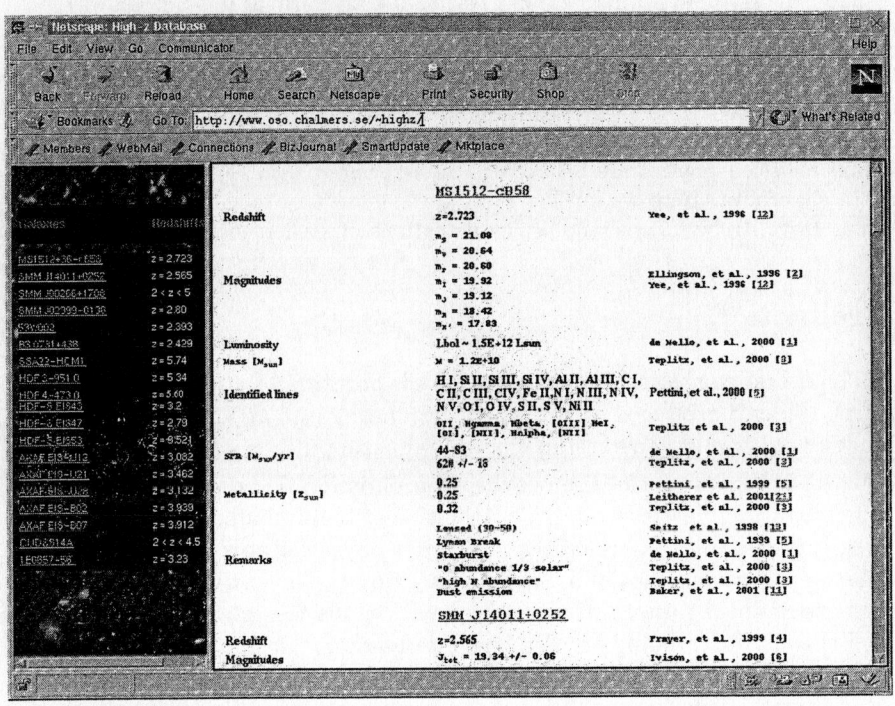

Fig. 1. MS1512-cB58, one of the galaxies in the database

Serendipitous Detection of Radio Pulses from Evaporating Black Holes, GRBs and Extragalactic Supernova Using SETI@home

Paul Demorest[1], Aaron Golden[2], Eric Korpela[1], Dan Werthimer[1], and Ron Ekers[3]

[1] Space Sciences Laboratory,University of California, Berkeley, U.S.A.
[2] Computational Astrophysics Laboratory, National University of Ireland, Galway
[3] Australia Telescope National Facility, CSIRO, Epping, Australia

Abstract. There are strong theoretical grounds for the generation of an intense radio pulse of intrinsically short duration during the evaporation of a primordial black hole. Similar emission physics are believed to be associated with Gamma Ray Bursts and certain supernova. Here we outline a program underway to mine the 50 Tbyte SETI@home Data Archive for serendipitous detections of such events.

1 Introduction

Rees (1977)[1] first suggested that during the evaporation of a primordial black hole (PBH), the pair produced e^+/e^- 'shell' expanding into an ambient interstellar magnetic field would generate an electromagnetic pulse of duration $1/\nu$ sec, with ν the 'peak' frequency of the synchrotron emission. Here $\nu \sim \frac{\gamma^2 c}{r_{max}}$, γ the Lorentz factor of the blastfront and r_{max} the size of the fireball when the energy of the swept up field finally decelerates it. Similar physics is applicable to certain classes of supernova, and that of a Gamma Ray Burst. The emission is expected to peak in the radio regime at flux densities within the range of existing observational facilities. Intervening interstellar/intergalactic plasma would disperse such pulses. For evaporating PBH within the galactic halo, the Taylor & Cordes [2] galactic electron density model allows one to compute the dispersion measure (DM) as a function of distance & (l, b), and so de-disperse & recover the original pulse signal. Typical values range from 5 to 30 cm^{-3} pc (from about 0.5 kpc to 40 kpc), increasing near the arms, with 5 to 250 pc cm^{-3}. For extragalactic GRBs & supernovae events, the DM is a function of the Hubble constant, the deceleration parameter, the mass of baryonic intergalactic matter and the distance to the source object in question [3]. Thus for GRB 970228 ($z \sim 0.7$), $\rightarrow$ DM ~ 420. Critically, computation of the DM for a GRB of known z would allow one to invert the DM relationship to constrain the other cosmologically relevant parameters. Such pulse events are however sporadic, random occurrences and their detection requires receiver systems that operate at the highest temporal resolution yet guarantee high sensitivity. UC Berkeley's SETI@home receiver system is located at the 305m Arecibo radiotelescope in Puerto Rico observing in a 'piggyback' fashion, operating at 1.4 GHz with a bandwidth of 2.5 MHz.

Its 4 year programme will sample $\sim$ 25% of the sky, re-visiting on average 3 times each location within its 0.1 degree beam. Whilst better known via the SETI@home screensaver initiative, the data archive provides tremendous opportunities for collateral research, such as the production of a unique spectral map of galactic HI [4], and to search for existing and undiscovered pulsars. We propose to analyse the entire dataset for evidence of dispersed pulses of intrinsically short duration consistent with an astrophysical origin as outlined above.

2   Mining the SETI@home Archive

Time domain data is obtained in the following manner: a 30 MHz band from the receiver is converted to baseband using a pair of mixers and low pass filters. The resulting complex signal is digitized and then filtered to 2.5MHz using a pair of 192 tap FIR filters in the SERENDIP IV instrument [5]. One bit samples are recorded on 35 GByte DLT tapes (one bit real and one bit imaginary per complex sample). These tapes are shipped to Berkeley and form the SETI@home Data Archive. The pulse search algorithm under active development involves use of a coherent de-dispersion and thresholding technique, with the latter typically set at between 15 - 20 σ. Each set of raw 1-bit data is converted to complex floating point, a 2k chunk is FFT'd, multiplied by a chirp function encapsulating a specific DM, the inverse FFT obtained and the time series thresholded accordingly. The process is repeated for a range of DM values, and then on to the next 2k chunk. At a sample rate of 2.5 MHz, each 2k chunk represents 0.8 msec of data, and overlapping is done for completeness. Chunks in excess of 0.8 msec will be required for accurate de-dispersion in the case of extragalactic sources We estimate that to analyse the 3 years worth of data to date in real time would require a computational throughput of $\sim$ 500 GFLOPS/sec. As such the problem is eminently parallel in nature, and a distributed approach similar to the SETI@home screensaver system is a likely solution, although more specific distributed computing paradigms are under examination, such as Grid and BOINC[1]. Such an optimised search algorithm could be of some relevance as regards a 'real time' or pipeline application in the next generation of radio telescope facilities, such as SKA and the ATA.

References

1. Rees, M. (1977) Nature **266**, 333
2. Palmer, D.M. (1993), ApJ **417L**, 25
3. Taylor, J.H., Cordes, J.M. (1993) ApJ **411**, 674
4. Korpela, E.J., et al. (2002) in "Seeing Through the Dust: The Detection of HI and the Exploration of the ISM in Galaxies", Eds. R. Taylor, T. Landecker, & A. Willis (ASP: San Francisco), 2002
5. Werthimer, D., et al. (1997) in "Astronomical and Biochemical Origins and the Search for Life in the Universe", IAU Colloq. 161, Bologna, 163.

---

[1] Berkeley Open Infrastructure for Network Computing

High Energy Photon Flux Prediction from Neutralino Annihilation in M 31

Alain Falvard[1], Edmond Giraud[1], Agnieszka Jacholkowska[1],
Karsten Jedamzik[2], Julien Lavalle[1], Gilbert Moultaka[2], Eric Nuss[1],
Frederic Piron[1], Pierre Salati[3], Mariusz Sapinski[1], and Richard Taillet[3]

[1] GAM, Univ. Montpellier II, Place E. Bataillon, 34 095 Montpellier Cedex, France
[2] LPMT, Univ. Montpellier II, Place E. Bataillon, 34 095 Montpellier Cedex, France
[3] LAPTH, Chemin de Bellevue, BP 110, Annecy-le-Vieux Cedex, France

Abstract. Considering a NFW halo of M 31 made of neutralinos, predictions for the γ fluxes are computed for different sets of SUSY parameters using the DarkSuspect MC simulation [1] [2]. Then they are compared with the efficiency of CELESTE Cerenkov telescope [3] which has a threshold energy of 30 GeV.

1 Introduction

Most cosmological models studied today are variant of a scenario where perturbations start in a medium of WIMPS, the so-called CDM scenario. An overview of this scenario, including some of its possible drawbacks, was recently presented in [4]. Since a disk and a bulge are not enough to model the rotation curve of M 31, we assumed the presence of an additional spherical halo. We considered mass models in which the M/L ratios of the bulge and the disk are in the ranges $3.5 \leq \Upsilon_{bulge} \leq 4.2$ and $2.5 \leq \Upsilon_{disk} \leq 4.2$, to be consistent with population synthesis models, and fitted the rotation curve by adding a halo with mass density distribution given by $\rho_\chi(r) = \rho_0 \left(\frac{r_0}{r}\right)^\gamma \left\{\frac{r_0^\alpha + a^\alpha}{r^\alpha + a^\alpha}\right\}^\epsilon$. We found that for M 31 a NFW profile ($\gamma = 1$), for the remaining halo, provides a better fit to the rotation curve than either a Moore profile ($\gamma = 3/2$) or a inner profile with $\gamma = 0.5$.

Physics beyond the standard model could be supersymmetry. These issues are reviewed by J. Ellis in this book. If R-parity is conserved, the lowest mass supersymmetric particle (LSP), most probably the neutralino, is stable, has a very small cross-section, and is a natural candidate for CDM. Being a Majorana particle, the neutralino is its own antiparticle which can then annihilate through various channels, several of them ending with high energy γ-rays. We have assumed that the halo of M 31 is made of neutralinos and we have derived high energy photon fluxes by exploring the parameter space of neutralino models.

2 Theoretical Physics Models and Tools

Unification and universality are assumed at some GUT scale for the gauge couplings, gaugino masses, soft-supersymmetry breaking scalar masses for squarks and leptons, trilinear coupling, and Higgs fields. Thus, starting from the common scalar soft supersymmetry breaking (SSB) mass $m_{\tilde{Q}}^2 = m_{\tilde{U}}^2 = m_{\tilde{D}}^2 = m_{\tilde{L}}^2 =$

$m_{\tilde{E}}^2 = m_{\tilde{H}_1}^2 = m_{\tilde{H}_2}^2 \equiv m_0^2$, SSB gaugino mass $M_1 = M_2 = M_3 \equiv m_{1/2}$, SSB trilinear coupling $A_t = A_b = A_\tau \equiv A_0$ and gauge couplings $\alpha_1 = \alpha_2 = \alpha_3 \equiv \alpha_{GUT}$, all taken at the GUT scale, the relevant low energy quantities are obtained from the renormalization group evolution of these parameters from the GUT scale down to the EW scale. We use a code made of two parts: 1) the SUSPECT code [2] which starts from the 5 free parameters (m_0, $m_{1/2}$, A_0, $\tan\beta$, $\text{sgn}\mu$) at the GUT scale and evolves the SUSY Lagrangian to the EW scale, 2) the Dark-SUSY code [1] which calculates masses and couplings at the EW scale using the SUSPECT values as input. Accelerator bounds and relic densities are taken into account. Then the γ-ray fluxes are derived from the neutralino mass, annihilation cross-sections and branching ratios, for various halo distributions of M 31. Several benchmark models have been proposed for the purpose of comparing various experiments and codes [5]. We compare our results with 5 of the benchmarks (parameters given in the Table) located in the so-called "bulk" region.

3 Predicted Fluxes Above 1 GeV from M31 for the Benchmark Models

We call "ground-level" models of M 31, halo models with the above M/L ratios, which ignore the black hole and the CDM clumpiness. The predicted fluxes for the "ground-level" astrophysical model, and the SUSY benchmark models BCGIL are given in the lower part of the Table. They are integrated for $E_\gamma > 1\ GeV$ and $\Theta = 10^{-3}\ sr$. The Galactic Center (G.C.) values are also given. The fluxes are in units of $10^{-12} cm^{-2} s^{-1}$.

model	B	C	G	I	L
$m_{1/2}$	255	408	383	358	462
m_0	102	93	125	188	326
$tan\beta$	10	10	20	35	45
$sign(\mu)$	+	+	+	+	+
EWSB scale [GeV]	492.8	721.2	719.1	657.4	836.1
GUT scale [GeV]	2.17×10^{16}	0.30×10^{16}	1.90×10^{16}	1.99×10^{16}	1.86×10^{16}
m_{χ^0}	99.4	162.3	155.1	144.8	190.2
R_g	0.968	0.988	0.987	0.985	0.990

astrophysical source and model	B	C	G	I	L
G.C. NFW ($J = 1214$) [this paper]	304.5	31.4	185.3	1095	934.0
M31 NFW ($J = 8$) [this paper]	2.1	0.2	1.3	7.7	6.5
G.C. from [5] ($J = 500$)	84.29	10.19	63.90	535.0	992.4
G.C. from [5] renormalized with $J = 1214$	204.8	24.76	155.3	1300	2412

4 Integrated Flux Predictions from M31 for the CELESTE Experiment

CELESTE is a sampling atmospheric Cerenkov telescope which uses 53 large mirrors of a former solar plant to explore the γ-ray energy range 30 GeV <

440　A. Falvard et al.

E < 300 GeV [3]. The predicted fluxes for the "ground-level" model of M 31 at threshold energy $E_\gamma > 30 \, GeV$ and angular aperture $\theta = 1.4 \times 10^{-4} sr$ (the aperture of CELESTE) are shown in the figure below.

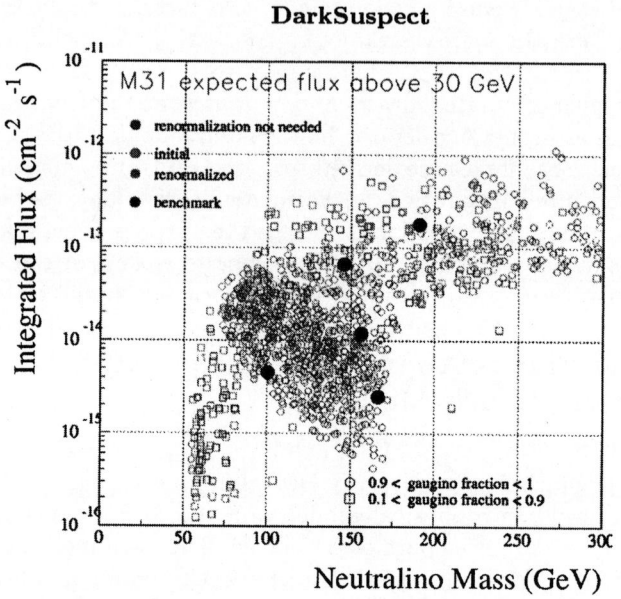

Fluxes derived for the benchmark models BCGIL are also shown. Models providing the largest fluxes, which have the largest cross-sections, lead to $\Omega_\chi h^2 \simeq \frac{1.07 \times 10^9 x_f}{g_*^{1/2} M_{pl}(GeV)(a+\frac{b}{2x_f})} < 0.1$ where $x_f \equiv \frac{m}{T_f}$; T_f being the freeze-out temperature. For models with low relic density we show both fluxes: bared and "normalized" as if they populate a fraction of the halo given by $\Omega_\chi/0.1$. We have computed the flux convolved with CELESTE efficiency and acceptance for various halos, with or without clumps, and ignoring or not the central black hole. In the most favourable neutralino models, one may expect between 0.3 and 3×10^{-3} photon/min depending on astrophysical parameters. Exploring a significant fraction of the SUSY parameter space would require 100 times higher sensitivity.

References

1. http://www.physto.se/~edsjo/darksusy/
2. http://www.lpm.univ-montp2.fr:7082/~kneur/suspect.html
3. http://doc.in2p3.fr/themis/CELESTE
4. J. Primack, 2001, astro-ph/0112225
5. J. Ellis, J.L. Feng, A. Ferstl,, K.T. Matchev, & K. Olive, 2001, astro-ph/0110225

A New Unifying Field Spacetime Structure

Bruce Feigler and Margaret Joseph

MDD Research, 13 Ronaki Rd, Auckland, New Zealand

Analysis of the data and equations for the solar system and the hydrogen atom identified a unifying spacetime structure intrinsic to the fields. The spacetime structure provides a means of connecting disparate field information, explains unknowns and provides new insights into fields. Since the solar system can represent a gravitational field and the hydrogen atom a quantum electromagnetic field, the structure opens a new route towards a quantum gravitational electromagnetic (GEM) field and unification of forces. The structure has advantages of not requiring complex mathematics as it is based on the natural simple unities to be found in fields and linkage with the tangible physical world.

The starting point is Kepler's 3rd law ($d^3/t^2 = k$) that represents an intrinsic spacetime ratio connecting distance and time in the field. Since this law applies to both the solar system gravitational field ($GM/4\pi^2 = d^3/t^2$) and the hydrogen atom electromagnetic field ($c^2 e^2/m_e 4\pi^2 = d^3/t^2$), it suggests the existence of an underlying common spacetime structure which is scale invariant and independent of mass, charge or type of force in inverse square law fields.

1 Field Kinematic Connections

The normalised format of Kepler's 3rd law ($t^2 = d^3$) can be extended to include volume and orbital velocity when they are also normalised. The volume and d^3 terms have physical world meaning given the 3 dimensional spherical cloud model of the electron around the proton.

$$\text{volume}^1 = \text{t}^2 = \text{d}^3 = 1/\text{velocity}^6$$

From the normalised $t^2/d^3 = 1$ format, the following array can be developed which gives an interlocking framework for the kinematic terms and represents an intrinsic field spacetime structure based on d/t. A method deriving this array is $(d/t)^n \times t^2/d^3$.

$$(d/t)^1 = t/d^2 = 1/\text{angular velocity} \quad = \text{velocity} \qquad = v^1$$
$$(d/t)^2 = 1/d = 1/\text{distance (wavelength)} = \text{kinetic velocity} = v^2$$
$$(d/t)^3 = 1/t = 1/\text{time(period)} \qquad = \text{frequency} \qquad = v^3$$
$$(d/t)^4 = d/t^2 = 1/\text{area} \qquad\quad = \text{acceleration} \quad = v^4$$

There are more levels to the d/t field spacetime structure, evidenced by $(d/t)^6 = 1/\text{volume}$, and $(d/t)^8 = \text{kinetic velocity/volume} = \text{kinetic velocity}^4$ indicating this d/t space structure encompasses the ideal gas law $P = T/V$ and thermodynamic radiation pressure fields, for example the sun's radiation pressure field. The useful $v^3 = 1/t = f$ relationship can be seen in the $v^3 = 2\pi GMf$ and $v^3 = 2\pi c^2 e^2 f/m_e$ equations.

The kinematic terms represent different aspects of the same underlying fundamental spacetime (d/t) ratio in the field, as they can all be defined as a power of $(d/t)^n$ field space time and there is a constant d/t gap between adjacent levels.

2   Field Constants Connections

As the kinematic terms represent different aspects of the same field spacetime structure it is to be expected the fundamental c, G and h constants also do, with simple connections to the fundamental field d/t spacetime constant and each other. Linkage of the constants to the field spacetime structure gives the opportunity to develop theoretical values for them and new insights into their meaning in the physical world. Analysis of the solar system and hydrogen atom data and equations identified the following proportional relationships between field d/t spacetime and the constants.

$$(d/t)^2 = (v/2\pi)^2 \propto (1/G)^2 \propto (\pi/h)^2 \propto (c/2)^2$$
$$\text{becomes } (2d/t)^2 = (v/\pi)^2 \propto (2/G)^2 \propto (1/\!\!\!h)^2 \propto (c)^2 \propto 8.99$$

The magnetic constant is $4\pi \times 10^{-7} Hm^{-1}$ supporting the idea of basic constants like π re-emerging at different orders of magnitude as fractal constants. The magnetic constant can be included in relationship $(4\pi/Gh)^2 \propto c^4$.

A major surprise is that d and t in the equations represent earth's orbital radius and time, which is due to light speed being measured in seconds. This makes a simple but fundamental connection between the straight line behaviour of light on the radial, the curved field behaviour and a whole unit of field spacetime. $C = $ constant $\times$ earth$(2d/t)$, giving a constant of 31621. This constant represents a $\sqrt{10}$ ratio so disappears when terms are squared. The hydrogen atom data is also measured in seconds which again accounts for the surprising result of $\pi/\text{earth } v = \pi/29.79 = 0.105458 \propto \!\!\!h$, a whole spacetime unit. As light is a physical phenomenon and force, c is the best candidate as the base for deriving theoretical numbers for other spacetime constants. eg the G value would be $6.67128 \ 10^{-11} N.m^2/kg^2$.

3   Field Quantum Connections

Analysis of the normalised hydrogen atom data identified that the d/t spacetime structure and kinematic terms are quantum; with $(t/d)^n = n^n$. This matches with the known stable quantum states of the Bohr hydrogen atom model.

Term	term	n	1st	2nd	3rdstate
1/velocity	angular velocity	n^1	1	2	3
1/kinetic velocity	distance(wavelength)	n^2	1	4	9
1/frequency	time(period)	n^3	1	8	27
1/acceleration	area	n^4	1	16	81

The kinematic terms and spacetime of the hydrogen atom being quantum suggests spacetime of fields in general are quantum, independent of their size, type or orbital bodies. This would allow gravitational fields such as the solar system to have a stable quantum d/t spacetime field structure, particularly since earth's orbit is connected to quantum $\cancel{k}$. Such an extension of the quantum field concept could give an explanation for the simple mathematical expansion series $(0.85)^n \times 47.88$ which contains numbers matching the velocities of the supposedly randomly located planets.

4 Field Force Connections

When the mass force of the gravitational field and the charge force of the electromagnetic field are equal their spacetime structures are identical. Since the G, c and h constants are equivalents in the field spacetime structure it suggests that the different fields they operate in are also equivalent, with the gravity and electromagnetic force representing different aspects of the same underlying force and field. The spacetime structure also provides a framework within which to arrange mass and charge related terms and fields.

Level	kinematic term	mass x term	m/q x term	formula	light
$(d/t)^1$	velocity	momentum			c^1
$(d/t)^2$	kinetic velocity	energy	voltage	md^2/qt^2	c^2
$(d/t)^3$	frequency		magnetic field	m/qt	c^3
$(d/t)^4$	acceleration	force field	electric field	md/qt^2	c^4

The structure places the electric field at the acceleration c^4 level and the magnetic field at the frequency c^3 level, preserving the known $E/B = c$ relationship. This supports the validity of the field d/t spacetime structure and the field d/t to c connection. As well it identifies c gaps between other electromagnetic terms and the potential concept of an interlocking momentum, energy, magnetic and electric field with its spacetime structure controlled by c; electromagnetic radiation. This would represent a unified quantum gravitational electromagnetic (GEM) field.

References

1. R. Serway: *Physics For Scientists And Engineers* 4th Edition (Saunders College Publishing, Philadelphia 1996)
2. E. Chaisson, S. McMillan: *Astronomy Today* 3rd Edition (Prentice Hall, New Jersey 1999)
3. B. Feigler, M Joseph: *Code of the Universe* (Rutherford Books, Auckland 2001)
4. M. Bakich: *The Cambridge Planetary Handbook* (Cambridge University Press, Cambridge 2000)

Visualization of Gravitational Lenses

Francisco Frutos-Alfaro

Escuela de Física, Universidad de Costa Rica, San José, Costa Rica

1 Introduction

In scientific research the visualization of the natural phenomena is becoming important. Gravitational lenses are presently a topic of intense research, therefore a program to visualize them is very useful, not only for research, but also for teaching. The visualization program that we are presenting, was written in C and runs under Unix and Linux platforms with previously installed libraries (XForms, Mesa or Open GL and Imlib). These libraries can be found freely on the Internet.

2 The Program

The program creates a control panel, which allows the user to choose the lens model and the source. The images produced by the chosen lens model appear on a window by clicking the image window button on the control panel. When the user chooses the lens model all corresponding sliders will appear on the control panel. These sliders represent the parameters of the lens model and they can be varied interactively. All changes are shown in realtime in the image window. Other features can be presented on this window with just a click on the control panel buttons. There is an old Linux version of the program on the Internet under following addresses:

```
http://lia.efis.ucr.ac.cr/~frutos/
http://www.tat.physik.uni-tuebingen.de/~frutto/
http://www.linmpi.mpg.de/~frutos/
```

These websites will be updated. The sources of the program are available on request. More information about the program can be found in the references ([1], [2]).

3 Applications

3.1 Animations

A window appears by clicking the track button on the control panel and it enables the user to fix the trajectory of the source. To see the animation the user just has to click on the animate button that is on this track window.

3.2   Visual Modeling of Gravitational Lenses

Modeling a gravitational lens is not an easy task. With our program it is easy to find out the model parameters that approaches the images of an observed lens system. The user can test the already modeled and fitted observed gravitational lens systems, for example, the Einstein Cross. The OBIP button on the control panel allows the user to enter the observed image positions and to draw circles that represent them (click set to save the position values). After that, the user just has to vary the parameters of the model to find out a best visual fit. The model images can be considered well-fit when they overlap with most of the area of the observed images (represented through circles), as visually estimated. The parameter can also be fitted using Brent's optimization method which is included in the program (click on the fitting button to get the optimization window). First of all, the user has to choose the parameters that must be fitted (click on model parameter button to do that). Then the user can optimize the model parameters (click the optimize button in the fitting window) and the parameter values will appear on the parameter window. On the image window one can see if the results of the fit are good or not. At the moment we can optimize the model parameters taking into account the observed positions, not the observed flux ratios.

4   Future Work

The program can be improved by the inclusion of some additional subroutines:

- Contour subroutine for the isochrones (time delay)
- Light curves subroutine (dependence of brightness on time)
- Subroutine for computing the image magnification
- Subroutine to calculate critical curves and caustics
- Fitting subroutine (not completely finished yet)
- Subroutine to load images of observed gravitational lenses
- Subroutine with more complex (elliptical) models
- Subroutine for superposition of models in different lens planes
- Subroutine with cosmic string lens models
- Subroutine for non-parametric reconstruction
- Kaiser-Squires subroutine

An improvement of the fitting subroutine to include the optimization of the observed flux ratios is needed to have an even better tool to model gravitational lenses. The author is working on some of these improvements and wishes to invite the interested reader to participate in the process.

References

1. F. Frutos-Alfaro, *Die interaktive Visualisierung von Gravitationslinsen*, PhD Thesis, Eberhard-Karls-Universität Tübingen, 1998.
2. F. Frutos-Alfaro, *A computer program to visualize gravitational lenses*, Am. J. Phys., **69**(2), February 2001.

Evolution of Density Fluctuations in a Universe Dominated by $su(2)$-Valued Yang-Mills Fields

André Füzfa and Dominique Lambert

Facultés Universitaires Notre-Dame de la Paix, B-5000 Namur, Belgium

1 Introduction

According to the standard model of elementary particles, the intermediate bosons of the weak interaction were massless before the Higgs transition at approximately $10^{-11}s$ after the Big Bang[1], making this interaction long-ranged. We present here a new hamiltonian formulation of the Einstein-Yang-Mills (EYM) system that classically couples gravity and the weak force. These equations have been widely studied in the last decade but very few results are known for inhomogeneous dynamical gravitational fields. Our purpose is thus to use this formulation to study, e.g. the evolution of density fluctuations before the Higgs transition that could lead to those of the Cosmic Microwave Background. Here, we focus on Yang-Mills fields only, but in a near future we intend to add scalar (Higgs) and spinor (leptons) fields to our model.

2 Spherically Symmetric EYM System for $SU(2)$ Gauge Group

The spherically symmetric gauge 1-form $\mathbf{A} = A_\mu^m \tau_m dx^\mu$ can be written as the so-called *Witten ansatz* [1] (without the pure gauge potential)[2]:

$$\mathbf{A} = a(r,t)\tau_3 dt + b(r,t)\tau_3 dr + c(r,t)\tau_1 d\theta + (\cot\theta\tau_3 + c(r,t)\tau_2)\sin\theta d\varphi \quad (1)$$

where the τ_i's are the usual basis of $su(2)$ ($[\tau_m, \tau_n] = i\epsilon_{mn}{}^l \tau_l$; m, n, l are gauge indices). The Yang-Mills field strength tensor is given by $F_{\mu\nu}^l = \partial_\mu A_\nu^l - \partial_\nu A_\mu^l + \epsilon_{mn}{}^l A_\mu^m A_\nu^n$ and its associated stress-energy tensor by $T_{\mu\nu} = -2F_{\mu\alpha}{}^m F_\nu{}^{\alpha m} + \frac{1}{2}g_{\mu\nu}F_{\alpha\beta}{}^m F^{\alpha\beta m}$. The covariant formulation of the EYM system includes the Einstein equations of general relativity[3] $R_{\mu\nu} - \frac{1}{2}Rg_{\mu\nu} = \kappa T_{\mu\nu}$ together with the Yang-Mills equations $\mathcal{D}_\mu F^{\mu\nu} = \partial_\mu F^{\mu\nu} + [A^\mu, F^{\mu\nu}] = 0$. In the case of $su(2)$-valued, spherically symmetric Yang-Mills fields, it consists of a system of nine non-linear partial differential equations (5 for Einstein and 4 for Yang-Mills), which contains many crossed derivatives. Thus, in order to integrate numerically this complicated problem, it would be more convenient to break the explicit covariance of the equations and to move to a Hamiltonian formalism of the EYM system.

---

[1] If we consider a vacuum expectation value of the Higgs field of $246\,GeV$.

[2] The weak gauge coupling constant g_W has been put to 1.

[3] The cosmological constant Λ has been put to 0, $\kappa = 8\pi$ and $G = c = 1$.

3   Hamiltonian Formulation of EYM Equations

The ADM (Arnowitt-Deser-Misner, [2]) formulation of general relativity is the starting point of all methods in numerical relativity. We refer the reader to the impressive amount of work done by this community for further information while, here, we will restrict ourselves to the hamiltonian formulation of the Yang-Mills part of the EYM system. The ADM equations for spherically symmetric space-times are well known and can be found for example in [3,4]. The basic idea of this formalism is to write the usual lagrangean density for Yang-Mills fields:

$$\mathcal{L}_{YM} = \sqrt{-det(g)}\, g^{\mu\alpha} g^{\nu\beta} F^l_{\mu\nu} F^l_{\alpha\beta} \tag{2}$$

under hamiltonian form

$$\mathcal{L}_{YM} = -\pi^{i,l}_A F^l_{0i} + \frac{g_{00}}{8\sqrt{(^3g)}}\pi^{i,l}_A g_{ij}\pi^{j,l}_A + g_{00}\sqrt{(^3g)}\, g^{ij} g^{mn} F^l_{im} F^l_{jn} \tag{3}$$

(where the indices i, j, m, n are spatial, l is a gauge indice and (^3g) is the determinant of the spatial 3-metric). Doing so, we have defined the canonical variables as the components of the gauge potentials A^l_μ and their conjugate momenta were set to

$$\pi^{i,l}_A = 4\sqrt{(^3g)}\, g^{tt} g^{ij} F^l_{0j} \,. \tag{4}$$

This gives for the $su(2)$-valued case (cf. (1)) in an isotropic space[4]:

$$\mathcal{L}_{YM} = -\pi_b \left(\dot{b} - a' \right) - 2\pi_c \dot{c} - 4g_{rr} g^{tt} a^2 c^2 + \frac{g_{tt}}{8}\sqrt{g_{rr}}\, g^{\theta\theta} \pi_b^2 + \frac{g_{tt}}{4}\sqrt{g^{rr}}\, \pi_c^2$$
$$+ 4g_{tt}\sqrt{g^{rr}} \left(c'^2 + b^2 c^2 \right) + 2g_{tt}\sqrt{g_{rr}}\, g^{\theta\theta} \left(c^2 - 1 \right)^2 \,, \tag{5}$$

where a prime denotes a derivative w.r.t. the radius r and a dot a derivative w.r.t. time t. The variation of this lagrangean density w.r.t. the canonical variables b, c and their conjugate momenta π_b, π_c provides the Hamiltonian equations while the variation w.r.t. the time component a gives the constraint, all those relations constitute the hamiltonian counterpart of the covariant Yang-Mills equations. From the resulting equations, it is easy to check that the equilibrium points of this hamiltonian system correspond to the static Yang-Mills equations given in the pioneering work [5]. First numerical results on the full system of EYM equations are looking rather promising and are still under consideration.

References

1. E. Witten: Phys. Rev. Lett. **38**, 121 (1977)
2. R. Arnowitt et al. in *Gravitation: An Introduction to Current Research*, edited by L. Witten (John Wiley, New York 1962)
3. A. Moussiaux et al.: Gen. Rel. Grav. **15**, 209 (1983)
4. B. K. Berger et al.: Phys. Rev. D **5**, 2467 (1972)
5. R. Bartnik, J. McKinnon, Phys. Rev. Lett. **61**, 141 (1988)

---

[4] For which the metric is diagonal and $g_{\theta\theta} = g_{\varphi\varphi}$

High Energy Photon Flux Prediction from Neutralino Annihilation in the Globular Cluster Palomar 13

Edmond Giraud[1], George Meylan[2], Mariusz Sapinski[1], Alain Falvard[1], Agnieszka Jacholkowska[1], Karsten Jedamzik[3], Julien Lavalle[1], Eric Nuss[1], Gilbert Moultaka[3], Frederic Piron[1], Pierre Salati[4], and Richard Taillet[4]

[1] GAM, Univ. Montpellier II, Place E. Bataillon, 34 095 Montpellier Cedex, France
[2] STScI, 3700 San Martin Drive, Baltimore, MD 21218, USA
[3] LPMT, Univ. Montpellier II, Place E. Bataillon, 34 095 Montpellier Cedex, France
[4] LAPTH, Chemin de Bellevue, BP 110, Annecy-le-Vieux Cedex, France

Abstract. The distant globular cluster Palomar 13 has been found to have a very high mass-to-light ratio. Its profile can be well fitted either by using a King model with a tail, or by a NFW model [1]. This cluster may be the first case of the many clumps predicted by CDM simulations that would not be disrupted by the galactic halo potential. We assume that Pal 13 is made of neutralinos and we run the DarkSuspect code for five benchmark models. This code allows to estimate the gamma-ray flux produced in annihilations of neutralinos. These fluxes may be used as targets to be reached in proposals for future ground-based high altitude Cerenkov telescopes.

1 Introduction

The distant globular cluster Pal 13 has been found to have a very high M/L ratio of 40 $M_\odot/L_\odot$. Its profile can be well fitted either by using King model with a power-law tail or by NFW model [1] with scale radius 2.4 ± 0.2 pc and central density 80 $M_\odot pc^{-3}$. A possible explanation is that this distant cluster ($D = 24.3$ kpc) is one of the numerous dark clumps predicted by CDM scenarios, which has not been destroyed by the galactic tidal field. Pal 13 may also be a disrupted cluster, out of dynamical equilibrium. Here we assume that the NFW profile is the signature of a halo made of cold particles.

Physics beyond the Standard Model could be Supersymmetry. The lowest-mass Supersymmetric particle called neutralino, is a natural candidate for CDM. If R-parity is conserved then the neutralino is stable and has a very small cross-section for annihilation. We assume that the halo of Pal 13 is made of neutralinos and calculate the flux of high energy γ-rays due to their annihilation.

2 Theoretical Physics Models, Tools and Benchmarks

Theoretical physics beyond the Standard Model is reviewed by J. Ellis in this book. The 13 SUSY benchmark models have been proposed to provide a common way of comparing the discovery potential of future accelerators [2]. The models fulfill the conditions imposed by LEP measurements, $g_\mu - 2$ result and relic density constraint $0.1 < \Omega_\chi h^2 < 0.3$. We calculate the γ fluxes for five of the benchmark models, designated as BCGIL and placed in the "bulk" region of parameter space, with our current MC simulation programs: DarkSUSY [3] and SUSPECT [4]. The SUSPECT code performs RGE evolution from GUT scale to EWSB one. The DarkSUSY package derives the relic density and fluxes.

3   Predicted Fluxes for Palomar 13

The fluxes integrated above energy threshold and within solid angle of 10^{-3} sr in function of the threshold are shown in the figure below.

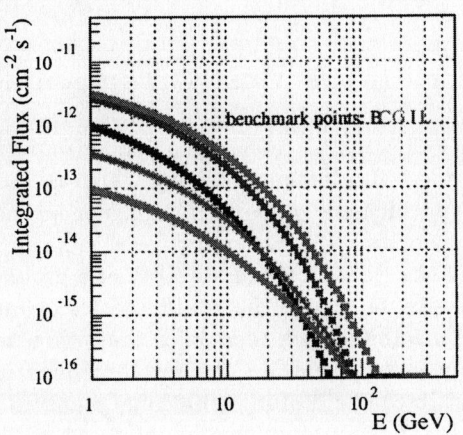

4   An Array of 16-20 Cerenkov Telescopes at High Altitude

From the above results one can conclude that we must work at low energy threshold and with a very low fluxes ($10^{-12.5}$ $cm^{-2}s^{-1}$). These fluxes are not out of reach but the sensitivity of the existing ground-based instruments will have to be improved by one order of magnitude for that purpose. An array of 5 HESSes (each including four 15-m class telescopes like in HESS), operating in adjacent areas at 5000 m altitude, would reach a flux limit of 2.5×10^{-13} $cm^{-2}s^{-1}$ at 25 GeV in 400 h. *With that flux limit most of the SUSY parameter space could be explored by measuring the Galactic Center.* This is the instrument needed to demonstrate or disprove Supersymmetry in astrophysics.

References

1. P. Coté, S.G. Djorgovski, G. Meylan, S. Castro, J.K. McCarthy, 2002 astro-ph/0203410
2. J. Ellis, J.L. Feng, A. Ferstl,, K.T. Matchev, & K. Olive, 2001, astro-ph/0110225
3. http://www.physto.se/~edsjo/darksusy/
4. http://www.lpm.univ-montp2.fr:7082/~kneur/suspect.html

Connections Between Quantum Theory and Cosmology: An Unconventional View

Thomas Görnitz

Inst. f. Didaktik der Physik, FB Physik der J.W. Goethe-Universität,
D-60486 Frankfurt/M., Germany

Abstract. Usually cosmology is seen as a special part of general relativity. Quantum theory is normally understood as responsible for the small and the very small. Here another perspective will be presented.

1 Quantum Theory is the Most Fundamental Theory of Nature

Quantum theory is the most fundamental theory which is known today. There is not any part in nature which suggests its failure. For an introduction to the philosophy of quantum theory see [5]. For many cases in nature classical physics is sufficiently good. Only the existence and stability of a ground state is beyond any reach of classical concepts. From the mathematics of quantum theory it follows by its first principles that it is a theory for the "wholeness' – irrespective of the point of extensions. At present we can see that quantum theory has its most fundamental aspect in the field of information. Quantum computing gets a rising importance, but also at the interconnection of quantum theory and general relativity, in case of the Bekenstein-Hawking-Radiation, the informational aspect is central.

2 Can Cosmology Be Treated Like Other Parts of Physics?

The huge success of physic has rooted in the distinction between a universal law and contingent initial conditions. The meaning of the law of nature is that it holds everywhere and at any time – and in all – at least in many – cases. The Universe is the wholeness of all from which it is not impossible to get empirical knowledge. Therefore it is unique and applying an universal law to it is a mistake. General relativity gives excellent descriptions of situations inside the cosmic space. But as a system of non-linear partially differential equations it has the consequence that small deviations in the initial conditions may result in absolutely different solutions in the global range. Because every solution of GRT is a whole cosmic space, almost all of them cannot be a description of the physical reality.

An equation's content of information is the same as the amount of all its solutions, therefore GRT stated to much. Since the 20th century cosmology has claimed to be an empirical science. Many data allow to check the models and to sort out many of them. If one is willing to keep the prerogative of an empirical evidence, models with an infinite volume must be excluded or all our empirical knowledge would cover exactly 0% of the whole Universe – not a safe ground in any science. Accepting the finite volume the cosmic space will be an $\mathbf{S}^3$.

3   Towards an Ontological and Universal Quantum Theory of Information

Quantum theory has transformed differences in fundamental scientific and philosophical concepts into mere useful distinctions. So the difference between matter and movement or between force and substance disappeared. In physics movement is expressed as kinetic energy and matter by rest mass, and both are equivalent. In second quantization force becomes an ensemble of particles. Weizsäcker[6] and his co-workers developed the mathematical frame to express quantum objects as aggregates of quantum information. The construction of massless and massive particles from abstract quantum information can be found in [1]. So all concepts of matter and energy can be traced back to the concept of quantum information.

To bring the qubits from philosophy to physics they require a cosmic representation. The symmetry group of a qubit is $SU(2)$. It can be represented as a sub-representation of its regular representation, the representation space of it is the Hilbert space of functions over the largest homogeneous space – an $\mathbf{S}^3$. This allows to characterize an abstract qubit by a "wave function" which is extended over the whole $\mathbf{S}^3$, possessing only a two-dimensional "knot-plane". The $\mathbf{S}^3$ is interpreted as cosmic space and a single qubit divides it in two parts.

If there are more qubits present, they constitute by its tensor product a higher dimensional representation of the symmetry group $SU(2)$, which can be decomposed into Clebsch-Gordan-Series. Each irreducible component can be represented by functions over $\mathbf{S}^3$ having higher frequencies. Therefore, with more qubits it is possible to get a higher spatial resolution. By group theoretical arguments a smallest length, the Planck length λ can be defined. Then a cosmic space containing N qubits gets a diameter of $N^2\lambda$ [2]. It is possible to derive a cosmological model which is an $\mathbf{S}^3$, expanding linearly in time and is as well solution of GRT. Quantum concepts and the first law of thermodynamics require a cosmic pressure. The pressure can be interpreted as an effective cosmological term and is naturally of the right scale. In the consequence, the model fits to the actual data better then to the data believed in the time of its first invention. It is shown that the Bekenstein-Hawking entropy can be explained as a special form of the cosmic quantum information. [3] [4] If mass is set in a cosmic correlation it becomes equivalent to the square of this cosmic information and in consequence we can state: Also Quintessence is abstract quantum information.

References

1. Th. Görnitz, D. Graudenz, C.F. v.Weizsäcker: Intern. J. Theoret. Phys. **31**, 1929 (1992)
2. Th. Görnitz: Intern. J. Theoret. Phys. **27**, 527 (1988)
3. Th. Görnitz: Intern. J. Theoret. Phys. **27**, 659 (1988)
4. Th. Görnitz, E. Ruhnau: Intern. J. Theoret. Phys. **28**, 651 (1989)
5. Th. Görnitz: *Quanten sind anders*, (Spektrum, Heidelberg 1999)
6. C.F. v.Weizsäcker: *Aufbau der Physik*, (Hanser, München 1985)

The Prototype Synchrotron Radiation Detector

Oliver Grimm

ETH Zürich, Laboratory for High-Energy Physics, 8093 Zürich, Switzerland

On behalf of the PSRD Collaboration*

The upgraded version of the Alpha Magnetic Spectrometer (AMS-02) is planned to be installed for a three-year mission on the International Space Station [1]. A new component, the Synchrotron Radiation Detector (SRD), has been proposed that allows the precise determination of the differential energy spectrum of high-energy electrons and positrons above 1 TeV. Due to inverse Compton and synchrotron losses, their lifetime is limited and the distance they can travel from their origin is believed to be below 1 kpc. Current theory states that the acceleration of cosmic rays takes place in supernova shock fronts through stochastic acceleration. Within kpc distance from Earth, only a small number of supernova remnants are identified, so, following [2], precise knowledge of the spectra at the upper end will allow definitive deductions about the responsible source(s) and their characteristics.

The SRD will identify high-energy electrons and positrons by their emission of synchrotron radiation in the Earth's magnetic field [3]. The concept plans to use an array of YAP(Ce) inorganic scintillators and photomultipliers to detect synchrotron photons in the energy range from 2.5 keV to about 100 keV.

The identification is complicated by the presence of a diffuse charged particle and photon background in the same energy range. The means to suppress this background is a very good time resolution, since the synchrotron photons arrive in coincidence with the leading charged particle.

The photon background has been measured well by, for example, the ASCA and HEAO-1 experiments, and amounts to 8 Photons/(cm^{2}s sr). To suppress protons, which are expected to be several orders of magnitude more abundant, to a level of 10^{-7}, a time resolution of the order of 10 ns is needed due to this background.

However, low-energy charged particles will also contribute to this background since they are indistinguishable from photons in the scintillator. The charged particle rate at these low energies has been measured only coarsely (for example by the NOAA-12 satellite), with insufficient detail for the development of a large-scale SRD. It was therefore necessary to measure the background rates in space.

* PSRD Collaboration: RWTH Aachen, III. Physikalisches Institut B, Germany; Lockheed Martin Company, Houston, USA; Johnson Space Center, NASA, Houston, USA; MIT, Laboratory for Nuclear Science, Cambridge, USA; ETH Zürich, Institut für Quantenelektronik, Switzerland; ETH Zürich, Labor für Hochenergiephysik, Switzerland; Institute of Physics, Academia Sinica, Taipei, Taiwan; CSIST, Lung-Tan, Tao Yuan 325, Taiwan; Institute of High-Energy Physics, Kyungpook National University, Taegu, Korea

The Prototype Synchrotron Radiation Detector (PSRD) has been built for this purpose.

The PSRD was designed to fly as a secondary payload on the Space Shuttle, participating in the Hitchhiker program of the NASA Shuttle Small Payloads Project. The PSRD employs an array of 12 YAP(Ce) crystals of $18 \times 18 \times 1 \, \text{mm}^3$ coupled to R5900U photomultipliers from Hamamatsu to measure the background spectrum up to about 100 keV. The crystals are shielded against sunlight by beryllium windows of thicknesses between 25 μm and 100 μm and their output is digitized by 8-bit flash ADCs at 20 MHz. Four YAP(Ce) crystals of $30 \times 30 \times 30 \, \text{mm}^3$ are used to extend the range to several MeV. These crystals are embedded in charged-particle veto counters and shielded by a 25 μm thick TOR-LM plastic foil. The event rate above three thresholds (10 keV, 40 keV, 100 keV) is integrated by 16-bit scalers in 10 ms time-bins. Furthermore, in collaboration with the Institute of Quantum Electronics at ETH Zürich, a novel type of CdTe/CdS solar cell is tested under actual space conditions.

A second objective of the PSRD, besides the background measurement, is a realistic space-test of the APV read-out chip proposed for the SRD. This analog-pipeline chip, originally developed for the CMS experiment at the LHC accelerator (CERN), is a favourable option for the read-out due to its low power consumption and compactness. Since the APV is not suited for the background measurement, it was included as a separate component. A silicon macrostrip detector that is planned to be employed by the GLAST experiment was found as a convenient detector to interface with the chip, available off-the-shelf. The read-out trigger for the APV chip is generated by a two-fold coincidence between two plastic scintillators, providing a trigger on charged particles passing through the PSRD.

The detector components are controlled by two 80x486 computers. The science data is stored directly on four hard disks of in total 100 GByte that are located in pressurized containers. Health and status information of the detector is transmitted to ground via a simple data link. Basic commands can also be uplinked to the experiment.

The PSRD was successfully flown with the Space Shuttle Endeavour on mission STS-108 in December 2001. A total of 38 GByte of data were taken during the flight. Data analysis is currently underway. Details on the experiment can be found in [4].

References

1. J. Alcaraz et al.: Nucl. Instr. Meth. A **478**, 119 (2002)
2. T. Kobayashi et al.: 'High energy cosmic ray electrons beyond 100 GeV'. In: *Proceedings of the 26th International Cosmic Ray Conference, Salt Lake City, USA, August 17-25, 1999*, ed. by B.L. Dingus, D.B. Kieda, M.H. Salamon (AIP, Melville 2000) Vol. 3, pp. 61-64
3. H. Hofer, M. Pohl: Nucl. Instr. Meth. A **416**, 59 (1998)
4. O. Grimm: Design and Construction of the Prototype Synchrotron Radiation Detector. PhD Thesis, ETH Zürich (2002)

Cosmic Shear from STIS Pure Parallels –
II. Analysis

H. Hämmerle[1,2], J.-M. Miralles[1,3], P. Schneider[1,2], T. Erben[2,4,5],
R.A.E. Fosbury[3], W. Freudling[3], N. Pirzkal[3], and S.D.M. White[2]

[1] Institut für Astrophysik und Extraterrestrische Forschung der Universität Bonn,
Auf dem Hügel 71, 53121 Bonn, Germany
[2] MPA, Karl-Schwarzschild Str. 1, 85748 Garching, Germany
[3] ST-ECF, Karl-Schwarzschild Str. 2, 85748 Garching, Germany
[4] Institut d'Astrophysique de Paris, 98bis Boulevard Arago, 75014 Paris, France
[5] Observatoire de Paris, DEMIRM, 61 Avenue de l'Observatoire, 75014 Paris, France

Abstract. The tidal gravitational field due to the inhomogeneous distribution of matter in the Universe causes the distortion of light bundles from distant sources. This yields a small but observable imprint on the distribution of galaxy ellipticities, an effect called cosmic shear. The statistical properties of the shear field reflect the statistical properties of the (dark) matter distribution in the Universe. Comparing the results with theoretical predictions can constrain cosmological parameters.

1 Data

With the initiation of a public parallel program with the STIS camera on board HST in June 1997 we decided to use these archival data for a cosmic shear study. The data set we have analysed consists of 498 coadded frames (which are available at http://www.stecf.org/projects/shear/).

The individual frames were coadded and drizzled to make full use of the subpixel dither information (see [9]). Catalogues were produced using both SExtractor [2] for the position and magnitude of objects and IMCAT [6] for size and shape information, since this software was designed specifically to measure robust ellipticities for faint galaxy images.

2 Cosmic Shear Measurement

Since the expected gravitational distortion of galaxy ellipticities on the STIS angular scale of $50''$ is a few percent, any instrumental distortion and other causes of PSF anisotropy need to be well understood. We analyse the PSF on 51 star fields and find that the mean ellipticity of stars (PSF anisotropy) is about 1%, which is sufficiently small to not affect the cosmic shear analysis.

To get an unbiased estimate of the shear the ellipticities of the galaxies also have to be corrected for the smearing of the isotropic part of the PSF. The fully corrected ellipticities are then used to estimate the cosmic shear dispersion in each of the 121 galaxy fields. By averaging over these results one obtains an unbiased estimate of the cosmic shear on the STIS angular scale.

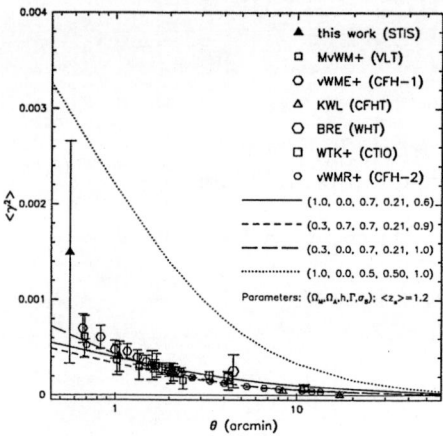

Fig. 1. Comparison of our cosmic shear result with measurements at larger angular scales and with model predictions. The lines show the theoretical predictions if one uses different cosmological models, which are characterized by Ω_m, Ω_A, h, Γ and σ_8. The redshift distribution is taken from [3], with a mean source redshift of $\langle z_s \rangle = 1.2$

In Fig. 1 we show our result [5], in comparison with those obtained by other groups on larger scales and with theoretically expected values for different cosmological models. The data agree very well with each other and with expectations from several cosmological models. However a high density universe with COBE normalization is clearly ruled out.

The parallel observations with STIS are currently being continued in the frame of a GO cycle 9 parallel proposal (8562+9248, PI: Schneider). Multicolour observations with the VLT are presently being carried out to determine the redshift distribution of the galaxies in the STIS fields using photometric redshifts.

Acknowledgements: We thank Y. Mellier and L. van Waerbeke for fruitful discussions. This work was supported by the TMR Network "Gravitational Lensing: New Constraints on Cosmology and the Distribution of Dark Matter" of the EC under contract No. ERBFMRX-CT97-0172 and by the German Ministry for Science and Education (BMBF) through the DLR under the project 50 OR106.

References

1. D. Bacon, A. Refregier, R.S. Ellis: MNRAS **318**, 625 (2000) [BRE]
2. E. Bertin, S. Arnouts: A&AS **117**, 393 (1996)
3. T.G. Brainerd, R.D. Blandford, I. Smail: ApJ **466**, 623 (1996)
4. A.S. Fruchter, R.N. Hook: PASP **114**, 144 (2002)
5. H. Hämmerle, J.-M. Miralles, P. Schneider, et al.: A&A **385**, 743 (2002)
6. N. Kaiser, G. Squires, T. Broadhurst: ApJ **449**, 460 (1995)
7. N. Kaiser, G. Wilson, G.A. Luppino: submitted to ApJ (2000), astro-ph/0003338 [KWL]
8. R. Maoli, L. van Waerbeke, Y. Mellier, et al.: A&A **368**, 766 (2001) [MvWM+]
9. N. Pirzkal, L. Collodel, T. Erben, et al.: A&A **375**, 351 (2001)
10. L. van Waerbeke, Y. Mellier, T. Erben, et al.: A&A **358**, 30 (2000) [vWME+]
11. L. van Waerbeke, Y. Mellier, M. Radovich, et al.: A&A **374**, 75 (2001) [vWMR+]
12. D.M. Wittman, J.A. Tyson, D. Kirkman, et al.: Nature **405**, 143 (2000) [WTK+]

Radio Ice Cherenkov Experiment

Pauline Harris for the RICE Collaboration

Department of Physics and Astronomy, University of Canterbury, Private Bag 4800, Christchurch, New Zealand

1 Introduction

The Radio Ice Cherenkov Experiment(RICE), located at the South Pole, is designed to detect coherent broad-band radio Cherenkov radiation emitted by electromagnetic cascades caused by the charged current interactions of high energy (10^{15} to 10^{18} eV) electron neutrinos interacting with nucleons in the ice. As a cascade develops, atomic electrons in the target medium are swept into the forward moving shower in which positrons annihilate causing a net negative charge to develop [1]. The shower front is a compact pancake with transverse dimensions of 20cm. At distances far from the shower and at wavelengths long compared to its transverse spread, the emitted radiation is coherent. For electromagnetic cascades in ice one gets coherence at radio wavelengths.

The RICE experiment currently consists of an array of 20 radio receivers distributed in an approximate 200m×200m×200m volume, at 100m-300m depths. The signal from each receiver is boosted in the ice by a 36-dB amplifier and carried via coaxial cable to the surface observatory. The signal is then filtered (suppressing noise contributions below 200MHz), re-amplified (either 52- or 60-dB gain), and fed into a CAMAC crate. After initial discrimination (using a LeCroy 3412E discriminator), the signal is routed into a NIM crate where the trigger logic resides. A valid trigger signal initiates readout of receiver waveforms, as recorded on HP54542 digital oscilloscopes. Also deployed are three large TEM surface horn antennas which are used to veto surface-generated noise.

2 Expected Signal Strength

We use a Monte Carlo simulation to determine the detection efficiency of RICE. There are three basic stages to the Monte Carlo. In the first stage the neutrino flux, its interaction with the nucleon, the cascade and the generation and propagation of the radio Cherenkov radiation are modelled resulting in an electric field at each antenna. The characteristics of the radio-frequency electric field produced by the electromagnetic cascade has been studied using a GEANT-based simulation code [1]. The second stage takes account of the antenna response and signal propagation through the electronics to form the output voltage at each channel. Finally the event trigger is generated, and tests against the various veto conditions applied in hardware and software.

3 Event Triggers, Source Reconstruction and Flux Limits

Background signals can be received by antennas from other sources. To discriminate valid events from random hits on the receivers we require that any one of three criteria is satisfied within a time window of $1.2\mu s$. The time window corresponds to the propagation of an electromagnetic signal through the array. The three criteria are: a)$\geq$ 4 under-ice antennas register signals above threshold, b) $\geq$ 1 under-ice antenna registers a signal above threshold in coincidence with a high-amplitude SPASE event, c)$\geq$ 1 under-ice antenna registers a signal above threshold in coincidence with a 30-fold PMT AMANDA-B event.

To eliminate background noise, two primary vetoes are used. The first veto rejects signals if one of the surface horn antenna registers a signal in co-incidence with the first of the three trigger criteria above. The second veto eliminates data if the under-ice antenna register signals consistent with that from a surface generated background.

Reconstruction of an event and its source is based on knowing the antenna array geometry and ice properties. Using the timing which one expects an electromagnetic signal to propagate through the ice and hit each antenna, the events can be reconstructed and the most likely source, location and direction can be identified by performing a χ^2 minimisation analysis.

The number of events detected can be used to calculate a limit on the electron neutrino flux. A report on the results of this upper limit study is currently in preparation.

Acknowledgements

We gratefully acknowledge the logistical support of the AMANDA Collaboration, the National Science Foundation Office of Polar Programs, the University of Kansas Graduate Research Fund, the University of Canterbury Marsden Grant, and the Cottrell Research Corporation.

References

1. S. Razzaque, et al.: Phys Rev. **D65** ,103002 (2002)

First High-Resolution Observations of the Lyman α Forest at $0.9 < z < 1.7$

Eckart Janknecht, Robert Baade, and Dieter Reimers

Hamburger Sternwarte, Gojenbergsweg 112, 21029 Hamburg, Germany

Abstract. Spectroscopy with *HST/STIS* Echelle and *VLT/UVES* of the bright QSO HE 0515-4414 ($z_{em} = 1.73, B = 15.0$) offers for the first time the opportunity to study the Lyman α forest in the redshift range $0.9 < z < 1.7$ at a resolution ≤ 10 km s^{-1}. The number density evolution of the Lyman α lines is well described by the power law approach $dn/dz \propto (1 + z)^\gamma$. We derive $\gamma = 0.01 \pm 0.64$ for the weak lines ($13.10 \leq \log N_{HI} \leq 14.00$) consistent with a constant comoving density of these objects, while the strong lines ($13.64 \leq \log N_{HI} \leq 16.00$) show evolution with $\gamma = 2.23 \pm 1.21$. The expected slow-down in their evolution does not appear earlier than $z \sim 1$.

1 Observations and Data Analysis

HE 0515-4414 was observed with *STIS* (S/N $\simeq$ 10, FWHM $\simeq$ 10 km s^{-1}) and with *UVES* at the *VLT*/Kueyen telescope (S/N $\simeq$ 10 - 50, FWHM $\simeq$ 6 km s^{-1}). The combination of the *HST* and *VLT* data contains information about the Lyman α forest from $z = 0.87$ up to $z = 1.73$. To avoid the proximity effect we exclude a region of about 5000 km s^{-1} from the quasar leading to an investigated redshift range $z = 0.87 - 1.68$.

The detected Lyman α lines (235 altogether) were fitted with the FITLY-MAN code in the MIDAS package [3] using Voigt profiles convolved with the instrumental profile. FITLYMAN adjusts three independent parameters per line by χ^2 minimizing: The redshift of an absorption line z, its H I column density N_{HI}, and its Doppler parameter b.

2 Results

The differential column density distribution function $f(N_{HI})$ is defined as the number of Lyman α absorption lines per unit column density and per unit absorption distance path as a function of N_{HI} [9]: $f(N_{HI}) = \frac{n}{\Delta N \, \Sigma_i \Delta X_i}$.

Fitting this quantity by a power law of the form $f(N_{HI}) = A \, N_{HI}^{-\beta}$, we derive $\log A = 9.62 \pm 0.58$ and $\beta = 1.61 \pm 0.04$. This result is in accordance with other analyses in comparable redshift ranges (for example [2]; [4]; [6]).

The evolution of the number density per unit redshift of Lyman α clouds can be well approximated by the power law

$$\frac{dn}{dz} = \left(\frac{dn}{dz}\right)_0 (1 + z)^\gamma. \tag{1}$$

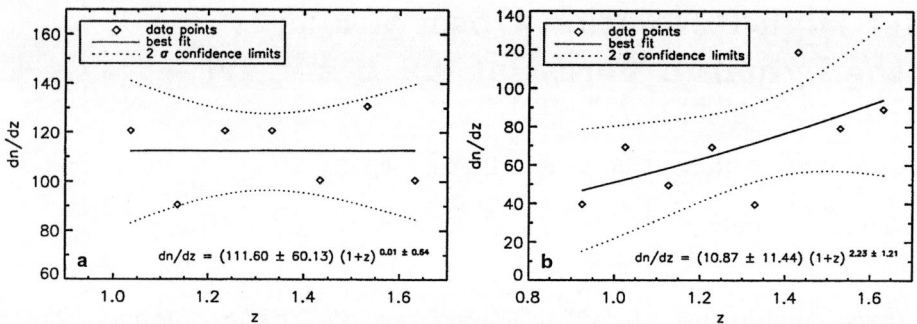

Fig. 1. The number density evolution of the Lyman α forest for weak lines ($13.10 < N_{HI} < 14.00$) (**a**) and for strong lines ($13.64 < N_{HI} < 16.00$) (**b**). The data points are binned with $\Delta z = 0.1$. The best fits were obtained by χ^2 minimization. The dotted curves represent the 95% confidence band

The exponent γ includes the cosmological evolution as well as the intrinsic evolution of the absorbers.

Because there is some evidence that weaker and stronger Lyman α clouds evolve differently, it is convenient to distinguish between two subsamples. In Fig. 1 we present the line numbers per unit redshift plotted over the redshift for the weak and for the strong lines, respectively. The weak absorbers show no evolution ($\gamma = 0.01 \pm 0.64$), in accordance with, e.g., [1] (for medium resolution data) or [7] (for $z > 2$). In contrast, there is a clear dependence on z for the strong Lyman α lines. We detect an obvious correlation between the evolution and the line strength, i.e., the high column density absorbers evolve with $\gamma = 2.23 \pm 1.21$. This is in contradiction to [8] and [2] who found no or only marginal evidence for a different evolution, respectively.

We cannot recognize a slow-down in the evolution of the stronger absorbers. Therefore, we conclude that this break does not occur earlier than $z \sim 1$ rather than at $z \sim 1.5 - 1.7$ as previously claimed ([2]; [5]; [10]).

A full discussion of the analysis is submitted for publication in Astronomy & Astrophysics. To extend the statistical basis we are going to include further QSOs covering the relevant redshift range.

References

1. J. Bechtold: ApJS **91**, 1 (1994)
2. A. Dobrzycki, J. Bechtold, J. Scott, M. Morita: ApJ, in press (2002)
3. A. Fontana and P. Ballester: ESO Messenger **80**, 37 (1995)
4. E.M. Hu, T.-S. Kim, L.L. Cowie, A. Songaila, M. Rauch: AJ **110**, 1526 (1995)
5. C.D. Impey, C.E. Petry, M.A. Malkan, W. Webb: ApJ **463**, 473 (1996)
6. T.-S. Kim, S. Cristiani, S. D'Odorico: A&A **373**, 757 (2001)
7. T.-S. Kim, E.M. Hu, L.L. Cowie, A. Songaila: AJ **114**, 1 (1997)
8. S.V. Penton, J.M. Shull, J.T. Stocke: ApJ **544**, 150 (2000)
9. D. Tytler: ApJ **321**, 49 (1987)
10. R.J. Weymann, et al.: ApJ **506**, 1 (1998)

Finding Atmospheres of Extra-Solar Planets in High-Dispersion Near-Infrared Spectra

Hans Ulrich Käufl

European Southern Observatory, D-85748 Garching bei München, Germany

Abstract. In the wavelength regime of 950-5500nm CRIRES, ESO's Cryogenic Infrared Echelle Spectrograph will offer a spectral resolution $\frac{\lambda}{\Delta\lambda} \approx 10^5$ in combination with a spatial resolution of $0.2''$. This makes it well suited to search for spectral signatures of atmospheres of extra-solar planets. Sensitivity estimates for the detection of the non-thermal OH glow in oxygen-bearing atmospheres are given. With the VLT such a search is still sensitivity limited, but a dedicated spectrograph at the projected ESO 100m OWL telescope could detect Earth-like planets at a distance of ≈ 5 parsec.

1 Rationale for High-Resolution Infrared Spectroscopy

Direct detection of terrestrial extra-solar planets is a question of contrast. Tabulated below is the flux ratio of a G2-star (Sun) to typical orbiting planets (Earth and Jupiter) as a function of wavelength. In the column for $0.5\mu m$ the ratio of stellar flux to re-emitted scattered light is given. In the other columns the ratio of stellar flux to thermal infrared flux of the planet is calculated[1]. From this table the sensitivity advantage of infrared observations it obvious (c.f. [1]), while also straylight problems are strongly reduced in the IR.

$\lambda[\mu m]$	0.5	2.3	4	10	20	∞
$\frac{F_{Jupiter}}{F_{Sun}}$	$1.0 * 10^{-9}$	$4.4 * 10^{-19}$	$5.9 * 10^{-12}$	$6.3 * 10^{-7}$	$2.1 * 10^{-5}$	$3.0 * 10^{-4}$
$\frac{F_{Earth}}{F_{Sun}}$	$5.0 * 10^{-10}$	$7.0 * 10^{-14}$	$3.0 * 10^{-10}$	$1.7 * 10^{-7}$	$1.0 * 10^{-6}$	$4.1 * 10^{-6}$

The presence of infrared rotational-vibrational transitions of suitable molecules in planetary atmospheres can enhance the contrast substantially, provided the level population exceeds thermodynamical equilibrium and the lines have sufficient optical depth. Examples for such lines are the well known non-thermal lines of the OH radical in the near-infrared[2] or of hydro-carbonates (see e.g. for resolved spectra of C_2H_6 on Jupiter[3])[2]. While in case of the hydro-carbonates the peak-brightness at line center can be expected to exceed the thermal continuum radiation by a factor of 10, one can expect for the case of OH that the peak brightness may exceed the thermal continuum by factors of 10^3. This in turn

---

[1] The optical (scattered) flux is estimated from the numbers given in the *Astronomical Almanac* assuming quadrature; for the infrared flux ratios black-body spectra with T=5770, 290 and 170 K have been assumed for Sun, Earth and Jupiter.

[2] Even stronger is the transition of C_2H_2 at $13.7\mu m$ [4], which however, cannot be observed with ground-based telescopes due to telluric absorption.

implies, that when spectrally resolved, the OH lines in the $2.2\mu m$ atmospheric window may approach or even exceed the radiation level of scattered light from the central star.

2   Significance of OH-Transitions, Prospects for Detection

The identification of the near-IR telluric air glow spectrum is given in [5]. It results from the reaction of O_3 with H_2 ([6]) and has been studied with great precision in the laboratory [7]. The occurrence of an OH spectrum in a planetary atmosphere hence can be considered as a rather strong indication of the presence of oxygen, the parent molecule of O_3, which in turn is the signature of life.

The most suitable instrument for these observations appears to be CRIRES, presently under construction for ESO's VLT. Intrinsic line-widths of rot-vib atmospheric lines are $\approx 0.5\frac{km}{s}$ while CRIRES has a nominal resolution of $3\frac{km}{s}$. The presently estimated sensitivity for CRIRES ($m_{2.2\mu m} \approx 13$ for a 100σ-detection), however, precludes to detect OH-line flux from an "Earth" at 5pc (peak-flux in line center may equal $m_{2.2\mu m} \approx 26$. Using a 100m telescope rather than an 8m and a spectrograph with resolution of $0.2\frac{km}{s}$ would reduce the sensitivity gap to 6-7 magnitudes. In other words, at 5pc a single OH-line of an Earth-like planet could be detected at the 100σ-level, in case this planet would have 10-15 times the Earth diameter!

3   Conclusions

A thorough assessment of the brightness of the telluric OH-lines with CRIRES in day time appears valuable. This will allow to better extrapolate the future use of these transitions for the identification of oxygen-bearing atmospheres of potential life-bearing planets in the Solar neighborhood. Due to the sensitivity limits imposed by the 8m VLT unit telescope CRIRES will most likely be restricted to searches for hydro-carbonates of extra-solar *Jupiter*-like planets.

References

1. H.U. Käufl, G. Monnet: 'From ISAAC to Goliath or better not!? Infrared Instrumentation Concepts for 100m Class Telescopes'. In: *Bäckaskog Workshop on Extremely Large Telescopes, Sweden, June 1–July 2, 1999*, ed. by T. Andersen, A. Arneberg, R. Gilmozzi (ESO, Garching b. München, 2000) pp. 282–289
2. P. Rousselot, C. Lidman, J.-G. Cuby et al.: A&A **354**, 1134-1150 (2000)
3. G. Wiedemann, G.L. Bjoraker, D.E. Jennings: ApJ **383**, L29-L32 (1991)
4. V.G. Kunde, A.C. Aikin, R.A. Hanel et al.: Nature **292**, 686-688 (1981)
5. A.B. Meinel: ApJ **11**, 555-564 (1950)
6. K.G. Anlauf, R.C. MacDonald, C.J. Polanyi: Chem. Phys. Lett. **1**, 619 (1968)
7. M.C. Abrams, S.P. Davies, M.L.P. Rao, R. Engleman, J.W. Brault: ApJ Supp.**93**, 351-395 (1994)
8. A.F.M. Moorwood, et al.: 'CRIRES: a high-resolution infrared spectrograph for the VLT'. In: *Astronomical Telescopes and Instrumentation Waikoloa, Hawaii USA, Aug. 22 - 28, 2002*, (SPIE, in press)

A New X-Ray Probe for Supermassive Black Holes in Non-Active Galaxies

Stefanie Komossa[1], Dawei Xu[2,1], and Jianyan Wei[2]

[1] Max-Planck-Institut für extraterrestrische Physik, Postfach 1312,
   85741 Garching, Germany; skomossa@mpe.mpg.de
[2] National Astronomical Observatories, Datun Road 20A, Beijing 100012, China

Abstract. In this contribution, we review the previous observations of giant X-ray flares from normal, non-active galaxies, and present multiwavelength follow-up observations. These are used for a comparison of different methods (optical, radio, X-rays) to estimate the masses of the SMBHs at the centers of the flaring galaxies.

1 The Search for Supermassive Black Holes at the Centers of Galaxies, and Flares from Tidally Disrupted Stars as Probes

The search for and study of supermassive black holes (SMBHs) in non-active galaxies is of great interest in the context of galaxy/AGN formation and evolution: how and when do SMBHS form; how do they grow; did *all* galaxies pass through an active phase; why, then, are most of the SMBHs 'dark'?

It was suggested (e.g., Rees 1988) that SMBHs in non-active galaxies might be tracked down by occasional tidal disruptions of stars captured by supermassive black holes, producing a flare of electromagnetic radiation when the stellar remains are accreted onto the black hole.

In X-rays, several excellent candidates of such tidal disruption flares in *optically non-active* galaxies have been discovered in the last few years (see Komossa 2002 for a review), starting with the case of NGC 5905 (e.g., Komossa & Bade 1999). The galaxy exhibited a huge-amplitude X-ray flare which reached high peak luminosity, yet high-quality ground-based optical spectroscopy did not reveal any signs of AGN activity.

Below, we compare different methods (X-ray and non-X-ray) to estimate the masses of the black holes in the X-ray flaring galaxies. In this contribution, we concentrate on NGC 5905. A program to derive improved blue magnitudes, also for the other flare galaxies, is presently in progress.

2 Black Hole Mass Estimates

In X-rays, a *lower limit* on the mass of the black hole at the center of NGC 5905 can be estimated via the Eddington luminosity, using the observed X-ray peak luminosity of NGC 5905 of at least several 10^{42} erg/s. This gives $M_{\rm BH} \gtrsim$ a few 10^4 M$_\odot$.

In order to obtain non-X-ray estimates of the BH mass of NGC 5905, we first used the correlation between bulge properties and BH mass. NGC 5905 has a total blue magnitude of $m_{B,0} \approx 12.1^m$. Using the bulge-to-disk luminosity ratio generally valid for galaxies of the Hubble-type of NGC 5905 (SBb), $k = 0.25$ (Salucci et al. 2000), gives the absolute bulge blue magnitude, $B_{T,0}^{bulge} = -20.5$. We then compared with two recent studies that correlate BH mass and bulge blue luminosity: (i) the work of Ferrarese & Merritt (2000; FM00) which concentrates mostly on elliptical galaxies, and (ii) the work of Salucci et al. (2000; S00) on spiral galaxies.

Using the relation between $B_{T,0}^{bulge}$ and M_{BH} of FM00 (their 'sample A', Tab. 2) gives a BH mass of a few times 10^8 $M_\odot$ for NGC 5905. This is close to the limiting BH mass for tidal disruption of a solar-type star to work; atmosphere stripping of giant stars will still occur for even larger BH masses. It has to be kept in mind, that the $B_{T,0}^{bulge}$ - M_{BH} relation shows a large scatter (in contrast to the M_{BH} - σ relation), and that the few spirals in the sample of FM00 tend to be located below the relation followed by ellipticals. Therefore, in a second step, we used the results of S00 on BH masses in (late-type) spiral galaxies. We then obtain an upper limit for the BH mass of NGC 5905 of $M_{BH} \lesssim 10^7$ $M_\odot$.

Finally, the relation between radio luminosity and BH mass of Franceschini et al. (1998; see also Wu & Han 2000) was employed. The measured radio upper limit for the nucleus of NGC 5905 (Komossa & Dahlem 2002) then translates into an upper limit on black hole mass of $M_{BH} \lesssim 2.5\, 10^8$ $M_\odot$.

3   Outlook

Future X-ray sky surveys – like those planned with *ROSITA*, *LOBSTER-Eye* and *MAXI* – will be valuable in finding new X-ray flares, and then trigger *rapid multi-wavelength follow-up observations.*

Such observations will allow to scrutinize the favored outburst scenario, and are expected to open up a new window to detect and investigate SMBHs and their *immediate environment* in *non-active* galaxies.

References

1. Ferrarese L., Merritt D., 2000, ApJ **539**, L9
2. Franceschini A., Vercellone S., Fabian A.C., 1998, MNRAS **297**, 817
3. Komossa S., 2002, in: *Lighthouses of the Universe. The Most Luminous Celestial Objects and their Use for Cosmology*, ESO Astrophysics Symposia, R. Sunyaev et al. (eds), in press [astro-ph/0109441]
4. Komossa S., Bade N., 1999, A&A **343**, 775
5. Komossa S., Dahlem M., 2002, in: *MAXI workshop on AGN variability*, ISAS Report, in press
6. Rees M.J., 1988, Nature **333**, 523
7. Salucci P., Ratnam C., Monaco P., Danese L., 2000, MNRAS **317**, 488
8. Wu X.-B., Han J.L., 2001, A&A **380**, 31

EUSO: From Raw Data to Scientific Results

Maria Concetta Maccarone, Osvaldo Catalano, Andrea Santangelo, and
Livio Scarsi
(on behalf of the EUSO Collaboration)

Ist. Astrofisica Spaziale e Fisica Cosmica, IASF-CNR,
Via Ugo La Malfa 153, 90146 Palermo, Italy

Extended Abstract

The Extreme Universe Space Observatory *EUSO* [1] will be accommodated on
board the International Space Station as an external payload for the Columbus
module. During a period of three years, *EUSO* will collect a statistically significant number of Extreme Energy Cosmic Ray and Neutrino (EECR/ν) events
together with other Atmosphere Physics events. This multidisciplinary aspect
must be taken into account in the event data analysis and requires an efficient
Ground Data Handling system [2,3].

In a schematic view, the following main topics are to be considered as relevant
for the project:

- **Target** - The primary objective of *EUSO* is to obtain a detailed description
  of the EECR/ν energy spectrum together with a map of the arrival directions, as unique global result from 'certified events'. The certification phase
  for each single detected event obviously requires the knowledge as complete
  as possible of the conditions of the primary detector involved, i.e. the Earth
  atmosphere
- **Interaction Medium** - The Earth atmosphere constitutes the variable and
  not-modifiable environment for *EUSO*. Basic environmental parameters that
  can influence the response of the instrument are, for example: the ozone level,
  responsible for the UV absorption; the presence of clouds, that can partially
  hide the UV fluorescence track or the Čerenkov signal; the Earth surface
  (land, sea, ice-region, desert, ...) that can influence the Čerenkov signal
  itself; the atmospheric temperature and pressure, responsible for variations in
  the fluorescence yield; and the winds that, although in a smaller measure, can
  transport dusts, aerosol and micro particles. These parameters are variable
  with time and with geographical coordinates
- **Background discrimination** - The sources of UV background for *EUSO*
  are several; they can be roughly divided into two broad categories, namely
  man-made (city lights, lights from ships and airplanes, ...) and 'natural'
  (reflected starlight and moonlight, atmospheric near-UV emission light from
  chemical reactions, low energy cosmic ray air showers, ...). Ephemeris for
  the moon phase, meteorological and geographic maps and databases, will be
  the necessary completion for the knowledge of the observation conditions

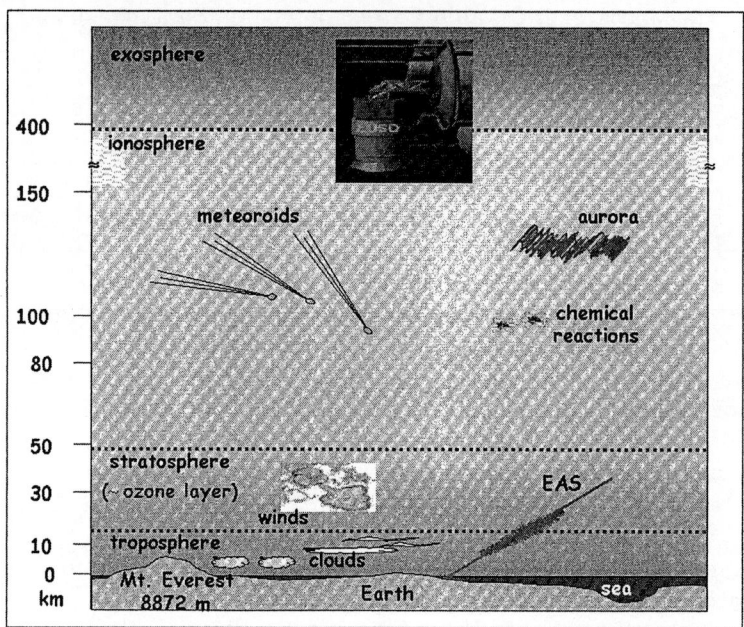

Fig. 1. Artistic view of an EECR signal together with the basic environmental parameters influencing the *EUSO* response

- **EECR and Neutrino event discrimination** - In the class 'other events' we can enumerate the natural phenomena as aurora, lightning, meteors. These can be detected (and studied) by *EUSO* and discriminated against the primary cosmic events making use of their very different propagation speed in the Earth atmosphere and of the event duration.

The *EUSO* Scientific Data Centre will control the scientific payload performance and functional status, monitor the ambient background, perform on flight calibration, make on-line quick-look analysis and produce the related results database. Certified events will be classified and collected in an archive as support for the global result of determination of energy spectra and drawing of maps for the arrival directions.

References

1. O. Catalano, M.C. Maccarone, A. Santangelo, L. Scarsi: 'EUSO - Extreme Universe Space Observatory'. In: *these Proceedings*
2. M.C. Maccarone, T. Mineo: 'EUSO - Ground Data Handling and Outreach'. In: *Proc. 27th ICRC, Hamburg, Germany, August 2001*, ed. by K.-H. Kampert, G. Hainzelmann, C. Spiering (Copernicus Gesellschaft, 2001), Vol. HE, pp.852-855
3. *"The Official EUSO Web Site"*, http://www.euso-mission.org

Cosmic Alignment Towards the Radio Einstein Ring PKS 1830–211

G. Meylan[1], F. Courbin[2], J.-P. Kneib[3], and C. Lidman[4]

[1] Space Telescope Science Institute, 3700 San Martin Drive,
    Baltimore, MD 21218, U.S.A., gmeylan@stsci.edu
[2] Institut d'Astrophysique et de Géophysique, Université de Liège,
    17 Allée du 6 Août, Liège 1, Belgium, Frederic.Courbin@ulg.ac.be
[3] Observatoire Midi-Pyrénées, Laboratoire d'Astrophysique, UMR5572,
    14 Avenue Edouard Belin, 31000 Toulouse, France, kneib@ast.obs-mip.fr
[4] European Southern Observatory, Casilla 19001, Santiago 19, Chile,
    clidman@eso.org

Abstract. Optical and near-IR Hubble Space Telescope and Gemini-North adaptive optics images, further improved through deconvolution, are used to explore the gravitationally lensed radio source PKS 1830–211. The line of sight to the quasar at $z = 2.507$ appears to be very busy, with the presence, within 0.5 arcsec, from the source of: (i) a possible galactic main-sequence star, (ii) a faint red lensing galaxy visible only in H-band and (iii) a new object whose colors and morphology match those of an almost face-on spiral galaxy. The $V - I$ color and faint I magnitude of the latter suggest that it is associated with the molecular absorber seen towards PKS 1830–211, at $z = 0.89$ rather than with the $z = 0.19$ HI absorber previously reported in the spectrum of PKS 1830–211. While this discovery might ease the interpretation of the observed absorption lines, it also complicates the modeling of the lensing potential well, hence decreasing the interest in using this system as a mean to measure H_0 through the time delay between the images. This is the first case of a quasar lensed by an almost face-on spiral.

1 Are Two Different Galaxies Acting Both as Lenses?

The HST observations used in our study are publicly available from the STScI archives. They were obtained as part of the CfA-Arizona Space Telescope LEns Survey (CASTLES). Lehár et al. (2000) identified a possible lensing galaxy and a point source (labelled G and P, respectively, in Figs. 1 and 2 below) from their H-band data. The deeper data we deconvolve here allow to unveil an additional spiral galaxy located between the two images of the quasar source. We argue (Courbin et al. 2002) that the point source P is a galactic star, while Winn et al. (2002) propose that it is the bulge of the new spiral structure. Given the present observations we cannot discriminate between these two interpretations, however spectra to be taken with AO on 10-m class telescopes will settle the issue.

References

1. Courbin F., Meylan G., Kneib J.-P., Lidman C, 2002, ApJ, in press.
2. Lehár J., Falco E.E., Kochanek C.S., et al., 2000, ApJ, 536, 584
3. Winn J.N., Kochanek C.S., McLeod B.A., et al. 2002, ApJ, in press.

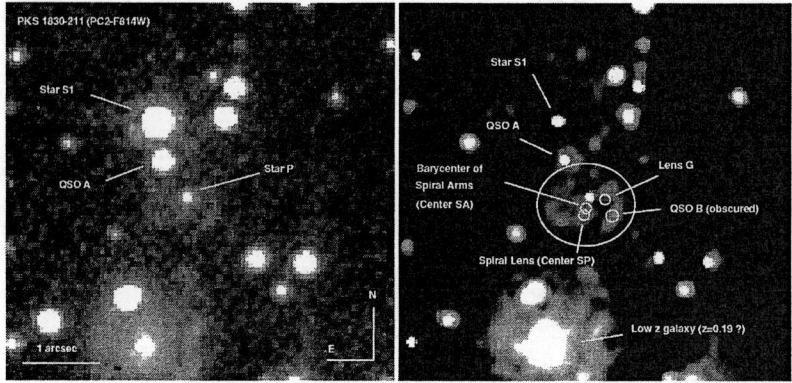

Fig. 1. Left: Part of the HST/WFPC2 image of PKS 1830-211. This image is a combination of all available frames in the F814W filter (8 in total), for a total exposure time of 6,400 sec. **Right:** Simultaneous deconvolution of the 8 frames, reaching a resolution of 0.046 arcsec, and sampled with a pixel of 0.023 arcsec. The various positions of the two probable lenses and of the very obscured (and hence invisible here) QSO B are indicated with circles. Note the obvious spiral shape between the quasar images, with one spiral arm passing right onto the line of sight to QSO B. We identify this galaxy as the probable source of absorption at $z = 0.89$. The center of the ellipse shown in this figure defines the barycenter SA of the spiral. It is one of three centers, together with SP and G, considered in the modelling. There is no significant trace in this image of the much redder lens G, but we plot its position as measured from the near-IR F160W image, in overlay on the F814W image

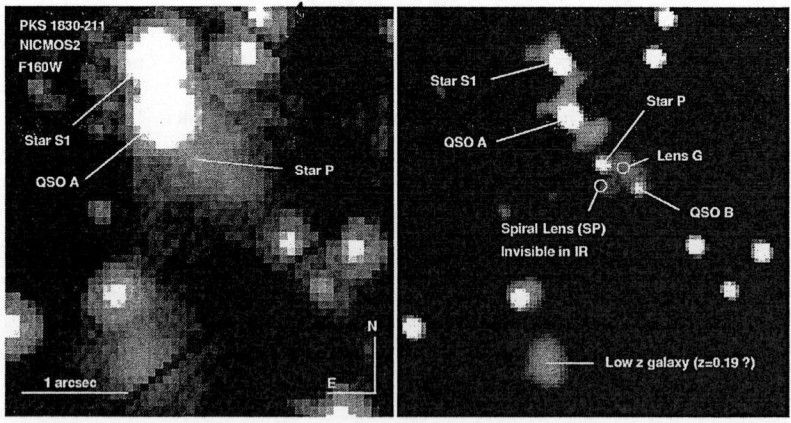

Fig. 2. Left: Part of the HST/NICMOS2 image of PKS 1830-211 obtained in the F160W filter (4 individual frames). **Right:** Simultaneous deconvolution of the four F160W images, with a final resolution of 0.075 arcsec. Note the object right to the West of star P. Lens G is visible in this frame only. There is no trace of the spiral galaxy SP, whose center is marked with a circle

Gauge-Ready Formulation of the Cosmological Kinetic Theory and CMBR Anisotropy

Hyerim Noh

Korea Astronomy Observatory, Korea

Abstract. We present cosmological perturbations of the kinetic components based on the relativistic Boltzmann equations in the context of generalized gravity theories. Our general theory considers an arbitrary number of scalar fields generally coupled with the gravity, arbitrary number of mutually interacting hydrodynamic fluids, and components described by the relativistic Boltzmann equations like massive/massless collisionless particles and the photon with accompanying polarizations. We also include direct interactions among fluids and fields. We consider three different types of perturbations, and all the scalar-type perturbation equations are arranged in a gauge-ready form so that one can implement easily the convenient gauge conditions depending on the situation.

1 Classical Formulation

We consider the most general perturbation in the FLRW (Friedmann-Lemaître-Robertson-Walker) world model. As a metric we take

$$ds^2 = -a^2 \left(1 + 2A\right) d\eta^2 - 2a^2 B_\alpha d\eta dx^\alpha + a^2 (g^{(3)}_{\alpha\beta} + 2C_{\alpha\beta}) dx^\alpha dx^\beta, \qquad (1)$$

where $a(t)$ is the cosmic scale factor and $dt \equiv ad\eta$. $A, B_\alpha, C_{\alpha\beta}$ are generally spacetime dependent perturbed order variables.

1.1 Gauge Strategy

We present all scalar type perturbation equations in a "gauge-ready method". This has a strong advantage that any gauge condition can be chosen depending on the situation. Also, if a solution is known in one gauge, all solutions in every other gauge can be obtained through the gauge transformation.

2 Kinetic Theory Formulation

The evolutions of the collisionless particles and the photon are described by specifying distribution functions which are governed by the relativistic Boltzmann equation. We consider the massless, massive collisionless particles and the photons.

3 CMBR Anisotropy

The temperature anisotropies can be derived by expanding the observed temperature in terms of the spherical harmonic function as

$$\Theta(\mathbf{x}, \eta_0, \hat{\gamma}) \equiv \sum_{\ell} \sum_{m=-\ell}^{\ell} a_{\ell m}^{\Theta}(\mathbf{x}) Y_{\ell m}(\hat{\gamma}). \tag{2}$$

The polarization anisotropies are expanded using the spin-weighted harmonic functions as

$$Q(\mathbf{x}, \eta_0, \hat{\gamma}) \pm iU(\mathbf{x}, \eta_0, \hat{\gamma})$$

$$\equiv \sum_{\ell} \sum_{m=-\ell}^{\ell} \left[a_{\ell m}^{E}(\mathbf{x}) \pm i a_{\ell m}^{B}(\mathbf{x}) \right] {}_{\pm 2}Y_{\ell m}(\hat{\gamma}). \tag{3}$$

Then, the angular power spectra is given by

$$C_{\ell}^{XY} \equiv \frac{1}{2\ell + 1} \sum_{m=-\ell}^{\ell} \langle a_{\ell m}^{X}(\mathbf{x}) a_{\ell m}^{Y*}(\mathbf{x}) \rangle_{\mathbf{x}}$$

$$= \frac{1}{(2\ell + 1)^2} \frac{2}{\pi} \int n^2 dn \sum_{m=-2}^{2} X_{(\ell)}^{(m)}(n, t_0) Y_{(\ell)}^{(m)*}(n, t_0), \tag{4}$$

where X and Y can be any one of Θ, E and B.

4 Numerical Implementation and Summary

We present the evolution of the cosmological perturbations of the kinetic components based on relativistic Boltzmann equations. We include the baryon, CDM, photon, massless and massive neutrino species, the spatial curvature, and the cosmological constant. Our formulations are presented in the context of generalized gravity theories and are arranged in gauge ready form. For the numerical calculation, we implemented four different gauge conditions: comoving gauge, the zero shear gauge, the uniform curvature gauge, and the uniform expansion gauge. Comparing the solutions obtained from these different gauge conditions we can check the numerical accuracy. The detailed equations and numerical results including the power spectra for both scalar and tensor type perturbations are shown in our recent paper ([1]).

References

1. J. Hwang and H. Noh: Phys. Rev. D **65**, 023512 (2002)

APPLES: A Parallel Slitless Imaging Survey for ACS

Anna Pasquali[1], Norbert Pirzkal[1], Jeremy R. Walsh[1], James E. Rhoads[2], Sangeeta Malhotra[2], Zlatko Tsvetanov[3], and the APPLES Team

[1] ESO/ST-ECF, Karl-Schwarzschild-Str. 2, D-85748 Garching bei München, Germany
[2] STScI, 3700 San Martin Drive, Baltimore MD 21218, USA
[3] Department of Physics and Astronomy, JHU, 3701 San Martin Drive, Baltimore MD 21218, USA

Abstract. We present APPLES, the ACS Pure Parallel Lyα Emission Survey for Cycle 11, which is aimed to collect the largest and most uniform sample of high redshift Lyα emitters ever pursued with HST. By using the Wide Field Channel of the Advanced Camera for Survey (ACS), we estimate to find 10 to 20 high redshift Lyα emitters per ACS pointing down to a flux limit of 2×10^{-17} ergs cm^{-2} s^{-1} in two orbits, for a final sample of about 1000 Lyα emitters, which will provide: a) robust statistics on the population and evolution of Lyα emitters at $4 \leq z \leq 7$; b) a measurement of the reionization redshift; and c) spatial clustering information for Lyα emitters as a function of redshift.

1 Introduction

State of the art instrumentation has pushed observational cosmology to $z = 6.56$ with the detection of ~ 10 galaxies at $z > 5$ (cfr. Hu et al. 2002 and Dey et al. 1999) whose spectra indicate on-going star formation in a relatively non-dusty environment (Malhotra & Rhoads 2002). This first generation of stars is believed to have re-ionized the intergalactic medium at $z > 5$ (Becker et al. 2001, Djorgovski et al. 2001). The lack of bright Lyα emitters at $z < 3$ implies a strong redshift-evolution of this class of galaxies, and thus a statistically-significant survey of Lyα emitters at high redshift is needed to understand their evolution in redshift.

2 The Challenge

The Wide Field Channel (WFC) of ACS is an ideal instrument for such a survey: the spatial sampling ($0''.05$/pix) of its field of view ($3'.4 \times 3'.4$) allows to resolve galaxies as compact as $0''.25$. Its grism (working in the range 0.55 - 1 μm at 40 Å/pix resolution in the first order) allows at the same time the detection and redshift determination for Hα at $0 < z < 2$ and Lyα at $4 < z < 7$. We used SLIM 1.0 (Pirzkal et al. 2001) to produce grism images of template galaxies and derive the grism limiting magnitude and S/N ratio as a function of galaxy luminosity and redshift. Figure 1 shows the simulated spectra of a Lyα and a Lyα-break

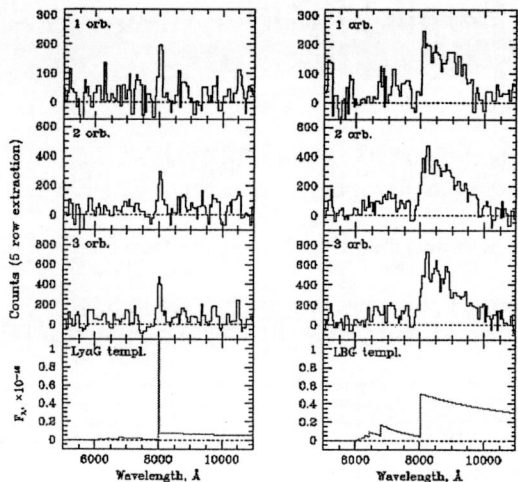

Fig. 1. Simulated grism 1st order spectra of a Lyα and a Lyα-break galaxy at $z = 5.6$, for different exposure times (1 orbit is assumed to be 40 min)

galaxy at $z = 5.6$. For Seyfert II galaxies, the S/N ratio in the Lyα decreases from 20 to 2 as the redshift increases from 4.5 to 7 and the object brightens from m(F850LP) = 26 to 25 as Lyα moves through the passband. QSO Lyα emission can be detected up to $z = 7$ and as faint as m(F850LP) = 25 with S/N $\simeq 5$. Lyα and Lyman-break galaxies at $z = 5.6$ can be already detected with S/N ≥ 5 in one orbit exposure. On the basis of our simulations, APPLES has been designed to acquire direct images in the F775W and F850LP filters and grism images with G800L. With a minimum of three assigned orbits, one will be devoted to photometry and two orbits will be used for spectroscopy.

A number of cosmological issues can be addressed with APPLES data: *census of Lyα galaxies* to study their evolution with redshift; *measurement of the reionization redshift* independent of the Gunn-Peterson effect; *constraint on hierarchical galaxy formation models* from the number distribution of Lyα emitters with redshift; *star formation in the local Universe* using the APPLES sources at $z < 1.7$ from [OII]; *clustering of Lyα and Lyman-break galaxies* to estimate the dark halo masses associated with these sources.

References

1. R.H. Becker et al.: AJ 122, 2850 (2001)
2. A. Dey, H. Spinrad, D. Stern et al.: ApJ 498, L93 (1998)
3. S.G. Djorgovski, S. Castro, D. Stern, A.A. Mahabal: ApJ 562, 55
4. E.M. Hu, L.L. Cowie, R.G. McMahon et al.: ApJ 568, L75 (2002)
5. S. Malhotra, J.E. Rhoads: ApJ 565, L71 (2002)
6. N. Pirzkal, A. Pasquali, J.R. Walsh et al.: ST-ECF ISR ACS2001-03 (2001)

Radiative Effects in the Modelling of Accretion onto Stellar Magnetospheres *

Nikolai Pogorelov[1,2], Igor Kryukov[1,2], Ulrich Anzer[2],
Guennadii Bisnovatyi-Kogan[3], and Gerhard Börner[2]

[1] Institute for Problems in Mechanics, Russian Academy of Sciences,
Vernadskii Avenue 101-1, Moscow 119526, Russia
[2] Max-Planck-Institut für Astrophysik, Karl-Schwarzschild-Str. 1,
D-85741 Garching, Germany
[3] Space Research Institute, Russian Academy of Sciences,
Profsoyuznaya St. 84/32, Moscow 117810, Russia

Accretion onto neutron stars and black holes is the main energy supply in galactic X-ray sources. The angular momentum captured by the X-ray star from the optical companion's wind with velocity V_∞ and the binary period P is proportional to $V_\infty^{-4} P^{-1}$ [4]. If V_∞ is high or the binary components are widely separated, the angular momentum is not sufficient for the accretion disk formation at distances of the Alfvén radius where the magnetic pressure of the star is balanced by the dynamic pressure of infalling matter, and the accretion at large distances from the star becomes nearly spherically symmetric. This occurs, for example, in some X-ray sources with a massive companion star where long-periodic pulsars are observed [1], [3].

To be accreted onto a magnetized star, the matter should penetrate through the stellar magnetosphere. In [6] the results have recently been presented of the magnetohydrodynamic computation of the quasi-spherical Bondi-like accretion onto a magnetic dipole and several solutions with infinitely expanding shock waves were obtained. The penetration of the matter beneath the surface of the magnetosphere in those calculations occurred due to substantial, artificially introduced, resistivity of the plasma. In [5], a somewhat different implementation of the accretion scenario [2] was used. According to it, the plasma flow is initially decelerated by the interaction with the stellar magnetic field, with cusps forming in the polar regions of the magnetosphere. Further on, as a result of the Rayleigh–Taylor (interchange) instability, clumps of plasma penetrate inside the magnetosphere and, threaded by the magnetic field, freely fall onto the poles along unperturbed magnetic field lines under the action of gravity. In [5] we assumed that, on entering inside the magnetosphere, these clumps of plasma become homogenized within a layer adjacent to the inner side of the initially impermeable surface of magnetosphere and soon after that one can again consider the gas flow in the continuum approximation. Thus, the final stage of the scenario was investigated, in which the flow occurs around some modified shape of the inner boundary that can be interpreted as an Alfvén surface possessing polar

* This work was partially supported by the Russian Foundation for Basic Research under Grant No. 02-01-0948.

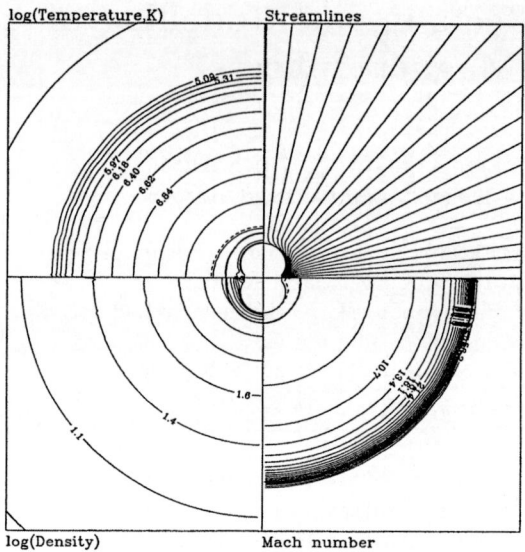

Fig. 1. Parameter distribution for the quasi-spherical accretion onto the star with the mass $\mathcal{M} = 1.4\mathcal{M}_\odot$ and the characteristic size of the magnetosphere $R_* = 3.52 \times 10^8$ cm for $\dot{\mathcal{M}} = 10^{-9}\mathcal{M}_\odot/\text{yr}$, the outer boundary radius $R_0 = 100R_*$, velocity $V_0 \approx 515$ km/s, temperature $T_0 = 10^4$ K, and the accretion efficiency $E = 0.1$

holes. Although somewhat simplified, this model allows us to simulate the averaged picture of the accretion. It is important that the accreted plasma attains rather high temperatures and its consideration in the polytropic approximation used in [5] and [6] is hardly acceptable. We show that by taking into account sufficiently realistic (optically thin medium) cooling and heating terms in the energy equation and the radiation pressure term in the momentum equation, implying that the latter can be important if the luminosity of the star approaches the Eddington limit, one can obtain solutions with physically adequate distributions of parameters. These solutions involve additional discontinuities and/or layers of steep increase in temperature. We obtained accretion regimes that admit sufficiently high accretion rates, thus making possible the existence of steady flows. There are also unsteady accretion regimes with the shock wave propagating away from the star.

References

1. U. Anzer, G. Börner: A&A **299**, 62 (1995)
2. J. Arons, S.M. Lea: ApJ **207**, 914 (1976)
3. G.S. Bisnovatyi-Kogan: A&A, 245, 528 (1991)
4. I.F. Illarionov, R.A. Sunyaev: A&A **39**, 185 (1975)
5. I.A. Kryukov, N.V. Pogorelov, G.S. Bisnovatyi-Kogan, U. Anzer, G. Börner: A&A **364**, 901 (2000)
6. Yu.M. Toropin et al.: ApJ **517**, 906 (1999)

The Dynamic Steady State Universe:
A Qualitative Unification Model

Conrad Ranzan

Cosmic Research Center, 5145 Second Avenue, Niagara Falls, Canada L2E 4J8
email: CozmicResCenter@aol.com

The *Dynamic Steady State Universe (DSSU)* is a comprehensive model of our Universe. Although it is described as a steady-state model, it is *not* a static model. It is an integrated model which readily explains all the major observed phenomena, including structures and their functions, using existing physics – without resorting to inflation, dark matter, infinite density singularities, nor higher dimensions. Its main feature is that it has made possible the interpretation of the large scale structure of the Universe. Its main postulate is the synchronic expansion and contraction of 'space' resulting in the continuous expansion of 'space' within a non-expanding universe. The following is a list of selected features of the model.

As a theoretical model the DSSU is characterized by the expansion of *space* within a non-expanding universe. A simple yet elegant concept. It is derived from the static but unstable Einstein universe (1917) and the expanding Lemaitre universe (1927), and the steady state Bondi-Gold model (1948); it incorporates aspects of the early *Big Bang model*. It defines *space* and describes the two *space postulates* of expansion and contraction. It postulates the contraction of *space* within mass and the contraction of *space* within *space-contracting fields* which surround mass bodies. It describes how the dual geometry (spherical and hyperbolic geometry) of the universe combines to form the large scale structures. It postulates the *two primary functions of black holes* and their essential role in the dual geometry universe. It explains the significance of the two known properties of black holes. It provides a new and unambiguous meaning of the term *the gravitational distortion of space*. Most remarkably it answers the question: *how can space be distorted?* It contributes two new key ideas to *Olbers' paradox*. It explains the relationship between the *lambda force* and each of: vacuum energy, space curvature, hyperbolic space, and anti-gravity. It is cellular. It is isotropic and homogeneous on the largest scale. It is infinite in time and space (volume). It does not violate conservation and physical laws in its creation and annihilation processes.

As a practical working model the DSSU explains the shape of elliptical galaxies as being the result of the tidal effect in hyperbolic space. Itxplains the formation of all non-rotating structures such as giant elliptical galaxies, normal ellipticals, dwarf galaxies, globular clusters, star clusters, and gas-dust clouds. It explains the mechanism which induces *angular momentum* in galaxies. It explains why the *intrinsic velocity*of galaxies cannot exceed + or − 3000 km/sec. It explains random motions of galaxies. It explains the orientation of the plane of galaxy rotation. It explains the randomness of the sense of galaxy rotation. It explains the source of the central bulge of galaxies. It explains how the evolution of spiral galaxies is affected by the DSSU postulates. It explains the distinct pattern found in the original Geller-Huchra wedge of galaxies map (from the 1980s). Explains the chains and webs of galaxy clusters. It describes how the dual geometry creates the mechanism for galaxy clustering and maintains galaxy cluster cohesion (without the use of exotic dark matter). It explains *anomalous redshifts* of galaxies; and describes various sources of spectral redshift errors. This provides justification for using a *variable* Hubble's constant. It provides a reasonably good value for the rate of space expansion. This will make it possible to calculate the *unit-universe* cosmological constant. Unequivocally, it explains the *Cosmic Background Microwave Radiation*.

As a predictive model the DSSU predicts the destination and ultimate fate of the Milky Way galaxy. It predicts that the magnitudes of the lambda force and gravitational force are equal within each unit-universe. It explains why there is no danger of a *gravitational collapse of the Universe*. By using a basic fact of physics it can be shown that the geometric structure of the DSSU cancels gravity on a large scale. It predicts that a small cluster of galaxies exists in the core of each unit-universe void. It predicts that the *Higgs* is a process and not a particle. Gravitation and *Higgs* represent two aspects of the same process. It explains why *lambda* can have a distinct (positive) value and yet result in an observable value of zero over cosmic distances.

The DSSU is the only model of the Universe which uses the *simultaneous expansion and contraction of space* in spatially separated large scale regions; while the currently accepted standard model (the big-bang inflationary model) uses only expansion.

C. R. (2002/08/01) email: Ranzzan@aol.com

ARCHEOPS: A Large Sky Coverage Balloon-Borne Experiment for Mapping the Cosmic Microwave Background

Cécile Renault[1] for the Archeops collaboration

[1] Institut des Sciences Nucléaires, 53 av. des Martyrs, F-38026 Grenoble cedex, France

1 The Archeops Experiment

Archeops is an international collaboration gathering laboratories from France [CRTBT, ISN, LAOG (Grenoble); IAS, CSNSM, LAL (Orsay); SPP (Saclay); APC, IAP (Paris), CESR, OMP (Toulouse)], Italy [Univ. La Sapienza (Roma), IROE-CNR (Firenze)], UK [QMW (Cardiff)] and USA [CALTECH, JPL, the University of Minnesota].

Here are summarised the main features of the instrument, the observations and the expected sensitivity to cosmological parameters *via* the C(l) curve; a complete description of the experiment and of its performances is given in [1].

In order to have very sensitive detectors, Archeops uses spider-web bolometers cooled down at 100 mK by an open-cycle dilution based on He^3-He^4 mixture. Fourteen bolometers at 143 GHz or 217 GHz are optimised for CMB physics, one bolometer at 545 GHz monitors atmospheric -and galactic- emission while six polarised bolometers at 353 GHz are dedicated to the observation of polarised foregrounds. The resolution of 10 arc-minutes is obtained by a 1.5 m mirror.

A main feature of Archeops is the scanning strategy: the gondola rotates on itself around a vertical axis in 30 seconds and the telescope points at a fixed elevation of 41°, describing large circles on the sky. Taking benefit of the Earth rotation, it allows to cover 30 % of the sky in 12 hours. This strategy, combined with the design of the focal plane allows redundancy at 3 scales: a few ms between two bolometers at the same elevation, a few minutes between bolometers located on different rows of the focal plane and a few hours between large circles crossing each other. It leads to a higher signal over noise ratio and an efficient track of systematic effects.

In order to avoid perturbations from the sunlight, the launch is performed above the polar circle by CNES people from the Swedish base of Kiruna: a 24 hour night flight is then possible if the stratospheric winds are cooperative - it was rarely the case.

2 The Archeops Observations

After a technical flight from Trapani (Sicilia) in July 1999, two successful scientific flights were performed from Kiruna (Sweden): the January, 29th 2001 with 7.5 hour night data and the February, 7th 2002 with 12 hour night data (+

7 hour day data). Figure 1 shows the map obtained from the first flight data, although it is measured at very low frequency with respect to the peak emission. Anisotropies are studied at high north galactic latitude, far from intense dust emission.

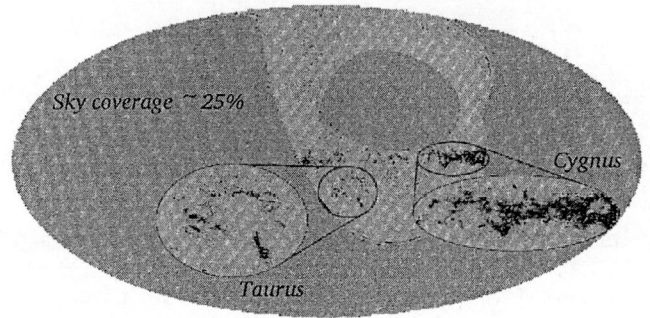

Fig. 1. Archeops map at 217 GHz (01-01-29 flight)

Figure 2 shows the expected sensitivity for the second flight with 12 hour night data and 10 bolometers splitted at 143 GHz and 217 GHz. Archeops is sensitive to many angular scales with some emphasis from l=10 to l=400. It will allow connection with the results obtained by COBE [2] to check the shape of the *plateau* before the first acoustic peak.

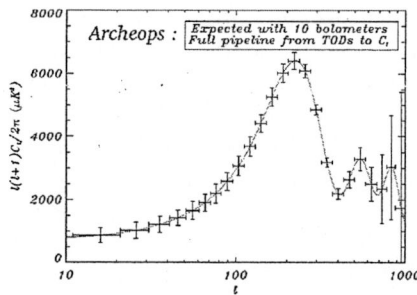

Fig. 2. Simulation of C(l) obtained with 12 hours of data and 10 bolometers using Archeops sensitivity, sky coverage, calibration and noise measured during the flight

References

1. A. Benoit et al.: Astropart. Phys. **17**, 101 (2002)
2. D. J. Fixsen et al.: Astroph. Journal **486**, 623 (1997)

The European Ultra-High Precision Stellar Photometry Road Map for Asteroseismology and Planet Finding

Ian Roxburgh[1,2], Fabio Favata[3], Annie Baglin[2], and Jørgen Christensen-Dalsgaard[4]

[1] Astronomy Unit, Queen Mary, University of London, London E1 4NS, UK
[2] LESIA, Observatoire de Paris, Place Jules Janssen, 92195 Meudon, France
[3] Astrophysics Division, European Space Agency, PO Box 299, 2200 AG, Noordwijk, The Netherlands
[4] Teoretisk Astrofysik Center, Danmarks Grundforskingsfond, and Institut for Fysik og Astronomi, Aarhus Universitet, DK 8000 Aarhus C, Denmark

Abstract. We give a brief description of the 3 European space missions COROT, MONS and Eddington, devoted to asteroseismology and planet finding.

1 Introduction

Since the early 1980's European scientists have studied space projects in stellar seismology, using high precision photometry to extend the powerful tools of seismology from the Sun to stars. The first such project, EVRIS, developed by the French space agency CNES, was launched on the Russian mission Mars96 but was lost when Mars96 failed. More recently, the discovery of extrasolar planets has led scientists to propose using the same technique to find planets around other stars by detecting transits across the stellar disc. This ultra-high precision photometric approach implies very specific instrumental requirements, high stability of all the elements of the detection chain and long uninterrupted observations in exactly the same direction.

Three missions, MONS, COROT and EDDINGTON are being developed and will be launched from 2005 onwards, in a step by step approach, with complementary objectives and different level of budget, leading to slightly different designs.

2 COROT

COROT [1] is a mini satellite with a telescope of 28cm aperture, and is led by the French Space Agency CNES. The scientific programme combines seismology and planet finding, with equal priority. Its low earth orbit will allow continuous observations of 150 days duration on stars in 5 target fields of 8 square degrees close to the galactic plane. This mission profile will give high accuracy on the frequencies of oscillation of stars and the detection of a reasonable number of telluric planets, slightly larger than the Earth and orbiting slightly closer than an a.u. around their parent star.

3 MONS

MONS [2], with a telescope of 32cm aperture, is the scientific mission on the Rømer minisatellite being developed in Denmark. The satellite will fly in a elliptical orbit and the primary goal of the mission is to observe nearby bright stars whose oscillations properties are expected to be similar to the Sun. Each star will be observed for 1-2 months. Observations of a large number of other stars will be made at lower precision using an auxiliary telescope and the star trackers.

4 Eddington

The next step will be the Eddington mission [3] being developed in the framework of the ESA scientific programme. Eddington is a medium class mission and is devoted to both planet finding and stellar seismology. With its large collecting area of close to 1 square meter, and its position at L2, it will determine the seismic parameters of a very large number of stars and especially members of nearby clusters. In its planet finding mode Eddington will stay on a single field for 3 years and its wide field of view will permit the observation of several 100 000 stars during the 5 year mission lifetime. The photometric accuracy and sky coverage will be sufficient to detect a significant number of Earth-like planets in the habitable zone around solar-like stars.

References

1. COROT: see http://www.astrsp-mrs.fr/www/corot.html
2. MONS: see http://astro.ifa.au.dk/Roemer
3. Eddington: see http://sci.esa.int/home/eddington/index.cfm

BaR-SPOrt: Balloon-Borne Radiometers for Sky Polarization Observations

C. Sbarra[1], S. Cortiglioni[1], G. Bernardi[1], E. Carretti[1], S. Cecchini[1], C. Macculi[1], G. Ventura[1], M. Baralis[2], O. Peverini[2], R. Tascone[2], G. Boella[3], S. Bonometto[3], M. Gervasi[3], G. Sironi[3], M. Tucci[3,12], M. Zannoni[3], V. Natale[4], R. Nesti[4], R. Fabbri[5], J. Monari[6], S. Poppi[6], L. Nicastro[7], R. Di Raffaele[7], A. Boscaleri[8], E. Pascale[8], P. de Bernardis[9], M. De Petris[9], S. Masi[9], M.V. Sazhin[10], and E.N. Vinyajkin[11]

[1] IASF-CNR, Sez. di Bologna, Via Gobetti 101, 40129 Bologna, Italy
[2] IRITI-CNR, C.so Duca degli Abruzzi 24, 10129 Torino, Italy
[3] Dip. di Fisica, Università di Milano Bicocca, P.zza della Scienza 3, 20126 Milano, Italy
[4] IRA-CNR, Sez. di Firenze, largo E. Fermi 5, 50125 Firenze, Italy
[5] Dip. di Fisica, Università di Firenze, via Sansone 1, 50019 Sesto Fiorentino, Firenze, Italy
[6] IRA-CNR, Sez. di Bologna, Via Gobetti 101, 40129 Bologna, Italy
[7] IASF-CNR, Sez. di Palermo, via U. La Malfa 153, 90146 Palermo, Italy
[8] IROE-CNR, via Panciatichi 64, 50177 Firenze, Italy
[9] Dip. di Fisica, Università La Sapienza, P.le A. Moro 2, 00185 Roma, Italy
[10] Schternberg Astronomical Institute, Moscow State University, Moscow 119899, Russia
[11] NIRFI, 25 B. Pecherskaya st, Nizhnij Novgorod, 603600/GSP-51, Russia
[12] Instituto de Fisica de Cantabria, Fac. de Ciencias, Avda. Los Castros s/n, 39005 Santander, Spain

Abstract. BaR-SPOrt, funded by ASI (Italian Space Agency), is a 32 (90) GHz balloon-borne correlation polarimeter for direct measurements of the Q and U Stokes parameters, with an angular resolution of $0.6°$ ($0.2°$). Aim of the experiment is the detection of the polarized emission of the diffuse Galactic Background and the Cosmic Microwave Background (CMB). The most likely launch site is Antarctica (2 to 4-week flight). Kiruna (Sweden, 1-week flight) and Svalbard (Norway, > 1-week flight) are possible launch site to observe the Northern sky.

1 The Science

Main scientific goal of BaR-SPOrt is measuring the linear polarization level of the sky emission on small sky patches [1]. Good observing targets, characterised by low emission from dust and synchrotron, exist in both the Southern and the Northern hemisphere (the area already observed by BOOMERanG [2] and, e.g., the area centered at RA=11h, DEC=45°, respectively).

The expected polarization level ($P_{rms} = \sqrt{< Q^2 > + < U^2 >}$) of the CMB is maximum at small angular scales and is only weakly dependent on the cosmological model. If flown for 2 weeks or more, the instrument at 90 GHz, having a beam of 0.2°, is expected to detect CMB polarization irrespective of the presence of a reionization period.

The instrument at 32 GHz, with a beam of 0.6°, will at least be able to improve current upper limits on both CMBP and synchrotron polarized emission.

2 The Instrument

The polarimeter design has been developed to minimize instrumental effects and to increase long-term stability [4], as to reduce 1/f noise effects. The instrument shares most of the SPOrt [3] know-how. The main instrumental characteristics are:

- Low cross-polarization (< -40 dB) on-axis optics providing HPBW$\simeq 0.6°$ at 32 GHz and HPBW$\simeq 0.2°$ at 90 GHz;
- Correlation Unit based on custom design waveguide Hybrid Phase Discriminator, with unpolarized component rejection > 30 dB [5];
- Custom design OMT with high isolation (> 60 dB) to limit contamination from the unpolarized component;
- Custom design internal calibrator for polarized signals [6];
- A Cryostat to cool (< 80 K) LNAs, circulators, polarizer and OMT by a closed-loop cryocooler, and a thermal shield, temperature regulated, located inside the cryostat to increase the thermal stability.

Table 1. BaR-SPOrt technical characteristics

Frequency	Bandwidth	Angular resolution	Instantaneous sensitivity
32 GHz	10%	0.6°	0.5 mKs$^{1/2}$
90 GHz	20%	0.2°	0.5 mKs$^{1/2}$

References

1. M. Zannoni et al., in *Astrophysical Polarized Backgrounds, Bologna, Italy, October 9-12, 2001*, ed. by S. Cecchini et al. (AIP Conf. Proc. Vol. 609) pp. 115-121
2. P. de Bernardis et al.: Nature, **404**, 955 (2000)
3. E. Carretti et al., in *Astrophysical Polarized Backgrounds, Bologna, Italy, October 9-12, 2001*, ed. by S. Cecchini et al. (AIP Conf. Proc. Vol. 609) pp. 109-114
4. E. Carretti et al.: New Astronomy, **6**, 173 (2001)
5. O.A. Peverini et al., in *Astrophysical Polarized Backgrounds, Bologna, Italy, October 9-12, 2001*, ed. by S. Cecchini et al. (AIP Conf. Proc Vol. 609) pp. 177-182
6. M. Baralis et al., in *Astrophysical Polarized Backgrounds, Bologna, Italy, October 9-12, 2001*, ed. by S. Cecchini et al. (AIP Conf. Proc Vol. 609) pp. 257-260

Effects of Gravitational Lensing on the Hubble Diagram in Quintessence Cosmology

Mauro Sereno[1,2]

[1] Dipartimento di Scienze Fisiche, Università degli Studi di Napoli "Federico II",
Compl. Univ. Monte S. Angelo, Ed. G, Via Cinthia, 80126 Napoli, Italia

[2] Istituto Nazionale di Fisica Nucleare, sez. Napoli,
Compl. Univ. Monte S. Angelo, Ed. G, Via Cinthia, 80126 Napoli, Italia

Abstract. We discuss the lensing dispersion on the Hubble diagram of standard candles in a flat Universe filled in with dark energy with negative pressure. This effect must be considered when estimating cosmological parameters from observations of Supernovae (SNe) of type Ia

1 Introduction

Observational and theoretical evidences [1] support a nearly flat Universe dominated by a new type of energy component, called dark energy or quintessence; evidences, coming from SNe Ia, of the acceleration of the Universe's expansion demand a strongly negative pressure for the dark energy ($w_Q \equiv p_Q/\rho_Q < -1/3$, where p_Q and ρ_Q are, respectively, the pressure and energy density of the dark energy).

The importance of these observations makes necessary a complete study of all systematics. The effect of amplification dispersion by gravitational lensing must be accurately considered. Observational data are taken in the inhomogeneous Universe. The effect of gravitational lensing results in the appearance of shear and convergence in images of distant sources according to the different amount and distribution of matter along different lines of sight. Sources most likely appear to be de-magnified relative to the standard Hubble diagram. In fact, light bundles, propagating far from clumps, experience a matter density less than the average matter density of the Universe. On-average Friedmann-Lemaître-Robertson-Walker (FLRW) models [2] address this problem. It is assumed that the relations on a large scale are the same of the corresponding FLRW Universe, while inhomogeneities only affect local phenomena like the propagation of light. The smoothness parameter α_M represents, in a phenomenological way, the magnification effect experienced by the light beam, that is the effective fraction of matter density in the beam connecting the observer and the source. In this scenario, the luminosity distance–redshift relations in terms of the cosmological parameters must be corrected for the effects of inhomogeneities in the pressureless dark matter (DM) [3].

2   Lensing Dispersion

Since the amplification of a source at a given redshift has a statistical nature, the smoothness parameter is direction-dependent. At a given source redshift z, there is an unique mapping between the magnification μ of a standard candle and the direction-dependent α_M [3].

The magnification probability distribution function (pdf) of standard candles is strongly dependent on the equation of state, w_Q, of the quintessence [3]. With no regard to the nature of DM (microscopic particles, such as axions or WIMPS, or macroscopic compact objects, such as black holes and MACHOs), the dispersion increases with z and is maximum for dark energy with very large negative pressure. The main features of the magnification pdf are a mode biased towards de-amplified values ($\alpha_M < 1$) and a long tail towards large magnifications ($\alpha_M > 1$). The pdf is characterized by the parameter $\Delta\mu$, defined as the difference in amplification between the mean FLRW value ($\alpha_M = 1$), and the magnification in the empty beam case ($\alpha_M = 0$). When $\Delta\mu$ increases, the mode value moves towards greater de-magnification: to preserve the total probability and the mean value, the pdf must both reduce its maximum and enlarge its high amplification tail. From the properties of the angular diameter distance in a clumpy Universe [3], it follows that lensing dispersion increases with the redshift of the source and with dark energy with large negative pressure, being maximum for the case of the cosmological constant. Both microscopic DM and quintessence with an intermediate w_Q partially attenuate the effect of the clumpiness.

The noise in the Hubble diagram due to gravitational lensing strongly affects the determination of the cosmological parameters from SNe data. The errors on the pressureless matter density parameter, Ω_M, and on w_Q are maximum for quintessence with not very negative pressure, since in these models the luminosity distance is less sensitive to the cosmology [3]. The effect of the gravitational lensing is of the same order of the other systematics affecting observations of SNe Ia. Due to lensing by large-scale structures, in a flat Universe with $\Omega_M = 0.4$, at $z = 1$ a cosmological constant ($w_Q = -1$) can be interpreted as dark energy with $w_Q < -0.84$ (at 2-σ confidence limit) [3].

References

1. A.G. Riess, A.V. Filippenko, P. Challis, A. Clocchiatti, et al.: AJ, **116**, 1009 (1998); S. Perlmutter, G. Aldering, G. Goldhaber, R.A. Knop, et al.: ApJ, **517**, 565 (1999); P. de Bernardis, P.A.R., Ade, J.J., Bock, J.R., Bond, et al.: Nature, **404**, 955 (2000)
2. C.C. Dyer, R.C. Roeder: ApJ, **174**, L115 (1972); P. Schneider, J., Ehlers, E.E., Falco: *Gravitational Lenses* (Springer-Verlag, Berlin 1992); S. Seitz, P. Schneider, J. Ehlers: Class. Quant. Grav., **11**, 2345 (1994)
3. M. Sereno, G. Covone, E. Piedipalumbo, R. de Ritis: MNRAS, **327**, 517 (2001); M. Sereno, E. Piedipalumbo, M.V. Sazhin: MNRAS, in press, (2002)

Problems of Cosmological Darwinian Selection and the Origin of Habitable Universes

Rüdiger Vaas

Zentrum für Philosophie und Grundlagen der Wissenschaft,
Justus-Liebig-Universität Gießen, Otto-Behaghel-Str. 10 C, 35394 Gießen, Germany
Ruediger.Vaas@t-online.de

Abstract. For the cosmologist and physicist Lee Smolin, our universe is only one in a much larger cosmos (the Multiverse), a member of a growing community of universes, each one being born in a bounce following the formation of a black hole [1,2]. In the course of this, the values of the free parameters of the physical laws are reprocessed and slightly changed in a random way. This leads to an evolutionary picture of the Multiverse, where universes with more black holes have more descendants. Hence, according to Smolin our universe is a product of mutation and selection analogous to the evolution of species described first by Charles Darwin in his seminal book "On the Origin of Species" in 1859 [3]. Smolin concludes, that due to this kind of Cosmological Natural Selection our own universe is the way it is. This is taken as an explanation for the so-called fine-tuning of physical parameters for the existence of earth-like life: They have the values we observe, because they make the formation of black holes much more likely than most other values. – This paper critically comments on some limits of Smolin's hypothesis. A more extended discussion can be found in [4].

1 No Necessary Link Between Black Holes and Life

In principle, life and Cosmological Natural Selection could be independent of each other. If it is possible that there are only short-"living" giant stars collapsing quickly into black holes or universes without stars at all but many primordial black holes, life as we know it could not evolve. Nevertheless, such universes might be very reproductive because of their giant stars or primordial black holes. On the other hand we can conceive a universe without black holes at all (if supernovae led to neutron stars only) but which could be rich in earth-like life nevertheless. Thus, there is no necessary relationship between black holes and life, but nevertheless there might be a contingent connection – namely via the role of carbon as the "molecule of life" due to its ability to make complex molecules (much more than any other element) and as an element accelerating star formation via shielding and cooling of birth clouds of gas and dust and thus providing the conditions for black hole production.

Smolin might explain the fine-tuning of physical constants, but life and intelligence remain an epiphenomenon, i.e. causally inert. Thus, there is no positive selection of life, even if the conditions of life would be exactly the same as the conditions for maximizing the numbers of black holes. But it could be possible that there is a hidden connection between the hospitability of universes for life and black hole formation. Black holes could be advantageous for life, or life could be advantageous for black hole formation. For instance, intelligent cosmic engineers could create universes via black holes [5]. Thus, there would be no self-organized evolution but a designed, preplanned development. Nevertheless, life could still be seen as a "tool" of the Multiverse to produce more universes.

2   The Darwinian Analogy Is Inadequate

Natural selection as described in biology depends on the assumption that the spread of populations (or genes) is mainly constrained by external factors (shortage of food, living space, mating opportunities etc.). In comparison with that, the fitness of Smolin's universes is constrained by only one factor – the numbers of black holes –, and this is an internal limitation. Furthermore, although Smolin's universes have different reproduction rates, they are not competing against each other (cf. [6]). There is no "overpopulation" and no selective pressure, hence no natural selection in a biological sense. Smolin's universes are isolated of each other (except maybe for their umbilical cords). Therefore, there couldn't be a quasi-biological evolution of universes.

3   There Is No Falsifiability of Smolin's Central Claim

Falsifiability of a hypothesis depends on holding fixed the auxiliary assumptions needed to produce the targeted conclusion. In practice, one tries to show that the auxiliaries are themselves well confirmed or otherwise scientifically entrenched. What should be falsifiable according to Smolin is his claim that our universe is nearly optimal for black hole production. However, this is not a necessary consequence of his premises. A consequence is only that most universes are nearly optimal. To move from this statistical conclusion to the targeted conclusion about our universe, Smolin ([2], p. 127) simply assumes that our universe is typical. This is an additional hypothesis as he admits. But this auxiliary assumption is neither confirmed nor otherwise scientifically entrenched. Thus, if changes in the values of our parameters did not lead to a lower rate of black hole formation – contrary to Smolin's prediction – we could always "save" Smolin's hypothesis by supposing that our universe is not typical! Hence, it is (at least for now) impossible to falsify Smolin's central claim that our universe is nearly optimal for black hole production (cf. [7]).

References

1. L. Smolin: Class. and Quantum Grav. 9, 173 (1992)
2. L. Smolin: The Life of the Cosmos (Phoenix, London 1998)
3. C. Darwin: On the Origin of Species (Cambridge: Harvard University Press 1964)
4. R. Vaas, http://www.vijlen.com/vip-projects/confs/mima/Vaas/VAAS.html and http://arXiv.org/abs/gr-qc/0205119
5. E. R. Harrison: Quart. J. R. astr. Soc. 36, 193 (1995)
6. J. Maynard Smith, E. Szathmáry: Nature 384, 107 (1996)
7. S. Weinstein, A. Fine: J. Phil. XCV, 264 (1998)

Big Bang Nucleosynthesis of Lithium-7 and the Baryon Density of the Universe

Elisabeth Vangioni-Flam[1], Alain Coc[2], and Michel Cassé[2,3]

[1] Institut d'Astrophysique de Paris, 98 bis Bd Arago, 75014 Paris, France
[2] CSNSM, IN2P3-CNRS and Un. Paris Sud, Bat. 104, 91405 Orsay Campus, France
[3] SAp, CEA, Orme des Merisiers, 91191 Gif sur Yvette CEDEX, France

Abstract. Thanks to recent nuclear physic compilations, we update Standard Big Bang Nucleosynthesis (SBBN) calculations. By a Monte-Carlo technique, we calculate the uncertainties on the light element yields related to nuclear reactions. The results are compared to astrophysical observations. The baryonic density obtained is confronted to other estimates deduced from recent independent approaches as the observations of the anisotropies of the Cosmic Microwave Background or the $Ly\alpha$ forest at high redshift. Lithium-7 could lead to more stringent constraints on the baryonic density of the universe than deuterium, because of a much higher observation statistics and an easier extrapolation to primordial values.

1 Observational Constraints from the Light Elements

The ^4He primordial abundance is derived from observations of metal poor extragalactic ionized hydrogen (HII) regions. The extreme values recently published cover the range: $0.231 < Yp < 0.246$. Deuterium is peculiar because, after BBN, this fragile isotope can only be destroyed in subsequent stellar or galactic nuclear processing. Hence, the primordial abundance should be represented in principle by the highest observed value. Remote cosmological clouds on the line of sight of high redshift quasars are the best candidates. However, these D abundance data are scarce and scattered and cast a doubt on the direct identification of these observed values with the primordial D abundance. Whatever, the extreme values deduced from the different observations lead to: $1.6 \times 10^{-5} < D/H < 4.65 \times 10^{-5}$. Compared to D, the determination of the ^7Li primordial abundance from observations seems easier. A plateau in the lithium abundance as a function of metallicity (Spite plateau) is interpreted as being representative of the primordial lithium abundance. Accordingly, we adopt the range: $0.9 \times 10^{-10} < ^7Li/H < 1.9 \times 10^{-10}$. (For the observational references see Coc et al 2002 [4]).

2 CMB and $Ly\alpha$ Forest Observations

CMB anisotropy measurements give independent estimates of the baryonic density of the universe. The most recent determinations from BOOMERANG [2]) and DASI [8] experiments are:
$0.019 < \Omega_b.h^2 < 0.026 \ (2\sigma)$.

The MAXIMA experiment reports a higher value [10]:
$0.02 < \Omega_b.h^2 < 0.045$ (2σ).

The Cosmic Background Imager (CBI)[7], ground based, has given preliminary results in marked contrast with above experiences: $\Omega_b.h^2$ is about 0.009.

On the other hand, the study of the baryon content of the intergalactic medium through the $Ly\alpha$ forest, in the redshift range $0 < z < 5$, leads also to an evaluation of $\Omega_b.h^2$:

$0.0125 < \Omega_b.h^2 < 0.03$ [9] (1σ)
$0.02 < \Omega_b.h^2 < 0.04$ [6] (1σ).

3 Nuclear Data and BBN Results

Most of the important reactions for ^7Li production (see Table 1 in [4]) are available in the NACRE compilation of thermonuclear reaction rates [1]. In this study, we ran Monte-Carlo calculations to obtain statistically better defined limits as in a previous work [11]. We get:

$0.006 < \Omega_b.h^2 < 0.016$ based on ^7Li only and
$0.012 < \Omega_b.h^2 < 0.018$ based on Li and D.

These results are to be compared with other BBN calculations:

$0.006 < \Omega_b.h^2 < 0.017$, based on He and Li, [5]
$0.017 < \Omega_b.h^2 < 0.023$, based on low D, [5]
$0.018 < \Omega_b.h^2 < 0.022$, [3].

Our results (see figures 4 and 5 in [4]) are in good agreement with those of [5] and in reasonable agreement with the study of [3] which is based on low D abundance. The BBN results are broadly consistent with the CMB ones and with the estimates obtained with the high redshift $Ly\alpha$ forest observations.

Big Bang nucleosynthesis has been the subject of permanent care since it gives access to the baryon density which is a key cosmological parameter. Though independent methods are now available, the SBBN one remains the most reliable because i) the underlying physics is well known and there is essentially one free parameter contrary to other methods. It is worth pursuing the improvement of nuclear reaction rates and abundance determination of light elements, essentially ^7Li and D.

References

1. C. Angulo, M. Arnould, M. Rayet et al., Nuclear Phys. A656, 3 (1999)
2. P. de Bernardis et al., ApJ, 564, 559 (2002)
3. S. Burles et al, ApJL 552, L1 (2001)
4. A. Coc, E. Vangioni-Flam, M. Cassé, Phys. Rev. D, 65, 043510 (2002)
5. R.H. Cyburt, B.D. Fields, and K.A. Olive, New. Astr. 6, 215 (2001)
6. J. Hui et al, ApJ, 564, 525 (2002)
7. S. Padin et al., ApJL, 549, L1. (2001)
8. C. Pryke et al, ApJ, 568, 46 (2002)
9. R. Riediger, P. Petitjean, and J.P. Mucket, AA 329, 30 (1998)
10. R. Stompor et al, ApJ, 561, L7 (2001)
11. E. Vangioni-Flam, A. Coc, and M. Cassé, AA, 360, 15 (2001)

Introducing Astronomical Topics at the University of Cooperative Education

Hans Weghorn

University of Cooperative Education, Rotebühlplatz 41, 70178 Stuttgart, Germany

1 Introduction

The University of Cooperative Education offers a dual study with focus on business and industrial applications. It can be considered to have a similar scope like the German Universities of Applied Sciences. The distinction to the latter are the shorter time of education - the final degree is attained after three years - and the joined education with partner companies. The main final degree is the German engineer with diploma, and in parallel for an international reputation a BSc (hons), which is spent from London Open University. After this, master degrees can be obtained on base of further studies at the same University or at other institutions. The question, which kind of astronomical work can be performed at our site, shall here be discussed especially on base of Fig. 1.

2 Possible Working Areas

Astronomy always depended and today even more depends on the intense use of technology. Therefore, astronomy offers an interesting and demanding working field for engineers in general, as especially for students of our technical courses. Dedicate areas of interest for possible projects can be identified: Instruments and control equipment for observations, development of general tools for image format conversion, display and presentation, and scientific image processing algorithms, specifically correlation or speckle processing methods [1]. For the work on correlation processing, raw data is available from former observation runs [2] at ESO on La Silla, Chile, not only for an improvement of processing methods, but also for a production of new scientific astronomical results.

3 First Projects on Interferometric Imaging

Figure 1a,b,c show a first interesting speckle processing sample of the relatively faint spectroscopic binary ADS1490: Despite the fact that the 500 speckle interferograms, which were recorded in V at the ESO Danish 1.5-m telescope, are dominated by camera noise (Fig. 1b), an output image is obtained by triple correlation processing of this data set (Fig. 1c), which has much higher an angular resolution than the classical long-time exposure of the same object (Fig. 1a). One project was started in 2001 to investigate, how the output of this noisy data set can be improved by employing projection techniques during the speckle image processing. The data set appears also being feasible for further investigations of

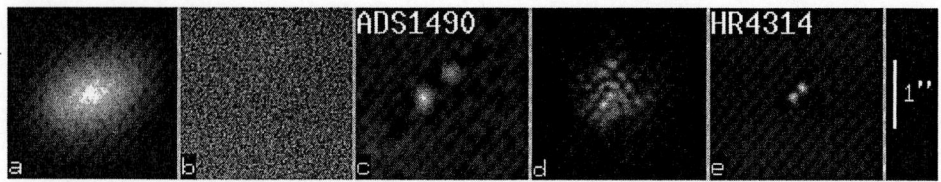

Fig. 1. Speckle processing of ADS1490 in V (a,b,c), and HR4314 in I (d,e)

the limiting magnitude for speckle observations with plain CCD detectors, since only 40 % of the theoretical angular resolution could be achieved due to the high noise in the input data.

The second sample demonstrates that speckle techniques can even obtain diffraction-limited angular resolution in the reconstructed images of the observed objects: 250 speckle interferograms of the binary HR4314 were recorded in I also at the ESO D1.5 (Fig. 1d), and triple correlation yields an output with the full theoretical spatial resolution of 0.13 arcseconds (Fig. 1e). For preparing speckle observation runs on the most recent telescope generation, e. g. an 8-m mirror of the VLT, last year a professional CCD camera (Photometrics CoolSNAP cf, $13\,e^-$ readout noise) was purchased. Around this, a speckle camera system shall be constructed, which is applicable in BV bands. With this system angular resolutions will be achievable in the optical, which compete to adaptive optics systems and interferometers, because these are operating at NIR and MIR wavelengths resp. [3]. According to previous investigations [4] and the CCD noise parameter, a limiting magnitude of 12^m can be assumed for this new camera system.

As completely different kind of project, the implementation of a graphical display library was started also last year, which will enable the porting of our proprietary interferometric processing package [5] to desktop computers operated by MS Windows. Due to the rapid hardware development in the recent years, it is nowadays possible to perform even power-consuming 4-D image processing on standard PCs providing a comfortable handling and environment to the user.

Concluding, it shall be stated here that our astronomical work will continue on the described areas with special focus on accuracy improvements in astrometric and photometric measurements of close binaries and stellar multiplets.

References

1. A. W. Lohmann, et al.: Appl. Optics **22**, 4028 (1983)
2. H. Weghorn, et al.: AG Abstr. Series **11**, ed. G. Klare (Weber, Leimen 1995), 158
3. C. Cesarsky: 'Introduction'. In: *Astronomy, Cosmology and Fundamental Physics, Garching, March 2002*, this book issue
4. H. Weghorn: AG Abstr. Series **9**, ed. G. Klare (Weber, Leimen 1993), 17
5. H. Weghorn: 'Software concept for four-dimensional bispectral analysis of stellar interferograms '. In: *Science with the VLT Interferometer, Garching, June 1996*, ed. by F. Paresce (Springer, Heidelberg 1997) pp. 397–398

Photometric Redshifts Using Submm Wavelengths

Tommy Wiklind

Onsala Space Observatory, Astronomy & Astrophysics, SE-43937 Onsala, Sweden

Abstract. Photometric redshifts using two wavelength bands in the submm region of dust spectral energy distributions are shown to work.

1 Introduction

Recent observations at sub- and millimeter wavelengths with sensitive bolometer arrays have revealed a population of objects believed to be dusty high redshift far-infrared luminous galaxies [12] [1] [9] [7]. Number counts indicate a large excess of sources compared to no-evolution models based on optical surveys [8] [3] and implies that the optically derived star formation density is significantly underestimated at high redshifts. However, despite detection of >200 sub-millimeter sources, only 2 (possibly 3) have secure optical or near-infrared counterparts [13]. This lack of secure redshift information constitutes a major problem when interpreting the submm observations in terms of galaxy formation and evolutionary models. Based on the tight correlation between non-thermal radio and thermal far-infrared fluxes for local galaxies [5], a technique using the radio-FIR spectral index as a redshift indicator for distant submm sources was devised by [4]. For a fixed observed wavelength the flux of synchrotron emission decreases with increasing redshift, while the opposite is true for the Rayleigh-Jeans part of the dust continuum. There are, however, several caveats with this radio-FIR relation. The spectral index depends not only on the redshift of the source, but also on the dust temperature, and the radio continuum spectral index, i.e. non-thermal vs. thermal contributions [6].

2 The Submm Method

A recent compilation of NIR-to-submm observations of 41 ultra-luminous infrared galaxies (ULIRGs) selected to have $L_{FIR} > 10^{12}$ $L_\odot$ was presented by [10]. These local ULIRGs are characterised by a dominating FIR luminosity, large dust masses, large molecular gas masses and signs of strong gravitational interaction and/or merging [11]. The heating source of the large FIR luminosity is in many cases pure star formation, but a significant fraction of all ULIRGs also contain an AGN. Local ULIRGs are the most likely low-z counterparts to the high redshift dusty submm objects seen in deep bolometer surveys. The sample obtained by [10] is therefore well suited as a template for submm detected galaxies. The SEDs as tabulated by Klaas et al. are plotted in Fig. 1a. No corrections

or normalisation has been done to the SEDs at this stage. In Fig. 1b the SEDs
have been normalised such that the overall dispersion is minimised. The 1σ deviations around the average SED are shown. The dispersion is remarkably small
for wavelengths $>50\mu$m. A least-square fit over $60-850\mu$m of a modified Planck
curve is also shown. In Fig. 1c a modified Planck curve has been fitted for 10^4
Monte Carlo realizations where the average SED values at each wavelength band
have been allowed to vary stochastically and uniformly $\pm1\sigma$. The fitted curves
fall within the grey band.

Integrating the fitted Planck functions over the filter function characteristic
for the SCUBA bolometer at 450μm and 850μm at different redshifts of the
emitting source, we get the flux ratio vs. redshift as shown in Fig. 1d. Here the
intensity if the grey scale represents the density of flux ratios from the 10^4 Monte
Carlo realizations. The average and $\pm95\%$ levels are indicated in the plot as well
as a few observed objects from a survey of gravitationally lensed AGNs [2].
The strongest lenses are marked and are more likely to suffer from differential
magnification, making the flux ratio unreliable as a redshift indicator.

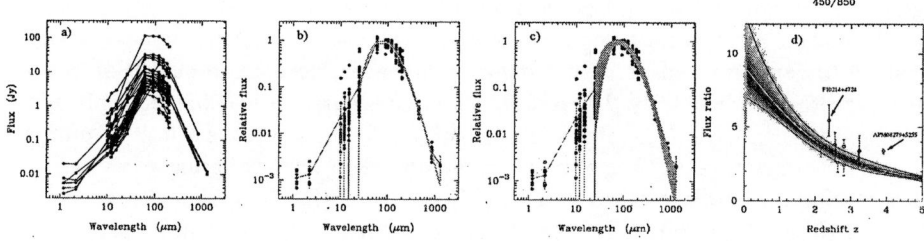

Fig. 1. a) The original SEDs for the local ULIRGs from [10]. **b)** Normalised SEDs with
a fit between 60-850μm. **c)** Monte Carlo simulation of the best fit. **d)** The $450/850\mu$m
flux ratio versus redshift with a few observed objects shown

References

1. Barger, A. J., Cowies, L. L., Sanders, D. B et al. 1998, Nature, 394, 248
2. Barvainis R., Ivison R. 2002, ApJ, 571, 712
3. Blain A. W., 1999, MNRAS, 309, 955
4. Carilli, C. L. and Yun, M. S. 1999, ApJ, 513, L13
5. Condon, J. J., 1992, ARA&A, 30, 575
6. Dunne, L., Clements, D. L. and Eales, S. A. 2001, MNRAS, 319, 813
7. Eales, S., Lilly, S., Gear, W., et al. 1999, ApJ, 515, 615
8. Guiderdoni, B., Hivon, E., Bouchet F. R. and Maffei, B. 1998, MNRAS, 295, 877
9. Hughes, D. H., Serjeant, S., Dunlop, J., et al. 1998, Nature, 394, 241
10. Klaas, U., Haas, M., Müller, S. A. H., et al., 2001, A&A, 379, 823
11. Sanders, D. B. and Mirabel I. F. 1996, ARA&A, 34, 749
12. Smail, I., Ivison, R. and Blain A. 1997, ApJ, 490, L5
13. Smail, I., Ivison, R. J., Blain, A. W. and Kneib, J.-P., 2002, MNRAS, 331, 495

Practical Problems with the Light Curves of GRB Afterglows

A. Zeh and S. Klose

Thüringer Landessternwarte Tautenburg, 07778 Tautenburg, Germany

Abstract. In many cases data gaps affect the interpretation of the optical light curves of GRB afterglows. We demonstrate this practical problem on two examples.

1   Introduction

The investigation of Gamma-Ray Bursts in general, and their afterglows in particular, is currently among the most active research fields in astronomy. From the observational point of view, one of the basic practical problems in this field is the transient, short-lived nature of the bursts and of the afterglow phenomenon. Observations of afterglows therefore usually require target-of-opportunity (TOO) requests, what can result in potential logistical problems (for example: the appropriate instrument is not mounted at the telescope, the required telescope is not available, the TOO time per night is strongly limited, the Moon does affect optical observations, the weather conditions are worse, etc,). Moreover, in general many observatories contribute to the observations of an afterglow, what can introduce additional complications due to potentially different photometric filters used by the various observers. The consequence of this situation is that very often afterglow light curves are affected by data gaps. In particular, this holds at later times, when the afterglow is faint and only accessible to larger optical telescopes. Based on two examples we demonstrate here in which manner incomplete light curves do affect the interpretation of afterglow light curves.

2   Data Gaps and Hidden Broken Power-Law Decays

The optical afterglow of GRB 991208 was in detail observed and investigated by Castro-Tirado et al. [1]. We fitted its R_c-band light curve by the analytical expression [2, 3] $F_\nu(t) = k \ \{(t/t_b)^{\alpha_1 n} + (t/t_b)^{\alpha_2 n}\}^{-1/n} + F_{\mathrm{gal}}$, which allows to search for a break in the light curve. Here, F_ν is the measured flux density, t the time after the burst trigger, t_b the break time, F_{gal} the flux density of the host galaxy, k is a constant, and n characterises the sharpness of the break. Figure 1a shows that evidence for a break in the R_c-band is apparent approximately 1 week after the burst, what indicates to a mildly collimated explosion. Since the knowledge of the break time allows an estimate of the beaming factor, finding a break in an afterglow light curve is of fundamental importance. A careful inspection of Fig. 1a shows that in this particular case t_b is finally determined by only one single data point. This, unfortunately, makes the error bar for α_2 relatively

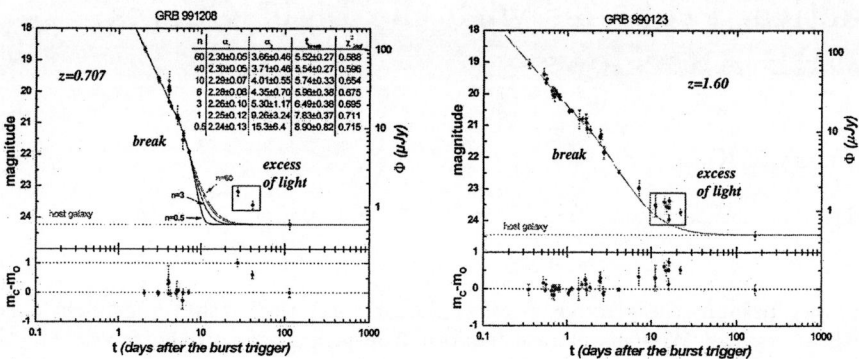

Fig. 1. R_c-band light curves of the afterglows of GRBs 991208 (a) and 990123 (b) based on data collected from the literature, corrected for interstellar extinction

large and the parameter n is ill-defined. Consequently, no tight constraints on the late-time behavior of this optical afterglow can be set.

3   Missing Late-Time Data and Hidden High-z Supernovae

The question of whether there is extra light in an afterglow light curve has gained considerable attention since the discovery of this phenomenon in the afterglow of GRB 980326. Such extra light could be due to an underlying supernova and hence it could be direct evidence for a high-mass star as a GRB progenitor. For the detection of such an excess, however, it is essential to have very accurate late-time measurements of the magnitude of the underlying host galaxy. This is demonstrated here for GRB 990123 which was at a redshift of $z=1.60$. In the literature there are two different R_c-band magnitudes reported for the host galaxy of this burst, $R_c = 24.07 \pm 0.07$ [4] and $R_c = 24.51 \pm 0.14$ [5]. If we take the latter, then there is an excess of flux between 10 to 30 days after the burst trigger (Fig. 1b). Within the context of the supernova model this excess could have revealed the most distant type Ib/Ic supernova ever seen (see also [6]). However, if we take the former magnitude [4], no clear excess is apparent [7] and, hence, there is no obvious evidence for a potential high-z supernova.

References

1. Castro-Tirado, A. J. et al., A&A 370, 398 (2001)
2. Beuermann, K. et al., A&A 352, L26 (1999)
3. Rhoads, J. E., Fruchter, A. S., ApJ 546, 117 (2001)
4. Sokolov, V. V. et al., A&A 372, 438 (2001)
5. Holland, S., & Hjorth, J., A&A 344, L59 (1999)
6. Dar, A., GCN #346 (1999)
7. Zeh, A. et al., in preparation (2002)

Dark Matter in Dwarf Elliptical Galaxies

Werner W. Zeilinger[1], Herwig Dejonghe[2], Sven De Rijcke[2], and
George K.T. Hau[3]

[1] Institut für Astronomie, Universität Wien
   Türkenschanzstraße 17, A-1180 Wien, Austria
[2] Sterrenkundig Observatorium, Universiteit Gent
   Krijgslaan 281, S9, B-9000 Gent, Belgium
[3] ESO, Casilla 19001, Santiago de Chile 19, Chile

Abstract. Results from an ESO Large Programme on the search for dark matter in
dwarf elliptical galaxies (dEs) are presented. Deep imaging and long-slit spectroscopy
was obtained for a sample of dEs ranging from dE0 to dS0. Only few dEs are found
to be flattened by rotation and the majority is supported by pressure anisotropy. The
typical M/L ratios are found to be in the range of 3 to 9 solar units which are consistent
with predictions based on cold dark matter cosmological scenarios for galaxy formation.

1 Introduction

Dwarf elliptical galaxies ($M_B > -18$ mag) dominate the nearby universe in numbers and are therefore of particular interest. A comprehensive review on the
subject is given by [1]. In the current cold dark matter scenarios they are
thought to be dark matter dominated systems and consequently a key factor
in understanding galaxy formation in general. dEs are supposed to form from
average-amplitude density fluctuations [2]. Supernova-driven winds are thought
responsible for expelling their gas and reshaping the galaxies. Alternatively, it is
argued that dEs can also form out of late-type galaxies that are stripped of their
disk material and dark matter by galaxy harassment [3]. However, the internal
dynamical structure of dEs is still largely unknown because to their very faint
surface brightness levels. The few dEs studied so far are found to be supported by anisotropy introducing a dichotomy in the otherwise linear sequence of
increasing rotational support with decreasing luminosity for ellipticals.

Here we present first results from an ESO Large Programme on dark matter
in dEs. We selected a representative sample of dE's with morphological types
ranging from dE0 to dS0. The aim of the programme is to study the dark matter
content through dynamical models and investigate the function of the environment.

2 Observations and Dynamical Modeling

The observations were carried out at the ESO 8.2 m telescopes VLT-UT2 and
UT4 using FORS2. Major and minor axis long-slit spectra were obtained with the
FORS2 grism GRIS_1028z+29 in the wavelength region $\lambda\lambda 7900 - 9300$Å achieving

an instrumental broadening of $\sigma_{instr} = 30$ km/s. In addition, VRI images of the sample were obtained during a periods of excellent seeing. The kinematic profiles, mean rotation velocity and the velocity dispersion as a function of radius, were obtained by fitting Gaussians to the line-of-sight velocity distributions. The spectra have a typical radial extent of 1-2 R_e. Ellipses were fitted to the isophotes of the calibrated VRI images in order to derive surface brightness, position angle and ellipticity profiles together with the Fourier coefficients that quantify the deviations of the isophotes from a pure elliptic shape.

By deprojecting the surface brightness profile the total spatial density was derived. The gravitational potential follows from Poisson's equation. The orbital structure is described by the distribution function $F(E, L, L_z)$, which was obtained by fitting a dynamical model directly to the spectra [4]. About 100 models were typically fitted to the spectra, with mass distributions varying between constant M/L and models with steeply outwardly rising mass-to-light ratios.

3 Results

At present, data of a sample of 9 dEs has been analyzed so far. Galaxies were observed in the Fornax cluster (dense cluster environment), NGC 5044 group (group environment) and NGC 5898 group (small group environment). The current sample of dEs suggests that probably not more than 10% of all dEs are flattened by rotation. The kinematic dichotomy seems therefore to be confirmed. A positive correlation (B-R) vs. M_R supports a scenario in which more massive dEs keep their processed material longer. The kinematic data are found to be largely in agreement with the scaling relations of Supernova-driven mass loss models, but the predicted $Z(L)$ relation [2] could not be verified. Dynamical models favour dark matter halos, although not dark matter dominated systems. A case study on the dE1 FS 76, member of the NGC 5044 group [5], yields $3.2 \leq M/L_B \leq 9.1$ $M_\odot/L_\odot$ within 1 kpc (≈ 1.5 R_e) [6]. An inversion of the velocity dispersion profile outside 1 kpc indicates the presence of a compact dark halo, possibly truncated by tidal stripping. Also the detection of a kinematically peculiar core in FS 76 could be a consequence of tidal torques accreting matter towards the galaxy center.

References

1. H.C. Ferguson, B. Binggeli: A&A Rev. **6**, 67 (1994)
2. A. Dekel, J. Silk: ApJ **303**, 39 (1986)
3. B. Moore, G. Lake, N. Katz: ApJ **495**, 139 (1998)
4. S. De Rijcke, H. Dejonghe: MNRAS **298**, 677 (1998)
5. H.C. Ferguson, A. Sandage: AJ **100**, 1 (1990)
6. S. De Rijcke, H. Dejonghe, W.W. Zeilinger, G.K.T. Hau: ApJ **559**, L21 (2001)

Cosmological Test of Brane Models

Houri Ziaeepour

Mullard Space Science Laboratory (MSSL), Holmbury St. Mary, Dorking RH5 6NT, Surrey, UK

Abstract. After obtaining a class of solutions for two-brane models we show that at the epoch of Primordial Nucleosynthesis, they have an unconventional cosmology which modifies the ^4He yield with respect to the BBN prediction. This reconciles the relatively high baryonic density suggested from CMB anisotropy observations and the observed low primordial ^4He yield. We also perform a second test using Ultra High Energy Cosmic Rays (UHECR). If at high energies particles penetrate into the bulk, the observed time coherence of the air showers from UHECR constraints the parameter space of the brane world models.

In 2-brane models with cold and hot matter on the visible brane, the evolution equation is not an exact FLRW and has the following form [1]:

$$H^2 \equiv \frac{\dot{a}_L^2}{a_L^2} = \frac{8\pi G}{3}(\alpha_{hot}\rho_{hot} + \rho_{cold} + \rho_{\Lambda_{obs}} + \mathcal{O}(\rho_m^2)). \qquad (1)$$

The relation between primordial yield of light elements depends on the temperature of the $p - n$ plasma when neutrinos decouple from weak interaction $p + e \leftrightarrows n + \nu$; $\Gamma_{pe\leftrightarrows n\nu}/H \approx 1/\alpha^{1/2}(T/0.8MeV)^3$. When the characteristic scale of the brane model $\mu L \gg 1$ (to solve the hierarchy problem) and the tension of branes $\rho_0 \approx \rho_L$, the asymptotic value of $\alpha \approx \frac{2}{3}$ [1]. With $(n/p)_{freez,FLRW} \approx 1/7 = 0.143$, $(n/p)_{freez,brane} \approx 0.125$. Consequently, the ^4He yield Y_p is $\sim 12\%$ less than standard cosmology (see Fig. 1).

Most of field localization mechanisms in brane models are evolutionary i.e. based on special field configurations with localized properties like topological defect [2] and a few QCD-like models (if any) leads to a consistent localization [3]. Therefore if large extra-dimensions exist, the physics at high energies must be effectively $D > 4$ and interactions in that energies can eject particles to the extra -dimensions. In this case, the arrival delay caused by propagation of remnant particles in the bulk can destroy the time coherence of the observed showers. This can be used to constraint parameter space of the brane models. We calculate [4] the minimum propagation time for particles ejected to extra-dimensions for a number of brane models and compare them with arrival time resolution of present Air Shower detectors. Our attention is mostly concentrated on the classical structure of the brane models and the results are mostly independent of the detail of their quantum field content and origin. The calculation is mainly based on the assumption that in the time scale of the propagation of a particle in the extra-dimension, the bulk and the branes are quasi-static. For

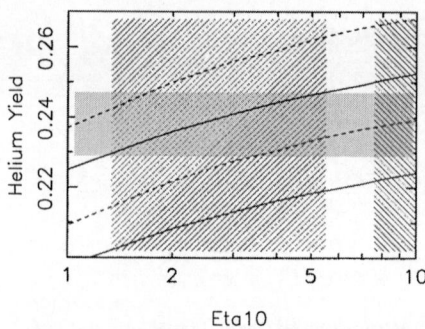

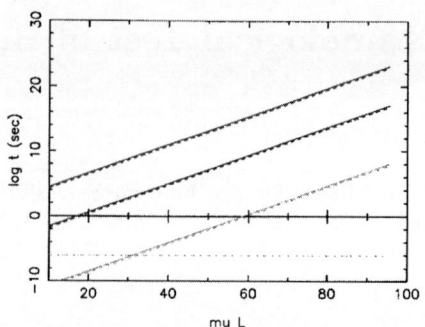

Fig. 1. Primordial Helium yield: FLRW (blue), 2-brane models (magenta). Full and dashed curves correspond respectively to $n_\nu^{light} = 3$ and $n_\nu^{light} = 4$. The hashed regions show a conservative range for the observed value of Y_p (yellow) and corresponding η_{10} according to BBN (green) and highest limit of η_{10} from CMB (red)

Fig. 2. Propagation time for relativistic particles with $u_L^0(t_0)/N^2 = 10^3$ (full line) and $u_L^0(t_0)/N^2 = 1.2$ (dashed line), in static RS model with $n^2(t,y) = e^{-\mu y}$, $a^2(t,y) = 1$ and $b = 1$. Red, magenta and light green curves correspond to $M_5 = 10^{13}eV$, $M_5 = 10^{15}eV$ and $M_5 = 10^{18}eV$ respectively. u_L^0 is the initial speed. The dark green line shows the time coherence precision of present Air Shower detectors.

most 2-brane models the acceptable range of μ is $\mu \gtrsim 1eV$ except for original RS model which needs smaller $\mu \lesssim 10^{-2}eV$. In Goldberger-Wise mechanism for stabilizing the distance between branes $\mu \sim m_{radion}$. With the mass range we obtain for radion, it seems that it is more similar to an axion than a Higgs and its coupling to other fields of the bulk or branes must be very small or the distance between two branes must be extremely small. Figure 2 shows as an example the propagation time in the bulk for static RS model.

References

1. Ziaeepour H., hep-ph/0010180.
2. Jackiw R. & Rebbi C. *Phys. Rev.* D **13**, 3398 (1976).
3. Dubovsky S., Rubakov V., *Int. J. Mod. Phys.* A **16**, 4331 (2001)hep-th/0105243.
4. Ziaeepour H., hep-ph/0203165.

Author Index

ESO ASTROPHYSICS SYMPOSIA
European Southern Observatory

Series Editor: Bruno Leibundgut

Series homepage – http://www.springer.de/phys/books/eso/

ESO ASTROPHYSICS SYMPOSIA
European Southern Observatory

Series Editor: Bruno Leibundgut

W. Hillebrandt, B. Leibundgut (Eds.),
From Twilight to Highlight:
The Physics of Supernovae
Proceedings, 2002. XVII, 414 pages. 2003.

P. A. Shaver, L. DiLella, A. Giménez (Eds.),
Astronomy, Cosmology
and Fundamental Physics
Proceedings, 2002. XXI, 501 pages. 2003.

Series homepage – http://www.springer.de/phys/books/eso/